INTRODUCTION
to *Animal Science*

Leland S. Shapiro
Los Angeles Pierce College

*Note: The cd that originally accompanied this book has been removed.

Prentice
Hall

Upper Saddle River, New Jersey 07458

Library of Congress Cataloging-in-Publication Data

Shapiro, Leland.
   Introduction to animal science / Leland Shapiro.
      p.cm.
   Includes bibliographical references (p.).
   ISBN 0-13-920992-1
      1. Livestock. 2. Animal industry. I. Title.

SF61.S397 2001
636—dc21   00-041668

Publisher: *Charles E. Stewart, Jr.*
Executive Editor: *Debbie Yarnell*
Associate Editor: *Kate Linsner*
Assistant Editor: *Kimberly Yehle*
Production Editor: *Lori Dalberg, Carlisle Publishers Services*
Production Liaison: *Eileen O'Sullivan*
Director of Manufacturing & Production: *Bruce Johnson*
Managing Editor: *Mary Carnis*
Manufacturing Manager: *Ed O'Dougherty*
Art Director: *Marianne Frasco*
Marketing Manager: *Jimmy Stephens*
Cover Design: *Lorraine Castellano*
Cover Photos: *Richard Hamilton Smith, Corbis (Holstein cows); Corbis Digital Stock (llama); Corbis Digital Stock (pigs at trough); PhotoDisc, Inc. (rabbit couple); Kim Steele, PhotoDisc, Inc. (turkey chicks); Alan and Sandy Carey, PhotoDisc, Inc. (Hereford cow and calf)*
Composition: *Carlisle Communications, Ltd.*

Prentice-Hall International (UK) Limited, *London*
Prentice-Hall of Australia Pty. Limited, *Sydney*
Prentice-Hall Canada Inc., *Toronto*
Prentice-Hall Hispanoamericana, S. A., *Mexico*
Prentice-Hall of India Private Limited, *New Delhi*
Prentice-Hall of Japan, Inc., *Tokyo*
Prentice-Hall Singapore Pte. Ltd.
Editora Prentice-Hall do Brasil, Ltda., *Rio de Janeiro*

ISBN 0-13-920992-1

To my parents: their love, dedication, and encouragement enabled me to succeed. And to my children, Ilana and Aaron, who deserve much credit for their patience and support while I researched, wrote, and edited this text.

Contents

Preface   xi

About the Author   xiii

CHAPTER 1

Overview of Animal Science: Statistics, History, and Future   1
Objectives   1
Key Terms   1
Introduction   1
The Beginning of Animal Science   2
The Introduction of Animals to North America   3
The Growth of Animals in the United States   3
Taxonomy   7
World Politics, Numbers, and Economics   8
Animal Science in the Twenty-First Century   12
Summary   15
Evaluation Questions   15
Discussion Questions   17

CHAPTER 2

Animal Breeding and Genetics   18
Objectives   18
Key Terms   18
Introduction   19
The Cell: Carrier of Genetic Material   19
Mendel's Experiments   20
Incomplete, Partial, or Codominance   22
Sex Determination   23
Sex-Limited Genes   23
Sex-Linked Traits   23

Hereditary Defects   24
Systems of Mating   25
Summary   27
Evaluation Questions   27
Discussion Questions   29

CHAPTER 3

Animal Reproduction   30

Objectives   30
Key Terms   30
Introduction   31
Anatomy   31
Estrous Cycle   42
Fertilization   44
Gestation   46
Artificial Insemination   51
Summary   60
Evaluation Questions   61
Recommended Reading   64

CHAPTER 4

Animal Nutrition   65

Objectives   65
Key Terms   65
Introduction   65
Anatomy of the Digestive System   68
Essential Nutrients   78
The Uses of Feed Nutrients   87
Summary   92
Evaluation Questions   92
Discussion Questions   95

CHAPTER 5

Dairy Industry   97

Objectives   97
Key Terms   97
Introduction   98
History of the Dairy Industry   98
The Dairy Industry Today   101
Dairy Farm Income and Expenses   103
Dairy Breeds   104

Feeding and General Management of Dairy Cattle   108
Common Dairy Metabolic Disorders   123
Breeding, Selecting, and Judging Dairy Cattle   128
Milk Secretion and Milking Machines   137
Herd Health Disorders   147
Miscellaneous Dairy Cattle Housing   166
Summary   166
Evaluation Questions   167
Discussion Questions   170

CHAPTER 6

Dairy and Meat Goat Industry   171

Objectives   171
Key Terms   171
Introduction   171
Dairy Goat Management   178
Characteristics of Goat Milk   179
Feeding Goats   180
Lactation   181
Reproduction in the Goat   181
Parturition (Kidding)   182
Management of Kids from Birth to Puberty   183
Herd Health   184
Summary   187
Evaluation Questions   187
Discussion Questions   188

CHAPTER 7

Beef Production   189

Objectives   189
Key Terms   189
Introduction   190
U.S. Regions of Beef Production   190
Beef Breeds   194
The Beef Carcass   205
Stocker Systems   217
Fattening Cattle   218
Beef Cattle Husbandry, Facilities, and Equipment   222
Selection of Beef Cattle   236
Summary   249
Evaluation Questions   249
Discussion Questions   252

CHAPTER 8

Sheep Production   253

Objectives   253
Key Terms   253
Introduction   254
Breeds of Sheep   256
Sheep Nutrition   278
Sheep Reproduction   286
Tail Docking and Castration   290
Diseases of Sheep   290
Predators   300
Sheep Selection, Wool Care, and Marketing   301
Slaughtering Process of Sheep   307
Summary   309
Evaluation Questions   310
Discussion Questions   312
Recommended Reading   313

CHAPTER 9

The Swine Industry   314

Objectives   314
Key Terms   314
Introduction   315
Brief History of Swine   315
Taxonomy   316
Breeds of Swine   316
Swine Selection   323
The Pork Carcass   330
Swine Management   341
Swine Reproduction   347
Swine Nutrition   349
Swine Health Management   360
Animal Welfare   373
Summary   374
Evaluation Questions   374
Discussion Questions   377
Swine Breed Associations   377
Sources of Boar Semen   378

CHAPTER 10

The Poultry Industry   379

Objectives   379

Key Terms   380
Introduction   380
Taxonomy and Breeds   382
Anatomy and Physiology of the Chicken   388
Reproductive System and the Egg   392
Nutrition   393
Diseases and Parasites of Poultry   400
The Broiler and Layer Industry   410
Special Considerations for Turkeys, Ducks, and Geese   417
Summary   423
Evaluation Questions   423
Discussion Questions   426
Recommended Reading   426

CHAPTER 11

The Equine Industry   427

Objectives   427
Key Terms   427
Introduction   427
Taxonomy   432
Breeds of Horses   433
Donkeys: Jacks and Jennets   457
Mules   457
Horse Reproduction and Management   458
Horse Selection   467
Teething   472
Nutrition   474
Diseases of Horses   487
Injuries of Horses   505
Summary   505
Evaluation Questions   505
Discussion Questions   508
Recommended Reading   508

CHAPTER 12

The Rabbit Industry   509

Objectives   509
Key Terms   509
Introduction   509
Domestication and Breeds of Rabbits   510
Selection of Breeding Rabbits   515
Pet Rabbit Industry   516
Rabbit Meat Industry   516

Laboratory Rabbits   517
Reptile Feeder Market   517
Rabbitry Management   517
Feeding the Rabbit Herd   520
Reproductive Management   523
Rabbit Fur or Wool Industry   525
Rabbit Health Matters   526
Summary   529
Evaluation Questions   529
Discussion Questions   531

CHAPTER 13

The Camelid (Lamoid) Industry   532

Objectives   532
Introduction   532
Llama Nutrition   534
Llama Reproduction   536
Male Berserk Syndrome   537
Herd Health   537
Uses of Llamas   540
Summary   542
Evaluation Questions   542
Discussion Questions   543
Recommended Reading   543

CHAPTER 14

Ostriches   544

Objectives   544
Introduction   544
Reproduction   545
Meat, Hide, and Feathers   546
Dangerous Birds   547
Nutrition   547
Diseases and Other Health-Related Disorders   549
Summary   550
Evaluation Questions   551
Discussion Questions   551

Glossary   553

Index   569

Preface

When I first came to L.A. Pierce College, almost 30 years ago, the introduction to livestock production course, as it was called back then, was being taught by Professor Lindsay Boggess. Its major emphasis was introducing animal, dairy, and equine science students, general agriculture students, and those pursuing careers in veterinary medicine to the basic terminology and management procedures associated with the various production animal agriculture enterprises in the United States. Most students taking the course were production oriented at that time and most came from families either directly or indirectly connected to animal agriculture.

In the early 1970s, the field of animal husbandry taught the *art* of raising livestock in a healthy, humane, and profitable manner. Today, it is animal *science* and not husbandry that is the main focus of introductory courses in livestock raising. A tremendous advancement in the use of chemistry, physiology, genetics, molecular biology, and nutrition, along with animal welfare, has changed the emphasis and interests in production agriculture. Although the art of animal production is still extremely important and is absolutely necessary for profitable and humane livestock enterprises, this text will concentrate primarily on the science. We suggest that students take additional laboratory hands-on courses to learn the art. It takes thousands of hours working with livestock and a trained master to really learn the art of animal husbandry.

In 1976, when I began teaching in the animal science department at Pierce College, we were using Blakely and Bade's *The Science of Animal Husbandry.* For more than two decades, it proved to be an excellent text to introduce students to this field. I was very fortunate in being asked to help rewrite and edit the sixth edition, published in 1994. Its use throughout the world at many colleges, universities, and high schools indicates its popularity.

This first edition of *Introduction to Animal Science* uses much the same format, photographs, and material that we included in our last edition of Blakely and Bade's text but adds additional emphasis on more modern husbandry, science, and welfare concerns of the twenty-first century. I would like to acknowledge and thank Drs. James Blakely and David H. Bade for their contributions to this text. We have included in this text some additional species that seem to have caught the interest of many new agricultural entrepreneurs around the United States.

Fourteen chapters covering introductory animal reproduction, genetics, nutrition, breeds, animal health, and general management of the various common livestock species are included. Some of the chapters are quite large and are not meant to be covered in one classroom setting but are simply divided into units based on animal species. At the end of each chapter an evaluation section will assist students in preparing for exams and quizzes. A glossary is found at the end of the text to enable students to comprehend new terms throughout their reading.

Dr. L. S. Shapiro

ACKNOWLEDGMENTS

I am greatly indebted to Drs. James Blakely and David Bade for their contribution to this text. Their original work, *The Science of Animal Husbandry,* formed the basic outline of this text. My longstanding collaborative relationship with the agriculture faculty at California Polytechnic State University, San Luis Obispo, California State Polytechnic University, Pomona, the University of California–Davis, and Oregon State University–Corvallis provided me with a clear understanding of the educational needs of undergraduate agriculture and veterinary students. It was this understanding that enabled me to recreate this first edition of *Introduction to Animal Science.* In particular, I want to acknowledge my former professors, the late Professor Harmon Toone, Dr. Herman Rickard, Dr. Joe Sabol, Professor Lindsay Boggess, Professor John Barlow, Professor Bernyl Sanden, Dr. Lloyd Swanson, Dr. Nancy East, Dr. Peter Cheeke, Dr. Dale Weber, and Dr. David Church.

Industry organizations and representatives such as the National Pork Producers Council, Rex Sprietsma of Westfalia-Surge, Inc., Tom Majeau and Dr. Craig Barnett of Bayer Agriculture Division, Coe Ann Crawford of VetLife, and Lori Wagner of Sport Horses of Color provided me with invaluable information and photographs that were used in this book.

Several of my colleagues provided meaningful criticism and added information from their areas of expertise. In particular I want to thank Professors Ronald Wechsler, Liz White, Rebecca Yates, Patrick O'Brien, Jana Smith, Russ Schrotenboer, and Bill Lander of L.A. Pierce College, Les Ferreira and Joe Sabol of California Polytechnic State University, San Luis Obispo, and Temple Grandin of Colorado State University, Fort Collins, for their encouragement and educational insight. I extend a special thank you to reviewers: John Mendes, Modesto Junior College; Murn Nippo, University of Rhode Island; and Brian J. Rude, Mississippi State University.

I thank my students, 63 of whom are now practicing veterinary medicine. These graduates help me stay in tune with the fast-moving trends of the new millennium. It is their energy and enthusiasm that drive me and guide me each semester.

I owe a tremendous appreciation to several individuals at Prentice-Hall who guided me through the development and editing stages of this text. I would like to especially thank Charles Stewart, Kate Linsner, and Debbie Yarnell for directing me through the various processes required in preparing this manuscript. Finally, I am particularly grateful to Lori Dalberg, my production editor, for catching all of my errors prior to publication. She has a tremendous amount of patience and talent.

About The Author

Dr. Leland S. Shapiro is the director of the preveterinary science program at Los Angeles Pierce College. He has been a professor of animal sciences for 24 years and is a member of the American Dairy Science Association, Dairy Shrine Club, Gamma Sigma Delta Honor Society of Agriculture, and Association of Veterinary Technician Educators, Inc. Dr. Shapiro was a dairy farmer for almost two decades, holds a California State pasteurizer's license, and for 14 years was a certification instructor for artificial insemination. Professor Shapiro is a member of the college's ethics committee and has completed two postdoctoral studies in bioethics. Dr. Shapiro is a University of California–Davis Mentor of Veterinary Medicine, and the recipient of several local college teaching awards as well as the prestigious The National Institute for Staff and Organizational Development Community College Leadership Program (NISOD) Excellence in Teaching Award, in Austin, Texas.

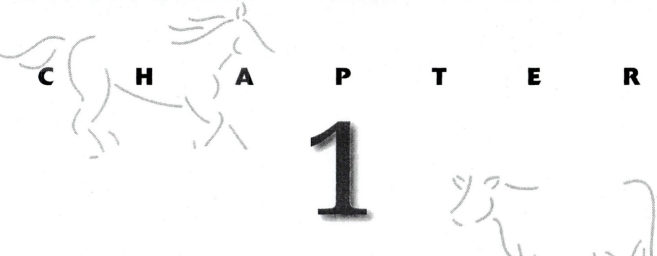

1

Overview of Animal Science: Statistics, History, and Future

OBJECTIVES

After completing the study of this chapter you should know:

■ The distribution of animal agriculture, both domestically and internationally.
■ The history of livestock domestication.
■ The history of livestock introduction and development in the United States.
■ Classifications of livestock and other domestic animal species according to their taxonomy.
■ How both local and world politics affect the price and availability of food.
■ Basic livestock names and physiological parameters, such as temperature, pulse, and respiration rates for each of the major common species.

KEY TERMS

Binomial nomenclature	Genus	Phylum
Class	Kingdom	Species
Classification	Order	Taxonomy
Family		

INTRODUCTION

A generation ago, the field of animal science was called animal husbandry. The production of healthy cattle, sheep, horses, poultry, and goats was more an art than a science. Skills needed in the feeding, breeding, and management of livestock were passed down from one generation to the next with little need for formal college or university education. Today, many ranchers and farmers are college educated. They rely on science-based, proven facts as to which, how, and when to feed, breed, medicate, and manage their stock. Ranchers and farmers rely on old wives tales less often today. Profit margins are slim, and detailed record keeping is necessary for profitable operations.

Farmers of the twenty-first century are the main stewards of the land. In addition to producing food and fiber, farmers have a separate requirement of

1

preserving what little is left of our natural resources. With only 2 percent of the voting public directly involved with production agriculture, farmers must constantly defend their use of limited natural resources (e.g., water, soil, and sometimes pastureland) for which city-raised citizens are also in competition.

THE BEGINNING OF ANIMAL SCIENCE

Today, we know of more than 1.5 million species of animals. New species are being discovered constantly. It is estimated that living animals represent less than 1 percent of all the animals that have ever existed.[1] There are more than 15,000 species of mammals known to exist today. Of those, there are hundreds of species of ruminants. Through natural selection and human domestication, certain species have become predominant in specific regions of the world. Humans and domestic animals have developed a synergistic relationship. The domesticated animals have flourished and in turn have provided humans with food, fiber, transportation, and draft power.

The early mountaineers likely domesticated the trailable species (cattle, sheep, horses, goats, and dogs) and followed them as they grazed or hunted. Plains farmers domesticated the other species (cats, pigs, chickens, geese, and ducks) by driving them into the farmers' villages during the night.

Domestication of animals occurred over thousands of years. Fossil remains of animals resembling cattle date back 3 to 4 million years. The prehistoric remains of an ancient oxlike animal have been found in Asia. It had horns similar to those of goats, with a spread of 6 to 7 feet, was about 6 feet tall, and grazed on twigs and shoots of trees. It was not until the New Stone Age (8000–6000 B.C.E.) that cattle were first domesticated in Europe and Asia. Modern-day cattle are believed to have descended from Aurochs, which roamed the forests of central Europe. Aurochs became extinct in the year 1627.

The ancestor of today's commercial chicken is believed to have originated in Southeast Asia. It is known today as the Red Jungle Fowl (*Gallus bankiva*). The earliest records of humans keeping poultry dates back to 3200 B.C.E. in India. The Egyptians, in 1400 B.C.E., bred artificially incubated chicken eggs and sold chicken meat and eggs.

Sheep were probably domesticated during the early Neolithic Age, making them among the first animals to feed the guiding hands of men and women. Probably first domesticated in central and western Asia, sheep are pictured on Egyptian monuments dated between 5000 and 4000 B.C.E. Biblical passages abound with mention of sacrificial lambs, so it is well documented that sheep were found in many civilizations. Wool fabrics have been found in the ruins of Swiss lake villages dating back 10,000 to 20,000 years ago.

Fossil remains of the horse have been discovered in the river valley clays, sands, and sandstone of the American West. The prehistoric horse was a four-toed animal. A million years ago, the "true horse" migrated throughout the world over existing land bridges. The horse in the Americas disappeared during the Pleistocene epoch. The horse was domesticated between 4500 and 2500 B.C.E. in ancient Mesopotamia. Ancient Egyptians, Greeks, and western Europeans used this "new horse" as a beast of burden, for transportation, and for warfare centuries before Columbus landed in North America.

---

[1]P. L. Stein and B. M. Rowe, *Physical Anthropology,* 5th ed. (New York: McGraw-Hill, 1993).

Although not native to America, swine are mentioned as early as 4900 B.C.E. in writings of Chinese historians and about 1500 B.C.E. in Biblical writings. Archaeological evidence demonstrates that swine may have been first domesticated as early as 9000 B.C.E. in the eastern part of Papua New Guinea. There is further evidence of swine domestication in Jericho (Palestine) around the year 7000 B.C.E. They were well-known in historical accounts of what is now Great Britain by about 800 B.C.E.

THE INTRODUCTION OF ANIMALS TO NORTH AMERICA

When Columbus reached America in 1492, he found no domestic animals. He brought horses, cattle, sheep, and hogs on his second voyage in 1493, and small importations continued periodically. Vera Cruz brought cattle from Spain to Mexico in 1521. Hernando De Soto introduced swine from Cuba to Florida in 1539. The horse, brought by the Spanish Conquistadors who explored much of the Americas in the early sixteenth century, soon became established. The animals imported from Europe were used mainly for milk, butter, hides, and draft. With wild game plentiful, meat was not the citizens' main concern. These animals, under the guiding hands of such notables as George Washington and Thomas Jefferson, multiplied, and purebred herds developed in eastern North America. After the American Revolution, livestock moved westward, and by the early nineteenth century it was distributed over most of the East, the South, and the Far West.

Similar movement of stock created large herds of cattle, sheep, goats, and horses at missions throughout the Spanish territory. Herds of 424,000 cattle at two missions were recorded. Thus large Mexican ranches developed, and many still exist today.

THE GROWTH OF ANIMALS IN THE UNITED STATES

Cattle in America

Prior to the American Civil War, large herds were marketed to the East for hides and tallow through the rendering plants in New Orleans. During the Civil War, the eastern and northern markets were cut off, and cattle multiplied in Texas. By the end of the war, cattle in Texas were selling for $3 to $6 per head, whereas in the North and East they were bringing $30 to $60 per head. The possibility of large profit led to the romantic period of the trail drives from Texas to meet the transcontinental railroad lines in Kansas. An estimated 10 million head of hardy, rough Longhorn cattle traveled the 600- to 1,700-mile trail to market (see figure 1–1).

The introduction of barbed wire in the 1870s, along with both the end of the Indian Wars and the killing of the buffalo, opened the Midwest for cattle production. When Columbus landed in 1492, it is estimated that there were 60 to 80 million American bison (buffalo). By 1900 there were only 2,000 left. American soldiers and others slaughtered them—mainly to destroy the spirit of the American Indian.[2]

---

[2]L. S. Shapiro, *Applied Animal Ethics* (Albany, NY: Delmar, 2000).

The cattle industry grew almost overnight, with the end of the open range announced by the coming of barbed wire. This closing of individual herds meant great strides in animal improvement. Closer supervision, a critical eye, and sound business practices dictated heavier concentrations of more efficient animals on dwindling lands. The quickest means of improvement lay in the practice of mating superior sires with existing herds. The purebred sire now made his contribution, demand outstripped supply, and replacements changed in type and temperament. The Longhorn was literally bred out of existence, ushering in the golden purebred era and marking the demise of one way of life and the birth of another.

Out of this romantic history arose the dynamic business industry of animal science that we know today. Research, technology, and improved management techniques are being developed and scrutinized to keep up with an affluent and ever-increasing population's demands for livestock goods and services. Environmental protection demands, limited resources, changes in animal welfare sensitivities, and an increasing number of hungry people to feed challenge this industry today.

Sheep in America

The only sheep native to North America are the bighorn, or Rocky Mountain, sheep. Columbus brought a few sheep in 1493. Cortez introduced Spanish Merinos to Mexico in 1519 and undoubtedly brought other types of sheep as the Spaniards moved into New Mexico, opening missions and establishing basic farming and husbandry practices. The multicolored sheep the Spanish imported to Mexico are thought to be the forerunners of the multicolored Navajo Indian sheep that exist today and produce a coarse blanket and rug wool. Although accounts of the exact type are not clear, the first direct importation, consisting of British breeds, was to Jamestown in 1609.

FIGURE 1–2 Sheep herds are still prevalent in the western states, where they utilize plants and land unsuitable for many other agricultural purposes.
Photo courtesy of James Blakely.

The first importations of sheep to North America were primarily for producing wool. Predators' attacks, poor management, and lack of shelter caused the quality and yield of wool of these early breeds to quickly deteriorate. In 1783, Merinos were imported from Spain to improve the quality and yield of wool. By 1810, the northeastern part of the United States was the main sheep-producing area of the country. In 1830, colonists moved westward over the Allegheny Mountains, where sheep progressed in quality and numbers.

Due to limited supplies of cotton during the Civil War, the price of wool increased dramatically. By 1884, there were 50 million sheep in the United States. This figure was not surpassed until 1942, when 56 million head of sheep were estimated in the United States. Since then, numbers have steadily declined because of competition from synthetic fibers and pressure from other meat products. Production shifted from the east to the Rocky Mountain region by the turn of the twentieth century and finally to the 11 western states and to Texas (see figure 1–2).

Swine in America

American breeds of swine come from two wild stocks: the East Indian pig and the European wild boar. Hogs were first brought to the New World, from the Canary Islands to what is now the island of Haiti, by Columbus in 1493. Swine multiplied rapidly on the island and served as a continuing supply of stock to accompany expeditions. In 1539, Hernando De Soto launched an expedition from Cuba and landed in Florida with 13 hogs, which numbered 700 only 3 years later. De Soto explored from Florida to Missouri, and it is thought that many razorback swine are descendants of these nondescript early hogs that wandered off. These same hogs served as foundation stock for later development into the many breeds in existence today. Mongrel sows were often trapped from the wild state and bred to imported boars, which began arriving as early as 1825. The early corn-growing areas of Tennessee, Kentucky, Ohio, and the Connecticut Valley became chief hog-producing centers and marketed their corn through swine.

As agriculture developed west of the Allegheny Mountains, so did the production of swine, a natural complement for corn and other grain farming. Pork was popular because it could be easily smoked, pickled, or otherwise preserved and it had a high level of energy because of its fat concentration—a fact not overlooked by the hard-working pioneers.

With the infusion of Chinese, Neapolitan, Russian, and European boars, swine-breeding programs began to evolve. They eventually produced thoroughly genuine American breeds, many of which are held in higher esteem than the ancestors from which they sprang. Most of the breeds raised in America originated from these early breeding programs and continue to draw on the existing genetic material for improvement within the breed. Swine have continued to make a steady, though fluctuating, climb in total numbers. Although styles and types have changed with human demands, both purebred and crossbreeding systems continue to have a substantial impact on the agricultural well-being of America.

Horses in America

Although Columbus brought horses to the West Indies in 1493, it was not until 1519, when Cortez landed on the shores of Mexico, that the initial herd of 16 head was reintroduced to the Americas. By 1521, more than 1,000 head had been imported to support the conquest by the ambitious Spaniards. Hernando De Soto, credited also with the introduction of swine to the United States, had 237 horses in his expedition to Florida in 1539. After the death of De Soto during the expedition, many of his followers abandoned their horses and returned by boat from the upper Mississippi River. Coronado followed in 1540, bringing more horses. The band led by Coronado no doubt followed a plan similar to that of De Soto, leaving stock that would thrive on the grassy plains, eventually to be prized by American Indians, master horsemen. During the early seventeenth century, the Spanish missionary movement into New Mexico also enhanced the rising horse population; it contributed further to wild bands because of lost, strayed, or abandoned animals.

While the wild horse population was busy developing under the guiding hand of the Indian tribes, 17 saddle horses for riding were imported to Jamestown between 1609 and 1611; at least two of these horses were stallions.

Because oxen were primarily used for plowing, the horse was neglected as a useful working animal until 1840, when the buggy appeared on the scene. Larger draft horses were imported from France to pull the larger wagons and freight-moving equipment that followed the buggy. The greater size of some horses led to their use as plow horses, a wise move for plantation owners. These large landowners then developed easy-riding saddle horses and flashy, high-stepping buggy horses, which they used to oversee their large holdings in comfort and speed.

The development of the western range, the trail drives after the Civil War, and the increasing affluence in America led to the breeding of cow horses, pleasure horses, and racing stock. By 1867, it was estimated that there were 7 million horses in the United States. This number increased to 21 million by 1913, but the coming of the automobile and the tractor paralleled the decline in horse numbers, which by 1960 had fallen to 3 million head. However, the love affair with the horse in America is far from over. The sport of racing and the renewed interest in recreation horses has spurred this proud, magnificent species into popularity once again. At the last census, it was estimated that there are up to 7 million horses in the United States.

Miscellaneous Animal Enterprises in America

In addition to the mentioned commonly raised livestock, many Americans are now raising exotic animals, endangered species, and culturally important alternatives to the standard livestock species. These animals include alligators, crawfish, alpacas, llamas, deer, buffalo, Nelore cattle, Waygu cattle, Chinese pigs, potbellied pigs, ostriches, emus, shrimp, and snails. Some of these enterprises will be discussed later in the text.

TAXONOMY

Classifying animals provides us with a means of comparing, discussing, and contrasting life forms. To understand the nature of living things, including their differences and similarities, we sort life forms into a number of groups that have similar physical characteristics or that are related to each other. The system of placing these life forms in order is called **classification** (see table 1–1). The science of classifying them into different categories is known as **taxonomy**.

Classification system of placing life forms in order based on similar physical characteristics

Taxonomy science of classifying life forms into different categories

TABLE 1–1	Classification of Domestic Animals						
Common Name	Phylum	Subphylum	Class	Order	Family	Genus	Species
Alligator	Chordata	Vertebrata	Reptilia	Crocodilia	Alligatoridae	*Alligator*	*mississippiensis*
Alpaca	Chordata	Vertebrata	Mammalia	Artiodactyla	Camelidae	*Llama*	*pacos*
Bison	Chordata	Vertebrata	Mammalia	Artiodactyla	Bovidae	*Bison*	*bison*
Cattle	Chordata	Vertebrata	Mammalia	Artiodactyla	Bovidae	*Bos*	*taurus*
Cattle	Chordata	Vertebrata	Mammalia	Artiodactyla	Bovidae	*Bos*	*indicus*
Chicken	Chordata	Vertebrata	Aves	Galliformes	Phasianidae	*Gallus*	*domesticus*
Donkey	Chordata	Vertebrata	Mammalia	Perissodactyla	Equidae	*Equus*	*asinus*
Duck	Chordata	Vertebrata	Aves	Anseriformes	Anatidae	*Anas*	*platyrhyncha*
Emu	Chordata	Vertebrata	Aves	Casuariiformes	Dromiceidae	*Dromiceius*	*novaehollandiae*
Goat	Chordata	Vertebrata	Mammalia	Artiodactyla	Bovidae	*Capra*	*hircus*
Goose	Chordata	Vertebrata	Aves	Anseriformes	Anatidae	*Anser*	*anser*
Horse	Chordata	Vertebrata	Mammalia	Perissodactyla	Equidae	*Equus*	*caballus*
Human	Chordata	Vertebrata	Mammalia	Primates	Hominidae	*Homo*	*sapiens*
Llama	Chordata	Vertebrata	Mammalia	Artiodactyla	Camelidae	*Llama*	*glama*
Ostrich	Chordata	Vertebrata	Aves	Struthioniformes	Struthionidae	*Struthia*	*camelus*
Rabbit	Chordata	Vertebrata	Mammalia	Lagomorpha	Leporidae	*Oryctolagus*	*cuniculus*
Sheep	Chordata	Vertebrata	Mammalia	Artiodactyla	Bovidae	*Ovis*	*aries*
Snail	Mollusca	—	Gastropoda	Sigmurethra	Helicacea	*Helix*	*aspersa*
Swine	Chordata	Vertebrata	Mammalia	Artiodactyla	Suidae	*Sus*	*scrofa*
Swine	Chordata	Vertebrata	Mammalia	Artiodactyla	Suidae	*Sus*	*vittatus*
Turkey	Chordata	Vertebrata	Aves	Galliformes	Meleagrididae	*Meleagris*	*gallopavo*

TABLE 1–2	Common Names for **Various Domestic Species**						
Species	Bovine	Equine	Ovine	Caprine	Porcine	Avian	Lapine
Common Name	Cattle	Horse	Sheep	Goat	Pig	Chicken	Rabbit
Male	Bull	Stallion	Ram	Buck	Boar	Rooster	Buck
Female	Cow	Mare	Ewe	Dow	Sow	Hen	Doe
Parturition	Calve	Foal	Lamb	Kid	Farrow	Hatch	Kindle
Name of young	Calf	Foal	Lamb	Kid	Piglet	Chick	Kitling
Young male	Bull	Colt	Ram lamb	Buckling	Boar	Cockerel	—
Young female	Heifer	Filly	Ewe lamb	Doeling	Gilt	Pullet	—
Castrated	Steer	Gelding	Wether	Wether	Barrow	Capon	—
Group name	Herd	Herd/band	Flock	Band	Drove/herd	Flock	Hutch

Source: L. S. Shapiro, *Principles of Animal Science* (Simi Valley, CA: Ari Farms, 1997).

Binomial nomenclature system developed using two names—a genus and a species—in classifying all life forms

Genus taxonomy category composed of a group of similar species

Species taxonomy category that is subordinate to a genus

The system, developed by Carolus Linnaeus in 1758, uses **binomial nomenclature.** Each species is given a two-part name; the first name represents the genus. A **genus** is a group of similar species, and this name is always capitalized. The second name is the **species,** and it is never capitalized. Both names should either be underlined or set in italic type. Both names are Latin based. Common names for various domestic species are found in table 1–2.

Species are classified into higher taxonomic levels based on their evolutionary relationships. In the past, all living things were placed into one of two large groups, called *kingdoms.* Some forms of life did not readily fit into either of these two kingdoms, animal or plant. Today, we have as many as five commonly accepted kingdoms (monera, protista, fungi, plantae, and animalia). Some researchers still use the original 2 and others have expanded the number of kingdoms to as many as 20.

There are seven basic levels within the taxonomy: *kingdom, phylum, class, order, family, genus* and *species.* In this text we discuss only animals found in the animal kingdom. The phylum Chordata includes humans and all domestic animals. Animals under this classification possess a notochord, a dorsal hollow neural tube, and gill slits (at least in the larval stages); are bilaterally symmetrical; and have a segmented body and a closed blood system with a heart located below the pharynx.

The two main classes discussed in this text are Aves and Mammalia. The class Aves includes small birds that lack teeth but have feathers, scales on the legs, well-developed lungs, and a completely subdivided heart with four chambers; birds maintain a constant body temperature of 41°C (106°F), and produce a small number of large eggs. The class Mammalia includes animals with mammary glands and hair, whose young develop in the uterus and (except in one order) are born alive; the animals have a greatly enlarged forebrain, metanephric kidneys, a diaphragm that separates the thoracic and abdominal cavities, and a four-chambered heart and are warm blooded, or homeothermic (which means they can maintain their own body temperature).

WORLD POLITICS, NUMBERS, AND ECONOMICS

For more than 10,000 years, people in various regions of the world have been using domestic animals as a source of food, to assist in labor, to assist in war, for sport, or as companions. Many nations have more than enough animals to feed their starving populations, but because of religious and/or cultural barriers cannot eat these animals; yet the animals utilize resources that could be used for harvesting other feeds. Many nations, because of dictatorial politics, refuse to allow available resources to be used for feeding their own people. Still others simply lack the basic knowledge necessary to produce abundant food supplies with available resources.

The United States has 5 percent of the world's population and supplies more than 20 percent of the world's total food. U.S. farmers produce 40 percent of the world's soybeans, 15 percent of the world's wheat, and 25 percent of the world's other grains. The U.S. exports 60 percent of its soybean crop, 46 percent of its wheat, and 27 percent of its feed grains. The exportation of these and other crops is very important in balancing U.S. trade.

The world's population is expected to almost double during the next half century, from 5.3 billion people in 1990 to more than 10 billion by 2050.[3] Food demand in many developing countries will more than double by 2025 and rise by another 50 percent by 2050.[4] Environmentalists and ecologists argue that to feed this larger population, farmers must continue to intensify agricultural practices. It is feared that these intensified practices will cause further ecological damage to limiting resources. The United Nations Environment Program reports that 17 percent of the land supporting plant life worldwide has lost value over the past 45 years. The loss is most widespread in Central America (25 percent) and least widespread in the United States (5.3 percent).

Realizing the importance of topsoil to the production of our food, clothing, and shelter, farmers in this country have adopted new methods of conservation tillage. This tillage system leaves just enough crop residue to adequately protect the soil from erosion, generating the least possible amount of sediment. Conservation tillage is the number one defense against sediment in the water, protects waters from sediment-associated chemical losses (fertilizers and pesticides), decreases fuel consumption, reduces machinery investment and repair costs, and increases the number of acres farmed per person hour. The production of these crops, whether to be consumed directly by people or indirectly through animals, is of great ecological concern to the planet. Yet we must eat.

Students of animal agriculture must become familiar with both ecological and animal welfare terminology. With only 2 percent of the American public involved with production agriculture, farmers must constantly defend their farming methods to a voting public capable of legislating farmers out of existence. Efforts of conservation should also be made for the farmers themselves. The land the farmers work is usually a lifelong residence and one passed on to future generations.

The amount of food purchased by average U.S. workers' pay has increased markedly. In 1950 the average American earned $1.34 per hour; by 1995 the average wage had increased to $11.46 per hour. Table 1–3 shows the increase in purchasing power for common food items.

Although many animal activists, ecologists, and environmentalists criticize American agriculture as being too factory styled, it has provided U.S. citizens with the cheapest supply of food in the world. In 1820, one American farmer fed four people. A very large percentage of the population thus needed to be

---

[3]J. Bongaarts, "Can the Growing Human Population Feed Itself?" *Scientific American* 270(3):36, 1994.

[4]Cargill Newsletter, 1995.

TABLE 1–3	Economics of Food Purchases in the United States	
Item	**Amount Purchased Per Hour of Labor**	
	1950	**1995**
Frying chicken (pound)	2.3	12.5
Milk (half gallon)	3.5	8.0
Eggs (dozen)	2.2	12.4
Pork (pound)	2.5	5.9
White bread (pound)	9.3	14.5
Ground beef (pound)	2.4	8.4

Source: USDA, 1997.

TABLE 1–4	Percentage of Personal Disposable Income Spent on Food in Other Nations
Nations	**Percentage**
Japan, England	26
Canada, Australia, Sweden	20–30
France	31
Poland, Korea	50
East India	60
Russia, China	65

Source: L. S. Shapiro, *Principles of Animal Science* (Simi Valley, CA: Ari Farms, 1997).

TABLE 1–5	History of Percentage of Personal Disposable Income Spent on Food in the United States
Year	**Percentage**
1950	30
1960	17.5
1970	13.9
1980	13.5
1990	11.8
1998	9.8[a]

Source: L. S. Shapiro, *Principles of Animal Science* (Simi Valley, CA: Ari Farms, 1997).
[a]Estimated.

directly involved with production agriculture. In 1998, the average American farmer fed 212 people.[5] On an ecological basis this increase in production efficiency is a plus. We can now feed more people, using less land and fewer resources (e.g.,water, fertilizer, fuel, and labor), than we ever could.

Whereas most Europeans spend from 30 to 65 percent of their personal disposable income on food, Americans spend less than 10 percent (see table 1–4). Consequently, Americans can use 90 percent of their disposable income on items other than food. They are the envy of the world. In comparing what the average American spends on food, the rest of the world is eating as Americans did before World War II (see table 1–5).

---

[5]Archer Daniels & Midland (ADM), 1999.

How do we feed a hungry world? Which animals are the most efficient? What kind of environmental interactions affect animals' ability to convert feed for nonhumans into food for humans? Where are all of these animals located? (See table 1–6.) Are Americans obligated to give up their eating practices and their methods of farming to help feed the rest of the world? These questions need to be discussed and answered, but in an animal ethics course. We will cover some of these questions later in the text.

TABLE 1–6	World Animal Numbers	
Species	**Leading Nations**	**Animal Numbers**
Cattle	Total	1,300,000,000
	India	271,437,000
	Brazil	130,700,000
	Russia	112,400,000
	United States	100,110,000
Sheep	Total	1.200,000,000
	China	205,000,000
	Australia	157,222,000
	Russia	128,000,000
	New Zealand	57,740,000
	India	48,178,000
	United States[a]	7,238,000
Goats[b]	Total	526,000,000
	India	—
	China	—
	Pakistan	—
	Nigeria	—
	United States[c]	2,000,000
Buffalo[b]	Total	140,000,000
	India	—
	China	—
	Pakistan	—
	Thailand	—
Camels[b]	Total	19,000,000
	Somalia	6,000,000
	Sudan	2,800,000
	India	1,500,000
	Ethiopia	—
Yaks[b]	Total	13,000,000
	Russia	—
	Tibet	—
Llamas[b]	Total	13,000,000
	South America	—
Chickens[b]	Total	11,300,000,000
	China	2,700,000,000
	Russia	—
	United States	1,500,000,000[d]
	Brazil	—

(continued)

TABLE 1-6	World Animal Numbers (continued)	
Species	**Leading Nations**	**Animal Numbers**
Swine	Total	803,753,000
	China	475,000,000
	Brazil	31,427,000
	United States	62,156,000
	Germany	24,795,000
Turkeys[b]	Total	362,000,000
	United States	285,000,000
	Russia	—
	Italy	—
Ducks[b]	Total	527,000,000
	China	—
	Bangladesh	—
	Indonesia	—
	Vietnam	—
	United States	6,000,000
Horses[b]	Total	60,000,000
	China	—
	Mexico	—
	Russia	—
	Brazil	—
	United States	5,000,000–7,000,000
Donkeys[b]	Total	43,000,000
	China	—
	Ethiopia	—
	Mexico	—
	United States	50,000
Mules[b]	Total	15,000,000
	China	—
	Mexico	—
	Brazil	—
	United States	30,000

Source: L. S. Shapiro, *Principles of Animal Science* (Simi Valley, CA: Ari Farms, 1997).
[a]Agricultural Statistics, 1999. National Agricultural Statistics Source, Livestock and Economics Branch, USDA.
[b]Current data not available for numbers of animals for each nation.
[c]U.S. numbers are given for comparison to world numbers even when not a leading country.
[d]Includes 281 million laying hens and a total of 6.4 billion broilers per year; 1.4 billion refers to numbers at any one time.

ANIMAL SCIENCE IN THE TWENTY-FIRST CENTURY

The optimum use of natural resources in the United States involves the use of both animals and plants to produce the nutrients humans need. Approximately half of the land area of the United States is grazing land that is not suitable for crop production. This land would be of no use for food production if it were not for the ruminant animals that graze this land. The United States has more than enough cropland to grow both feed grains (for livestock consumption) and food crops (for human consumption).

TABLE 1–7	Growth in Red Meat Consumption for the Chinese Middle Class		
Red meat	**Per Capita Consumption**		**Percentage of Increase**
	1995	**2005**	
Beef	19.7	36	82
Pork	100.9	108	7
Chicken	19.6	27	38

Source: Archer Daniels & Midland (ADM), 1999.

TABLE 1–8	Comparison of Chinese and U.S. Arable Lands
China	**United States**
286 million acres	426 million acres
9 percent of the world's arable land	13 percent of the world's arable land
22 percent of the world's population	5 percent of the world's population
<0.25 acre per person	1.64 acres per person

However, many areas of the world do not have this same luxury. For example, the most populous nation, China, does not possess the technology and expertise to feed its growing population. China is projected to import nearly 23.5 million tons of coarse grains (more than 18 percent of the total world trade) by the turn of the century.[6] The demand for red meat will also increase dramatically as China's middle class expands (see table 1–7).

The United States has twice the number of cultivable acres as China (see table 1–8). With China's much larger population, the Chinese will need to purchase food from other nations to feed its citizens. By 2010, it is estimated that China will have the largest economy in the world on a population basis. The United States could profit immensely by providing surpluses in plant and animal agriculture. Currently, the United States exports more than 2.5 billion pounds of meat per year. Much of its grain production (corn and soybeans) is exported through this meat.

The U.S. Department of Agriculture (USDA) Economic Research Service's 1998 outlook projected an increase in domestic per capita consumption of pork from 47.9 to 51.4 pounds. In addition, exports of pork were expected to grow from 1.1 billion pounds in 1997 to 1.21 billion pounds in 1998. Per capita consumption of chicken is expected to rise from 73.1 to 77.5 pounds, per capita consumption of beef is expected to drop from 67.2 to 65.6 pounds, and per capita consumption of turkey is expected to increase from 18.1 to 18.8 pounds.

Consumers are concerned about food safety and nutrition. The American public is concerned about cholesterol levels, fat, sugar and salt content, food additives, artificial coloring, caloric content, and pesticide and herbicide residues. In 1998, the USDA finally defined the term *organic* as used on food labels. Under the guidelines set by the USDA, livestock could be brought from

---

[6]Archer Daniels & Midland (ADM), 1999.

nonorganic sources into an organic operation but could not be moved in and out. Livestock have to be fed organically produced feeds, including pasture if appropriate. Animals can receive vitamin and mineral supplements but not hormones or antibiotics to stimulate growth. Under these guidelines, the use of any animal drug, except vaccines, is prohibited in the absence of illness. Manure must be managed to prevent water pollution and yet used to recycle its nutrients to the land. Agriculture Secretary Dan Glickman announced these standards and claimed they would "open the door to this exciting new market and economic opportunities for farmers."

The total number of farms in the United States will continue to decrease in number and increase in size. More efficient livestock production yielding a lower per unit cost of production will result. More automation, increased use of biotechnology in the production of medicaments and vaccines, biofriendly alternatives to herbicides and pesticides, superovulation, embryo splitting, sex-determined semen, and computer-managed livestock facilities will more than likely become the norm. Animal rights activists will continue to criticize the use of genetically engineered animals, xenotransplants, to produce organs for saving human lives, the use of animals in biomedical research, and the use of animals to test artificial organs. In addition, the nonagriculturally trained public will continue to try and pass laws that will prohibit American farmers from practicing traditional plant and animal agriculture.

It is important for beginning students of animal and veterinary science to appreciate the norms in basic animal physiology. Average temperature, pulse, and respiration rates are included in table 1–9. These values are only averages and vary depending on the age and activity level of the animal and the ambient temperature and humidity. Familiarity with these values will allow students to more easily spot a veterinary or management problem before it becomes critical.

Table 1–10 outlines leading states in livestock and livestock products. These states will change throughout the years. The costs of production greatly

TABLE 1–9	Normal Rectal Temperatures, Pulse, Respiration Rate, and Chromosome Number for Domestic Animals			
	Temperature (°F)	**Pulse**	**Respiration Rate**	**Number of Chromosomes**
Chicken	107.1	300	30	78
Goat	103.8	75	16	60
Rabbit	103.1	140–150	50–60	44
Pig	102.5	70	13	38
Sheep	102.3	75	16	54
Dairy cow	101.5	60	20	60
Dog	101.5	60–120	14–16	78
Cat	101.5	100–120	20–30	38
Beef cow	101.0	50	14	60
Mare	100.0	35	12	64
Llama	100.0	60–90	10–30	74
Stallion	99.7	35	12	64
Human	98.6	58–104	12	46

Source: L. S. Shapiro, *Principles of Animal Science* (Simi Valley, CA: Ari Farms, 1997).

TABLE 1–10	**Leading States in Livestock and Livestock Products**					
Beef Cattle	**Feeder Cattle**	**Dairy Cattle**	**Eggs**	**Broilers**	**Turkeys**	**Swine**
Texas	Texas	Wisconsin	California	Arkansas	North Carolina	Iowa
Kansas	Kansas	California	Ohio	Georgia	Minnesota	North Carolina
Nebraska	Nebraska	New York	Pennsylvania	Alabama	Arkansas	Minnesota
Oklahoma	Colorado	Pennsylvania	Indiana	North Carolina	Virginia	Illinois
Missouri	Iowa	Minnesota	Georgia	Mississippi	Missouri	Indiana

Source: USDA, 1996.

affect shifts in various agricultural enterprises. These costs include not only labor but also taxes (some states are more tax friendly to the farmer than others),
land values, competing enterprises, and transportation costs to marketing and
distribution centers.

SUMMARY

With only 2 percent of the voting public directly involved with production
agriculture, farmers must remain the main stewards of the land. Increased
property values, citizens wanting to move to the "country," and farmers need-
ing to capitalize on their real estate in order to survive will increase the likelihood of additional prime farmland being developed. Developing nations with
increasing populations will require more food and fiber from the United States
and other first world nations. The numbers of livestock will continue to increase to meet population demands.

With continued scientific innovations, fewer and fewer farmers are producing the feed and fiber to which we have become accustomed. The growth of
American agriculture has greatly contributed to Americans' availability of excess funds to buy primarily nonfood items (e.g., homes, cars, and televisions).
With the continued demand for cheaper and more abundant food supplies,
American agriculture will become less labor intensive, more mechanically
dependent, and more scrutinized for its effect on the environment and animal
welfare.

American farmers will need to justify their use of limited resources (such as
water, soil, oil, fuel, and power), management techniques that may require the
caging or discomfort of animals, and labor demands that may be different from
what has become the norm for nonfarm occupations. American farmers will
need to become diplomats, educators, and politicians as well as scientifically
suave entrepreneurs in the twenty-first century.

EVALUATION QUESTIONS

1. There are _____ species of animals in the world.
2. It is estimated that living animals represent less than _____ percent
   of all the animals that have ever existed.
3. Modern-day cattle are believed to have descended from _____ that
   roamed the forests of central Europe.

4. The ancestor of today's commercial chicken is believed to have originated in _____ .
5. The _____ were the first people to artificially incubate chicken eggs.
6. The prehistoric horse is thought to have originated from _____ .
7. When Columbus landed in North America, there were _____ head of American bison on the continent.
8. By 1900 only _____ bison were left in America.
9. _____ are the only sheep native to North America.
10. American breeds of swine come from two wild stocks: _____ and _____ .
11. The system of placing life forms in order is called _____ .
12. The science of classifying animals into different categories is called _____ .
13. _____ is a group of similar species.
14. The seven basic levels of taxonomy (in order) are _____ , _____ , _____ , _____ , _____ , _____ , and _____ .
15. The phylum with a notochord, dorsal hollow neural tube, and gill slits is _____ .
16. An animal's ability to maintain its own body temperature is called _____ .
17. The class that bears its young alive, has mammary glands, and has body hair is _____ .
18. A young rabbit is called a _____ .
19. Parturition for a sow is referred to as _____ .
20. A castrated rooster is called a _____ .
21. A group name for pigs is a _____ .
22. The United States has only _____ percent of the world's population and supplies more than _____ percent of the world's food.
23. _____ percent of the American population is directly involved with production agriculture.
24. Americans spend _____ percent of their disposable income on food.
25. The number of cattle in the world is _____ .
26. The number of sheep in the world is _____ .
27. The nation with the largest number of cattle is _____ .
28. The nation with the largest number of chickens is _____ .
29. The nation with the largest number of pigs is _____ .
30. The number of horses in the world is_____ .
31. The number of cattle in the United States is _____ .
32. What is the average rectal temperature of the major livestock species? Average pulse and respiration rates? Chromosome numbers?
33. Where might you find large numbers of camels?
34. What three areas (nations) have the highest population of cattle? Sheep? Goats? Chickens? Pigs? Turkeys? Horses?

DISCUSSION QUESTIONS

1. Why is the average American spending so much less (based on a percentage of disposable income) for food than those living in the rest of the world? Why is there such a difference?
2. Why do so many farmers readily adopt no-till, mulch-till, and ridge-till methods compared to conventional till? Why do so many farmers seek nonchemical means of ridding pests (plant and animal) on their farms?
3. What type of job opportunities do college graduates in agriculture have today that they did not have a generation ago?
4. How was the domestication of livestock important to the civilization of the world?

CHAPTER

2

Animal Breeding and Genetics

OBJECTIVES

After completing the study of this chapter you should know:

- The difference between animal breeding and animal propagation.
- The effect of the environment on the phenotypic expression of a genetic trait.
- The systems of mating used in animal agriculture.
- The differences between mitosis and meiosis.
- Generally, Mendel's theories and how they might relate to current methods of selection and breeding of animals.
- How the Punnett square is used to predict the offspring in the mating of two heterozygotes.
- How to fill out, read, and compare bracket pedigrees.
- Differences between sex-linked and sex-limited traits.

KEY TERMS

Allele
Autosomes
Bracket pedigree
Chromosomes
Conformation
Crossbreeding
Cytoplasm
Deoxyribonucleic
  acid (DNA)
Diploid
Dominant
Genes
Genotype

Grade
Haploid
Heterozygous
Holandric inheritance
Homozygous
Inbreeding
Linebreeding
Locus
Meiosis
Mitosis
Mutation
Nucleus

Outbreeding
Phenotype
Probability
Punnett square
Purebred
Recessive
Reduction division
Registered
Sex linked
Somatic
Type
Zygote

INTRODUCTION

Animal breeding is not animal propagation. The difference between the two is the increase in the quality of each generation with animal breeding versus the simple increase in animal numbers with animal propagation. The genetic makeup of an animal, its **genotype,** is greatly affected by environmental interactions. An animal's **phenotype,** its actual performance, thus depends not only on the genetic material passed down by each parent but also the effects of temperature, feed, stress, and management. Mathematically,

$$P = G + E$$

where P represents an animal's phenotype (e.g., milk yield, growth rate, wool production, and speed), G represents the animal's genotype or genetic makeup, and E represents the environmental effects to which the animal is exposed (e.g., temperature extremes, wind, nutrient deficiencies or toxicities, additional stressors from management, facilities, and competition with others).

Robert Bakewell (1760–1795) in England accomplished the first documented genetic improvement of cattle. Prior to Bakewell, natural selection was the main determinant of evolution for the bovine species. The Collings brothers used Bakewell's breeding principles to develop the Shorthorn breed of cattle and the first herd book in 1822. Soon thereafter, many new breeds and herd books were created. Breeders established ideal **type,** or *conformation,* for their respective breeds. These ideals were primarily color, size, and conformation that the farmers felt represented superior production traits. The opportunity for farmers to make real genetic progress, breeding animals, did not come about until the beginning of the twentieth century. The rediscovery of Gregor Mendel's research in 1903 by Walter J. Sutton and others and the first dairy cow–testing association in the United States in 1905 created new genetic opportunities for farmers.

The Austrian monk Gregor Mendel performed his now famous breeding experiments with peas and published his results in 1866. Mendel's laws of inheritance are the basis of all genetic and animal-breeding theory practiced since his time.

■ **Genotype** genetic makeup of an animal

■ **Phenotype** physical characteristics and performance of an animal

■ **Type** physical conformation of an animal

THE CELL: CARRIER OF GENETIC MATERIAL

We all learned through watching news reports of Dolly, the infamous cloned sheep of Dr. Willmut, that each one of our cells carries our entire genetic blueprint. Generally, cells have an outer membrane, inner *cytoplasm,* and a *nucleus.* The nucleus contains the hereditary material of the cell and controls the cell's growth, metabolism, and reproduction. The genetic material is found on rodlike structures called **chromosomes.** Each species has a characteristic number of chromosomes (see table 1–9). The mating of animals, even closely related animals (e.g., horse and donkey), with unlike chromosome numbers yields a sterile animal (mule or hinny). The resulting offspring is missing needed genetic material. The **genes** located on the chromosomes determine the animal's genotype or genetic makeup. Genes are composed of *deoxyribonucleic acid (DNA);* each of us has a slightly different combination. A pair of genes is necessary to determine a given trait. This pair of genes is located on the same site (**locus**) on homologous chromosomes.

All cells originate from other cells through a process of cell division. Two general kinds of cell division occur in the animal body: mitosis and meiosis.

■ **Chromosomes** rod-shaped bodies that occur in pairs in the nuclei of cells and that carry the genetic makeup of the animal in the form of genes arranged along the chromosome

■ **Genes** hereditary units located on a chromosome

■ **Locus** location on a chromosome where a particular gene is found

▓ **Mitosis** cell division of somatic cells that produces identical diploid daughter cells

▓ **Diploid** having two sets of homologous chromosomes; somatic cells are diploid

▓ **Meiosis** cell division that produces gametes, which contain the haploid number of chromosomes

▓ **Haploid** number of chromosomes found in the gametes, which is half the normal number of chromosomes found in the somatic cells (diploid condition)

▓ **Zygote** cell formed by the union of a sperm and an ovum

Mitosis refers to the cell division of *somatic,* or body, cells. Mitosis yields identical cells and is needed when the number of cells must increase to replace old, worn-out cells. Somatic cells have two of each type of chromosome (half from each parent). They are said to possess a **diploid** number, symbolized by $2n$ (n representing the number of chromosomes found in each of the gametes). Thus, mitosis maintains the parental number of chromosome sets in each of the daughter cells produced.

Meiosis refers to cell division in the gametes whereby the parental number of chromosome sets is reduced by half in each of the daughter cells. During meiosis, homologous chromosomes separate, carrying with them the members of the allelic pair. In other words, meiosis ensures that each parent contributes one-half of the genetic material of the progeny. A *reduction division* is necessary so that the union of a sperm and an ovum (during fertilization) restores the diploid number of chromosomes. This new reduced number is referred to as **haploid** and is symbolized by $1n$ or n.

The uniting of a sperm and ovum creates the **zygote,** or fertilized egg. The zygote is diploid. The chromosome pairs uniting are homologous. The first chromosome in the spermatozoa matches up with the first chromosome in the ovum and so on. This matching-up process results in many different possible combinations of traits in the offspring, contributing to animal variation. To calculate the possible variation in a population, one can use the formula $(2^n)^2$, where n is the haploid number. In humans, for example, the number of possible chromosomal combinations is $(2^{23})^2$, which equals a number greater than all of the people on the planet. This statistical phenomenon was used during the O. J. Simpson trial to demonstrate that a sample of blood could be proven to come from only one individual. Each person has a unique DNA makeup provided by the gene combinations on these chromosomes.

MENDEL'S EXPERIMENTS

Mendel experimented with the common garden pea plant. He wanted to see if there were any patterns in the way hereditary material was transmitted from parents to offspring. Before explaining Mendel's theories we need to define some basic genetic terms:

Gene Heredity unit consisting of a DNA sequence at a specific location on a chromosome

Allele An alternative form of a gene found on the same location of the homologous chromosome

Homozygous Pertaining to the condition in which both genes at a particular location, on homologous chromosomes, are the same allele or are identical (either dominant or recessive)

Heterozygous Pertaining to the condition in which two alleles at a given location, on homologous chromosomes, are not the same; although the appearance (phenotype) may be identical, the genetic makeup (genotype) is different

Dominant A gene whose effect masks the phenotypic expression of its allele; dominant alleles are indicated by capital letters

Recessive A gene whose expression is hidden by a dominant gene and only expresses itself when in the homozygous state; recessive alleles are represented by lowercase letters

Probability The likelihood that some event or outcome will occur

Mendel discovered that if a plant was homozygous dominant for a trait (TT), it had a dominant allele on each homologous chromosome. If a plant was homozygous recessive (tt), it had a recessive allele on each chromosome. Homologous chromosomes separate from each other during meiosis and end up in separate gametes and so do the two alleles of the pair. Thus, when gametes from a TT plant and a tt plant combine during fertilization, there is only one possible outcome. All of the offspring must be Tt (carry alleles for both forms of the trait). Mendel took tall homozygous dominant pea plants and crossed them with short homozygous recessive pea plants. All of the offspring were tall; they were in the heterozygous condition in which the dominant tall gene masked the expression of the recessive short allele.

Mendel crossed hundreds of plants and kept very meticulous records. The symbols he assigned to each of the generations are still used today.

P   *parental generation*
F_1 *first generation (filial means offspring)*
F_2 *second generation*

In the second generation, Mendel used mathematical models to predict the probability that a particular gene combination would take place. The **Punnett square** is one of the easiest methods of determination. Each square depicts the genotype of one kind of offspring. The outside of the square shows the possible gametes.

As you can see from table 2–1, a new pea plant will carry one or more dominant alleles and thus appear as a tall plant three out of four times. Only one offspring in four will phenotypically express itself as short (homozygous recessive). Although the phenotypic ratio is 3:1 (tall:short), the genotypic ratio is 1:2:1 (TT:Tt:tt).

Not all inherited traits are as simple as tall or short peas. Even simple qualitative traits, such as coat color and horned cattle versus polled, can have modifying genes that cause minor variations in phenotype. In cattle, at least nine major loci affect hair color and hair pattern. Four of these loci have at least three alleles each. Many breeds have color requirements that make it imperative to assess the probability offspring will exhibit these traits. In general, black is dominant to other colors. In some cases the dominance is partial or incomplete.

Only two alleles at a single location control the presence or the absence of horns in cattle. However, additional modifying genes can result in the formation of scurs (partial horns). The allele for polled is dominant to the allele for horns.

Animal breeders are often interested in more than one pair of genes per mating. For example, if we wanted to calculate the probability of obtaining a

▓ **Punnett square**
diagrammatic representation of gene segregation and their recombination at fertilization

TABLE 2–1	Punnett Square Predicting the Offspring of Two Heterozygote Pea Plants

		Female Gametes	
		T	t
Male Gametes	T	TT	Tt
	t	Tt	tt

T, tall (dominant) plant; t, short (recessive) plant.

TABLE 2-2	Results from Cross between Two Heterozygous Parents Carrying Dominant Black and Polled Genes			
		Cow		
		BP	Bp	bP	bp
Bull	BP	BBPP	BBPp	BbPP	BbPp
	Bp	BBPp	BBpp	BbPp	Bbpp
	bP	BbPP	BbPp	bbPP	bbPp
	bp	BbPp	Bbpp	bbPp	bbpp

B, black; b, red; P, polled; p, horned.
Genotype results: 1 (BBPP), 2 (BBPp), 2 (BbPP), 4 (BbPp), 1 (BBpp), 1 (bbPP), 2 (Bbpp), 2 (bbPp), 1 (bbpp).
Phenotype results: 9 black, polled; 3 black, horned; 3 red, polled; 1 red, horned. Ratio of 9:3:3:1.

black, polled animal in the second filial generation (F_2) we could use the Punnett square. Recalling our previous discussions of the F_1 generation, all are in the heterozygous state, we would start with all black, polled cattle (phenotypically) with a genotype of BbPp (see table 2–2).

Relatively few characters of economic importance in farm animals are inherited in as simple a manner as hair color or polled conditions. Most important traits, such as milk yield, growth rate, and speed, are due to many genes with multiple gene and environmental interactions. In quantitative inheritance, estimates of genotypes in offspring are much more complex and beyond the scope of this text.

INCOMPLETE, PARTIAL, OR CODOMINANCE

With some traits, the heterozygote is intermediate to the expressions of the homozygous genotypes. The roan color of Shorthorn cattle is an example of this phenomenon. When examined closely, it is clear that the hair coat is actually a mixture of red hairs and white hairs. The individual hairs are not a blend of red and white (as in a pink rose). If a red (RR) Shorthorn is mated with a white (rr) animal (both being homozygous), all of the F_1 offspring are roan. Genotypically they are expressed as Rr. In the mating of the F_1 animals, the F_2 generation expresses a phenotype of 1 red, 2 roan, and 1 white (1:2:1 ratio) with a genotype of 1:2:1 (see table 2–3).

Another example of partial dominance occurs in the equine condition known as hyperkalemic periodic paralysis (HYPP). HYPP causes periodic episodes of muscle tremors ranging from shaking to complete collapse. In many instances the horse dies. The mutated gene is inherited as a partial dominant gene. It is believed that the more serious clinical symptoms observed in some horses are due to their homozygous stage compared with the less severe partially dominant heterozygotes.

Gene expression in animals can also be expressed as an additive gene interaction. In this type of inheritance, there are many gradations between two extreme genotypes. Additive gene interaction contributes to the wide variance seen in the milk production rates of cows sired by the same bull. Polygenic traits

TABLE 2–3	F_2 Cross of Roan Shorthorn Cattle

		Cow	
		R	r
Bull	R	RR	Rr
	r	rR	rr

RR, Red; rr, White; Rr, Roan.
Genotype: 1 (RR), 2 (Rr), 1 (rr). Ratio of 1:2:1.
Phenotype: 1 red, 2 roan, 1 white. Also a ratio of 1:2:1.

are primarily determined by additive genes. Overdominance results in hybrid vigor. Here, an interaction between genes that are alleles results in the heterozygotes having production traits superior to those of either parent.

Lack of dominance can occur in such traits as the palomino color in horses. A dilution gene in the heterozygote condition results in the chestnut color. In the recessive condition there is no diluting effect (so the horse is chestnut in color). To get a palomino offspring 100 percent of the time, the breeder needs to breed a creamello horse (dominant homozygote, which is white in color) to a chestnut horse (recessive homozygote). The heterozygote condition will change the chestnut horse color to a dark cream color (the palomino).

SEX DETERMINATION

In mammals, females have a pair of chromosomes, XX, and males have a pair, XY, which determine the sex of the offspring. After meiosis, all of the female ova have an X chromosome. Half of the male sperm have an X and half have a Y chromosome. Thus, the sex of the offspring is determined by the male parent and the ratio is 1:1.

In birds, the determination of sex is reversed. The male bird has the homozygous pair (ZZ), and the female bird has the heterozygous pair (ZW). The predicted ratios in birds are the same as those in mammals, only the female is the determiner.

SEX-LIMITED GENES

The phenotypic expression of some genes is limited to one sex. These genes are known as sex-limited genes. An example of sex-limited genes is seen in milk production in dairy cattle. Both the bull and cow carry the genes, but phenotypic expression is limited to the cow. Roosters do not lay eggs, but they transmit genes for egg production. The presence or absence of one or a number of sex hormones determines the expression.

SEX-LINKED TRAITS

Autosomes are all chromosomes except the X and Y chromosomes (nonsex chromosomes). In addition to the genes located on the autosomes, many are found on each of the sex chromosomes. Some genes are carried on the nonhomologous

Autosomes chromosomes excluding the X and Y chromosomes

TABLE 2–4	Sex-Linked Condition in Ayrshire Cattle

MM = mahogany and white in either sex
Mm = mahogany and white (male)
Mm = red and white (female)
mm = red and white in either sex

M, mahogany (dominant); m, red (recessive).

TABLE 2–5	Sex-Linked Condition in Cats

YY = black in either sex
Yy = calico in the female
Yy = black in the male
yy = yellow in either sex

Y, black (dominant); y, yellow (recessive).

Sex linked genes that are carried on the nonhomologous portion of the X chromosome

portion of the X chromosome. In the male, a portion of the X chromosome does not pair with the Y. Genes that are carried on this portion are said to be **sex linked.** Usually sex-linked genes are recessive in their phenotypic expression. Because the male has only one X chromosome, a gene on the nonhomologous portion of the X chromosome will express itself even if it is a recessive gene. In the female, two recessive genes would be required to cause a recessive trait to appear.

Holandric inheritance genes carried on the portion of the Y chromosome that has no homologous portion on the X chromosome; traits represented by these genes are thus passed only from father to son

A portion of the Y chromosome has no homologous portion on the X chromosome. Genes carried here are thus transmitted only from fathers to sons; this inheritance is referred to as **holandric inheritance.** In general, however, sex-linked traits are determined by the genes carried on the X chromosomes.

Examples of sex-linked traits include color blindness and hemophilia in humans, hemophilia in dogs, barred feathers in chickens, and two of many color patterns in cattle and cats, as outlined in the following.

There are two color types of Ayrshire cattle, mahogany with white and red with white. These traits are sex linked. In the homozygous dominant condition, both sexes are mahogany and white. For the homozygous recessive condition, both sexes are red and white. With the heterozygous condition males are mahogany and white and females are red and white (see table 2–4). A similar condition exists in cats (see table 2–5).

HEREDITARY DEFECTS

Mutation chemical change in a gene such that it functions differently than before the change; this change is passed on to each succeeding generation

Sometimes something causes a change in a gene. When a new trait that did not exist in either parent appears we call it a **mutation.** Mutated genes are passed on to the offspring. Some mutations are beneficial, some mutations are harmful, and some are of little significance to the life or productivity of the animal. Exposure to various chemicals and exposure to radiation are two of the common causes of mutation. Hereditary defects are categorized as lethal, semilethal, or subvital. Some of the most serious (lethal) traits are discussed in the chapters concerning the individual species.

When deciding which animals to keep, to cull, to breed, to castrate, and to raise for meat, four sources of information must be analyzed. These four sources include the animal itself (its phenotype), the animal's progeny (if existing), the animal's ancestors (e.g., sire, dam, and maternal grandsire), and any collateral relatives (e.g., brothers, sisters, and half-brothers) that may provide information about the animal's genotype.

Depending on the age and production history of the animal one or a combination of information sources should be analyzed. For mature bulls, the most accurate information can be summarized from progeny performance. However, a young bull would be evaluated on its own performance and those of its pedigree (sire and maternal grandsire).

A **bracket pedigree** can be a helpful start. Three to five generations should be reviewed.

▍**Bracket pedigree**
three- to five-generation diagrammatic record of an individual's ancestors

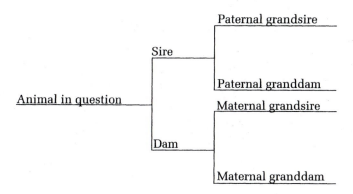

The two most important animals in a bracket pedigree are the sire and the maternal grandsire. Usually, more genetic information can be obtained through looking at these records than those of any other animals in the pedigree. These records are more informative largely because the males have more progeny and hence have a greater likelihood to demonstrate genetic worth through the additional information provided. A good pedigree includes genetic evaluations and information on the performance of the animal's ancestors. The accuracy of pedigree estimates of genetic transmitting ability is based on the relationship of each item of information to the animal's breeding value and the relative accuracy of the various items of information.

Only one-eighth of the genes from each great-grandparent is passed on to an animal. Therefore, other than checking for inbreeding, there is little value in tracing a pedigree beyond the grandparents.

SYSTEMS OF MATING

There are many different methods for mating animals. The system chosen depends on the kind of result desired. For example, to produce offspring with extreme breeding value to increase the rate of genetic change would require a much different system of mating than that required for producing hybrid vigor. The four basic systems of breeding are inbreeding, linebreeding, outcrossing, and crossbreeding.

▓ Inbreeding
breeding of animals
more closely related
than the average
relationship in the
population

Inbreeding is the mating of animals that are more closely related to each other than the average relationship in a population. Usually, inbreeding causes a decrease in performance in traits such as fertility, survivability, and hybrid vigor (inbreeding depression). With heavy culling, inbreeding can be useful in increasing homozygosity for the superior genes at a high proportion of loci. The undesirable genes also tend to become more homozygous. This deleterious effect requires severe culling to avoid inbreeding depression. Most dairy research shows an average 50-pound reduction in milk yield for every additional percentage of inbreeding.

▓ Linebreeding
breeding an animal to a
common ancestor; not
as closely related as
inbreeding

Linebreeding is the mating of several generations of offspring to a particular animal or its descendants. It is an attempt to concentrate the superior genes of a particular family or animal.

▓ Outbreeding
system of mating
animals less closely
related than the
average for the breed

Outcrossing, or **outbreeding,** is the opposite of inbreeding. It is the mating of animals with no common ancestors on the immediate pedigree within a breed. Outbreeding increases heterozygosity and thus, usually, hybrid vigor. Continuous outbreeding is defined as the yearly mating of the females in a population to unrelated males.[1]

▓ Crossbreeding
mating of animals of
different breeds

Crossbreeding is the mating of the sires of one breed to the dams of another breed. One of the main reasons to crossbreed is to increase heterosis or hybrid vigor. The offspring of these matings are usually more productive and more resistant to disease and have higher survivability and fertility rates. Heterosis lasts only one generation. For this reason, crossbreeding is not generally used with dairy cattle. Crossbreeding is primarily used for terminal crosses (where the offspring will be eaten). A second purpose of crossbreeding is to create a new breed. The Israeli Holstein (see figure 2–1), the highest-producing milk

FIGURE 2–1   Israeli
Holstein cow, the
highest-milk-producing
breed in the world.
Photo by author.

[1]R. Bogart, Oregon State University, Corvallis, Oreg.

cow in the world, the Jamaican Hope (a Jersey crossbred), and the Santa Gertrudis (a Brahman–Shorthorn cross) are three examples of new breeds created by crossbreeding.

Three terms are often misused by industry itself: *purebred, registered,* and *grade.* A **purebred** animal is one that has two registered parents of the same breed. Some breeds allow purebred classification for an animal whose parents are simply eligible for registry. A **registered** animal is a purebred that is recorded in the herd book of the breed. Most purebred animals are not registered simply because the animal's owners did not submit registration papers. A **grade** animal possesses the distinct characteristics of a particular breed but at least one of its parents cannot be traced to the registry (herd book). Most grades have one registered parent, usually the sire. The British use an additional term, *blood,* to denote a purebred racehorse. A half blood would indicate that the sire or dam was something other than a Thoroughbred.

At one time, the sheep industry also used this terminology to describe the amount of fine wool breeding (Merino and Rambouillet) represented in a particular ewe or ram. A full blood sheep produced fine wool, whereas a one-quarter blood sheep was noted for producing much coarser fiber.

Particular mating systems for each species are discussed in future chapters.

▓ **Purebred** animal with two purebred parents that is hence eligible for inclusion in the breed registry

▓ **Registered** animal whose name is recorded in the breed registry

▓ **Grade** animal that possesses the distinct characteristics of a particular breed but at least one of its parents cannot be traced to the registry (herd book)

SUMMARY

Animal breeding involves an improvement in quality during each subsequent generation and not simply an increase in offspring. The phenotype of an animal is greatly influenced by both its genetic makeup (genotype) and the environment. Earlier research by noted scientists such as Bakewell and Mendel contributed to the laws of genetics that we still use today in seeking to improve our livestock genetic base.

From Mendel's garden pea experiments to the more recent cloning of Dolly, the fundamental understanding of the DNA molecule, the replication of cells (mitosis and meiosis), and the carrying of genetic material on the individual chromosomes is essential. The splicing, removal, and addition of genes to an organism are evolving and increasingly controversial areas of animal and plant breeding that modern agriculture will face more frequently.

EVALUATION QUESTIONS

1. The genetic makeup of an animal is referred to as its _____ .
2. An alternative form of a gene located at the same locus on each of a pair of chromosomes is a _____ .
3. The mating of sires of one breed or breed combination to dams of another breed or breed combination is called _____ .
4. The basic physical unit of heredity consisting of a DNA sequence at a specific location on a chromosome is the _____ .
5. _____ refers to the first-generation progeny following the parental or P_1 generation.
6. A _____ is a cell formed from the union of a sperm and an ovum.
7. A _____ is a two-dimensional grid used to determine the possible zygotes obtainable from a mating.

8. _____ is an observed level of performance or physical trait of an animal.

9. The mating of animals in an attempt to concentrate the genes of a superior ancestor in animals of later generations is called _____ .

10. _____ is the mating of unrelated individuals.

11. The mating of relatives is called _____ .

12. A sex cell—a sperm or an ovum—is called a _____ .

13. _____ was the British livestock breeder who first successfully used estimates of breeding value, selection, and systems of mating to make genetic improvement.

14. _____ bodies, which occur in pairs in the nuclei of cells, carry the genetic material in the form of genes.

15. Animals that possess the distinct characteristics of a particular breed but that have at least one parent that cannot be traced to the registry are called _____ .

16. _____ is a condition in which both genes at a particular locus are the same allele.

17. _____ is a condition where the two alleles at a given locus in an animal are not the same.

18. _____ is another name for hybrid vigor.

19. Hybrid vigor is caused by the _____ mating system.

20. Cell division of somatic cells is called _____ .

21. Cell division of gametes is called _____ .

22. The removal from a herd of animals with lower genetic or phenotypic value is called _____ .

23. The system of mating animals that are less closely related than the average relationship in the population (no common ancestors in the immediate pedigree) is called _____ .

24. _____ is an animal's offspring.

25. _____ means naturally hornless.

26. A _____ is a list or diagram of an animal's ancestors (usually containing genetic and performance information).

27. A trait whose measurement would fall into a discrete classification such as color or the presence or absence of horns is referred to as a _____ .

28. An animal recorded in the herd book of the breed is _____ .

29. _____ refers to two sets of homologous chromosomes; represented by 2n.

30. _____ refers to having only one set of chromosomes; represented by n.

31. All chromosomes except the sex chromosomes are called _____ .

32. _____ describes a gene that when paired with its allele covers up the phenotypic expression of that gene.

33. _____ describes a gene whose phenotypic expression is covered or masked by its own allele.

34. _____ is a situation in which neither allele is dominant over the other, with the result that both are expressed in the phenotype.

35. _____ is the cross resulting from mating a mare with a jack (donkey).

36. An animal or a recognized breed that is eligible for registry in the official herd book of that breed is called _____ .

37. _____ designates the red-white color phase of Shorthorn cattle.
38. A trait limited to only one sex is referred to as a _____ .
39. _____ refers to genes carried on the nonhomologous portion of the X chromosome.
40. _____ is a gene carried on the nonhomologous portion of the Y chromosome and always transmitted from father to son.
41. The likelihood that some event or outcome will occur is the _____ .
42. _____ was an Austrian monk who worked with pea plants.
43. In birds the _____ sex is the determiner of the sex of the offspring.
44. _____ is a genetic disorder in horses that causes periodic episodes of muscle tremors.

DISCUSSION QUESTIONS

1. A roan, heterozygous polled bull is bred to a roan, horned cow. What are the possible genotype ratios and phenotype expressions (and ratios) in their offspring? Use a Punnett square to show your work.
2. Why would crossbreeding not be advisable for dairy cattle? Why would inbreeding not be advisable for dairy cattle?
3. A couple has just married. They plan to have three children. They want to have two boys and a girl. What is the probability their desire will be fulfilled?

CHAPTER 3

Animal Reproduction

OBJECTIVES

After completing the study of this chapter you should know:

- The anatomic features of both the male and female reproductive systems in each of the major livestock species.
- The definition of puberty and the effect the environment (including nutrition) has on its timely expression.
- About the estrous cycle and estrus lengths for the various livestock and other domestic animal species.
- The differences in fertilization of the ovum and development of the embryo between mammals and birds.
- The definition of differentiation and how Dolly changed our understanding of this process.
- Common methods of diagnosing pregnancy in livestock, reasons for performing these management procedures, and complications involved with difficult births.
- The current and future use of artificial insemination, embryo transfer, embryo splitting, and genetic manipulation in livestock management.

KEY TERMS

Albumen
Allantois
Amnion
Caruncles
Cervical plug
Cervix
Chimera
Chorioallantoic membrane
Chorioallantois
Chorion
Clitoris
Cloaca

Corpus luteum (CL)
Cotyledonary placental
   attachment
Cotyledons
Cowper's glands
Cryptorchidism
Cuticle
Diestrus
Differentiation
Endocrinology
Endometrial cups
Epididymis

Estrogen
Estrus
Fallopian tube
Fertilization
Fimbria
Flagellum
Funnel
Germinal disc
Gestation
Gonadotropin-releasing
   hormone (GnRH)
Graafian follicle

Gravid	Nidation	Prostaglandin
Gubernaculum	Nuclear transfer technology	Puberty
Heat		Scrotum
Infundibulum	Ovulation	Seasonal breeders
Inguinal canal	Ovum	Semen
Interstitial cells	Parenchyma	Seminiferous tubules
Involution	Parturition	Sheath
Isthmus	Penis	Sigmoid flexure
Labia majora	Placentation	Stocking rate
Labia minora	Placentome	Suburethral diverticulum
Magnum	Polytocous	Testosterone
Malpositions	Pregnant mare serum gonadotropin (PMSG)	Uterine milk
Malpresentation		Vagina
Metestrus	Proestrus	Vas deferens
Metestrus bleeding	Progesterone	Vent
Monotocous	Prolactin	Vestibule

INTRODUCTION

The reproductive process is the focal point of overall animal productivity and profit. Litter size, calving rate, services per conception, days open, eggs hatched, and so forth can be used to measure the efficiency of reproduction. Synchronization of some very complex processes in both the male and female is required. In the end, profitable reproductive efficiency is what counts. Because the profit margin per unit is so slim today, keeping breeding animals that are not pregnant, sows that farrow small litters, and mares that do not conceive at all are choices that will lead to economic failure.

Before one can understand the reproductive process, a basic review of reproductive anatomy and physiology is necessary. **Endocrinology,** or the study of the endocrine system (and its hormones), is a major factor in reproductive physiology. Thus, a basic understanding of the function of reproductive hormones is essential to the management of animal reproduction.

▮ **Endocrinology** study of the endocrine system and its hormones

ANATOMY

The first step in understanding reproduction is to understand basic reproductive anatomy.

The Male

The reproductive system of the male (see figure 3–1) can be divided into three major parts: the testes (also called gonads, testicles, or primary organs); the accessory, or secondary, sex glands; and the external copulatory organ, the *penis*.

Testes.   In mammals, viable sperm cannot develop at body temperature. Therefore, the *scrotum,* which houses the testes, is a heat-regulating structure. The testes will lower away from the body when the temperature outside the body is warm and will contract back toward the body at cooler times (see figure 3–2). By midpregnancy in cattle and just before birth in horses, the testes descend from the body cavity into the scrotum through a small opening known as the *inguinal canal.* Descent of the testes is caused by a shortening of the **gubernaculum,** a ligament that extends from the inguinal region to the tail of the **epididymis.**

▮ **Gubernaculum** ligament that extends from the inguinal region to the tail of the epididymis

▮ **Epididymis** tube that carries sperm from the seminiferous tubules to the vas deferens; final place for maturation of sperm in the male

FIGURE 3–1
The male reproductive
tract. (a), a bull and (b),
a boar. Note the large
sigmoid flexure in both
species.
Courtesy of Dr. James
Blakely.

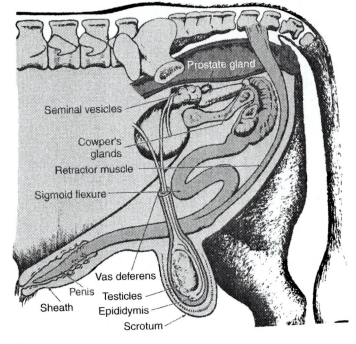

(a)

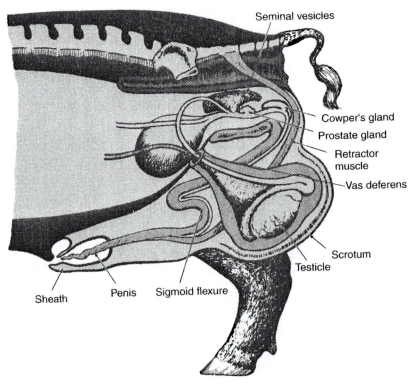

(b)

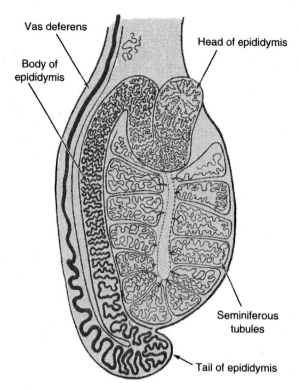

Vas deferens

Body of
epididymis

Head of epididymis

Seminiferous
tubules

Tail of epididymis

FIGURE 3–2 The testes produce both sperm and testosterone. Courtesy of Dr. James Blakely.

Sometimes one or both testes do not descend, a condition known as **cryptorchidism.** If only one testis descends, it is called unilateral cryptorchidism, and if both fail to descend, it is known as bilateral cryptorchidism. A unilateral cryptorchid may have impaired reproductive efficiency, and a bilateral cryptorchid is sterile. In addition, the testes differ from the ovaries in that spermatozoa are not present at birth (as are the ova in females). Correcting this disorder by surgical means would be considered unethical because it can be inherited, thus perpetuating this undesirable trait.[1]

Under normal development, the testes (plural for testis) function by producing sperm in the very small, convoluted tubules that make up a large part of the structure of the testes. If linked together to form one tube, these **seminiferous tubules** from one pair of bull testes have been estimated to stretch the length of 50 football fields. **Interstitial cells,** located in the spaces between the seminiferous tubules in the testes, produce the male hormone **testosterone,** which is responsible for secondary sex characteristics of the male (heavy chest, heavy muscling, and masculine appearance). On receiving the proper message from the brain, testosterone is released and causes sexual excitation in the male.

Scrotum. The scrotum is a two-lobed skin sac that encloses the testes. The scrotum has the same embryonic origin as the labia majora in the female. During hot weather, muscles lining the scrotum relax, permitting the scrotum to stretch and the spermatic cord to lengthen. More surface area is thus provided for cooling the testes.

▓ **Cryptorchidism** hidden testicle; a condition in which one or both testicles are retained in the abdominal cavity; if both are retained the animal is sterile

▓ **Seminiferous tubules** convoluted tubules located within the testes that are responsible for the production of spermatozoa

▓ **Interstitial cells** also known as Leydig cells; located in the spaces between the seminiferous tubules in the testes, they produce the male hormone testosterone

▓ **Testosterone** male sex hormone

---

[1]L. S. Shapiro, *Applied Animal Ethics* (Albany, NY: Delmar, 2000).

Epididymis.   The epididymis (see figure 3–2) serves four functions: transport, storage, maturation, and concentration of sperm. This structure, estimated to be 120 feet in total tubular length in the bull and up to 225 feet in the stallion, serves to transport the sperm from the testes to the accessory sex glands. Water is reabsorbed here to increase concentration; maturation is achieved because of cell excretions, and sperm is stored primarily in the tail of the epididymis.

Vas Deferens and Accessory or Secondary Sex Glands.   The **vas deferens** transports the mature semen from the tail of the epididymis past the accessory sex glands (commonly called secondary sex glands), the seminal vesicles, the **Cowper's glands** (also known as bulbourethral glands), and the prostate gland. These glands are responsible for producing the bulk of the fluid that we commonly refer to as *semen*. This seminal fluid may range from 5 to 10 cc in the bull and is ejaculated through the penis into the reproductive tract of the female. Seminal fluid is rich in fructose and sorbitol, major sources of energy for spermatozoa.

Penis.   In the bull, ram, and boar, sexual excitation causes blood to be pumped into the chambers of the penis, causing an erection by a straightening of the **sigmoid flexure;** this straightening allows for copulation. After copulation, the sigmoid flexure contracts because of a retractor penis muscle that retracts the penis into a protective *sheath* (prepuce). In the stallion, cavernous bodies containing large spaces fill with blood during excitation. This causes considerable increases in size that allows for penetration into the mare.

Reproductive Anatomy of Male Poultry.   The male reproductive organs in poultry are greatly simplified in comparison to the larger domestic species such as cattle, swine, sheep, and horses. The male fowl differs from other domestic animals in that the testes do not descend into a scrotum but remain in the abdominal cavity along the backbone close to the anterior portions of the kidneys (see figure 3–3). The testes of the fowl still produce live sperm and testosterone, which is responsible for secondary sex characteristics such as bright red combs, plumage, and the crowing response. Roosters crow because of testosterone.

As in other species, sperm is produced in the seminiferous tubules deep within the testes. In chickens the testes start producing sperm at about 100 to 250 days of age. The developing spermatozoa take about 30 days to mature. Sperm released from the tubules enter the *vasa deferentia* (ductus deferentes), small tubes that conduct the sperm along a pathway to the **cloaca.** The ductus deferentes do not open into a copulatory organ, as in other species, but rather into small papillae (fingerlike projections). These projections are located on the dorsal wall of the cloaca and serve as semen-transporting organs. The male fowl also has a rudimentary copulatory organ that has no connection to the ductus deferens and is located on the ventral part of the cloaca. Mating with the female is mostly a matter of joining cloacas long enough for semen injection.

Male fowl respond to light just as female fowl do. Males that are to be used for natural matings should receive the same form of light stimulation as females to produce the most viable semen in the largest quantities. Males will molt and decrease semen production as a result of excessive lighting. Birds respond best to light if first exposed to 3 or 4 weeks of increased darkness. A gradual increase in day length to 14 hours has been shown to improve fertility in males.

▓ **Vas deferens**   tube that transports mature semen from the tail of the epididymis past the accessory sex glands to the urethra

▓ **Cowper's glands** also known as bulbourethral gland; an accessory sex gland in the male that produces the seminal fluid needed to clean the urethra and neutralize its acidic environment

▓ **Sigmoid flexure** S curve in the penis of a bull, ram, or boar that allows for elongation or straightening of the penis during copulation

▓ **Cloaca**   common junction for the outlets of the digestive, urinary, and genital systems through the vent

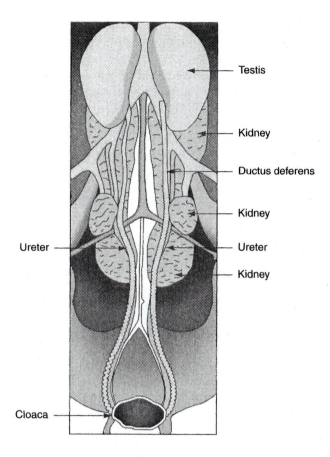

FIGURE 3–3 The reproductive and urinary organs of the male fowl.

Labels on figure: Testis, Kidney, Ductus deferens, Kidney, Ureter, Ureter, Kidney, Cloaca

TABLE 3–1	Stocking Rates for Common Farm Animals	

Species	Male–Female Ratio
Cattle	1:25–40
Horses	1:20–30
Sheep	1:35–60
Pigs	1:25–50
Goats	1:30
Rabbits	1:10
Chickens	1:12
Ducks	1:7
Turkeys	1:10–20

Note: The stocking rate largely depends on the size of the area (pasture versus range), flock versus pen mating, and time constraints (seasonal versus estrous cycle).

The correct ratio of males to females (**stocking rate**) depends on the type and the size of the birds involved (see table 3–1 for recommended stocking rates for all species). It is generally defined on the basis of the number of cockerels per 100 pullets. Most commercial operations place a few extra males in the pens at the time of breeding to allow for some early culling and mortality from fighting. A minimum of 8 to 11 males per 100 females is recommended in most cases.

▨ **Stocking rate** ratio of males to females, also refers to the number of animals per acre in a pasture or range setting

■ **Puberty** period
when the reproductive
system becomes
mature in form and
function

Puberty.   The male and female reproductive systems do not become fully functional until **puberty** (the age at sexual maturity). In the male, puberty is marked by the production of viable sperm and a desire for mating, influenced by the hormone testosterone. Follicle-stimulating hormone (FSH) is released by the pituitary in the male and is required for normal spermatogenesis (sperm production). Luteinizing hormone (LH) is also released by the male's pituitary and is needed to stimulate the production of testosterone by the testes.

Females are born with all of the ova that they will ever have. At about 1 month of age, follicles start to appear on the heifer's ovaries. However, estrous cycles do not begin until the heifer reaches 5 to 14 months of age. The age at which puberty occurs varies considerably with the species and breed (see table 3–2). In many species and breeds, puberty is more a function of weight than of age. Dairy cattle reach puberty when their body weight is 30 to 40 percent of their expected adult weight, whereas beef cattle[2] must be closer to 45 to 55 percent and sheep[3] 40 to 50 percent of their expected adult weight. As the level of nutrition increases, first estrus occurs at an earlier age. In heifers, restriction of energy, protein, or phosphorus will delay the onset of puberty.[4]

Although most breeds of sheep reach puberty at 5 to 7 months of age, the larger, slower-maturing wool breeds may be delayed to 16 months or more. Rams normally reach puberty about a month earlier than ewes in each case.

With *seasonal breeders* (horses, goats, and sheep), the season, as well as body weight, influence when the animal reaches puberty. Ewes born early in the season (January in the Northern Hemisphere) will reach puberty in September, as will a younger and smaller ewe lamb born in March or April. Gilts raised with other gilts reach puberty sooner than other pigs. The presence of an adult boar hastens puberty as well.[5]

TABLE 3–2	Age of Puberty in Farm Animals (in Months)		
Species	**Males**	**Females**	**Weight (kg) for Females**
Cattle	10–12	8–18	160–270
Sheep	4–9	5–10	27–34
Horse	13–18	12–15	a
Swine	5–8	4–7	68–90
Goat	3–5	5–7	10–30
Rabbit	4–8	3–4	—
Chicken[b]	22–26	22–26	—

[a]Varies considerably with the breed of horse. Males average 367 kilograms. J. D. Skinner and J. Bowen, "Puberty in the Welsh Stallion," *Journal of Reproduction and Fertility* 16:133, 1968.
[b]Age in weeks.

---

[2]J. H. B. Roy, C.M. Gilligs, and S.M. Shotton "Factors Affecting First Estrus in Cattle and Their Effects on Early Breeding," in *The Early Calving of Heifers and Its Impact on Beef Production,* ed. J. C. Taylor (Brussels: European Economic Community, 1975).

[3]H. J. Bearden and J. W. Fuquay, *Applied Animal Reproduction,* 3rd ed. (Upper Saddle River, NJ: Prentice-Hall, 1992).

[4]P. V. Rattray, "Nutrition and Reproductive Efficiency," in *Reproduction in Domestic Animals,* 3rd ed., eds. H. H. Cole and P. T. Cupps (New York: Academic Press, 1977).

[5]A. P. Mavrogenis and O. W. Robinson, "Factors Affecting Puberty in Swine," *Journal of Animal Science* 42:1251, 1976.

The Female

Although the emphasis in most breeding programs is on the male, the female reproductive system is much more complicated and important. It is therefore necessary to go into more detail to ensure full understanding of the anatomy of the female and the function of each organ or part at puberty (see figure 3–4).

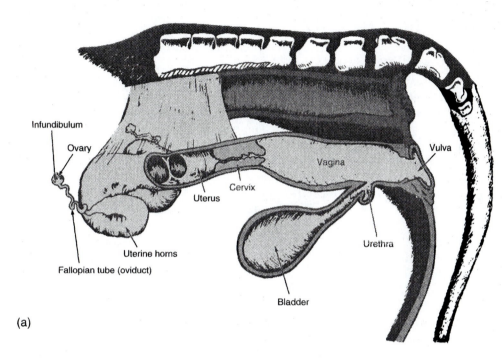

Infundibulum
Ovary
Vagina
Vulva
Cervix
Uterus
Urethra
Uterine horns
Bladder
Fallopian tube (oviduct)

(a)

FIGURE 3–4 The female reproductive system. The ovary and fallopian tubes come in pairs. Note the long uterine horns in (b), a sow's reproductive tract. This is very typical of polytocous (litter bearing) animals. Courtesy of Dr. James Blakely.

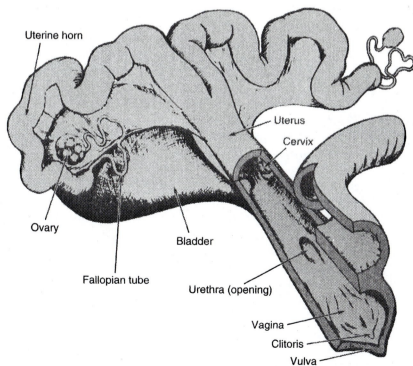

Uterine horn
Uterus
Cervix
Ovary
Bladder
Fallopian tube
Urethra (opening)
Vagina
Clitoris
Vulva

(b)

Ovum female reproductive cell or gamete; commonly called an egg

Ovaries.   The ovaries are homologous with the male testes. They remain in the body cavity near the kidneys and do not descend. The *ova* (eggs; singular is **ovum**), which when fertilized by the male spermatozoa begin the embryo, are present at birth. Ova are formed during the prenatal period. Although estimates of available ova run as high as 75,000 for both ovaries in cattle, relatively few—perhaps 20 to 30—are shed during an average cow's lifetime under natural conditions. Only 2,500 potential ova are left at the end of a cow's reproductive life. The remaining ova start development and become atretic. In humans, newborn girls have an average of 2 million oocytes at birth. By the time they reach puberty the number has decreased to 350,000.[6]

Monotocous normally giving birth to one young each gestation period

Polytocous litter bearing

The cow, mare, and ewe are **monotocous,** normally giving birth to one young each gestation period. Therefore, one ovum is produced each estrous cycle. The sow normally produces 10 to 25 ova in each cycle and gives birth to several young each gestation period. The sow and other litter-bearing animals are described as being **polytocous.**

The ovary of a cow is almond shaped and averages 10 to 20 grams in weight. The ovaries of the sheep and goat are also almond shaped but only half the size of those of the cow. The ovaries in the mare are kidney shaped and two to three times the size of those of the cow. Pig ovaries are shaped as a cluster of grapes.

Parenchyma functional layer

Graafian follicle mature follicle, ready to ovulate

Estrogen major female sex hormone produced by the ovaries; responsible for the outward signs of heat or estrus

In contrast to the male, in which the seed is developed deep within the seminiferous tubules, the tissue that produces the ovum lies very near the surface of the ovary. This functional layer is known as the **parenchyma.** It contains all of the ovarian follicles and the cells that produce the ovarian hormones. Potential ova, within primary follicles, are generally believed to be present at birth. These potential ova lie dormant until puberty. Following puberty, successive stages of maturity follow until a mature ovum called a **Graafian follicle** is produced. The Graafian follicle is filled with a fluid rich in **estrogen.** This protruding "blister" (see figure 3–5) on the surface of the ovary is produced under the influence of FSH from the anterior pituitary gland. The pituitary also produces LH, which causes a rupture (**ovulation**) of this follicle and a release of an ovum (egg). The hypothalamus (a small but important endocrine gland located above the pituitary) produces several controlling factors. One of them, *gonadotropin-releasing hormone (GnRH)*, influences the release of FSH and LH. Cows with a cystic follicle can be administered an injection of GnRH to cause an LH surge and ovulation of the follicle.

Ovulation release of the ovum from the ovary

Corpus luteum (CL) also known as yellow body; temporary structure formed on the ovary after ovulation that is needed to maintain pregnancy in most species and that secretes progesterone

Immediately after ovulation, the cavity left by the vacated ovum is acted on, and theca cells and granulosa increase to produce a scarlike structure, the **corpus luteum** (CL) (see figure 3–5). If fertilization of the ovum does not occur, *prostaglandin* ($PGF_{2\alpha}$) will be released from the uterus, causing a regression of the CL. The cycle can then repeat itself to form another Graafian follicle. If fertilization occurs, the CL retains its size under the influence of the LH and itself produces the hormone *progesterone*, responsible for suppressing further heats and maintaining pregnancy. In rodents, *prolactin* is a major anterior pituitary hormone responsible for CL maintenance.

Infundibulum funnel-shaped opening of the oviduct close to the ovary

Fallopian tube also called oviduct; site of fertilization in most species

Fallopian Tube (Oviduct).   The ovary is stimulated to release the ovum into the **infundibulum** of the **fallopian tube,** or oviduct (see figure 3–6). This action

---

[6]S. I. Fox, *Human Physiology*, 5th ed. (Dubuque, IA: Wm. C. Brown Publishers, 1996).

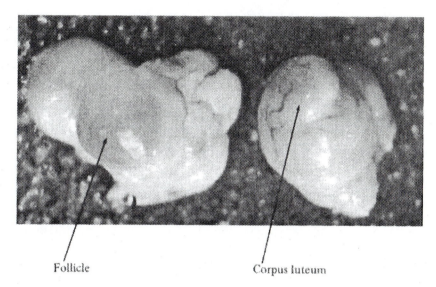

FIGURE 3–5 The ovarian follicle and corpus luteum. Courtesy of Dr. James Blakely.

Follicle

Corpus luteum

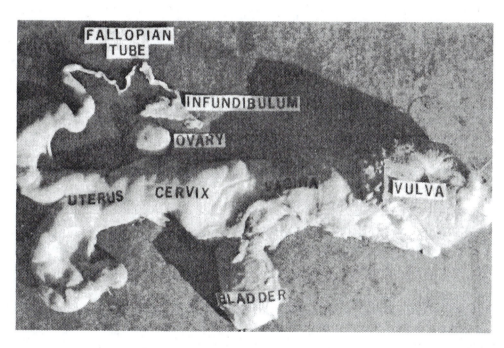

FIGURE 3–6 The female reproductive tract, illustrating dissected ovary and fallopian tube. Courtesy of Dr. James Blakely.

is actually delayed until 12 hours after the end of **estrus** (heat) in the cow. In the cow, doe, ewe, sow, and mare, the infundibulum is separate from the ovary. The ovum, swept into the infundibulum of the fallopian tube by ciliated action of the **fimbria** and muscular contractions, then makes its way into the horn of the uterus. **Fertilization** (union of the ovum and sperm) actually takes place in the upper third of the fallopian tube. This sequence of events can occur on either side of the paired mammalian system, initiated by either ovary.

Uterus.   The uterus (see figure 3–6) consists of two horns that curve like a ram's horn and a common body. The *cervix*, considered an integral part of the uterus, is discussed separately to simplify this study.

Estrus also called heat; period during which the female is receptive to the male for purposes of breeding

Fimbria hairlike projections on the infundibulum that help guide the ovulated ovum into the fallopian tube

Fertilization union of a sperm and an ovum to form a zygote

In cattle, the horns of the uterus make a complete spiraling turn before connecting with the fallopian tubes. These horns are usually well developed; the fetus develops in one horn or the other.

Blood and nerves reach the uterus through the supporting broad ligament. This ligament may stretch in older animals, allowing a lower carriage of the uterus and fetus.

Cows, does, and ewes have a *cotyledonary placental attachment.* Inside their uterus, the mucosa layer contains *caruncles.* These small projections, which enlarge to about the size of a U.S. half-dollar during pregnancy, are nonglandular and rich in blood supply. Arranged in rows extending into both horns, they have been estimated to number 70 to 120 in the cow. In the ewe and doe there are 88 to 96 points of attachment. The caruncles have a spongelike appearance because of small cavities that serve as attachment points for the opposite structure, the **cotyledon** from the placenta (membrane enclosing the fetus). The cotyledons and caruncles together, called the **placentome,** might be thought of as two buttons that snap together.

The mare has a unique placental attachment with the development of *endometrial cups* from the second to the fourth months of gestation. The endometrial cups are of fetal origin and are associated with the secretion of *pregnant mare serum gonadotropin* (PMSG).

The uterus has many functions. For example, it serves as a pathway for sperm at copulation, and motility of sperm to the fallopian tubes is aided through contractile actions. In the early weeks of gestation (pregnancy), the uterus is thought to sustain the embryo by secretions from the uterine glands (*uterine milk*). The uterus, capable of undergoing great changes in size and shape, serves as an attachment point through the placentome for the growing embryo during gestation. It plays a major role in expulsion of the fetus and membranes at **parturition** (birth) and is capable of regaining its posture quickly after parturition through *involution.*

Cervix.   The sphincterlike, thick-walled, inelastic structure that separates the uterine cavity from the vaginal cavity is the cervix (see figure 3–6). The basic function of the cervix is to seal off the uterus, protecting it against bacterial and other foreign matter invasions. The sphincter remains closed at all times except during birth. In cows, ewes, and does, the cervix additionally serves as a sperm reservoir after mating. Transverse interlocking rings help keep contaminants out of the uterus.

During estrus and copulation, the cervix serves as a passageway for sperm. If pregnancy results, a *cervical plug* develops, completely sealing off the uterine canal to protect the fetus. Removal of this plug increases the chance of abortion. Shortly before birth, the mucous plug liquefies, the cervix dilates, and passage of the fetus and membranes is allowed at parturition (birth).

Vagina.   Lowermost of the internal reproductive structures, the **vagina** serves as the female copulatory organ. It is here that semen is deposited by the male in most animal species (in the mare and sow semen is deposited in the cervix). Like the cervix, the vagina dilates to allow passage of the fetus and membranes.

Vulva.   The external opening to the female reproductive tract is the vulva. Its three main functions are the passage of urine, the opening for mating, and the terminal portion of the birth canal. It consists of the *clitoris,* homologous to the

Cotyledon fetal membrane that joins with the maternal caruncles of the uterus, forming placentomes (connection of the placenta)

Placentome combined cotyledon and caruncle "button" attaching the membrane that encloses the fetus

Parturition act of giving birth

Vagina birth canal and area of semen deposit during natural breeding in most species

glans penis; the *labia minora,* or inner folds, homologous to the prepuce; the *labia majora,* or outer folds, homologous to the scrotum; and the **vestibule,** the area common to both the urinary and reproductive systems. During estrus in the mare, the clitoris becomes erect and the labia have frequent contractions (winking). A *suburethral diverticulum* (blind pocket) is found just posterior to the external urethral orifice in the vestibule.

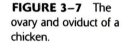

Vestibule area common to both the urinary and reproductive systems in mammals

Reproductive Anatomy of the Hen. Figure 3–7 illustrates the female reproductive system in the chicken. It should be noted that normally only the left ovary and oviduct are fully developed in a female fowl. During incubation, the right side fails to develop and by hatching time has degenerated to only a rudiment. There are five clearly defined regions of the oviduct:

1. The *funnel,* or *infundibulum,* which is responsible for picking up the yolk after it has been ovulated
2. The **magnum,** which secretes albumen, or the white of the egg (3 hours)
3. The *isthmus,* which secretes the shell membranes (1.25 hours)
4. The uterus (shell gland), which secretes the shell, primarily calcium carbonate (20 hours)

Magnum part of the oviduct in poultry that secretes the albumen, or white of the egg

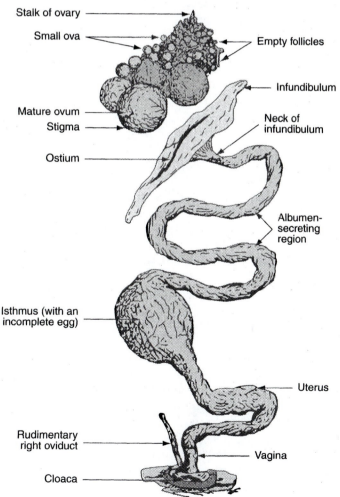

FIGURE 3–7 The ovary and oviduct of a chicken.

Stalk of ovary

Small ova

Empty follicles

Mature ovum

Stigma

Infundibulum

Neck of infundibulum

Ostium

Albumen-secreting region

Isthmus (with an incomplete egg)

Uterus

Rudimentary right oviduct

Vagina

Cloaca

5. The *vagina,* where the egg is temporarily stored and expelled when fully formed; here the **cuticle,** a very thin coat of albumen-like material, is deposited over the shell

The female reproductive system functions initially through stimulation by FSH, which causes development of the mature follicles (yolks). FSH production is normally stimulated by increasing periods of light brought about in wild birds by the lengthening days of spring. FSH production may be artificially induced by the lighting systems designed by people. The ovary is influenced by this stimulation to begin production of its own hormones, estrogen and progesterone. Estrogen causes an increase in the levels of blood calcium, protein, fats, vitamins, and other substances necessary for egg formation. Estrogen also stimulates separation of the pubic bones and an enlargement of the **vent** in preparing the hen for laying. Progesterone acts on the hypothalamus gland to produce LH from the anterior pituitary, which causes release of the mature yolk from the ovary to the funnel, or infundibulum. If sperm are present at this point, a fertile egg will be laid. Otherwise, production will continue, but the eggs produced will be infertile.

ESTROUS CYCLE

The estrous cycle varies among and within species. However, variability among individuals of a particular species may indicate a reproductive abnormality. Table 3–3 outlines the average estrous cycle and period of estrus (heat) for some of the major farm animal species.

The periods of the estrous cycle are proestrus, estrus, metestrus, and diestrus (see table 3–4). The periods occur in a cyclic manner for most species. Although most reproductive physiologists start the cycle with the estrus period, we begin with the proestrus period. *Proestrus* begins with the regression of the corpus luteum and an increased production of FSH. FSH stimulates follicular growth of one or more follicles. The follicles start producing increasing quantities of estrogen, which brings on the outward signs of estrus, or *heat.*

With the exception of humans and other primates, the female is only receptive to the male during the period of *estrus.* In the cow this period lasts 12 to 18 hours. Environmental temperatures, humidity, and stress can alter the length of this period considerably. The mare has the most variable estrus period, averaging 2 to 12 days (most mares average 4 to 8). As the mature Graafian follicle expands with estrogen-containing fluid, the anterior pituitary gland releases a surge of LH, triggering ovulation.

Metestrus starts as ovulation commences. *Metestrus bleeding* in cows occurs in about 90 percent of all heifers and 45 percent of all mature cows. It is caused by the breakage of uterine capillaries, which become more fragile with decreasing levels of circulating estrogen during the middle of the cow's cycle. It is not to be confused with menstrual bleeding in women, which occurs at the end of each cycle as the hormone-primed uterine lining (the endometrium) is sloughed off. Metestrus bleeding does not signify conception or failure to conceive in cattle, simply that estrus has passed. Typically, cows show blood-stained mucus on their tails and/or pin region. Once ovulation occurs, LH stimulates the formation of a new gland, the CL.

TABLE 3-3	Estrous Cycle and Estrus Lengths in Farm Animals	
Species	**Length of Cycle**	**Duration of Estrus**
Ewe	16–17 days	24–36 hours
Cow	18–24 (21) days	18–19 hours
Sow	19–20 days	48–72 (48) hours
Doe (goat)	21 days	32–40 hours
Mare	19–25 days	4–8 (5) days
Bitch	6–12 months	5–19 (9) days
Doe (rabbit)	15–16 days	12 days

Source: L. S. Shapiro, *Principles of Animal Science* (Simi Valley, CA: Ari Farms, 1997).

TABLE 3-4	Estrous Cycle in the Cow			
	Proestrus	**Estrus**	**Metestrus**	**Diestrus**
Time	3–4 days	12–18 hours	3–4 days	12–15 days
Hormones	FSH	Estrogen	LH	Progesterone
Features	Follicle growth	Signs of heat	CL formation	CL function

Source: L. S. Shapiro, *Principles of Animal Science* (Simi Valley, CA: Ari Farms, 1997).

TABLE 3-5	Description of Ovarian Follicles
Graded Follicle	**Physiological Characteristics**
Primary	Centrally located oocyte, surrounded by a single layer of granulosa cells
Secondary	Two or more layers of granulosa cells
Tertiary	Antrum separated from the vascular theca, filling with estrogen-rich follicular fluid
Graafian	Blisterlike structure filled with follicular fluid
Atretic	Regressing follicle
Corpus luteum	Yellow body made up of luteal tissue, secretes progesterone, and necessary to maintain pregnancy
Corpus albicans	White fibrous tissue forms from CL

The *diestrus* phase occurs when the CL is fully functional. An increase in circulating levels of progesterone can be detected by day 8 of the cycle and culminates with the regression of the CL on day 20 or 21.

Even though cattle are monotocous, two to three waves of follicular growth occur during each cycle. During each wave, a group of follicles start to mature. Most of these follicles become atretic. It is theorized that the estrogen from several developing follicles is required to initiate estrus. Usually only one follicle reaches the Graafian stage. The grading of follicles is summarized in table 3–5.

FERTILIZATION

In cattle, the average life of the ovum is 6 to 12 hours, and the average life of the sperm is 30 hours. Thus mating must take place at the latter part of estrus for fertilization to occur (figure 3–8). The deposited sperm, in natural mating, makes its way up the vagina through the cervix, into the uterus, and meets the ovum at the upper one-third of the fallopian tube. The sperm owes its motility to a *flagellum* (tail) that propels it toward the ovum, but it is moved mostly by the muscular activity of the uterus and fallopian tubes. The actual head of the sperm causes fertilization of the ovum. The first sperm that reaches the ovum causes a reaction that prevents all other sperm from uniting with the egg.

FIGURE 3–8 The breeding sequence of events.
Courtesy of Dr. James Blakely.

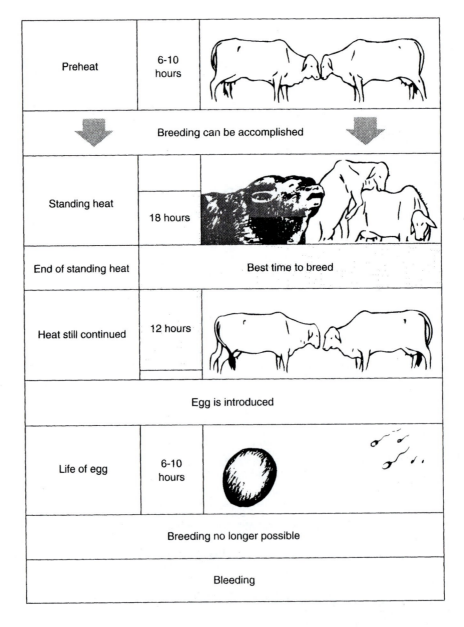

Preheat	6-10 hours	
	Breeding can be accomplished	
Standing heat	18 hours	
End of standing heat	Best time to breed	
Heat still continued	12 hours	
	Egg is introduced	
Life of egg	6-10 hours	
	Breeding no longer possible	
	Bleeding	

Fertilization and Development in the Fowl

Sperm from fowl retain their fertilizing capacity for a considerably longer time than mammals. Chicken sperm have been known to be viable for as long as 32 days after insemination, but weekly insemination is needed to ensure highest fertility. After mating, the sperm are stored in natural folds in the oviduct of the female. These folds are sometimes referred to as sperm nests. As the yolk enters the infundibulum, the walls of the oviduct are stretched, liberating the sperm to fertilize the egg. Fertilization takes place at the *germinal disc* on the yolk.

If fertilization takes place, the embryo begins development around the well-defined germinal disc. The area is clearly visible to the naked eye when broken out of the egg. Within 48 hours, a chick embryo has established an intricate type of blood circulation between itself and the life-sustaining yolk. Because there is no placenta as in the other species discussed in this book, the fowl embryo has to depend on this intricate blood vascular network to carry out the necessary functions of bringing in nutrients and ushering out waste products.

By the end of the third day, the embryo has a full complement of membranes, known as the *allantois, chorion,* and *amnion* (see figure 3–9). The allantois, which at first serves to store excretory wastes, later merges with the chorion to form the *chorioallantois.* The major part of this combined membrane is closely associated with the shell. This membrane serves as the respiratory organ for the developing embryo until the pulmonary type takes over about 24 hours after hatching. By the end of the first one-third of the incubation period, the general outline of the embryo is fully recognizable and most of the essential internal systems, such as the lungs and the nervous, muscular, and sensory systems, are developed. The sex of the chick embryo can be determined as early as the fifth day of incubation. By the middle of the incubation period, embryos of most domestic poultry species are fully covered with *down* (the first feathers).

As with other species covered in this text, the fowl embryo partly floats within fluid in the amniotic cavity. The fluid is necessary to protect the developing embryo and allow it free movement. This freedom of movement is especially important in the chick embryo and must prevail to within the last 3 or 4 days of hatching or malformations can occur that endanger the life of the newly hatched chick. The egg must be turned several times a day in the incubator to prevent the embryo from adhering to the chorioallantois membrane. Under

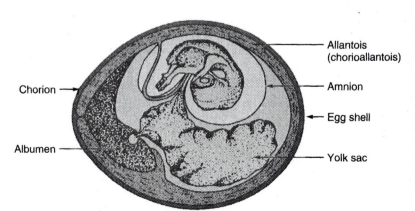

Chorion

Albumen

Allantois (chorioallantois)

Amnion

Egg shell

Yolk sac

FIGURE 3–9

Diagrammatic illustration of a 10-day-old chicken embryo, showing some important membranes. Reprinted with permission from J. Blakely and D.H. Bade, *The Science of Animal Husbandry,* 6th ed. (Englewood Cliffs, NJ: Prentice-Hall, 1994).

natural conditions, the hen shifts the egg several times a day out of instinct. The shell and the membranes also serve to further protect the developing embryo from harmful microorganisms or molds. Additional protection is provided by the mildly bacteriostatic action of the **albumen.**

▓ **Albumen** white of
the egg

Malpositions in fowl embryos may not be given much thought, but there is a natural position for poultry. About midway through the incubation period, the embryo assumes the normal position, which is to lie on its left side along the longest axis of the egg. The head should be tucked under the right wing facing the large end of the egg. Any other position is considered a *malpresentation.*

GESTATION

▓ **Gestation** period
of pregnancy

Gestation is the period of pregnancy. It starts with fertilization and concludes with parturition. The average length is summarized in table 3–6. There are both breed and individual differences, as well as other mitigating circumstances (such as twins). In cattle, gestation is 1 day longer when the cow is carrying a bull calf and 1 day shorter when the cow is carrying twins.

After fertilization, the embryo freely floats around the oviduct and then the uterine horns. Nutrients stored in the zygote (yolk) and uterine milk provide the developing embryo its energy. *Nidation,* or embryo implantation, takes days to weeks and is species dependent (see table 3–7).

Differentiation is the period during which the cells of the new embryo start the process of forming germ layers, organs, and extraembryonic membranes. Until Dolly, we thought this process was irreversible (see later section on cloning). As previously discussed in the section on poultry reproduction, three extraembryonic membranes surround the developing embryo soon after differ-

TABLE 3–6	Gestation Lengths in Domestic Animals	
Species	**Average Length (days)**	**Term Used (parturition)**
Cattle	282	Calving
Holstein	279	
Jersey	279	
Angus	279	
Brahman	293	
Hereford	284	
Sheep	145–150	Lambing
Goat	145–150	Kidding
Mare	337	Foaling
Belgian	335	
Morgan	342	
Arabian	337	
Thoroughbred	338	
Sow	114	Farrowing
Bitch	63–64	Whelping
Woman	266	Birthing
Queen	63	Queening
Elephant	600–660	Calving
Rabbit	28–35	Kindling
Llama	340–345	Birthing

TABLE 3–7	Nidation in Days Post Conception	
	Species	**Days**
	Cattle	30–35
	Sheep	30
	Pigs	18–24
	Horses	24–40

entiation begins. The amnion is the membrane closest to the embryo. The amniotic fluid bathes and protects the developing fetus. The middle membrane, the allantois, fuses with the chorion (outer membrane) in most species, forming the *chorioallantoic membrane.* This membrane attaches to the endometrium to form the placenta (*placentation*).

Pregnancy Diagnosis

It is very costly to the livestock operation to maintain breeding animals that are not pregnant. Timely culling decisions permit the animal manager to focus resources on reliable breeders. Diagnosing pregnancy early allows for proper nutritional demands of gestation, calving, and lactating, as well as considerations for rebreeding. In the past, the veterinarian did most if not all of the rectal palpations on the farm. Today, most progressive cattle managers are capable of diagnosing pregnancy in their animals.

The most common method of pregnancy diagnosis in large animals is by *rectal palpation.* Palpation is done by inserting the arm into the rectum and feeling the reproductive tract for pregnancy indications. The nongravid reproductive tract normally lies on the pelvic floor or against either pelvic wall. The cervix is the chief landmark, serving as a guide for locating all the other reproductive structures. The diagnosis is made chiefly based on the position and contents of the uterus. The horns are usually coiled on their front edge and in older cows may hang slightly into the abdominal cavity. In a **gravid** (pregnant) reproductive tract, the following changes can be felt:

■ Gravid   pregnant

1. *Fetal membrane slip:* At 30 to 35 days very experienced palpators can pick up the *embryonic vesicle* in the gravid horn and feel it as it gently slides through their fingers. The vesicle surrounding the developing embryo is about ¾ inch in diameter and filled with fluid. The chorionic membrane can be detected by gently grasping the uterine wall between the thumb and forefinger and lifting slightly. This method is *not* recommended for beginners because improper palpation can cause early embryonic death.

2. *Size increase:* Starting at about 45 days, the gravid horn is somewhat enlarged and has a thinner wall than the nongravid side. Slipping of the fetal membranes is much easier and safer at this point in the pregnancy. The fetus is only 1 to 1½ inches long at 45 days. By 120 days, the fetus will have grown to 12 inches in length.

3. *Cotyledons:* Cotyledons are the best indicators of a fetus being present. Starting at around 90 days, cotyledons can be palpated. They should measure ¾ to 1 inch across. By 120 days, they will have increased in size to 1½ inches.

4. *Miduterine artery:* After about 90 days, the miduterine artery to the gravid horn is about ⅛ to ³⁄₁₆ inch in diameter, or twice the size of the nongravid horn. A characteristic buzz or purr can be felt as blood is carried into the uterus to feed the developing fetus.

Radiography.   With smaller farm animals, such as sheep, goats, and pigs, rectal palpation is not physically possible. Therefore, other methods of pregnancy diagnosis must be used. X-rays can be used to identify fetal skeletons during the third trimester. This method is much too costly for production enterprises but is used occasionally in small animal veterinary practices for dogs and cats.

Ultrasound.   Ultrasound techniques have been used in human obstetrics for many years. Recently veterinarians have adopted this procedure to determine whether to breed a mare and as a means of pregnancy diagnosis in ewes, does, llamas, sows, and sometimes companion animals.

Hormonal Assays.   Concentration differences of various hormones in an animal's milk, blood, urine and perhaps saliva can be used to assist in diagnosis of pregnancy.

1. *Milk:* Progesterone levels in milk parallel those found in the blood. Prediction of pregnancy using a milk progesterone test is not 100 percent accurate. Cows with low levels of progesterone are open (not pregnant). Cows with higher levels (20 to 22 days postbreeding) may be pregnant or in the diestrus phase of estrous (if the cow was not bred during her estrus phase). Thus a false-positive test is possible.
2. *Blood:* PMSG can be detected in the pregnant mare's blood at 40 days postbreeding. Peak levels appear between day 50 and day 120. Progesterone levels in the blood of pregnant cattle, sheep, sows, goats, and mares can also be used to determine whether an animal is pregnant. However, animals in the diestrus phase of the estrous cycle can be confused with animals with high levels of progesterone.
3. *Saliva:* This test is still in experimental stages.
4. *Urine:* Urine hormone tests are available but not practical in livestock species.

Bouncing the Calf.   Bouncing or bumping the calf can only be done in the latter stages of pregnancy (after 5½ months). The fist is clenched and placed against the abdomen in the right flank. A short jab is made into the area and held in place to detect the swinging fetus returning to the area of pressure. This method was quite common in the United States before World War II and is still practiced in many developing countries today.

Vaginal Biopsy.   Vaginal biopsies have proven accurate in diagnosing pregnancy in laboratory animals, rabbits, and sheep. The tissue is fixed on a slide. The predominance of small cells with rounded nuclei is an indicator of pregnancy. [7] The technique is very time-consuming and not practical in most farm operations.

---

[7] A. M. Sorensen, Jr., *Animal Reproduction: Principles and Practices* (New York: McGraw-Hill, 1979).

Immunologic Diagnosis.   Immunologic assays depend on antibodies reacting with a hormone or product produced by the conceptus or extraembryonic membranes. In women, human chorionic gonadotropin (HCG) can be analyzed from urine samples 1 week into gestation. In cattle, pregnancy-specific protein B from the placenta can be detected in the blood of the cow as early as 24 days after conception. This method is much more accurate than evaluating milk progesterone levels because there are no false-positive results. However, there are no current cow-side tests for protein B. Cow-side progesterone testing has been available for many years. Similar protein B tests are available for sheep and goats but must be sent to a laboratory for analysis.

A more recently developed test, early conception factor (ECF), identifies 94 percent of open cows. The test is more commonly used to catch nonbreeders than to confirm pregnancy; it is advertised as a conception test and not a pregnancy test. ECF is a novel protein that appears in the blood serum and in the milk of certain mammalian species as early as 24 hours postbreeding. This test may eventually prove profitable in alerting farmers and ranchers to a large

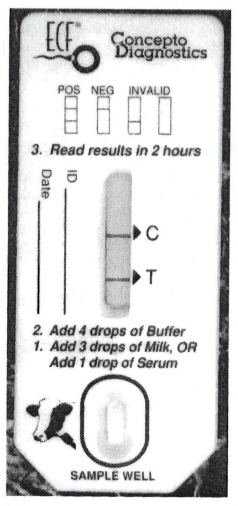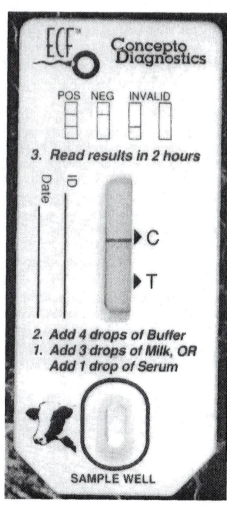

(a)   (b)

FIGURE 3–10
Early conception factor pregnancy tests. Test (a) indicating conception took place and test (b) indicating an open cow (no conception).
Courtesy of Concepto-Diagnostics.

TABLE 3–8	Cost Comparison of ECF Testing versus Traditional Methods		
	ECF Test	**Palpation**	**Ultrasound**
Days open	10	40	25
Revenue loss			
($2.00 per day open)	$20.00	$80.00	$50.00
Cost of method	$5.00	$5.00	$25.00
Income loss	$25.00	$85.00	$75.00
ECF net savings		$60.00	$50.00

Source: Concepto-Diagnostics, P.O. Box 6275, Knoxville, Tenn., 37914. 1999.

number of nonbreeders due to a ration or other herd problem. Currently, ECF testing can be conducted as early as 48 hours postbreeding but is more accurate between 6 and 15 days. Table 3–8 compares ECF testing and more traditional means of diagnosing open cows.

Palpation of Abdomen.   When palpating smaller animals, such as ewes, a plastic lubricated rod can be placed in the rectum to lift the fetus or fetuses against the abdominal wall, where they may be palpated. Ewes must be at least 65 days postbreeding for good accuracy. Rabbit does are easier to palpate, and fetuses can be felt as early as 10 days postbreeding (see later section on rabbits).

Parturition

Outward signs of approaching parturition vary slightly with each species, but the initiation is the same in all farm animals. In cattle, goats, and pigs a rise in fetal cortisol levels causes the placenta to release increasing amounts of estrogen, which initiates the release of $PGF_{2\alpha}$ from the uterus. $PGF_{2\alpha}$ causes the regression of the corpus luteum and thus the cessation of pregnancy. Release of relaxin, followed by oxytocin, causes relaxation of the pelvic area and contractions to assist with the delivery. In sheep, the placenta is the major source of progesterone production during pregnancy. Fetal cortisol in the lamb causes the ewe to convert placental progesterone to estrogen.

Appointment Births.   Appointment births, or planned times for delivery, can be initiated by injections of dexamethasone (a synthetic glucocorticoid), progesterone or estrogen (species dependent), and $PGF_{2\alpha}$. Increased levels of retained placenta and lowered subsequent fertility in the following breeding season have been seen with appointment births.

Three Stages of Birth.   The first stage of parturition is the dilation of the cervix and the positioning of the fetus in the birth canal. This stage is usually the longest and takes from 2 to 6 hours in the cow and ewe, up to 12 hours in the sow, and only 1 to 4 hours in the mare. The second stage of parturition involves the expulsion of the fetus. This part of the parturition cycle is usually very quick. Cows and ewes generally require less than 2 hours for delivery; mares deliver in 15 to 20 minutes. Once the chorioallantoic membrane separates from the endometrium, expulsion must be rapid or the fetus will suffocate.

ARTIFICIAL INSEMINATION

The Arabs are credited with the first modern use of artificial insemination. According to oral tradition, Arab sheiks used artificial insemination to spread poor genetics among neighboring sheikdoms. The beauty of his horses largely measured a sheik's wisdom. Traditional fireside stories describe collection of semen from stallions with poor confirmation and the depositing of this semen into mares belonging to a rival sheik.[8]

Dr. L. Spallanzani, an Italian physiologist, conducted the first scientific artificial insemination. In 1780, Dr. Spallanzani successfully collected and inseminated dogs. Dr. Spallanzani was also one of the first researchers to document that the cooling of semen (used stallion semen) did not necessarily kill the spermatozoa but simply held them in a motionless state until they were again exposed to heat.

In 1899 in Russia, Dr. E. I. Ivanoff began using artificial insemination to increase the quantity of high-quality horses for the chief of the Royal Russian Stud. Dr. Ivanoff trained some 400 technicians in the breeding of horses, sheep, and cattle.[9]

In 1937, Danish veterinarians developed the rectovaginal cervical fixation method of artificial insemination (see figure 3–11). It is still the most common method used today for cattle and horses. Using this method, the inseminator

FIGURE 3–11
Rectovaginal cervical fixation method of artificial insemination. Photo by author.

---

[8]As told to Dr. L. S. Shapiro by Bedouin leaders in the Sinai desert, 1979.
[9]E. J. Perry, *The Artificial Insemination of Farm Animals,* 4th ed. (New Brunswick, NJ: Rutgers University Press, 1968).

manipulates the cervix via the rectum, enabling him or her to guide the inseminating tube through the cervix and into the uterus. The semen is usually deposited at the junction of the cervix and the body of the uterus.

In 1949, semen was frozen for later use, using dry ice and alcohol (−79°C). With the addition of glycerolized diluters to protect sperm cells from ice crystals, liquid nitrogen (−196°C) soon replaced dry ice (see figure 3–12). Today, semen can be stored almost indefinitely and still maintain satisfactory conception rates. However, semen from bulls more than three or four generations back would most likely not improve the genetics of future generations. With each generation, the genetics should be better than the previous one, making older semen obsolete.

Although distributors in some nations still store semen in glass ampules, in 1972, American semen distributors began packaging bull semen in plastic French straws. The industry standard is 0.5 cc of semen containing 30 million live sperm cells. Half of these sperm cells are expected to die during the freezing and thawing procedures. The remaining 15 million live spermatozoa are those placed into the uterus of cows in heat. Conception rates have risen considerably with slight modifications of animal restraint, nutrition, and, most importantly, heat detection methods. Quality semen is collected by various methods; the most common by use of an artificial vagina (figures 3–13, 3–14). Semen quality is determined by concentration, using a spectrophotometer (figure 3–15), by motility and by morphology.

Distributors in the United States have been freezing boar semen since 1975. Most commercial companies freeze boar semen in pellets, which are thawed in dilutant. Additional research and work with frozen stallion, buck, and ram semen is necessary to reach the same acceptable conception rates we already have with frozen bovine semen. The economic incentives for developing this work are just beginning to surface as breeders start to understand the direct and indirect advantages of artificial insemination in other species.

FIGURE 3–12
A liquid nitrogen refrigerator can hold thousands of units of semen almost indefinitely and still maintain satisfactory conception rates. Photo by author.

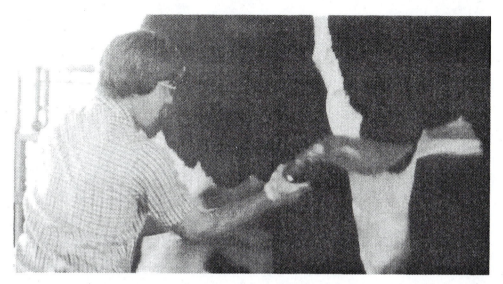

FIGURE 3–13
Collection of semen using an artificial vagina with handler and bull in an open barnyard.
Photo by author.

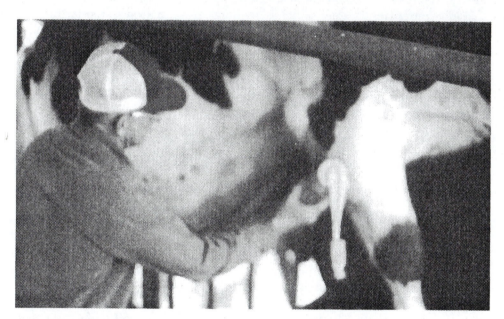

FIGURE 3–14
Collection of semen using an artificial vagina with bull in a protective breeding chute.
Photo by author.

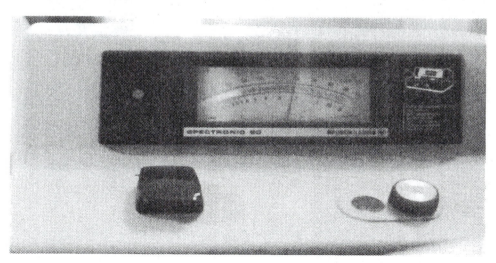

FIGURE 3–15   This spectrophotometer is used to measure the concentration of sperm cells before dilution and freezing.
Photo by author.

Embryo Transfer

The use of superior sires through artificial insemination, resulting in improved weight, quality, and disease resistance, is well-known. This method, of course, involves the freezing of semen to be later thawed and used to impregnate the female. The obvious advantage is to make the most extensive use of a superior male's genes. A similar method to take advantage of a superior female's genes is called embryo transfer and now embryo splitting. Embryo transfer is not new. It dates back to 1890, when Walter Heape, an English biologist, succeeded in making a rabbit produce a litter containing purebred young of two nonsimilar breeds. It has since been achieved in cattle, swine, horses, sheep, and zoo animals. There have even been cases in which a member of one species carries the young of another when compatible conditions exist. For example, there are records of a mouflon born to a sheep at Utah State University, and the Bronx Zoo delivered a rare Indian guar carried by a Holstein cow.

Embryo transfer, in a sense, takes artificial insemination a step farther. For instance, a cow with superior genetic characteristics is injected with a hormone that causes her to superovulate. Instead of releasing 1 mature egg into the reproductive tract, more typically she produces 6 to 8 but possibly as many as 20 or 30.

Through artificial insemination, the ova can then be fertilized by semen from a genetically superior bull. The superovulated cow then develops a number of fertilized embryos in 7 days. By inserting a long, thin, flexible latex catheter into the uterus (to seal off the cervix by an expandable bulb at the tip of the catheter), a saline solution can be passed into the uterus. Then, through a second channel in the same catheter, the microscopic embryos are floated out to collection cylinders. This technique, usually called embryo flushing (see figure 3–16), can produce many fertilized embryos (see figure 3–17); if used fresh, they can be transplanted within the next 24 to 48 hours into surrogate (recipient or substitute) mother cows.

Approximately 65 percent of fresh and 60 percent of frozen embryos result in pregnancies that will produce many genetically superior animals unrelated to the recipient mother.[10] Instead of getting 1 calf a year out of a genetically superior cow, it might be possible to get 100 calves. One superior donor cow is on record at Rio Vista Genetics, San Antonio, Texas, for producing 110 calves in 1 year, about 150 years of normal production.

The nonsurgical technique of recovery and transfer of embryos is considered the method of choice in cattle and accounts for >1 percent of cattle currently bred in the United States. More than 140 companies are doing over $20 million worth of business annually dealing in cattle embryo transfers. Although the cost still ranges from several hundred to several thousand dollars per embryo transfer, new technologies will undoubtedly bring the price down to benefit the average dairy farmer and beef rancher.

One such exciting technique recently developed by Rio Vista Genetics is a new transfer procedure for freezing and storing the embryos in much the same manner that semen has been stored. The technique involves a French straw (similar to a plastic soda straw) containing a few drops of protective fluid and the microscopic embryo. The straw is frozen in liquid nitrogen and can be

---

[10]Trans Ova Genetics, Sioux Center, Iowa, 1998. Personal Communication 1 (800) 372-3586.

(a)

FIGURE 3–16
Embryo transfer begins by nonsurgical flushing of fertilized embryos from the uterus of the donor cow. (a), By inserting a dual-channeled catheter with an inflatable head into the cervix, 200 to 300 ml. of saline solution can be introduced into the fallopian tubes, washing out the embryos into a collecting vial or filter. A technician, through rectal palpation, guides the catheter to its proper position and also manipulates the tract for maximum embryo collection. (b), Close-up view of tube arrangement for non-surgical flushing. Courtesy of Dr. James Blakely.

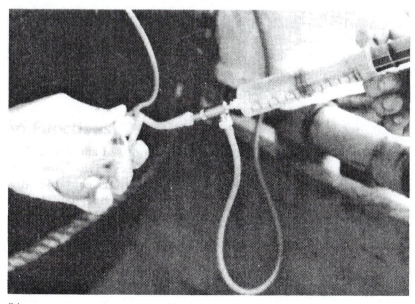

(b)

FIGURE 3–17
Fertilized embryos, either washed from a filter or simply discovered in the original flushing solution, are spotted in a grid plate and individually collected in a simple syringe (a). Embryos may then be transferred fresh into recipient cows or frozen for later use in a "French straw" (b). Photos by author.

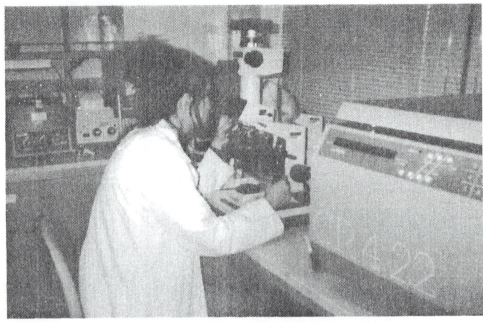

(a)

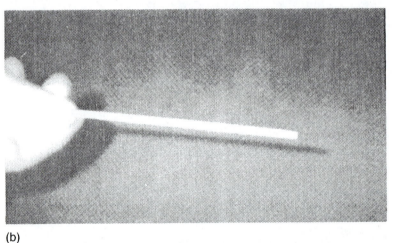

(b)

stored indefinitely. The process from thawing to implantation in the surrogate mother cow takes only about 10 minutes. Although the success rate has been about 50 percent, the exciting possibility of having embryo transfers in such a simplified and reliable state (no more complicated than artificial insemination) makes it possible for any practicing veterinarian and many skilled laypeople to perform the thawing and implantation successfully.

Embryo Splitting

Research scientists, taking a lesson from nature, long ago observed that identical twins were produced in a natural splitting of one fertilized embryo that produced two genetically identical offspring. Although the equipment necessary to create this natural occurrence artificially is very expensive, the microscopic

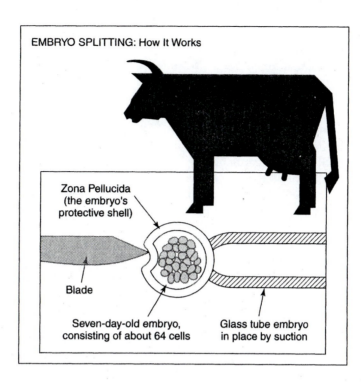

EMBRYO SPLITTING: How It Works

Zona Pellucida
(the embryo's
protective shell)

Blade

Seven-day-old embryo,
consisting of about 64 cells

Glass tube embryo
in place by suction

FIGURE 3–18 The calf embryo, held under a microscope on the tip of a tube, is split with a blade 7 days after fertilization. Both halves of the embryo will develop into identical calves.

surgery is relatively simple. Microsurgical embryo splitting and fusion have been applied to large domestic animals to increase the number of offspring with identical genotypes. One such method involves nuclear transplantation. Nuclei from 8- and 16-cell embryos are fused with enucleated halves of unfertilized eggs. This process results in the cloning of the animal and works up to the 16-cell stage.

Basically, a microscopic surgical blade is used to puncture the zona pellucida (the embryo's protective shell). The blade is then used to cut in half the 7-day-old embryo, consisting of about 64 cells; one-half of the embryo is then repackaged in a separate but natural shell (a zona pellucida from an unfertilized or inferior embryo). Embryos have been successfully divided into as many as four parts by microsurgery. It is most common, however, to limit embryo splitting to two halves. The embryo segments are referred to as demiembryos. The tough but pliable covering that surrounds the developing ball of cells allows the newly twinned embryos to be placed, generally one at a time, in the uteri of surrogate mongrel cows to gestate for the next 9 months (see figure 3–18).

The main advantage to embryo splitting is the increase in pregnancies per embryo collection. Pregnancy rates for fresh transferred split embryos are usually well over 50 percent. Relative to the whole embryo, this means better than 100 percent pregnancy rates. For example, if 7 whole embryos are split, yielding a total of 14 halved embryos and 7 pregnancy results, the result is a total of 8 pregnancies from the 7 whole embryos.

Chimeras

In the staggeringly complex world of reproductive manipulation, if calf embryos too small to be seen with the naked eye can be split and combined, it is reasonable to assume that the cells from numerous embryos might be recombined within one zona pellucida to produce an animal that is a combination of several animals in a single breeding cycle of just over 9 months. Such cattle have been produced in Australia using two, three, or even four individual breeds to produce offspring, joining the genetic material together surgically under a microscope. These creatures are called chimeras, after the mythical Greek monster that was part lion, part goat, and part dragon. Thus a **chimera** is simply an animal that possesses two or more cell populations. A chimera can also result from a transfer of cells between dizygotic twins within the placenta.

Although the process is still experimental, expensive, and time-consuming, the procedure is not much more difficult than embryo splitting because it involves basically the same microscopic surgical technique. Two or more embryos are taken at an early stage of development, the zona pellucida is removed surgically, and the two or more embryos are placed together, fusing a mixture of genetic material inside a common zona pellucida. The chimera can then be transferred into a recipient (surrogate) cow.

The reason for experimentation with chimeras is that it may be possible to transfer the milking ability of one cow, the muscular conformation of a second, the disease resistance of another, and numerous other qualities through the chimeric process. For instance, cows could be made to order for a certain function, to produce at maximum efficiency in a harsh climate, and to resist diseases that may have made cattle production impossible under natural conditions.

The drawbacks to the chimera are that it is still very experimental and expensive, and, at this point, any animal produced as a chimera does not pass on these outstanding qualities to natural offspring. Chimeras would have to be continually reproduced in the laboratory for replacements. Still, certain aspects of the process make it an exciting possibility.

▨ **Chimera** animal produced by the mixing of two or more cell populations or by the grafting of an embryonic part of one animal on to the embryo of another

Sex Control

The option to control the sex of animals within a species would have a great impact, and some successful manipulative research has been conducted in the area. The obvious positive examples of being able to control sex would be the production of only females in a laying-hen operation, only males in a broiler production, only females for the dairy trade, mostly males for the beef feedlot operator, and so on.

Sex-specific embryo transfer is now a reality. Through a biopsy technique, the sex of an embryo can be determined within hours, allowing same-day decisions regarding embryo disposition based on the embryo's sex prior to transfer.[11] Pregnancy rates for fresh-transferred, sexed embryos are slightly lower than for nonbiopsied embryos.

Identification of embryos for sex is not the only approach being researched; in addition, scientists have used several different methods to attempt to separate spermatozoa into the X and Y faction to produce either male or female embryos. Methods currently being investigated for sex determination of mammalian embryos include cytogenetic analysis (karyotyping) of some embryonic cells, immunological detection of a sex-specific factor (H-Y antigen), and measurement of differences in the metabolic activity between male and female

---

[11]Trans Ova Genetics, Sioux Center, Iowa, 1998. Personal Communication 1 (800) 372-3586.

embryos (glucose-6-phosphate dehydrogenase activity). In the case of the H-Y antigen method, 84 and 85 percent accuracy rates have been reported for cattle and sheep, respectively. Although the research is encouraging, repeatability in sperm cell separation is questionable for commercial application.

The method of sex determination showing the most promise involves the Y-specific DNA probes. This method involves the cloning of a DNA fragment with a nucleic acid sequence found only on the Y chromosome. The presence of this fragment in a sample of cells from an embryo indicates male, and its absence indicates female.[12] The accuracy with this procedure is greater than 95 percent. Economic feasibility has fallen short using this technique.

Another research project for sex control that holds promise is the immunological approach. The female is sensitized against the X or Y spermatozoa, causing the unwanted sex to be rejected before fertilization. In the same way, the unwanted fertilized embryo could be rejected by a sensitized female.

The obvious impact of sex control would be to improve efficiency of animal production beyond all imaginable techniques using natural breeding methods. To be able to control the exact numbers and sex of animals for production of meat, milk, work, sport, and replacements and to create these individuals made to order lends new credibility to the field of animal science through genetic engineering.

In June 1998, Switzerland became the first nation in the world to put the issue of genetic research to a ballot. The proposed law would have banned all research with genetically modified animals and crops, but Swiss voters overwhelmingly rejected the legislation. Not 1 of Switzerland's 26 cantons voted in favor of the restrictions. The Swiss are very environmentally conscious, but jobs and investment proved to be bigger concerns.

Cloning

Much has been written in the popular media about cloning, and many misconceptions have arisen as a result. The process is not that complicated or that new. Cloning means taking the cells from an individual and creating an environment in which those cells can reproduce an individual identical to the one from which the cells came.

Nuclear Transfer

Taking cloning one step further, researchers have been experimenting with nuclear transfer. Through microsurgery, several hundred cells are removed from a fertilized embryo. By removing the nucleus of each cell (which contains the chromosomes) and injecting it into an unfertilized egg that has had its own nucleus microsurgically removed, many hundreds of identical individuals might be possible as a result of embryo and nuclear transfer.

Furthermore, scientists are currently experimenting with nuclear fusion, a step beyond nuclear transfer. To greatly oversimplify the process, nuclei from two eggs or two spermatozoa are fused together to create an animal that has two mothers and no father or two fathers and no mother. This process would give the maximum amount of inbreeding and concentration of desired genes.

---

[12]H. J. Bearden and J. W. Fuquay, *Applied Animal Reproduction*, 5th ed. (Upper Saddle River, NJ: Prentice-Hall, 2000).

Dolly

In February 1997, Dr. Ian Wilmut of Scotland announced that he had success-fully cloned a sheep from a somatic cell. A first! Dr. Wilmut took cells from the udder of a Finn Dorset ewe and starved them in a culture dish. He then took an unfertilized ovum from a Scottish Blackface ewe and removed the nucleus, leaving only the zona pellucida and cytoplasmic organelles. Placing the two cells next to each other, he fused them using an electric pulse and a second pulse to jump-start cell division. Six days later, Dr. Wilmut transplanted this fused cell (embryo) into the uterus of another Blackface ewe. Five months later Dolly was born. One year later, Dolly gave birth herself, proving she was normal genetically. Similar *nuclear transfer technology* produced a 98-pound Holstein calf named Mr. Jefferson in February 1998.[13]

Recombinant DNA

DNA from one species is isolated and transferred to another species, usually a bacterium. For instance, DNA from a cow is transferred to *Escherichia coli* to culture and collect a large amount of a given substance under the direction of genetic instructions from DNA. DNA can be chemically modified to produce only one such substance, such as growth hormone. Therefore, natural hormones can be produced in very large quantities, using bacteria as the manufacturing agent. Hormones produced in this way can then be injected back into the cattle from which they came to increase production above what would be possible under normal conditions. Successful application of this technique at Cornell University has increased milk production in dairy cows and meat production in feedlot animals. Other products produced by recombinant DNA include interferon, used to combat shipping fever in cattle, and a vaccine against hoof-and-mouth disease. These are but a few examples of products that can be produced in large quantities by bacteria incubated in large vats. Scientists are looking for new ways to utilize these innovative procedures, others, such as animal activist Michael Fox of the United States Humane Society, condemn the production of these "super animals" as immoral acts.[14]

SUMMARY

The reproductive process is the focal point of overall animal productivity and profit. Success in fertility and reproductive efficiency relies on an understand-ing of seasonal variances, additional nutrient demands of the breeding animal, disease control and prevention, and marketing of projected offspring or prod-ucts. The mating of animals is more than just choosing a male and female of the same breed or species. It involves scientific analysis of each with statistically projected outcomes in the progeny.

---

[13]PPL Therapeutics Plc.

[14]M. W. Fox, "Genetic Engineering: Nature's Cornucopia or Pandora's Box?" *This Animal Agenda*, March 1987.

Modern veterinary products and management techniques are used to manipulate estrous cycles, improve heat detection, perform embryo transfer and splitting, and treat reproductive abnormalities. Many of these newer procedures are still undergoing industry and consumer criticism and evaluation. The twenty-first century will see continued improvement in reproductive performance with advantages to both the farmer and the consumer.

EVALUATION QUESTIONS

1. A mature follicle is called a _____ .
2. The funnel-shaped end of an oviduct is an _____ .
3. _____ is the exact location of fertilization.
4. _____ is the anatomical part of the reproductive tract that protects the uterus from infection and passage of foreign material.
5. The age of a mare when she reaches puberty is _____ .
6. The average gestation periods of a bitch, queen, cow, mare, sow, ewe, doe, and elephant (in days) are _____ , _____ , _____ , _____ , _____ , _____ , _____ , and _____ .
7. The average lengths of the estrous cycle in a cow, bitch, and mare are _____ , _____ , and _____ .
8. Nidation is complete _____ days after the cow is fertilized.
9. The pituitary hormone that stimulates ovulation is _____ .
10. The hormone that maintains pregnancy is _____ .
11. _____ is the milk letdown hormone.
12. _____ is an accurate method of pregnancy diagnosis in dogs used only during the last trimester.
13. In dairy cows, the hormone _____ is assayed from milk samples for pregnancy diagnosis.
14. A fetus releases the hormone _____ , which triggers parturition.
15. Difficult birth is called _____ .
16. ET is also called _____ .
17. _____ refers to an animal with testes abnormally remaining in the abdominal cavity.
18. A castrated stallion, chicken, boar, bull, and ram are called _____ , _____ , _____ , _____ , and _____ .
19. _____ , _____ , and _____ are three accessory sex glands in the male.
20. A beef heifer should reach puberty by _____ pounds and _____ months.
21. _____ is the main male sex hormone.
22. A heifer that is bred and who conceives on May 1 should calve on _____ .
23. Progesterone is primarily produced in the _____ . Its many functions include _____ .
24. In the United States most sheep are bred in the _____ season.
25. A sow does not return to heat until after _____ .
26. What is flushing and what animals practice it? _____
27. List three items you can check for in rectal palpation pregnancy diagnosis. _____ , _____ , _____

28. A sow that is bred on July 1 should farrow _____ .
29. Mating between the male and female chicken is mainly a matter of joining _____ .
30. In chickens, the _____ ovary and oviduct degenerate.
31. Turkey hens are commercially kept for breeding _____ times.
32. The recommended stocking rate for breeding turkeys is _____.
33. The drake–hen ratio should be _____ for breeding.
34. A young rabbit doe should first be bred at about _____ months.
35. Doe rabbits are usually rebred when the litter is _____ .
36. Dairy heifers should be bred when they _____ .
37. The tube that extends from the bladder to the end of the penis carrying semen and/or urine is the _____ .
38. The sensory and erectile organ of the female located just inside the vulva is the _____ .
39. _____ causes regression of the corpus luteum.
40. Waxing occurs prior to birth in this animal: _____ .
41. The milk secretion (production) hormone is _____ .
42. An infertile heifer born twin to a bull is called a _____.
43. The hormone _____ stimulates mammary duct development.
44. The hormone _____ has been shown to increase milk yield by as much as 20 percent when injected daily into the cow.
45. The cell formed by the union of a sperm and an ovum is a _____ .
46. An important reproductive characteristic in beef cattle is _____ .
47. Sperm is evaluated on the basis of _____.
48. The lengths of estrus in a cow, ewe, doe, mare, and bitch are _____ , _____ , _____ , _____ , and _____ .
49. Ages of puberty for a pig, cattle, horse, chicken, rabbit, and sheep are _____ , _____ , _____ , _____ , _____ , and _____ .
50. A doe kid born in the spring will be large enough to breed at _____ months and _____ pounds.
51. The typical stocking rates for goats, sheep, cattle, horses, ducks, turkeys, and chickens are _____ , _____ , _____ , _____ , _____ , and _____ .
52. For maximum fertilization, sows should be bred _____ and _____ hours after the onset of estrus.
53. Define the importance of the stigma, blood spot.
54. The eggshell is formed in the _____ .
55. Pregnancy testing of does by palpation can be done as early as _____ days after fertilization.
56. The reproductive life of rabbit does and bucks does not normally exceed _____ years.
57. Polyestrous seasonal breeders include _____ .
58. Roosters should be left with the flock for at least _____ .
59. Boar semen is stored in _____ .
60. What hormone or hormones induce superovulation? _____ .
61. What hormone causes the pelvic region to relax prior to birth?_____
62. When are rams most fertile? Least fertile? Why? _____
63. What keeps foreign matter out of the urinary tract? _____

64. The ligament that pulls the testes into the scrotum is the _____.
65. Sperm mature in the _____ .
66. Before the breeding season, ewes should be _____ .
67. A mature ram can handle _____ ewes.
68. GnRH controls the release of _____ and _____ from the anterior pituitary gland.
69. For best fertilization, gilts should be bred on the _____ heat.
70. Puberty in chickens occurs at _____ weeks.
71. In the male, FSH stimulates _____ .
72. In the first breeding season, turkey hens can be expected to produce _____ eggs.
73. After insemination, chicken sperm are viable for as long as _____ .
74. In the male LH stimulates the production of _____ .
75. Young female chicken _____ , horse _____ , and cow _____ .
76. The sensory and erectile organ of the female is the _____ .
77. The _____ is also known as the bulbourethral gland.
78. The S curve in the penis of the bull, ram, and boar is the _____ .
79. Roosters crow because of _____ .
80. Roosters produce the highest quality of semen when exposed to _____ hours of light.
81. The endometrial cups produce the unique hormone _____ .
82. _____ secretes albumen, or the white of the egg.
83. _____ is the shell gland in the hen.
84. _____ occurs in about 90 percent of all heifers and is caused by the breakage of uterine capillaries, which become more fragile with decreasing levels of circulating _____ during the middle of the animal's cycle.
85. _____ is the outermost of the fetal membranes. The fetal part of the placenta develops from it.
86. _____ is the temperature of liquid nitrogen in degrees Celsius.
87. _____ is a term designating animals that usually produce only one offspring at each pregnancy.
88. _____ is a term designating litter-bearing animals.
89. _____ is white fibrous tissue that formed from a regressing CL.
90. _____ successfully artificially inseminated a bitch that gave birth to a litter of puppies in the 1780s.
91. _____ is the most common method of artificial insemination in horses and cattle today.
92. Most bull semen in the United States is stored in _____ .
93. _____ live sperm cells are placed into one unit of semen prior to freezing.
94. _____ is from the Greek word meaning monster, a complex of recombined DNA from more than one animal source making up an entirely different embryo.
95. Bovine somatotropin that is used commercially today is made by a process called _____ .
96. An indication of estrus in the mare where the vulva opens and closes is called _____ .

RECOMMENDED READING

1. Bearden, H. J., and J. W. Fuquay. *Applied Animal Reproduction*. 5th ed. (Upper Saddle River, NJ: Prentice-Hall, 2000).
2. Hafez, E. S. E. *Reproduction in Farm Animals,* 6th ed. (Philadelphia, PA: Lea & Febiger, 1993).
3. Norris, D. O. *Vertebrate Endocrinology,* 2nd ed. (Philadelphia, PA: Lea & Febiger, 1985).
4. Shapiro, L. S. *Applied Animal Ethics* (Albany, NY: Delmar, 2000).

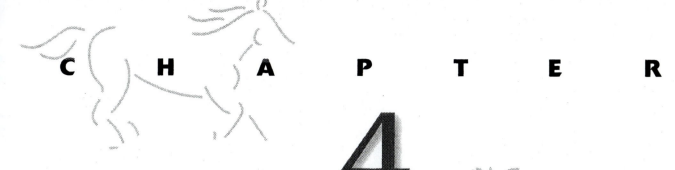

CHAPTER 4

Animal Nutrition

OBJECTIVES

After completing the study of this chapter you should know:

- The six basic nutrients (their importance, requirements, and deficiency and toxicity symptoms) of livestock.
- About photosynthesis and its importance to animals and humans.
- The anatomy of the four main digestive systems found in livestock.
- The basic physiology of the digestive systems found in livestock.
- The efficiency of feed conversion for the various species of livestock.
- The methods used to determine nutrient content in animal feed.
- The basic rules of thumb in feeding livestock.

KEY TERMS

Abomasum
Absorption
Acetic acid
Acidic detergent fiber (ADF)
Amino acids
Amylase
Anus
Arachidonic
Arginine
Avian
Bile
Bloat
Bloat Guard
Bolus
Bomb calorimeter
Butyric acid
Calorie
Calorie system
Camelids

Cannula
Carbohydrates
Carnivores
Cellulose
Chemical (digestion)
Chlorophyll
Chyme
Concentrate
Crude protein
Cud
Digestible energy (DE)
Digestion
Disaccharide
Duodenum
Emulsifier
Endocrine
Enzymatic (digestion)
Equine
Essential amino acids
Ether extract

Exocrine
Fat-soluble vitamins
Fatty acids
Fistulated
Fructose
Galactose
Gallbladder
Glandular stomach
Glucose
Glycine
Glycogen
Grit
Gross energy (GE)
Hardware disease
Hardware stomach
Herbivores
Hexoses
Histidine
Honeycomb
Hydrochloric acid

Hydrogenation
Hydrolysis
Hyperperistalsis
Hypoperistalsis
Ileum
Insulin
Isoleucine
Lactose
Large intestine
Least-cost ration
Leucine
Linoleic
Linolenic
Lipase
Lipids
Liver
Lysine
Maltose
Manyplies
Marbling
Masticates
Mastication
Maximizing profit
Mechanical (digestion)
Metabolizable energy
   (ME)
Methionine

Monogastric
Monosaccharides
Net energy (NE)
Neutral detergent fiber
   (NDF)
Nitrogen free extract
   (NFE)
Nonruminant herbivore
Oil
Omasum
Omnivores
Pancreas
Papillae
Paunch
Pentoses
Pepsin
Peptide bonds
Phenylalanine
Photosynthesis
Poloxalene
Polysaccharides
Propionic acid
Proximate analysis
Rancidity
Rectum
Reticulum
Roughage

Rumen
Ruminal tympany
Ruminant
Saturated fatty acids
Scours
Serine
Soft pork
Starch
Sucrose
Surfactant
Taurine
Threonine
Total digestible nutrient
   (TDN)
Traumatic gastritis
Trocar
Trypsin
Tryptophan
Unsaturated fatty acids
Uric acid
Valine
Villi
Volatile fatty acids (VFAs)
Water soluble vitamins
Weende system of
   proximate analysis

INTRODUCTION

To understand the proper science behind feeding livestock, a brief understanding of the basic digestive anatomy and physiology is essential. There are four distinct digestive systems among livestock species: *avian*, **equine** (or *nonruminant herbivore*), *ruminant,* and **monogastric.** More advanced nutrition courses go beyond discussion of the six basic nutrients, ration balancing, nutrient deficiencies, and toxicities that will be outlined in brief in this chapter.

Life on earth depends on **photosynthesis.** Without it, there would be no oxygen to breathe, no plants, no feed, no food, no animals, and no people. Photosynthesis is the process by which the *chlorophyll*-containing cells in green plants capture the energy from the sun and convert it into chemical energy **(carbohydrates).**

Sunlight

$$6CO_2 + 6H_2O \rightarrow C_6H_{12}O_6 + 6O_2$$

Chlorophyll

With very few exceptions, living organisms are linked together by a one-way flow of energy from the sun and by a cycling of materials such as CO_2 and O_2 on a global scale. Plants harness energy from the sun. Herbivores eat the plants. Omnivores and carnivores eat the plants and herbivores. When animals die, they become fertilizer for the plants and produce CO_2 for them to breathe;

Equine pertaining to a horse

Monogastric simple-stomached or nonruminant animal

Photosynthesis formation of carbohydrates and oxygen by the action of sunlight on the chlorophyll (green pigment) in a plant

Carbohydrates nutrient consisting of carbon, hydrogen, and oxygen with the general chemical formula $(CH_2O)_n$; includes sugars, starches, cellulose, and gums

in turn, the plants produce O_2 for the animals to breathe. We all depend on each other and must live in harmony.

Debate has increased on efficiency methods in feeding a starving world. What many refuse to acknowledge is that we are capable of feeding all that are currently hungry in the world. The problem is that no one wants to pay for it. As third world countries become more affluent, their consumption of poultry and red meat increases.

It is true that more hunger can be alleviated with a given quantity of grain by completely eliminating animals. For example, 2,000 pounds of concentrates must be supplied to livestock to produce enough meat and other livestock products to support one person for a year. We could feed that same person with just 400 pounds of grain in a year's time. Thus, a given quantity of grain eaten directly will feed five times as many people. In developing nations, where the population explosion is the greatest, virtually all grain is eaten directly by people. As people in these developing nations become more affluent, they tend to eat more and more grain indirectly through the increased consumption of animal products.

Animals provide more than just food for people. We have a continued need or use for animals in the following manner[1]:

- In developing countries animals are still used for power. They can be fueled on roughages. These animals include horses, water buffalo, and cattle.
- Animals provide needed nutrients not found in sufficient quantity or quality in available plants (e.g., essential amino acids, minerals, and vitamins).
- Animals produce protein of higher value than that of plants.
- Ruminants are capable of converting nonprotein nitrogen into protein.
- Animals can step up the protein content and quality of foods.
- Animals provide products that meet consumer preferences (choice).
- Much of the world's land is *not* cultivatable and thus would provide *no* food value to people unless they consume the animals that eat off of this land.
- Forages provide for most of the feed consumed by livestock (see table 4–1).
- Much of the feed grains and by-products consumed by animals are not used for human food (e.g., distillery wastes and fruit and vegetable wastes).

TABLE 4–1	Percentage of Feed for Different Species of Livestock Derived from Roughage versus Concentrates	
Species	**Roughages (%)**	**Concentrates (%)**
Sheep, goats	93.8	6.2
Beef cattle	84.5	15.5
Horses, mules	73.0	27.0
Dairy cattle	58.7	41.3
Swine	4.3	95.7
All livestock	61.7	38.3

Source: USDA, Economic Research Service, 1985.

---

[1]M. E. Ensminger, J. E. Oldfield, and W. W. Heinemann et al., *Feeds and Nutrition,* 2nd ed. (Clovis, CA: Ensminger, 1990).

TABLE 4–2	Feed to Food Efficiency Rating by Species of Animals				
Species	Unit of Production	Pounds Feed to Produce 1 Pound Product	Feed Efficiency[a] (%)	Total Digestible Nutrients Required per Pound Produced	Protein Required per Pound Produced
Broiler	Chicken	2.1	47.6	1.7	0.21
Fish	Fish	1.6	62.5	0.98	0.57
Dairy cow	Milk	1.11	90.0	0.90	0.10
Turkey	Turkey	5.2	19.2	4.21	0.46
Layer	Eggs	4.6	21.8	3.73	0.41
Hog	Pork	4.0	0.25	3.2	0.36
Rabbit	Fryer	3.0	35.7	2.2	0.48
Beef steer	Beef	9.0	11.1	5.85	0.90
Lamb	Lamb	8.0	12.5	4.96	0.86

Source: Modified from M. E. Ensminger, J. E. Oldfield, and W. W. Heinemann et al. *Feeds and Nutrition,* 2nd ed. (Clovis, CA: Ensminger, 1990).
[a]Feed efficiency based on pounds of feed required to produce 1 pound of product.

- Ruminants are capable of utilizing low-quality roughages (high in cellulose) that would provide no nutritive value to people.
- Animals are also capable of utilizing by-products, such as corncobs, citrus pulp, oilseed meals, gin trash, and cottonseed hulls, that would provide no value for human consumption.
- Animals provide elasticity and stability to grain production.
- Animals provide medicinal and other products (e.g., more than 100 medicines, cosmetics, and candles).
- Animals are an effective method of storing food for later consumption.
- Animals maintain soil fertility (1 ton of manure contains 10 pounds of nitrogen, 5 pounds of phosphorus [P_2O_5], and 10 pounds of potassium [K_2O]).

Organic farming would be impossible if it was not for large numbers of animals maintained and the manure they contribute to provide soil fertility without the use of chemicals.

How much feed does it take to produce 1 pound of animal product? The answer to this question is highly variable with the species, the country, the region, and the genetics and management of the particular animal in question. Table 4–2 summarizes feed to food efficiency by species. Clearly, the dairy cow is the most efficient animal on an energy and on a pound of feed per pound of product produced basis. On a protein basis, a broiler is more efficient.

ANATOMY OF THE DIGESTIVE SYSTEM

We can classify animals by both the type of food they eat and the anatomy of their digestive systems. There are three basic groupings of animals based on their food consumption:

▓ Carnivores
animals that eat
primarily flesh (meat)

1. **Carnivores** are animals that eat animal tissues. Examples of carnivores include cats and members of the mink family. Restricting meat from true carnivores can result in illness and/or death.

2. **Omnivores** are animals that normally eat both plant and animal products. Examples of omnivores include people, pigs, and dogs. A healthy combination of animal and plant products should be provided to animals that fall in this category.

3. **Herbivores** are animals that eat feed of plant origin. Examples of herbivores include cattle, sheep, goats, and horses.

There are four groupings of animals based on classification by anatomical differences of the digestive tract. They include

- Monogastric with nonfunctional cecum
- Monogastric with functional cecum
- Ruminant
- Avian

Monogastric (Simple-Stomached) Digestive System

The digestive systems of pigs, dogs, and people are very similar. What we learn from one species can often be applied to the others. All of these species are omnivores. All have simple stomachs (with nonfunctional ceca) (see figure 4–1). All have limited ability to digest fiber. The order of food passage and the mechanisms of digestion are basically the same.

Food enters the digestive system via the mouth, where partial digestion takes place. **Digestion** is defined as the breakdown of feed particles by *mechanical, chemical,* and/or *enzymatic* means to a size suitable for absorption.

▧ **Omnivores** animals that eat both plants and meat

▧ **Herbivores** animals that eat feed of plant origin

▧ **Digestion** process of breaking down feed into particles small enough to be absorbed

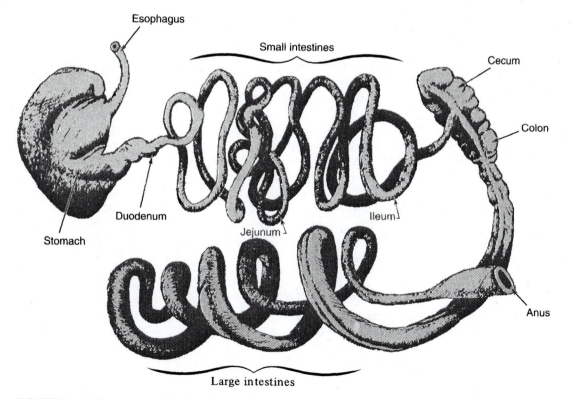

FIGURE 4–1 The monogastric digestive system.

Reprinted with permission from J. Blakely and D. H. Bade, *The Science of Animal Husbandry,* 6th ed. (Englewood Cliffs, NJ: Prentice-Hall, 1994).

▓ **Absorption**
process by which
digested food particles
pass through the
digestive tract and into
the circulatory system

Absorption is the passage of feed nutrients from the digestive system into the circulatory system. In many simple-stomached animals, salivary glands contain digestive enzymes that initiate digestion. The feed is also broken down mechanically in the mouth by *mastication* (chewing).

Food then passes down the esophagus and into the stomach. The stomach can temporarily store ingested food. Its primary function, however, is to further digest what was swallowed. *Hydrochloric acid* and various digestive enzymes (*pepsin*) continue the digestive process in the stomach. The fluid substance (partially digested feed) leaving the stomach is called *chyme.* The chyme passes into the small intestine. The first part of the small intestine, called the **duodenum,** is where most digestion (in monogastrics) takes place. Many additional enzymes are produced and secreted (from the *pancreas*) here to help with this process. Most nutrients are absorbed in the last part of the small intestine, the **ileum.** The wall of the small intestine is lined with many small, fingerlike projections called *villi.* Villi increase the absorption area of the small intestine. Damage to this area (by internal parasites) greatly reduces an animal's ability to utilize the nutrients consumed.

▓ **Duodenum**   first
segment of the small
intestine, following the
stomach

▓ **Ileum**   third portion
of the small intestine

▓ **Bile**   fluid produced
by the liver that emulsi-
fies fats in the small
intestine

The *liver* produces an *emulsifier* of fats, called **bile.** In most animals, bile is stored in the *gallbladder* until it is needed and then is secreted into the duodenum. In the horse and rabbit there is no gallbladder. Bile is still produced in these animals (by the liver), but it is simply deposited directly into the small intestine in small quantities. The percentage of fat in the diet of these animals is normally quite low.

The pancreas is both an *endocrine* (produces hormones) and an *exocrine* (produces enzymes) gland. Like all endocrine glands, its products—insulin and glucagon—are deposited directly into the blood. The enzymes of the pancreas flow through two ducts into the duodenum and are responsible for assisting with the digestion of feedstuffs.

Nondigested and/or nonabsorbed ingesta leaving the ileum pass into the *large intestine.* The large intestine is much shorter than the small intestine but is larger in diameter. The main functions of the large intestine are to store undigested feedstuffs and to absorb water. Large numbers of microorganisms present in the large intestine help further digestion in this area. However, because most nutrients in monogastric animals are absorbed in the small intestine, further digestion in the large intestines is of little benefit to the animal itself.

Nonabsorbed feed that continues to pass through the *rectum* and vacates through the *anus* is at this point called feces. Mucus added in the large intestine aids as feces pass through the terminal portions of the digestive tract. When feed goes through too quickly *(hyperperistalsis),* **scours** (diarrhea) results because not enough water was absorbed in the large intestine. When feed goes through too slowly (*hypoperistalsis*), constipation results because too much water was removed.

▓ **Scours**   diarrhea

Functional Cecum Digestive System (Nonruminant Herbivore)

Examples of animals with a functional cecum digestive system include the horse, rabbit, guinea pig, and hamster. The cecum and colon are extremely large in this group of animals and provide an area for microbial digestion of fiber (see figure 4–2). In an average horse, the large intestine (made up of the colon, cecum, and rectum) is approximately 25 feet long. It takes about 65 to 75 hours for

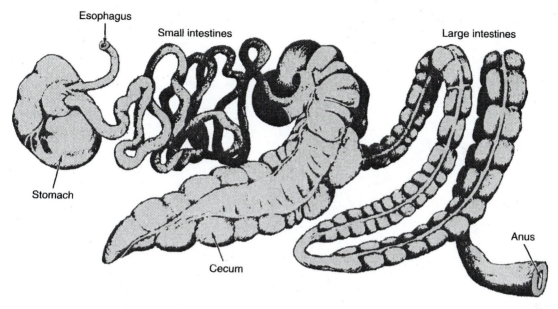

Esophagus

Small intestines

Large intestines

Stomach

Cecum

Anus

FIGURE 4–2 The horse's digestive system. Note the size of the cecum.
Reprinted with permission from J. Blakely and D. H. Bade, *The Science of Animal Husbandry,* 6th ed.
(Englewood Cliffs, NJ: Prentice-Hall, 1994).

feed to pass from the mouth to the anus in a horse.[2] In ruminants, the rumen precedes the small intestine. Thus, any amino acid produced by the microbes can be absorbed in the ileum. In the equine (and other nonruminant herbivores), the ileum precedes the microbial digestion of most of the fiber consumed. In other words, the cecum of the horse is *not* as valuable as the rumen of the cow in providing nutrients for the animal.

Volatile fatty acids (VFAs) produced in the cecum by fiber breakdown can be absorbed by and provide a considerable amount of energy for horses. Microbes in the colon and cecum also synthesize vitamins needed by the equine. However, not all of the required vitamins that are synthesized by the horse are absorbed in the hindgut. Thus, it is more important to check and balance vitamin requirements of horses than those of beef steers, for example.

> **Volatile fatty acids (VFAs)** referring to acetic, propionic, and butyric acids produced in the rumens of ruminants and the ceca of horses and rabbits; provide extensive energy to the animal

Ruminant Digestive System

Cows, sheep, goats, deer, and giraffes all fall under the classification of *ruminants* (see figure 4–3). Ruminant animals have no upper incisors or canine teeth. They do not normally chew or masticate their food the first time it is placed in their mouths. Instead, ruminants swallow their food in large quantities and then find a comfortable spot to lay down, throw up their food, and chew it (ruminate).

Llamas and other *camelids* are not true ruminants. Camelids have two toes on each foot but have three and not four compartments to their stomach. They also have two upper incisor teeth. These pseudoruminants have no horns and run with a pacing gait (move the front and rear legs on one side in unison).

---

[2]T. J. Cunha, *Horse Feeding and Nutrition,* 2nd ed. (San Diego, CA: Academic Press, 1991).

FIGURE 4–3 The ruminant's digestive system.
Reprinted with permission from J. Blakely and D. H. Bade, *The Science of Animal Husbandry,* 6th ed. (Englewood Cliffs, NJ: Prentice-Hall, 1994).

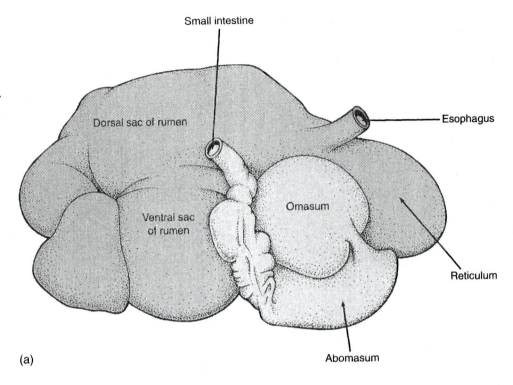

(a)

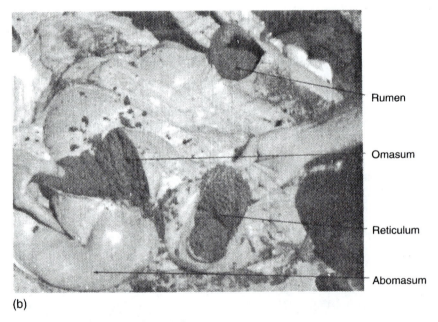

(b)

The four-compartment stomach of the ruminant includes the following:

1. **Rumen:** Also known as the *paunch*, the rumen is a large fermentation vat that can hold as much as 50 gallons in a mature Holstein cow. It is very muscular and is lined with fingerlike *papillae* that increase the absorptive area of the rumen (see figure 4–4). The ingested feed is mixed and partially broken down by microorganisms in the rumen. An average beef

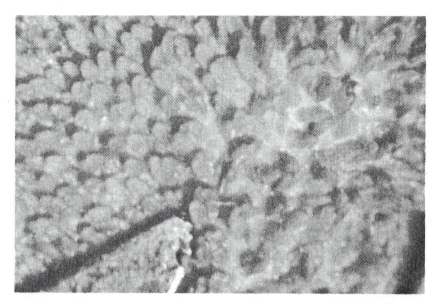

FIGURE 4–4 Villi line the wall of the rumen. Reprinted with permission from J. Blakely and D. H. Bade, *The Science of Animal Husbandry,* 6th ed. (Englewood Cliffs, NJ: Prentice-Hall, 1994).

cow ruminates for 8 hours per day and chews (*masticates*) 42,000 times as she casts up (regurgitates) undigested feed repeatedly throughout the day.[3]

2. **Reticulum:** Also known as the *hardware stomach* or *honeycomb;* because ruminants normally do not chew before swallowing their food, they occasionally swallow foreign objects, such as wire, nails, screws, and so forth. These objects become lodged in the reticulum, which borders the heart. A small magnet is usually placed into the reticulum of most commercially raised cattle to prevent the disorder known as **hardware disease.** Hardware disease (**traumatic gastritis**) occurs when the pieces of metal lodged in the reticulum work their way into the heart and/or surrounding tissues. Symptoms of this disorder include anorexia, emaciation, arched back, fever, and pain when defecating.

3. **Omasum:** Also known as the bible stomach or *manyplies,* the main purpose of the omasum is to grind up feed ingesta into smaller particles and remove some of the water out of the feed. This water removal helps with digestion in the abomasum by providing a more concentrated solute when the feed is mixed with hydrochloric acid and digestive enzymes there.

4. **Abomasum:** Also known as the true or *glandular stomach,* the abomasum makes up only 7 percent of the stomach of the cow. The digestion that occurs here is similar to the type that occurs in human stomachs. Hydrochloric acid and digestive enzymes assist in the breakdown of feedstuffs prior to their entry into the small intestine.

The microorganisms of the rumen have a true symbiotic relationship with the host; 22 genera and 63 different species of bacteria have been described in the rumen.[4] As many as 25 to 80 billion bacteria and 200,000 to 500,000 protozoa

■ **Reticulum** second compartment of the ruminant stomach; also known as the honeycomb or hardware stomach

■ **Hardware disease** caused by the ingestion of small pieces of baling wire, nails, screws, and so forth and their puncturing and/or scratching of the walls of the digestive system in ruminants; also known as traumatic gastritis

■ **Omasum** third compartment of the ruminant stomach; also known as the manyplies or bible stomach

■ **Abomasum** fourth or true stomach of a ruminant

---

[3]L. S. Shapiro, unpublished research (Corvallis, Oreg., 1984).

[4]D. C. Church, *The Ruminant Animal: Digestive Physiology and Nutrition* (Englewood Cliffs, NJ: Prentice-Hall, 1988).

are estimated to be in 1 milliliter of ruminal fluid. The large microbial population enables the ruminant to produce needed VFAs—*acetic, propionic,* and *butyric acids.* These VFAs provide 60 to 80 percent of the ruminant's energy needs. The microorganisms also produce B complex vitamins and essential amino acids. Microbes provide the means for the ruminant to utilize nonprotein nitrogen (NPN) by combining with nitrogen free extract (NFE) to produce amino acids.

A large amount of gas, primarily methane (CH_4) and carbon dioxide (CO_2), is produced as a result of bacterial fermentation in the rumen. The process of eructation (belching) generally removes this gas.

When more gas is produced than can be removed, a potentially lethal condition known as **bloat** (*ruminal tympany*) results. Common causes of bloat include consumption of

- Legume pastures
- Succulent forages high in cell content material
- Wheat pasture that has been heavily fertilized
- High levels of grain
- Low-roughage diets

The primary symptom of bloat is a swelling to abnormal proportions of the left side of the animal. Severe cases cause pressure on the diaphragm and lungs, inducing gasping for breath. It is thought that toxic substances produced during this period are inhaled or absorbed through the lungs, causing death.

Two types of bloat are recognized: a gaseous form and a gas bubble (frothy) form. The gaseous type may be relieved by having the animal walk, passing a tube down the esophagus, or, as a last resort, puncturing the rumen using a **trocar** and *cannula.* The bubble type must be burst like a balloon. A *surfactant* (something to prevent or break the tension needed to produce a bubble) is usually quite effective. Vegetable oils and even detergent soap powders have been used in emergency situations. However, a prepared surfactant such as *poloxalene* is recommended. It may be used as a drench or in critical cases injected through the wall into the rumen.

Prevention of bloat is preferred over treatment. Filling cattle on dry hay prior to grazing legume pastures, a heavy stocking rate to prevent rank growth, and use of surfactants are common procedures that have been very effective in preventing bloat. Bloat Guard blocks (containing poloxalene) have been used successfully to prevent bloat. Animals lick the blocks on a regular basis, maintaining sufficient concentration of poloxalene in their rumen, thus preventing the frothy bloat.

Normal Flow Pattern of the Ruminant Stomach.   Feed floating on the rumen fluid moves in a circular pattern, becoming heavier and slowly sinking. This circular, tidelike motion becomes most active after grazing is completed. The resting cow then begins rumination (**cud** chewing), which is a common sign that a cow is feeling well. Early cattlemen observed that cows not regurgitating were often ill and explained it as having "lost their cud." Not understanding the function of ruminants, they forced artificial cuds (balls of grass, even rags) into the mouth, trying to simulate nature and initiate a return to health. A **bolus** (cud) is formed through muscular action of the reticulum (pacesetter) from material delivered to it by the tide. This cud is regurgitated (forced up the esophagus) back to the mouth, where it is more thoroughly chewed, reswallowed, and

Bloat metabolic disorder in ruminants in which gas is produced in excess compared to what can be removed; also known as ruminal tympany

Trocar sharp-pointed instrument equipped with a cannula that is used to puncture the body wall and rumen to relieve the gaseous pressure in bloat

Cud bolus of regurgitated feed that a ruminant animal chews during rumination

Bolus large pill for dosing animals; sometimes used interchangeably with *cud*

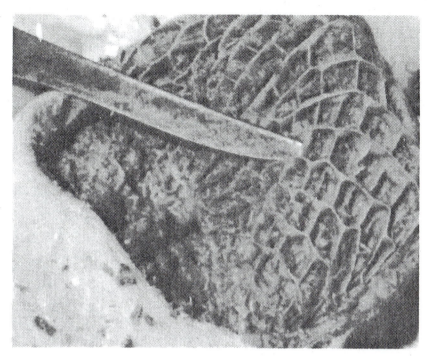

FIGURE 4–5 The wall of the reticulum (honeycomb). Reprinted with permission from J. Blakely and D. H. Bade, *The Science of Animal Husbandry,* 6th ed. (Englewood Cliffs, NJ: Prentice-Hall, 1994).

this time goes through the reticulum that due to its honeycomb structure (see figure 4–5) prevents most foreign objects, such as wire, from further travel. These sharp objects, creating the condition known as hardware disease, often puncture the reticulum, as previously mentioned. Prevention of this disorder is vital, because damage to the nearby heart generally leads to death.

Today, modern farmers keep *fistulated* cows (cows with artificial windows surgically placed into their rumen). They can then remove a bolus of feed from a healthy cow to place back into a cow "off feed." Microbes can also be purchased. Stomach pumps are occasionally used to remove ruminal fluid from healthy cows to use in a drench form to medicate the cow not ruminating.

The omasum (manyplies) receives the slurry mixture and removes most of the water through its large absorptive area (see figure 4–6). Most digestion is completed by the abomasum (true stomach). The various nutrient building blocks (e.g., amino acids, sugars, and fatty acids) are produced here through the action of the digestive juices on the bacteria and protozoa and absorbed through the wall of the small intestine. The undigested material is moved through the cecum and large intestine and is excreted as feces (manure) through the anus.

Esophageal Groove (Reticular Groove).   In a young ruminant (preruminant) animal, the abomasum is the only part of the four-chambered stomach that is functional. Therefore, these animals are not capable of digesting or utilizing feeds high in fiber. The primary diet of a preruminant is milk. Given access to pasture, a newborn ruminant will begin to graze within the first week to 10 days of life. The consumption of small amounts of forage will stimulate forestomach growth and development. The maximum rate of growth of the rumen occurs during the first 2 months after birth. Calves do not start ruminating until about 3 months of age, and even then only slight amounts of roughage are digested.

FIGURE 4–6 Folds
of tissue make up the
omasum (manyplies).
Reprinted with
permission from
J. Blakely and D. H. Bade,
*The Science of Animal
Husbandry,* 6th ed.
(Englewood Cliffs, NJ:
Prentice-Hall, 1994).

The system continues to increase in size and efficiency so that by weaning it should be functioning completely. By then, the VFAs produced by the microbes can be absorbed across the wall of the rumen via the villi, yielding energy to the animal. Fatty acids are the only nutrients absorbed in appreciable quantities in the rumen. Ammonia and some minerals are also absorbed in the rumen.

While the preruminant is nursing (consuming milk), all of its liquid ingesta is diverted directly into the omasum via the reticular groove (esophageal groove). The reticular groove is a structure that begins at the lower end of the esophagus and, when closed, forms a tube from the esophagus into the omasum. When a preruminant is nursing, milk will then bypass bacterial fermentation in the rumen. Closure of this groove is stimulated by the normal sucking reflexes.

Avian Digestive System

The monogastric (simple stomach) digestive tract of the fowl is divided into the various parts shown in figure 4–7. A brief discussion of the function of each part will provide the basic knowledge necessary to understand introductory poultry nutrition.

Mouth.   Chickens do not have teeth or serrated edges on their beaks, so they cannot chew. The chicken is basically a scavenger. Its beak is designed to rip or tear flesh, to scratch for seeds and insects, and to consume its feed in large pieces. The tongue provides some assistance in eating because of the forked section at the back, which forces food down into the gullet. Little saliva is secreted in the mouth to aid in swallowing.

Esophagus (Gullet) and Crop.   The gullet is a canal that leads to the crop and continues on to the proventriculus. It has great ability to expand. Food is stored in the crop temporarily. Some softening and predigestion take place here due to the action of enzymes.

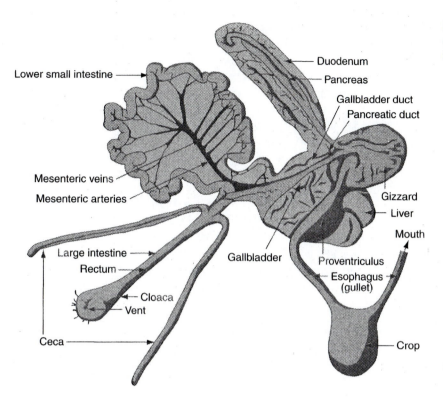

FIGURE 4–7 The digestive system of a fowl.

Glandular Stomach (Proventriculus).   The stomach secretes gastric juice, principally hydrochloric acid, and the enzyme pepsin, which acts to break down proteins to amino acids. There is very little storage space here.

Gizzard (Ventriculus).   The gizzard is a horny, lined structure that is heavily muscled. The involuntary grinding action of the muscular gizzard has a tendency to break down food products much as teeth would do. Feeding of **grit** is controversial but often practiced to assist the gizzard. It is especially beneficial when birds consume whole grains. Crushed granite, oyster shell, or other hard, insoluble material may be used as a free choice additive (made available, not mixed in the ration). When the gizzard is not used heavily, as in the absence of sufficient grit, its muscle becomes flaccid and weak.

Grit small particles of granite added to poultry rations to help in their grinding of feed in the gizzard

Small Intestine.   Most digestion takes place in the small intestine. Intestinal juices are enzymes secreted to split sugars and other nutrients into more simple forms so they can be transported to the bloodstream. Vitamins and minerals are absorbed here as well. Peristaltic action (contraction of the smooth muscle) also takes place here to move materials through the system to the ceca and rectum.

Ruminants have the ability to digest cellulose. However, because the passage of food through the digestive system of the chicken is very rapid, and the only location where bacteria occur to any extent is in the ceca, little digestion of fibrous feed can occur.

Ceca.   Similar to the appendix in humans, the ceca have no known function. Not all ingested feed passes through the ceca. Although some bacteria are found in these two blind pouches, so little digestion of fiber occurs that, for all practical purposes, it can be disregarded.

Large Intestine and Cloaca.   The large intestine is a continuation of the digestive tract leading from the junction of the ceca to the cloaca. The major function of the large intestine is to store undigested waste material and to absorb water from the intestinal contents. The cloaca is a common junction for the outlets of the digestive, urinary, and genital systems through the vent.

Accessory Organs.   The liver and pancreas contribute to digestion, although food does not pass through them. As previously mentioned in the section on simple-stomached animals, the liver secretes bile, which is stored in the gallbladder and used by the body to emulsify fat in preparation for digestion. The liver also stores readily available energy **(glycogen)** and breaks down protein waste by-products into *uric acid* for excretion by the kidneys. The white portion of chicken feces is mostly uric acid and represents the fowl's system of urinary excretion. The pancreas secretes enzymes (*amylase, trypsin,* and *lipase*) to aid in the digestion of carbohydrates, proteins, and fats. Sugar metabolism is also regulated by production of the pancreatic hormone **insulin.**

▓ Glycogen   animal starch

▓ Insulin   hormone secreted by the pancreas that is involved with the regulation and utilization of blood glucose

ESSENTIAL NUTRIENTS

Feed is a substance eaten and digested by an animal that provides the essential nutrients for maintenance (body repair), growth, fattening, reproduction (estrus, conception, and gestation), and lactation (milk production). Feeds can be divided into concentrate and roughage. **Concentrate** (grain-type product) and **roughage** (hay or grass-type product) make up the basic structure of a ration. A combination of the feeds contains essential elements (nutrients) for the well-being of an animal. All classes of livestock require the following six essential nutrients: water, protein, carbohydrates, fats, minerals, and vitamins.

▓ Concentrate   feed that is relatively high in energy and low in fiber (<18 percent crude fiber)

▓ Roughage   feedstuff that is relatively high in fiber (>18 percent crude fiber) and low in energy

Water

The cheapest and most abundant nutrient, water, is necessary for adequate growth, fattening, or lactation-type rations. A beef cow consumes about 12 gallons of water per day (see table 4–3 for comparison of consumption by other

TABLE 4–3	Daily Water Consumption by Livestock	
Species	**Condition**	**Water Consumption**
Cattle (beef)	Fattening	8–9 (gallons)
	Lactating	10–25
	Grazing	4.5–9
Cattle (dairy)	Lactating	25–80
Sheep	Grazing	0.5–2.1
Swine	Maintenance	1.5–3.5
	Pregnant	4.0–5.0
	Lactating	5.0–6.5
Chickens (per 100 birds)	Growing	4–5
	Nonlaying	5
	Laying	5–9
Turkeys (per 100 birds)	Mature	17
Horses	Mature	12

Source: L. S. Shapiro, *Principles of Animal Science* (Simi Valley, CA: Ari Farms, 1997).

species). It can live for weeks without food but only days without water. Water regulates body temperature, dissolves and carries other nutrients, eliminates wastes, and constitutes up to 80 percent of the cow's body. Water is necessary for the digestive processes, both as a medium and as a participant in body chemical reactions. Some enzymes are more effective when diluted in water, and the ions OH^- and H^+ (which come from water) are involved in the process of *hydrolysis*, the primary method of protein, fat, and carbohydrate digestion.

Circulating throughout the body, water carries dissolved nutrients to the cellular level and waste products through the excretory system. As a vital part of the blood, water has a temperature-regulating function because of this circulation, just as temperatures are regulated in water-cooled engines by a circulatory system.

Protein

Protein is the major component of tissues such as muscle, and it is a fundamental component of all living tissues. It contains carbon, hydrogen, oxygen, nitrogen, and sometimes sulfur and phosphorus. The protein molecule is composed of a number of smaller units linked together. These building blocks are called **amino acids.** The name stems from the characteristic chemical structure that combines an amino group (a base) with an acid. The simplest amino acid, glycine, illustrates the structure in figure 4–8. More complex amino acids are also illustrated by a chemical group represented by R. Amino acids are linked together at the junction CO—NH by a peptide linkage, as represented by figure 4–9. Glycine and another simple amino acid, alanine, are combined to form alanylglycine, liberating a molecule of water. The reverse process, hydrolysis, effects separation by taking up a molecule of water.

▓ **Amino acids** building blocks of proteins; they contain oxygen, carbon, hydrogen, and nitrogen

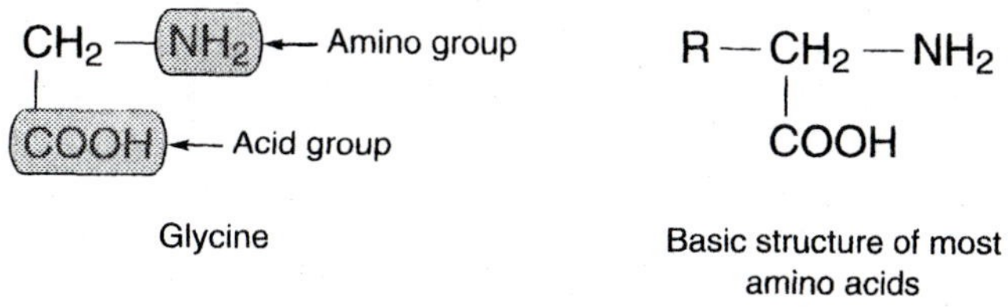

FIGURE 4–8 The simplest building block, or amino acid, is glycine. The presence of amino and acid groups gives rise to the name.
Courtesy of Dr. James Blakely.

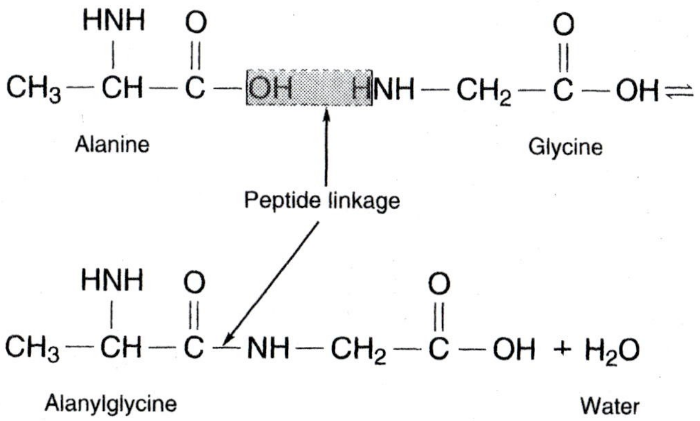

FIGURE 4–9 The combination of amino acids by peptide linkage is the basis of protein molecule formation. The reverse process, hydrolysis, is the basis of separation.
Courtesy of Dr. James Blakely.

Twenty-five different amino acids have been identified as constituents of the protein molecule, although most proteins contain from as few as 3 or 4 to as many as 14 or 15 different ones. The average protein contains 100 or more amino acid molecules linked together by *peptide bonds.*

Nonruminants (swine, poultry, and humans) need specific amino acids. Therefore, a high-quality protein containing a large variety of amino acids is desired. Ruminants do not need a wide variety because of their ability to synthesize needed amino acids through the actions of microorganisms in the rumen. The concern in cattle feeding is more for total protein and protein solubility (undegraded intake protein [UIP] versus degraded intake protein [DIP]) than quality protein. Protein that is broken down in the rumen is generally less available for the cow itself and has been attributed (when in excess) to decreasing fertility in high-milk-producing dairy cows.

Amino acids vary slightly in composition but average about 16 percent nitrogen. Because this figure is fairly constant, protein estimates are most often made by analyzing for nitrogen chemically and dividing by 16 percent or multiplying by 6.25. Feed samples are treated chemically to release ammonia, a form of nitrogen. Trapped ammonia is titrated for a nitrogen calculation expressed as a percentage. This number is multiplied by 6.25. The resulting figure is called **crude protein** (see figure 4–10).

▌**Crude protein** estimated chemically derived protein content of a feed based on its percentage of nitrogen

Plants bind amino acids in a designated pattern to make plant protein, which, through digestion, provides the amino acids that serve as building blocks for animal protein. Some plant protein sources are soybean meal, cottonseed meal, linseed meal, and peanut meal (concentrates) and legume hays (roughage).

FIGURE 4–10 The Kjeldahl apparatus is used to determine nitrogen for protein estimates.
Photo courtesy of the U.S. Department of Agriculture.

Plant protein is required to build animal protein, which comprises the bulk of muscle tissue and milk solids. All types of growth and production require this nutrient. Just as manufacturers convert huge quantities of raw materials into desirable products, a cow converts lush, green grass into steaks and fresh milk. Younger, faster-growing animals require large amounts, as do pregnant and lactating cows. Individual amino acids function as a necessary part of enzymes that aid in digestion, hormones that regulate body functions, hair and skin pigmentation, and metabolic body cell reactions, just to name a few examples.

Protein quality for simple-stomached animals (e.g., swine, dogs, humans, and chickens) refers to the amino acid content. The higher the level and the more the variety, the higher the protein quality. Pigs require 10 **essential amino acids,** cats require 11, and poultry require 12 for normal growth and development. The 10 essential amino acids are *arginine, histidine, isoleucine, leucine, lysine, methionine, phenylalanine, threonine, tryptophan,* and *valine.* In addition, cats need the amino acid *taurine* and poultry require *cystine* and *tyrosine* in their diets. Because swine are omnivores, their protein should come from a variety of sources (such as soybean meal, tankage, and dried milk). Protein supplements sometimes need to be added to the diets of nonruminants to supply the essential amino acids that are lacking in the grain portion of their diets.

▓ **Essential amino acids** amino acids that cannot be synthesized in the body at a rate needed for normal growth

Carbohydrates

The building blocks of carbohydrates are sugars, which provide most of an animal's energy requirements. The sugar in a candy bar gives quick energy and may be compared with the carbohydrates in concentrates (grain) and roughages. This energy is used for maintenance, growth, fattening, reproduction, and lactation. It is the fuel that is used to drive the reactions necessary to maintain the fire of life.

Carbohydrates are so called because these compounds are composed of carbon, hydrogen, and oxygen; the hydrogen and oxygen are in the same proportion as they are in water. The formula for water is H_2O. In both water and carbohydrates, there are two atoms of hydrogen for each atom of oxygen. This group of compounds includes sugars, starch, cellulose, and other more complex substances.

The sugars are the simplest of the carbohydrates. Without a doubt, *glucose* is the most important sugar in carbohydrate metabolism, because it is the sugar in blood, and all organisms seem to be able to utilize it. Glucose occurs in combination with other substances in the body. Figure 4–11 provides one example: the combination of two simple sugars, glucose and galactose, to form a compound sugar, **lactose.** Note that both the glucose and galactose units have six carbon atoms. They have the same general formula, $C_6H_{12}O_6$, but differ in arrangement and grouping of the atoms. The simplest sugars (*monosaccharides*) contain either six carbon atoms (*hexoses*) or five carbon atoms (*pentoses*). The most common hexoses are *glucose* (blood sugar), *fructose* (found in ripe fruit, honey, and seminal plasma), and *galactose* (found mostly in dairy products). The pentoses are seldom found free in nature, but they are found as part of the complex carbohydrates.

▓ **Lactose** milk sugar

A union of two molecules (**disaccharide**) of various simple hexose sugars forms three important sugars: *sucrose* (cane sugar), formed by a molecule of glucose and a molecule of fructose; *maltose* (malt sugar), formed by two

▓ **Disaccharide** union of two molecules of simple sugars

FIGURE 4–11   The building blocks of carbohydrates are sugars. The linkage of two simple sugars—glucose and galactose—forms the compound sugar lactose.
Reprinted with permission from J. Blakely and D. H. Bade, *The Science of Animal Husbandry*, 6th ed. (Englewood Cliffs, NJ: Prentice-Hall, 1994).

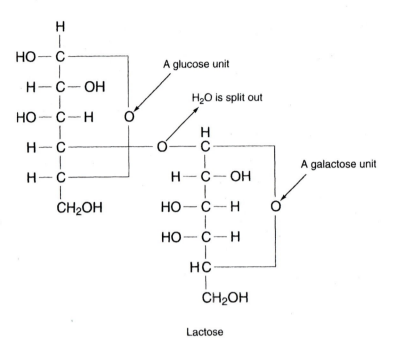

Lactose

molecules of glucose; and *lactose* (milk sugar), formed by a molecule of glucose and a molecule of galactose. Most of the carbohydrates in plants and feeds consist of the more complex carbohydrates formed by the union of many simple sugar molecules, with the splitting out of water. Called *polysaccharides* (many sugars), they include starch and cellulose.

Starch, composed of many molecules of glucose, is the principal form of stored energy in grain (especially corn), wheat, and grain sorghum. Digestion of starch by animals reverses the formation process to yield glucose and is therefore very important in livestock rations because of ease of digestion and availability. The feeding value is thus relatively high.

Cellulose and related compounds, which comprise the bulk of plant cell walls and form the woody fiber parts, are less completely digested. However, the end product is glucose. Cellulose yields energy the same as starch but not as efficiently because some energy must be expended (lost) in the work of digestion. Hay and related feeds typify cellulose-containing products. Animals cannot digest cellulose. The enzymes produced by large microbial populations found in the rumen and cecum enable ruminants and nonruminant herbivores to digest fiber and utilize the energy stored within.

For ease of analysis, discussion, and communication, chemists have divided the often confusing array of carbohydrates into fiber and NFE. A crude fiber digestion apparatus (see figure 4–12) simulates digestion by washing a feed sample alternately with dilute acid and dilute alkali (base). The material left behind roughly indicates the more poorly digested carbohydrates, fiber, or crude fiber. The chief fibers are represented by cellulose and other carbohydrates not easily dissolved.

Beginning in 1963, Van Soest of Cornell University developed a better method to estimate the fiber content of forages. Predicting the nutritive value of forages from the traditional **proximate analysis** proved to be inaccurate. Van Soest showed that the portion of the plant material that is soluble in neutral de-

Cellulose   a carbohydrate (polysaccharide) found in the fibrous portion of plants, and one which is digested by microbial enzymes produced in the rumen and cecum of herbivores

Proximate analysis   chemical evaluation of a feedstuff

FIGURE 4–12
Fiber is determined as part of a carbohydrate analysis by a crude fiber digestion apparatus.
Courtesy of Dr. James Blakely.

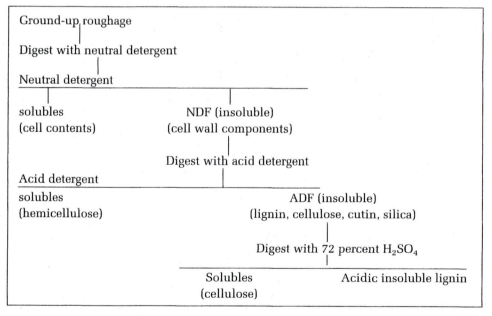

Ground-up roughage
|
Digest with neutral detergent
|
Neutral detergent
|
solubles   NDF (insoluble)
(cell contents)   (cell wall components)
|
Digest with acid detergent
|
Acid detergent
solubles   ADF (insoluble)
(hemicellulose)   (lignin, cellulose, cutin, silica)
|
Digest with 72 percent H_2SO_4
|
Solubles   Acidic insoluble lignin
(cellulose)

FIGURE 4–13
Detergent feed analysis summary.

▓ **Neutral detergent fiber (NDF)** measures the cell wall of a plant (cellulose, hemicellulose, and lignin) and has an inverse relationship with dry matter intake; a minimum level of NDF is required, however, to prevent ruminal disorders such as acidosis, rumen parakeratosis, and low milk fat percentage

▓ **Acid detergent fiber (ADF)** measure of cellulose and lignin in forages and feeds analyzed by boiling the sample in an acid detergent solution; it consists primarily of cellulose, lignin, and silica, which are the least digestible parts of the plant; as the plant matures ADF increases and digestibility decreases

tergent, cell contents (proteins, sugars, minerals, starch, and pectins), is virtually completely digested by the ruminant. The insoluble residue is called **neutral detergent fiber (NDF)** and is chiefly made up of the cell wall constituents. **Acid detergent fiber (ADF)** involves a further step. The NDF is boiled in an acid detergent solution. The acid detergent dissolves hemicellulose. Thus, NDF − ADF = hemicellulose content of the feed. The remaining ADF fraction consists largely of cellulose and lignin. The ADF can be treated with sulfuric acid to dissolve the cellulose, leaving the lignin with some residue of silica and cutin (see figure 4–13).

▓ Nitrogen free extract (NFE) readily available carbohydrates of a feed (primarily the sugars and starches)

Nitrogen free extract (NFE) represents the more soluble carbohydrates. It is extracted during the fiber determination by the weak acid and weak alkali. Starch, the sugars, the more soluble pentosans, and the other complex carbohydrates are included. NFE is derived by subtracting the sum of the other proximate components from 100:

% NFE = 100 − (% moisture + % crude protein + % ash + % ether extract + % crude fiber)

Fats

The building blocks for fats are *fatty acids,* which are also used as a source of energy. Fats are concentrated: they have 2.25 times as much energy as carbohydrates. The customary uses are to raise the energy level and to improve flavor, texture, and palatability (animal acceptance) in a ration. Fat choices fall almost entirely in the concentrate category and include by-products of the oilseed crops (vegetable oils) and rendered animal fats. *Fat-soluble vitamins* (discussed later in this section) are also associated with this nutrient, but they are not a part of the molecular structure. Fat serves as a body insulator and protector of various body parts and organs.

The term *oil* is often used interchangeably with *fat.* They are alike in composition except that fats are solids at ordinary room temperature whereas oils remain liquid. Fats and oils are soluble in ether and certain other solvents. Because of this characteristic (analysis is affected by chemically extracting with ether), all the substances dissolved are included under the classification fats or *ether extract.* The term *lipids,* often used by chemists, also means fats.

Like carbohydrates, fats are also made up of carbon, hydrogen, and oxygen, but the oxygen proportion is much lower, making the energy value greater. It is generally accepted that 1 pound of fat has 2.25 times the oxidation energy value of 1 pound of carbohydrate. Fat is composed of three fatty acids linked chemically to glycerol (see figure 4–14).

The fatty acids that are linked to glycerol vary with the type of fat from which they originate. Generally, there are two types: unsaturated and saturated. The term *unsaturated fatty acids* means the fatty acids are able to absorb oxygen or other chemical elements. Some unsaturated fatty acids are believed to be necessary for normal life. These essential fatty acids (which need to be in the feed or ration) include *linoleic, arachidonic,* and *linolenic.* Deficiencies in these fatty acids can result in dermatitis, reduced growth, increased water consumption, impaired reproduction, and increased metabolic rate. *Saturated*

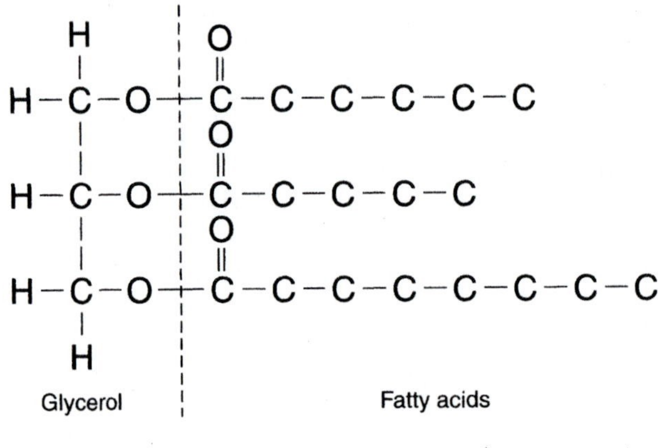

FIGURE 4–14 Fats are composed of three fatty acids linked chemically to glycerol. Courtesy of Dr. James Blakely.

Glycerol   Fatty acids

fatty acids have a single bond joining each pair of carbon atoms and carbon atoms within the chain that have two hydrogen atoms joining them. When the carbon atoms are joined by a double bond, the carbon atoms within the chain are able to have only one hydrogen bond to them. These fatty acids are referred to as unsaturated and are more prone to *rancidity* (through decomposition or oxidation of the fat).

Chemists, trying to prevent oxidation of unsaturated fatty acids, developed a process known as **hydrogenation,** or hardening, of the fats. With this process, vegetable fats are converted into a semisolid form used in shortenings and margarines. Hydrogenation of unsaturated fats increases the trans fatty acids, which are known to be unhealthy in the diets of humans because they contribute to heart disease.

Digestion of fat liberates the fatty acids from the glycerol molecule, allowing them to move through the bloodstream to muscle tissue or adipose (fatty) tissue to be united again to form fat. This fat deposit in muscle tissue is called **marbling.** In other forms it is a storehouse of energy.

Soft Pork.   Dietary fat can influence the properties of the body fat of an animal. For example, when pigs are fed fats high in unsaturated fatty acids, their body fat is softer than if they are fed fat from saturated sources. The condition *soft pork* is commonly produced when pigs glean peanuts, soybeans, and other oil-rich seeds that are high in unsaturated fats.[5] From a consumer standpoint, this condition is a highly undesirable carcass trait.

> **Hydrogenation** chemical process that saturates fats with hydrogen ions and increases the trans fatty acids in a feed in an attempt to prevent the oxidation of unsaturated fatty acids
>
> **Marbling**   intermingled fat in the lean muscle that contributes to its tenderness and palatability

Minerals

What are minerals? Simply stated and easily visualized, minerals are the ashes of a burned tree or other organic matter, including feed. To analyze for minerals, feed is burned at 600°C until a constant weight is reached. All that is left is ash, often seen on feed tag information. This ash residue is the inorganic portion of feed, minerals. Eighteen minerals are required for normal body function. Table 4–4 gives a condensed appraisal of functions of these minerals. Minerals are conveniently divided into two groups based not on their relative importance

TABLE 4–4	Minerals Required by Livestock
Mineral	**Function**
Ca, P, Mg, Cu, Mn	Skeletal formation and maintenance
P, S, Zn	Protein synthesis
Fe, Cu	Oxygen transport
Na, Cl, K	Fluid balance
Na, Cl, K	Acid–base balance
Ca, P, K, Mg, Fe, Cu, Mn, Zn	Activators and/or components of enzyme systems
Ca, P, Co, Se	Mineral-vitamin relationship

Source: L. S. Shapiro, *Animal Nutrition,* 3rd ed. (Simi Valley, CA: Ari Farms, 1997).

---

[5]R. L. West and R. O. Myer, "Carcass and Meat Quality Characteristics and Backfat Fatty Acid Composition of Swine as Affected by the Consumption of Peanuts Remaining in the Field after Harvest." *Journal of Animal Science* 65:475–480, 1987.

but on their respective quantities required in the daily diet. Those required in larger amounts are termed macro, or major, elements and include Ca, P, Ma, S, Cl, Mg, and K. Elements required in smaller amounts (but that are equally important) are termed micro, or trace, elements and include Co, Cu, Fe, Mo, Zn, F, I, Mn, Se, Cr, and Si.

When feeds are grown on fertile land with no deficiencies, they will supply sufficient mineral elements. Some deficiencies exist (see figure 4–15); therefore, a common practice is to supply a supplement of one part trace mineralized salt and two parts dicalcium phosphate or bonemeal to pastured or feedlot animals. Providing such supplements would ensure against common deficiencies.

FIGURE 4–15 Phosphorus deficiency in feedlot cattle. The steer at the bottom received the same ration as the one at the top plus 1/2 pound of steamed bonemeal daily to correct the phosphorus deficiency.
Reprinted by permission from J. Blakely and D. H. Bade, *The Science of Animal Husbandry,* 6th ed. (Englewood Cliffs, NJ: Prentice-Hall, 1994).

(a)

(b)

Vitamins

Vitamins are organic substances that are essential for the health, reproduction, lactation, growth, and general maintenance of animals. They are required in very minute quantities. The name is derived from the French, who in the early part of the twentieth century thought these small quantities of nutrients were amines. The term *vital amine* was reduced to *vitamin* and is used to this day.

There are two divisions of vitamins, *fat soluble* and *water soluble*. The fat-soluble vitamins are obtained from the ration or from synthetic supplements. They include vitamins A, D, E, and K. Ruminants and animals with a functional cecum (equine) can synthesize the water-soluble group. These vitamins include all of the B complex vitamins (riboflavin, thiamin, niacin, biotin, choline, folic acid, pyridoxine, pantothenic acid, vitamin B_{12}) and vitamin C. Vitamin C is only required in the diets of humans, monkeys, fruit-eating bats, and guinea pigs. Most other animals can convert glucose into vitamin C.

THE USES OF FEED NUTRIENTS

Although each of the six essential nutrients has a specific function, they are required by the animal in specific combinations to perform a desired body function. Although all six are needed, the amount of each varies depending on the use of the feed by the animal. In general, nutrients from feed are utilized by the animal in two or more of the body functions, described in the following sections.

Maintenance of the Body

Maintenance requires those nutrients necessary to maintain life without a gain or loss in body weight and without any production (milk, fattening, and reproduction). Maintenance includes heat to warm the body, energy for normal body work (heartbeat and breathing), and protein and minerals for repair and replacement of worn-out body tissues. Because maintenance requirements are based on body size, the larger the animal the more nutrients needed to maintain the body. On the average, from one-third to one-half of the nutrients in a ration go toward maintaining the body. This requirement must be met first; the excess of nutrients above maintenance are used for desired production.

Production Functions

Production body functions use those nutrients left after maintenance requirements have been satisfied. Production functions include one or more of the following: growth, fattening, milk production, reproduction, fiber production (e.g., wool and hair), or work. The main nutrients for each vary depending on the type of production desired. The amount of nutrients required also depends on the amount of production desired. For example, more nutrients are required for growth of 1.5 pounds per day than 0.5 pound per day. The total requirement of an animal is the sum total of all the desired body functions. For example, the requirement of a pregnant heifer growing 1.0 pound per day is the total of the requirements of body maintenance, growth, and fetal development.

The amount of each of the six nutrients in a feed is important in selecting and mixing the right feeds. Essential feed nutrients are obtained by a chemical analysis of the feed. In general, the main system used is the *Weende system of proximate analysis.* This system, devised more than 100 years ago, is still used. By the chemical processes of the Weende system (discussed briefly for each class of nutrient) the nutrients in a feed are separated into the fractions shown in table 4–5).

Feed Digestibility

Although the methods discussed earlier indicate the total amount of nutrients present, they do not provide information about the amount of the nutrient that will be available for use by the animal. For a feed nutrient to be utilized by an animal it must be consumed and digested. Undigested parts of the consumed feed are not used by the animal but excreted as feces (manure). Thus the term *digestible nutrient* is more accurate than *total nutrient,* such as digestible protein versus total or crude protein.

The Weende analysis works very well for concentrate feeds but has limitations in forages, which vary greatly in digestibility of the fiber content. Table 4–6 compares two systems currently used in analyzing feed nutrients. Students of animal science should recognize what each system represents as well as their limitations. Some nutrient deficiencies are readily recognized early on, whereas others are not noticeable without a trained eye or until severe deficiencies exist (see table 4–7).

TABLE 4–5	Weende Analysis
Corresponding Nutrient	**Analysis**
Dry matter	Feed minus dry matter equals water
Ether extract	Fats
Crude protein	Total proteins
Crude fiber	Hard-to-digest or undigested carbohydrates
Nitrogen free extract	Easy-to-digest carbohydrates
Ash	Minerals

Note: Vitamins are not included because they are contained in very small amounts and must be analyzed separately. In addition, as previously mentioned, there are some newer, more exact methods of analyzing digestible fiber in forages (ADF and NDF).

TABLE 4–6	Nutrient System Comparison		
Weende System	**Nutrient Component**	**Van Soest Analysis**	
Ash	Minerals		
Ether extract	Fats		
Crude protein	Proteins	Neutral detergent solubles (cell contents)	
Nitrogen free extract	Carbohydrates		
	Starch		
	Sugar		
	Hemicellulose		
Crude fiber	Lignin	Acid detergent fiber	Neutral detergent fiber
	Cellulose		

TABLE 4–7	Common Nutritional Disorders
Disorder	**Chief Causes**
Bloat	Legume, succulent forages
Hardware disease	Wire or nails lodged in reticulum
Ketosis	Sudden need for extra energy
Milk fever	Sudden need for Ca (lactation)
Acidosis	Excess grain consumption (ruminants)
Nutritional muscular dystrophy	Se or vitamin E deficiency
Blind staggers	Se toxicity (equine)
Grass tetany	Mg deficiency (ruminants)
Parakeratosis	Zn deficiency (swine)
Night blindness	Vitamin A deficiency
Goiter	Iron deficiency
Rickets	Ca, P, or vitamin D deficiency (young animals)
Anemia	Fe, Cu, vitamin B_{12}, or folic acid deficiency

Source: L. S. Shapiro, *Animal Nutrition,* 3rd ed. (Simi Valley, CA: Ari Farms, 1997).

Feed Energy

The energy in a feed usually comes from carbohydrates and fats. Its importance in animal feeding has resulted in expression of the energy of a feed in one of two methods.

The **total digestible nutrient (TDN)** system is the oldest. It is based on the digested fractions of the Weende analysis and their energy contribution. It is still widely used by many people in animal science.

The *calorie system* is based on the actual energy **(calorie)** content of the feed as measured by the heat given off in burning a small pellet of feed, usually in a measuring device such as the *bomb calorimeter*. Based on this system the energy content of a feed will be expressed in one of the following terms:

- **Gross energy (GE):** The total amount of energy in a feed
- **Digestible energy (DE):** The amount of energy in a feed digested by the animal, expressed in megacalories (1,000,000 calories of energy) or kilocalories (1,000 calories of energy)
- **Metabolizable energy (ME):** A measure of the energy in a feed used by the animal
- **Net energy (NE):** The amount of energy in a feed used for specific body functions

Example:

DE = GE − energy found in the feces
ME = GE − (fecal energy + gaseous and urinary energy)
NE = GE − (fecal energy + gaseous and urinary energy + heat increment)

It should be noted that only about 20 percent of the energy intake is actually available for use by a cow for the production of body weight gain or milk yield. Approximately 20 percent is lost in heat increment, 30 percent in feces, 5 percent each for urinary and gaseous production, and 20 percent for maintenance.

▓ **Total digestible nutrient (TDN)** method of estimating amount of energy in a feed by adding up the digestible crude protein, crude fiber, nitrogen free extract, and ether extract × 2.25 (fats)

▓ **Calorie** unit of heat energy defined as the amount of heat needed to raise the temperature of 1 gram of water from 14.5°C to 15.5°C

▓ **Gross energy (GE)** total heat of combustion of a feed burned in a bomb calorimeter

▓ **Digestible energy (DE)** energy that was absorbed by the animal after accounting for the energy lost in the feces

▓ **Metabolizable energy (ME)** energy absorbed and utilized by the animal after subtracting for energy lost in the feces, urine, and gases

▓ **Net energy (NE)** most commonly used energy term for determining lactating dairy cows' energy needs; the energy available after removing energy lost in feces, urine, gas, and heat increment (heat produced during digestion and metabolism)

Ration Balancing

The type of ration selected for an animal depends on the goal intended—fattening a steer for slaughter, feeding show animals, feeding dairy cows for maximum lactation, wintering cattle, feeding sheep for quality wool production versus lamb growth, and feeding horses for speed versus for draft. Regardless of the type of feeding desired, some rough rules of thumb can be used for developing basic rations. Biblical stories of cattle raising have been interpreted as "the eye of the master fattens his cattle." This is an excellent idea to keep in mind when developing rations. No matter what type of feeding system is used, any signs of unthriftiness should indicate a change in the ration if parasitism and illness can be ruled out. Total mixed rations have become very popular on the larger western style dairies (figure 4–16). On these large operations, dairy farmers store commodity feeds that they can then use to mix specific rations needed for their individual milk production strings (figure 4–17). Basic rules of thumb can be used as guidelines (see table 4–8).

For the more particular feeder or student, rough rules may not suffice. Researchers in the United States have accumulated a massive amount of data on nutritional requirements to provide information wherever needed. The National Academy of Sciences National Research Council (NRC) has made available a summary of this information. Using these requirements, a ration may be scientifically balanced for the specific purpose desired.

To balance a ration, the content of the various feed ingredients should be analyzed. Because such analysis is expensive and not always practical, the NRC has compiled average values for numerous feeds. The animal's age, sex, production level desired, activity level desired, and economics all must be considered when balancing a ration. In addition, the balancing of rations is

FIGURE 4–16 Total mixed rations (TMRs) are very popular on large California dairies. Photo by author.

FIGURE 4–17
Commodity storage and feeding saves money and is used to more precisely balance the ration. At this dairy the feeding of whole cottonseed is an economic way of meeting the herd's protein, energy, and fiber needs.
Photo by author.

TABLE 4–8	Sample of Basic Rules of Thumb in Feeding Livestock
Species	**Rules of Thumb**
Beef cattle	2 percent of body weight fed as hay
	3 × amount fed if fed as silage
Dairy cattle	3.5 percent of body weight fed in dry matter content
	50 percent of dry matter should come from roughage
Sheep	1.5–2 × maintenance diet should be fed to nursing ewes at peak of lactation
Horses	2 percent of body weight per day in dry matter
	1 percent of body weight per day in roughage
Pigs	3–4 percent of body weight per day in dry matter

Source: L. S. Shapiro, *Animal Nutrition,* 3rd ed. (Simi Valley, CA: Ari Farms, 1997).

time-consuming. Costs and availability of feed ingredients change frequently. Combinations will continue to shift with supply and demand, making the business of animal nutrition worthy of study. Years ago, farmers fed to produce a *least-cost ration.* Today, a *maximizing profit* is first priority when feeding and breeding animals. Maximizing profit is not always maximum yield or output. Rations have and will continue to change to keep pace with the times (e.g., producing leaner meats and lower-cholesterol-containing eggs and dairy products).

SUMMARY

There are four distinct digestive systems among livestock species: avian, non-ruminant herbivore, ruminant, and monogastric. Although there are many similarities between systems and even between species within each system, every animal species has its unique and special nutritional needs. Because feed constitutes the single largest expense in livestock production, a basic understanding of the six primary nutrients, the physiology and anatomy of the digestive system, and the common feeds used for each species is essential.

EVALUATION QUESTIONS

1. _____ is the trapping of solar energy and its conversion to chemical energy, which is used in the manufacturing of food molecules from carbon dioxide and water.
2. The classification of animals with stomachs divided into four compartments is _____ .
3. Examples of animals described in question 2 are _____ , _____ , and _____ .
4. _____ is the external opening of the lower end of the digestive system in poultry.
5. _____ is a metabolic disorder caused by excessive accumulation of gases in the rumen.
6. _____ is used as a last resort to treat the disorder in question 5.
7. The portion of a feed that contains mainly cellulose, lignin, and silica is _____ .
8. The greenish yellow pigment produced by the liver and used in emulsifying fats is _____ .
9. _____ is the organic compound containing carbon, hydrogen, and oxygen in which the hydrogen and oxygen are in the same proportion as that found in water.
10. The blind pouch located at the point where the small and large intestines meet is the _____ .
11. _____ is the gross energy of a feed minus the energy remaining in the feces.
12. Partially digested feed that moves from the stomach to the small intestine is called _____ .
13. Animal tissues and bones from animal slaughterhouses and rendering plants that are cooked, dried, and ground and then used as an animal protein supplement are called _____ .
14. The practice of ingesting feces is _____ .
15. _____ is a classification of animals that eat plants as the main part of their diet.
16. _____ is a classification of animals that eat meat as the main part of their diet.
17. _____ is a classification of animals that eat both meat and plants to balance out their dietary needs.
18. The process of breaking down feed into simple substances that can be absorbed by the body is called _____ .

19. _____ is the muscular stomach that crushes and grinds the feed and mixes it with digestive juices in birds.
20. _____ is also called a gullet.
21. The tubelike passage from the mouth to the stomach is the _____ .
22. Ash is also called _____ .
23. Ether extract determines the _____ content of a feed.
24. _____ is the presence and distribution of fat and lean in a cut of meat.
25. GE − (fecal energy + gaseous energy + urinary energy) = _____ .
26. GE − (fecal energy + gaseous energy + urinary energy + heat increment) = _____ .
27. The portion of a feed that is of lower digestibility, consisting of the more insoluble material found in the cell wall (cellulose, lignin, silica, and some protein), is called _____ .
28. _____ is the third compartment of the ruminant stomach.
29. _____ is the total of the digestible protein, digestible nitrogen free extract, digestible crude fiber, and 2.25 times the digestible fat.
30. The most commonly fed nonprotein nitrogen (NPN) is _____ .
31. The last part of the large intestine in mammals (before the anus) is the _____ .
32. _____ is also called honeycomb stomach.
33. _____ is also called manyplies stomach.
34. A _____ is also referred to as a paunch.
35. _____ is the process of chewing the cud.
36. _____ is the true stomach of a cow.
37. _____ is the passage of feed nutrients across the intestinal wall into the circulatory system.
38. _____ is a feed category that is high in fiber and low in digestible energy.
39. _____ is a feed category that is high in digestible energy and low in fiber.
40. _____ is regurgitated food or a large pill for dosing animals.
41. The first milk given by a female after delivery of her young is called _____ .
42. A groove in the reticulum between the esophagus and omasum that directs milk in young nursing animals away from the rumen/reticulum is _____ .
43. The deficiency disorder caused by insufficient iodine consumption is _____ .
44. Milk sugar is called _____ .
45. Table sugar is called _____ .
46. Another word meaning simple sugar is _____ .
47. A word meaning two-sugar molecule is _____ .
48. A word meaning many-sugar molecule is _____ .
49. _____ means to chew.
50. The process whereby the animal picks up feed or food is _____ .
51. _____ is the metabolic disorder occurring at or near the time of giving birth caused by a sudden drop in blood calcium levels.
52. The percentage of body weight eaten per day by beef cattle is _____ percent.
53. _____ is caused by a deficiency of iron and copper.
54. _____ is a metabolic disorder caused by Ca or P deficiency in young growing animals.

55. _____ and _____ are the two main gases produced in the rumen.

56. The normal wavelike muscular contractions that propel ingesta through the digestive tract are _____ .

57. _____ is a storage pouch off of, but continuous with, the esophagus in poultry; it is used for storage of food materials hastily eaten by the bird.

58. _____ is used by the chicken as a grinding material to help the gizzard break down large particles of feed into smaller ones.

59. _____ is a digestive disorder in the horse, brought on by overeating, excessive drinking while hot, moldy feeds, and/or infestation of roundworms.

60. Nutrients are absorbed in the simple-stomached animal primarily in the _____ .

61. Monogastrics cannot easily digest _____ .

62. Three common vegetable protein supplements are _____ , _____ , and _____ .

63. Three common animal protein feeds are _____ , _____ , and _____ .

64. List common high-energy feeds for various species of livestock: horses _____ , cattle _____ , and pigs _____ .

65. A sample of alfalfa hay containing 3 percent nitrogen would mean a protein content of_____ percent.

66. If 1 pound of corn yields eight units of energy, 1 pound of rendered animal fat would yield _____ units.

67. _____ is a mineral required by bacteria to synthesize vitamin B_{12}.

68. _____ vitamins are synthesized by the cow and thus not needed in the ration.

69. A common condition known as grass tetany is caused by a lack of _____ .

70. An element necessary for bone growth and found primarily in forages is _____ .

71. A deficiency of _____ shows up as a paleness of the blood vessels under the eyelid.

72. Sun-cured hay, green in color, supplies vitamins _____ and _____ .

73. The disease rickets is associated with vitamin _____ deficiency.

74. Night blindness results from a vitamin _____ deficiency.

75. A dairy cow will normally consume _____ percent of its body weight per day in dry matter.

76. Hardware disease results when sharp objects puncture the walls of the _____ .

77. A 600-pound steer being full fed a fattening ration would eat about _____ pounds feed per day.

78. Calcium gluconate is used to treat _____ .

79. Swine saliva differs from that of the cow because it contains _____ that begin the digestion process.

80. Too much salt in poultry rations can cause the undesirable management of _____ manure.

81. Some assistance is provided chickens in swallowing because of the unusual shape of their _____ .

82. Some softening and predigestion of feed, in poultry, takes place in the _____ .
83. The ventriculus, or _____ , serves as the chicken's _____ .
84. In poultry, the digestive system shares a common junction with the _____ and _____ system.
85. _____ _____ causes the white portion of chicken manure.
86. The main digestion of carbohydrates and proteins is regulated by enzymes secreted by the _____ and eventually absorbed in the _____ .
87. A 900-pound horse eats about _____ pounds of dry matter (feed) per day.
88. The rabbit's digestive tract is much like that of a _____ .
89. Challenge feeding is used to prevent _____ .
90. _____ is the main cause of soft pork.
91. _____ is the cheapest and most abundant nutrient.
92. Iron dextran injections are given to _____ to prevent _____ at _____ age.
93. _____ is the true stomach of a chicken.
94. The closest chickens can come to having teeth is the action simulated by the _____ .
95. _____ is completely indigestible by all animals.
96. _____ is caused by a deficiency of selenium.
97. _____ measures the readily available CHO of a feed.
98. Excess Se in the diet causes _____ in horses.
99. The percentage of N in most proteins is _____ percent.
100. What are the fat-soluble vitamins? _____
101. An average horse will consume _____ gallons of water per day.
102. _____ are the building blocks of carbohydrates.
103. _____ is blood sugar.
104. A polysaccharide that forms the bulk of plant cell walls is _____ .
105. _____ is a term, often used by chemists, meaning fat.
106. Chemists, trying to prevent rancidity of fats, developed the _____ process.
107. _____ is a fat deposit in the muscle tissue.

DISCUSSION QUESTIONS

1. What are the basic elements and/or building blocks of the six essential nutrients?
2. Give examples of a mineral–vitamin interaction; that is, how could an excess or a deficiency of one create signs of imbalance of either? For example, even though cattle do not normally require B vitamins, a deficiency of cobalt could create deficiency signs similar to those of B complex deficiencies.
3. Discuss the movement of feed through the ruminant digestive system and the function of each portion from the lips to the anus. Compare and contrast the anatomy and physiology of the ruminant digestive system with that of the other three types of digestive systems.

4. Calculate the total roughage, grain, and supplement you would recommend stocking as the new manager of a 30,000-acre ranch in a southwestern range environment to winter 900 cows and 35 bulls.

5. What is meant by "quality" of protein supplements for swine? Do swine actually need protein supplements? Why?

6. Describe deficiency signs of and give the nutrients responsible for correcting the following conditions:
   a. Anemia
   b. Goiter
   c. Parakeratosis
   d. Xerophthalmia
   e. Rickets

7. Why should prepared cattle feeds not be fed to horses? Are there possible exceptions to this rule? If so, what are they?

8. Define cribbing. Include common causes.

9. Match up basic deficiency and toxicity disorders for the vitamins, minerals, and other nutrients discussed in class or listed in your text (e.g. Zn deficiency with parakeratosis).

10. What is the difference between silage, haylage, soilage, hay, straw, fodder, stover, pasture, and so forth?

11. Define the following: NE, NFE, GE, TDN, ether extract, VFA (give examples), and NPN.

12. Give examples of monosaccharides, disaccharides, and polysaccharides. What is the difference between them?

CHAPTER 5

Dairy Industry

OBJECTIVES

After completing the study of this chapter you should know:

- The history and future trends of the dairy industry in the United States.
- The reasons for the size and scope of the U.S., California, and Wisconsin dairy industries.
- Why Israel and Japan have the highest-producing dairy cows in the world.
- The five commonly recognized dairy breeds in the United States.
- Special considerations in feeding and managing dairy cattle.
- Body condition scoring and its importance to animal reproduction, nutrition, and overall well-being.
- The steps used in balancing a ration for a typical dairy cow.
- The effects of negative energy balance on reproductive performance and overall performance of the dairy cow.
- The importance of milk urea nitrogen (MUN) levels in dairy cattle.
- Common metabolic disorders in dairy cattle.
- Methods of selection, breeding, and judging of dairy cattle, including the use of genetics and artificial insemination.
- Common genetic disorders of cattle.
- The anatomy and physiology of lactation, methods of milk collection, and functions of the milking machine and parts.
- Causes, prevention, treatments, and diagnosis of mastitis.
- The importance of bovine somatotropin (BST) and its use in the dairy industry.
- Common bovine health disorders and their causes, symptoms, treatments, and prevention.

KEY TERMS

Actinomycosis
Alveoli
Annular ring
Babcock test

Body condition score
Bovine leukocyte
  adhesion deficiency
  (BLAD)

Brown stomach worm
Challenge feeding
Clinical
Colostrum

Condensed milk	Linear score	Residual milk
Degraded intake protein (DIP)	Lobule	Resting phase
Dried poultry litter (DPL)	Lumen	Roundworms
Dry cows	Mange	Scabies
Electrolyte	Mastitis	Serological test
Fiddleneck poisoning	Mechanical refrigeration	Sphincter muscle
Furstenberg's rosette	Milk fever	Springers
Gland cistern	Milk letdown	Streak canal
Grass staggers	Milk urea nitrogen (MUN)	Subclinical
Grass tetany	Milking phase	Surfactant
Groundsel poisoning	Mule foot	Syndactylism
Intradermal test	Nymphs	Tall fescue toxicosis
Judging	Polydome™	Teat cistern
Ketosis	Predicted transmitting ability (PTA)	Tuberculin testing
Larkspur poisoning	Psoroptic mange	Undegraded intake protein (UIP)
Limber legs	Rectovaginal constriction	Zoonotic

INTRODUCTION

The dairy industry is based on the ability of mammals to produce milk in excess of that which is required to nourish their young. The excess milk, which has a high nutritional value, is used for human consumption. Cattle, the primary producers of milk in the world, produce more than 91 percent of the world's supply. In some countries, sheep, goats, and water buffalo are the principal producers of milk. Camels and reindeer are also used for milk production. Over the centuries, cows and some other mammals have been selectively bred for their ability to produce large quantities of milk.

Milk is very important in the human diet because of its high protein and calcium content. Protein provides many of the essential amino acids often deficient in the cereal grains used for food. Calcium from dairy sources has become increasingly important in preventing osteoporosis in humans. Scientists recently started recommending that people drink milk to reduce their chances of getting colon cancer.[1] Milk is also a flexible food source, and its fat and protein content can be modified easily to meet today's consumer demands.

HISTORY OF THE DAIRY INDUSTRY

Domestication of cattle and the use of their milk for human food began somewhere in Asia or northeast Africa between 6000 and 9000 B.C.E. Ancient Sanskrit writings indicate that milk was one of the most essential of all foods thousands of years ago. In India, around 2000 B.C.E., cattle were raised and used to make holy offerings to the gods. Butter was also used at that time. Egyptians used milk, cheese, and butter as early as 3000 B.C.E. The Old Testament contains at least 40 references to cattle. Milk and milk products are mentioned at least 50 times. Hippocrates recommended milk as a medicine 2,500 years ago.

---

[1]P. Holt, "Calcium May Reduce Colon Cancer," *Journal of the American Medical Association*, Vol. 2, 80, September 1998.

The first importation of dairy cattle to the New World took place in 1611 at the Jamestown Colony. In 1624, Plymouth Colony received its first cows, brought over to lessen the high death rate of the colonists (particularly the children) due to poor nutrition. From colonial times until about 1850, there was very little change in the dairy industry in the United States. Every family had a cow or two to supply milk products for the family. If not, a close neighbor did. Only small amounts of milk were sold. The availability of milk was largely seasonal; large amounts were produced when the pastures were lush.

The first rail shipment of milk, from Orange County, New York, into New York City (a distance of 80 miles), took place in 1841. There was no refrigeration of dairy products at this time. The perishable nature of milk and problems in transporting fluids limited early growth of the industry. In 1850, excess milk was pooled for the manufacture of cheese. It was poorly sanitized, and the milk and products were of poor quality. Milk started to be brought into the cities by boat or wagons. Milk was dipped into receptacles supplied by the consumer, was diluted in most cases, and was sold without any health regulations.

In 1851 the first cheese factory in the United States was established in Oneida, New York. This factory, and Gail Borden's first patent on *condensed milk* in 1856, provided dairy farmers with additional outlets for their milk.

There was no organized selection in breeding dairy cattle until approximately 1860. At this time, a number of dairy farmers became interested in organizing national registry associations and started to record the ancestry of animals and promote individual breeds. The associations established official production-testing programs, type classification, and the exhibition of purebred cattle.

Mechanical refrigeration was developed in 1861. Refrigeration of milk eventually enabled farmers to extend the shelf life of their product and to ship milk and other dairy products longer distances.

The first college of agriculture was established in Michigan in 1855. Seven years later, in 1862, Congress passed the Land Grant Act. This act established colleges of agriculture in every state. Dairy and other farmers have profited by being able to apply the scientific principles of breeding, feeding, and general management established at these universities. The husbandry of dairy animals eventually changed to the science of dairy cattle and processing. In 1887, the land grant universities received additional congressional support with the passage of the Hatch Act. This act created experiment stations in each state and provided millions of dollars to fund research on dairy cattle production and milk processing.

In 1871, Gustav DeLaval invented the centrifugal cream separator. Prior to this invention, cream was separated from the milk by allowing the cream to rise to the top of milk pans and then manually skimming off the cream. The economic importance of fat production became apparent early in dairy industry history. However, it was not until Dr. S. M. Babcock's fat determination test of 1890 that accurate determinations of milk fat content became possible. The **Babcock test** is still used today for on-farm testing of butterfat in small dairy goat and cow herds and to check the more modern electronic automated machines used by all large processing centers.

Also in 1890, the first *tuberculin testing* of dairy cows in the United States began. Tuberculosis is a **zoonotic** disorder, one that can easily spread from

■ **Babcock test**   test for butterfat quantity in milk and milk products

■ **Zoonotic**   capable of being transmitted between animals and humans

cow to human through the consumption of raw milk from a cow that tests positive for tuberculosis. The first commercial pasteurizing machines for the destruction of tuberculin microbes and other pathogenic bacteria did not become available until 1895.

In 1905, what eventually became the Dairy Herd Improvement Association (DHIA) began the first U.S. cow test program in Newaygo County, Michigan, and Humbolt County, California. The idea of herd testing and recording of complete lactation records gradually gained industry acceptance. It soon became the norm to test all cows twice a month for milk and fat levels. Cow test records are used to evaluate sire genetic values based on daughter performances as well as to determine the profitability of one's own cows. In addition, dairy farmers use official records to market the genetics within their herds. The use of computer banks of updated milk production records on each cow allows the U.S. Department of Agriculture (USDA) to calculate a national sire summary. Not all dairy farmers want to pay to have their cows tested, but more than 30 percent of the nation's dairy population is in on Dairy Herd Improvement (DHI) testing.

In 1914, Congress passed an additional law, the Smith-Lever Act. This federal legislation established an extension service at each land grant university. Its purpose was to provide a more direct means of passing on new research ideas to dairy farmers and milk processors and eventually consumers. Also in 1914, tank trucks became an alternative to wagons for transporting milk from the farms into the cities.

Some additional inventions and developments that helped shape the dairy industry include the following:

1919	Homogenized milk was sold successfully.
1932	Vitamin D was added to milk.
1938	Bulk tanks began replacing milk cans on farms.
1939	Artificial insemination was available commercially.
1948	Ultrahigh-temperature pasteurization was introduced.
1948	The first plastic-coated paper milk cartons were commercially available.
1968	Electronic testing of dairy products for milkfat was officially accepted.
1970s	Automated milking unit detachers were developed.
1974	Fluid milk products were labeled with nutrition information.
Mid-1970s	Nonsurgical embryo transfers were introduced.
1980s	Farmers began to backflush milking units between cows.
1980s	Somatic cell counts were accepted nationwide to monitor subclinical mastitis.
1980s	Computerized feeding was introduced.
1990s	Microimplants began to be used to monitor health, production, and identification of cows.
1990s	The use of biotechnology, such as BST, increased.

During the twenty-first century we can expect to see rapid advances in the following areas:

1. Sex preselection of both semen and embryos
2. Genetic engineering to select for specific traits or to select against specific conformations, illnesses, or other genetic abnormalities by the simple addition or removal of key genes.

3. Cloning dairy animals, engineered with human proteins, to make biopharmaceuticals,[2] including orphan drugs (drugs that work on diseases and illnesses that affect only a small percentage of the human population); it will prove much more economical to produce these drugs using transgenic technology and dairy cows.

THE DAIRY INDUSTRY TODAY

Today milk is produced and processed in all 50 states. However, the major areas of milk production are near dense urban populations. Major producing states are California, Wisconsin, New York, Minnesota, Pennsylvania, Michigan, Ohio, and Texas. The United States ranks third behind Israel and Japan in total milk production per cow. Although milk is used as a basis for many products, about 44 percent of the milk produced in the United States is still consumed as fresh milk and cream. Milk by-products make up 54 percent: cheese, 24 percent; butter, 18 percent; ice cream products, 10 percent; and evaporated and condensed milk, 2 percent. The remaining 2 percent is still used on the farm where it is produced. The percentages vary slightly from year to year depending on the production of milk and consumption trends. Table 5–1 lists some typical dairy products available to today's consumer.

Total milk production has kept up with population demands in the United States. This parity in supply and demand has been made possible by increased milk production per cow and by decreased per capita consumption of milk and milk products (from 730 pounds per person in 1950 to about

TABLE 5–1	Modern Dairy Products
Fluid milk products	Solid products
Homogenized milk (3.25% fat or more)	Dried skim milk
Lowfat milk (1 or 2% fat)	Butter
Skim milk (0.5% or less fat)	Cheese
Half and half (10.5% fat)	Cheddar (Colby, Monterey Jack)
Whipping cream (32% fat)	Swiss
Flavored milk products	Italian types
Chocolate milk (3.25% fat)	Brick
Chocolate drink (less than 3.25% fat)	Limburger
Eggnog (6% fat)	Blue
Fermented milk products	Cottage
Buttermilk	Frozen Desserts
Yogurt	Ice cream (10% fat or more)
Sour cream	Sherberts
Sour half and half	Ice milk (2–7% fat)
Sour cream dips	

Source: Reprinted with permission from J. Blakely and D. H. Bade, *The Science of Animal Husbandry,* 6th ed. (Englewood Cliffs, NJ: Prentice-Hall, 1994).

---

[2]Infigen, Inc. (sister company of ABS Global), DeForest, Wisc.

TABLE 5–2	Cow Numbers by State (1998)
United States	9,245,000
Wisconsin	1,449,000
California	1,264,000
New York	702,000
Pennsylvania	644,000
Minnesota	598,000

Source: Western Dairyman, Vol. 79, Issue 1, January 1998.

TABLE 5–3	Leading States in Annual Milk Production per Cow (in Pounds)
California	20,458
Arizona	20,083
Washington	19,996
Colorado	19,440
New Mexico	19,221
United States	16,498

Source: Western Dairyman, Vol. 78, Issue 5, May 1997.

540 pounds today). Per capita fluid milk consumption is only 21 gallons per year, compared with 47 gallons each year for soft drinks. Since 1950, milk production per cow has more than tripled, with 21,000 pounds of milk per cow per year now being common. Many top cows today produce more than 40,000 pounds of milk per year. Several cows have exceeded 60,000 pounds per year. Muranda Oscar Lucinda-ET, of Marathon, Wisconsin, produced 67,914 pounds of milk (twice-daily milking) in 365 days (November 1997). One cow in Stellenbosch, South Africa, Rendal Superstar (#9405), milked 245 pounds of milk in one day.[3] The cow was milked four times per day and was not being treated with BST.

The number of dairy cows in the United States reached its maximum of about 25 million in 1945 and has since steadily declined to 9.2 million in 1998 (see table 5–2). Whereas the average national herd size is 93 cows per farm, herds in southern California average greater than 700 milking cows per farm in 1999.

Although California holds the record for milk production per cow at 20,458 pounds and 685 pounds of butterfat, the United States averages only 16,498 pounds of milk and 613 pounds of butterfat per cow (see tables 5–3 and 5–4). This amount falls far behind Israel's average production of 23,405 pounds of milk and 789 pounds of butterfat and Japan's 18,425 pounds. See table 5–5 for international comparison values. Israeli Kibbutz herds average 26,539 pounds of milk, 830 pounds of butterfat, and 768 pounds of protein per cow per year.[4]

---

[3]Dr. Christian W. Cruywagen, Department of Animal Sciences, University of Stellenbosch, South Africa, 1998.

[4]Israel Cattle Breeders Association, *1997 Holstein Herdbook.*

TABLE 5–4	California Dairy Statistics (1997)

Number one dairy state in the nation
14,000 people employed on California dairy farms
14,500 people employed in California dairy-processing plants
Milk had a $3.7 billion value at the farm
2,324 dairy farms in the state
Average number of cows per farm was 575
Dairy production is state's leading agricultural enterprise
16 percent of U.S. milk produced in California
Number one milk-producing state; number one state in production per cow

Source: California Department of Food and Agriculture, 1997.

TABLE 5–5	International Milk and Butterfat Averages

Nation	Milk per Cow (lb.)	Butterfat per Cow (lb.)	Protein per Cow (lb.)	Cows
United States	16,498	613	527	9,200,000
Israel	23,405	789	730	102,776[a]
Japan	18,425	—	—	1,035,000[b]
India	—	—	—	99,360,000
Australia	10,838	—	—	1,800,000
Russia	4,460	—	—	17,400,000
New Zealand	7,527	—	—	3,200,000

[a]Amir Ben Yehoshua, Beer Tuvia, Israel, June 5, 1998.
[b]*Hoard's Dairyman,* August 25, 1997.

DAIRY FARM INCOME AND EXPENSES

Managing a dairy enterprise requires tremendous academic as well as hands-on experience. Most large dairy farm managers are college educated; many have graduate degrees as well as years of practical experience. Although the income from dairying is considered on the high end for agriculture income, the hours are very long. Typical expense and income sheets from dairies in California and Texas are provided in table 5–6 and in the following:

Income (1995 average California income per hundredweight of milk: $11.63)
   Milk: $12.02 per hundredweight × 650 cows × 184.61 per hundredweight of milk: = $1,442,357.90 per dairy per year
   Cull cows, replacement heifers, bull calves
   Manure (average cow produces 86 pounds of manure per day; 45 pounds N, 9.4 pounds P, and 29 pounds K)[5]
   Equipment rental and labor
Expenses (1995 average California cost per hundredweight of milk: $10.99)
   Feed, labor, mortgage, and depreciation on equipment
   Chemicals, utilities, and veterinary and related bills

[5]Deanne Meyer, Livestock Waste Management Specialist, Department of Animal Science, University of California–Davis, December 16, 1996.

| TABLE 5–6 | Top 25 Percent of Dairy Farms' Average Incomes and Expenses for the 9 Months Ending September 30, 1997 |

	California		Central Texas	
	Amount	Per Hundredweight	Amount	Per Hundredweight
Income				
Milk	$3,495,991	$12.35	$2,006,289	$13.45
Calves	22,869	0.08	43,192	0.29
Patronage dividend	56,016	0.20	6,633	0.04
Other	38,161	0.13	815	0.01
Total income	$3,613,037	$12.76	$2,056,929	$13.79
Expenses				
Feed	$1,839,217	$6.49	$1,057,299	$7.08
Herd replacement	366,821	1.30	244,233	1.64
Labor	267,835	0.95	144,179	0.98
Other operating expenses	585,032	2.08	437,042	2.94
Total expenses	$3,058,905	$10.82	$1,882,753	$12.64
Interest and rent expenses	$274,998	$0.97	$123,838	$0.83
Net Income	$279,134	$0.97	$50,338	$0.32
Average number of milking cows	1,599		874	
Average daily production per cow	65		63	
Average butterfat test	3.56 percent		3.49 percent	
Annualized herd turnover rate	35.47 percent		42.56 percent	

Source: Genske, Mulder & Co., LLP, Certified Public Accountants.

| TABLE 5–7 | Characteristics of Major Dairy Breeds |

	Ayrshire	Brown Swiss	Guernsey	Holstein	Jersey
Cow size (lb.)	1,200	1,400	1,100	1,500	1,000
Bull size (lb.)	1,850	2,000	1,700	2,200	1,500
Calf size (lb.)	75	90	75	95	60
Percentage of registered	2	3	5.5	78.5	10
Percentage of grade	0.5	0.5	1.5	95	2.5
Color	Red or mahogany and white	Brown	Fawn and white	Black and white, red and white	Fawn
Grazing ability	Excellent	Excellent	Fair	Fair	Good
Value for beef	Good	Excellent	Poor	Excellent	Poor
Protein (%)	3.4	3.57	3.57	3.3	3.79
Milk fat (%)	3.9	4.02	4.55	3.65	4.75
Solids not fat (%)	9.0	9.2	9.5	8.5	9.5
Annual average Milk (lb.)	14,365	15,615	13,456	19,569	14,443
Fat (lb.)	566	638	611	720	671
Protein (lb.)	485	558	481	464	539
Origin	Scotland	Switzerland	Isle of Guernsey	Holland	Isle of Jersey

DAIRY BREEDS

In the United States today there are five commonly recognized dairy breeds: Ayrshire, Brown Swiss, Guernsey, Jersey, and Holstein. American dairy cattle are generally taller, longer, and more angular than their European counterparts. Selection programs in the United States have concentrated their efforts on high milk yield. In Europe, many dairy breeders continue to select for beef characteristics so that they can use their cattle for dual purposes. Table 5–7 lists some qualities for which each breed is noted.

Ayrshires

The Ayrshire breed was developed in the county of Ayr in southwestern Scotland. This area is cold and damp, with relatively little forage available. Hence animals were selected for hardiness and excellent grazing ability. The color pattern of the Ayrshire breed varies from red and white to mahogany and white (see figure 5–1). The breed is more nervous than other dairy breeds. Early breeders were careful about selection in type, and the breed is still noted for its style, symmetry, strong attachment of udder, and smooth, clean "dairy-type" bodies. The Ayrshire has only a fair rating for beef and veal. The number of Ayrshires registered ranks them as fifth among the top five dairy breeds. Currently, registrations are less than 60 percent of what they were 40 years ago.

Brown Swiss

The Brown Swiss breed (see figure 5–2) was developed on the mountain slopes of Switzerland. It is perhaps the oldest of the dairy breeds. Bones found in the ruins of the Swiss Lake dwellers date back to 4000 B.C.E. They were grazed from the foot of the mountains in the spring to the highest slopes during summer. Such terrain produced hardy animals with excellent grazing ability. Their large size and white body fat make them also desirable for beef and veal.

FIGURE 5–1
Ayrshire cow.
Courtesy of Agri-Graphics.

FIGURE 5–2 Brown Swiss cow.
Courtesy of Agri-Graphics.

The Brown Swiss color varies from light to dark brown. The breed is noted for being very docile, with a slight tendency to be stubborn at times. Brown Swiss were developed for cheese and meat production, and they produce large volumes of milk with a relatively high percentage of milk fat and solids.

Guernseys

The Guernsey breed (see figure 5–3) was developed on the island of Guernsey, one of the Channel Islands between France and England. The island is noted for luxuriant pastures; hence, grazing ability was not a trait that was selected in early breeding.

Guernsey cattle are fawn colored with clear white markings. They are very docile, but their yellow body fat and small size make them undesirable for veal production. They are noted for producing milk with a deep yellow color indicating high carotene (vitamin A precursor) content. Guernsey milk is high in milk fat and total solids content.

Jerseys

The Jersey breed was developed on the island of Jersey, located only 22 miles from the island of Guernsey. This island also has luxuriant pastures, making it unnecessary to stress selection for grazing ability. Jerseys are the smallest and most refined of the five major dairy breeds. Butter was the major product produced on the island; hence, the Jersey was bred for large amounts of milk fat, a trait for which it is still noted today. During the development of the breed, only animals of excellent type were allowed on the island, and Jerseys are still known for uniformity in type (see figure 5–4).

FIGURE 5–3
Guernsey cow.
Courtesy of Agri-Graphics.

FIGURE 5–4
Jersey cow.
Courtesy of Agri-Graphics.

Milk from the fawn-colored cattle is yellow due to carotene content and is very high in percentage of milk fat and milk solids. The body fat of the Jersey is yellow, and this feature, combined with its small size, makes the Jersey undesirable for veal and beef. There are an estimated 325,446 Jersey cattle in the United States.[6] Jersey record holders include

Queen-Acres Boomer Celeste, EX-90 percent.
> 6 years, 7 months, 365 days, 3×, 38,030 Milk, 4.6 percent, 1,732 Fat, 1,333 Protein

Barbs MBSB Dayetta-ET, EX-90 percent
> 1,451 Protein in 365 days, 2× 35,910 Milk, 4.0 percent, or 1,439 Fat

BW Champs Lou W546, VG-84 percent
> 26,610 Milk, 7.5 percent, 1,990 Fat, 4.5 percent, 1,206 Protein, 365 days, 3×, 4 years, 11 months

Holsteins

The Holstein (see figures 5–5a, b, c) is the most prevalent dairy breed in the United States, accounting for 90 to 95 percent of the dairy cattle. Its origin was in the Netherlands in the provinces of North Holland and western Friesland, an

(a)

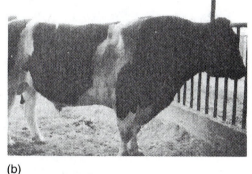

(b)

(c)

FIGURE 5–5 Black and White Holstein cow, (a). An outstanding Holstein bull used in artificial insemination, (b). Red and White Holstein cow, (c).
Photos (a) and (c) courtesy of Agrigraphics. Photo (b) by author.

---

[6]C. Covington, Executive Secretary, American Jersey Cattle Association, January 1998.

TABLE 5–8	Comparison of Holsteins and Jerseys	
	Holstein	**Jersey**
Average age at first calving	27 months	26 months
Average calving interval	13.6 months	13.2 months
Calving ease	Occasional	Very rare
Number of cows supported on a 55-acre farm	80.9	114.6
Productive herd life	38.4 months	39.4 months
Average calvings per lifetime	3.0	3.5
Average annual production per cow	19,569 pounds	14,443 pounds

area noted for excellent pastures. The breed was not forced to develop good grazing abilities during its early stages. Milk was used for cheese, so selection for animals with large volumes of milk was important. There are an estimated 8.8 million Holstein cows currently in the United States.[7] A separate Red and White Dairy Cattle Society started in 1964 (Figure 5–5c). A comparison between Holsteins and Jerseys is summarized in Table 5–8.

Milking Shorthorn

The Shorthorn breed originated in England and evolved into a triple-purpose animal (milk, meat, and power). In 1967, the introduction of Australian Illawara blood into the Shorthorn cow sent the breed in a new direction. In 1969, the milking Shorthorn breeders declared their cattle separate and distinct from those being raised for beef. They were finally accepted into the Purebred Dairy Cattle Association (PDCA) in 1972. About 1 percent of all registered and 0.25 percent of all grade dairy cattle are of this breed. Milking Shorthorns are white to all red, and many exhibit a roan color. The 1997 averages were 14,040 pounds of milk, 500 pounds of fat, and 464 pounds of protein. The Shorthorn cow weighs between 1,400 and 1,800 pounds. The bull averages 2,000 to 2,500 pounds.

Other Dairy Breeds

Other breeds of cattle are also used for milk production in the United States. These are usually dual-purpose breeds for which selection was based more on milk production than on beef. The Red Poll, Brahmanstein, and Dutch Belted cows are such examples.

FEEDING AND GENERAL MANAGEMENT OF DAIRY CATTLE

Feed more than any other factor determines the productivity and profitability of dairy cows. Within a herd, approximately 25 percent of the difference

---

[7]Holstein Association, January 1999.

in milk production between cows is due to heredity; the remaining 75 percent is determined by environmental factors, with feed making up the largest portion.

Care of Newborn Dairy Calves

Although some dairy operations obtain replacements from sales, most dairy farmers still raise their own herd replacements. Care of the newborn calf from birth to weaning, therefore, becomes a necessary part of dairy herd management. The dam should be removed from other cows and placed in an area where it can be carefully watched in case it needs assistance in calving, but it should not be disturbed unnecessarily. Soon after birth, any mucus from the calf's nostrils and mouth should be removed if necessary. In cold weather, the calf is dried or the cow is allowed to lick the calf dry so that the calf does not become chilled. Tincture of iodine is added to the calf's navel to prevent entry of bacteria into the body through the umbilical cord.

Because the object of the dairy is to produce milk, little is wasted on nursing calves from birth to weaning. In addition, from an animal welfare perspective, it is considered more humane to remove the newborn from its dam prior to any nursing. This early separation is required for the health of both the calf and the cow and is much more humane than later separation.

There is a tremendous difference in stress levels between calves that are allowed to nurse and those that are bottle-fed the first feeding. When a calf is allowed to stay with its dam for even 1 day and then forced to separate, the calf bawls excessively and refuses to feed and the dam refuses to feed or to let-down her milk. In addition, dairy cows have been bred for generations to produce much more milk than any one calf needs. Calves are not smart enough to stop nursing when they have had enough. Like all young children, they will eat, and eat, and eat—so long as it tastes good. Calves often overeat and develop scours (diarrhea) from that overconsumption. Uncorrected scours is the most common cause of death in young dairy calves.

It is important for dairy calves, and for all newborn animals, to receive some **colostrum** (first milk) within an hour of birth. The colostrum provides antibody protection against diseases. Dairy calves must receive about 4 pounds of colostrum as soon after birth as possible. Calves continue to receive colostrum for another 2 to 3 days, usually hand-fed from a mail pail or nipple. Extra colostrum cannot be sold as milk, so it is usually diluted with water (two parts colostrum to one part water) and fed to other young calves. See table 5–9 for suggested dilution rates and feeding levels. It can also be frozen or stored in the fermented state and used for other calves or sick calves at a later date. Most dairy farmers regulate the quantity, quality (monitor the colostrum immunoglobulin G [IgG] levels), temperature, and sanitation of the colostrum and the milk that follows to each and every replacement heifer (calf being raised).

Calves are raised in isolation pens (see figure 5–6) until they are consuming enough grain and hay to be weaned; then they are placed into group housing. Group housing is much cheaper from a labor standpoint but is associated with an increase in mortality rates, in competition for feed, in injuries resulting from horseplay and bullying, and in a greater increase in the spread of infectious diseases in very young animals. For these reasons, group housing should be avoided until after weaning.

▌**Colostrum** by legal definition, the lacteal secretion 15 days before and 5 days after parturition; it is rich in the antibodies needed to provide passive immunity to the newborn

TABLE 5-9	Dilution Rate for Various Liquid Diets for Calves Fed Once or Twice Daily		
Type of Milk (Ingredient)	Percent Dry Matter	Once-Daily Feeding[a] Ingredient + Water = lb. Dry Matter (lb./Feeding Once Daily)	Twice-Daily Feeding[a] Ingredient + Water = lb. Dry Matter (lb./Feeding Twice Daily)
First milk colostrum	28	3.5 + 3.5 = 1.0	2.0 + 2.0 = 1.1
Pooled excess colostrum	16	6.0 + 0 = 1.0	3.0 + 1.5 = 1.0
Whole milk, Holstein	12	7.0 + 0 = 0.8	4.0 + 0.0 = 1.0
Milk replacer	88	0.8 + 5.0 = 0.7	0.5 + 3.5 = 0.9

Source: Reprinted with permission from J. Blakely and D. H. Bade, *The Science of Animal Husbandry,* 6th ed. (Englewood Cliffs, NJ: Prentice-Hall, 1994).

[a]Use 75 to 80% of these amounts for Jersey or Guernsey calves.

FIGURE 5-6
Polydome calf nursery.
Courtesy of Poly Tank, Inc.

Feeding Dairy Calves to Weaning

Although there are numerous calf-rearing systems, the recommended feeding practices are generally as follows:

1. Keep calves in dry, draft-free housing (see figure 5–6).
2. Make sure calves receive colostrum for at least 1 day and preferably 2 to 3 days.
3. Change to whole milk or milk replacer at 2 or 3 days of age. If using a milk replacer, make sure the replacer is animal based and not plant based (i.e., soybean) for the first 3 weeks.

4. Do not overfeed; overfeeding is a frequent cause of scours and loss of calves.
5. Feed milk or mixed milk replacer daily at a general rate of 10 percent of body weight (10 pounds for a 100-pound calf).
6. Feed regularly, usually twice daily. Once-daily feeding has been successful but requires more attention to health problems, especially scours, at an early age.
7. Use clean feeding utensils and make sure sanitary conditions exist around pens at all times.
8. Feed milk or milk replacer at the same temperature each feeding, usually 95 to 100°F (35 to 38°C).
9. Feed milk or milk replacer until calves are at least 3 to 8 weeks of age.
10. Wean when the calves develop functional rumina and can eat from 1.5 to 2 pounds of calf starter per day (usually at age 6 to 8 weeks).
11. Feed calf starter and high-quality hay starting when calves are 7 days of age. Feed these items free choice, keeping fresh feed available at all times. Fresh, clean water must be available. See table 5–10 for example starter rations.

TABLE 5–10   Example of Calf Starter Rations

	Ration[a]					
Ingredients	A	B	C	D	E	F
Corn, coarse grind (lb.)	50	39	54	44	34	24
Oats, rolled or crushed (lb.)	35		12	22	34	24
Barley, rolled or coarse grind (lb.)		39				
Beet pulp, molasses (lb.)						20
Corncobs, ground (lb.)					15	
Wheat bran (lb.)		10	11			
Soybean meal (lb.)	13	10	8	26	15	25
Linseed meal (lb.)			8			
Molasses, liquid (lb.)			5	5		5
Dicalcium phosphate (lb.)	1	1	1	1	1	1
Trace mineral salt (lb.)	1	1	1	1	1	1
Vitamin A (IU)	200,000	200,000	200,000	200,000	200,000	200,000
Vitamin D (IU)	50,000	50,000	50,000	50,000	50,000	50,000
Total (lb.)	100	100	100	100	100	100
Protein (% of DM)	16	16	16	20	16	20
Fiber (% of DM)	6	5	5	5	11	9

Source: Reprinted with permission from J. Blakely and D. H. Bade, *The Science of Animal Husbandry,* 6th ed. (Englewood Cliffs, NJ: Prentice-Hall, 1994).

[a]Rations A, B, and C recommended for calves weaned after 4 weeks of age and receiving forages.
Ration D recommended for calves weaned before 4 weeks and receiving forage.
Ration E recommended for calves weaned after 4 weeks and not consuming forage.
Ration F recommended for calves weaned before 4 weeks and not receiving forage.

Calf-Rearing Systems

Calf-rearing systems to be used after feeding colostrum are as follows:

1. *Nurse cow system:* Two or more calves are allowed to nurse one cow. This is the most expensive system because of the loss of income from one cow. It is usually only recommended for veal production and is no longer widely practiced in the United States.
2. *Whole milk system:* Milk is hand-fed to calves at a rate of 6 to 8 pounds per day. Calves are usually weaned at about 60 days and will eat 300 to 500 pounds of whole milk. Many of the nation's large dairies are pasteurizing "mastitic milk" and feeding it to replacement calves.
3. *Milk replacer system:* This method can cut the cost of the whole milk system in half. Calves are hand-fed milk replacer instead of whole milk at a rate of 6 to 8 pounds per day. They are weaned at 28 days (early weaning) to 60 days, depending on how fast they can eat solid concentrates and forage. Milk replacer should include animal protein, because young calves do not have the digestive enzymes at birth to utilize nonmilk ingredients.
4. *Combination of milk replacer and whole milk systems:* This system reduces the amount of whole milk used by gradually switching from whole milk to milk replacer when the calves are 2 weeks of age.

Feeding Dairy Replacements from Weaning to Puberty

The main management purpose of the period between weaning and puberty is to produce, as inexpensively as possible, large, fast-growing heifers that are capable of beginning lactation at an early age. Concentrates are fed at a rate of 3 to 4 pounds per day up to 1 year of age, with hay and/or high-quality pasture fed free choice. More grain is fed if roughage quality is fair or poor. Animals are kept in good condition for proper growth but are never fat or overly conditioned. Tables 5–11 and 5–12 give specific ration suggestions.

Care at Breeding and during Gestation

Well-cared-for heifers should be of sufficient size for breeding at 13 to 15 months of age so they can enter the milking herd at about 2 years of age. Most dairy farmers breed according to weight or size, not age, with the following minimum weights given as a guide for breeding:

Holstein and Brown Swiss	750 pounds
Ayrshire	600 pounds
Guernsey	550 pounds
Jersey	500 pounds

Recent research has shown that breeding at 13 months of age regardless of weight is a simple, effective method, and it is being used by large dairies. In breeding heifers, attention must also be given to planned calving time of replacement heifers to fill low-milk-production gaps. Breeding time is correlated with desired calving time and entry of replacement heifers into the milking herd.

The bred heifers are kept in good condition on pasture and/or hay, with concentrates supplied only if needed. Heifers are expected to continue growth

TABLE 5–11	Growers' Rations for 400-pound Calves

Ration 1
6 lbs. alfalfa-grass hay, free choice (16–18% CP)
4 lbs. grain mix (9.8% crude protein):
    1500 lbs. coarsely ground shelled corn
    455 lbs. rolled or ground oats
    20 lbs. trace mineral salt
    20 lbs. monosodium phosphate
    5 lbs. vitamin premix

Ration 2
5 lbs. alfalfa-grass hay, free choice (12–16% CP)
5 lbs. grain mix (12.8% crude protein):
    900 lbs. rolled barley
    1000 lbs. rolled oats
    55 lbs. dry molasses
    20 lbs. trace mineral salt
    20 lbs. dicalcium phosphate
    5 lbs. vitamin premix

Ration 3
5 lbs. grass hay, free-choice (10–14% CP)
5 lbs. grain mix (10.9% crude protein):
    1800 lbs. corn and cob meal
    100 lbs. soybean meal
    55 lbs. dry molasses
    20 lbs. trace mineral salt
    20 lbs. dicalcium phosphate

Ration 4
6 lbs. corn silage (8–9% CP)
3 lbs. grass hay (12–14% CP)
4 lbs. grain mix (17% crude protein):
    1000 lbs. coarsely ground shelled corn
    655 lbs. rolled or ground oats
    300 lbs. soybean meal
    20 lbs. trace mineral salt
    5 lbs. limestone
    15 lbs. dicalcium phosphate
    5 lbs. vitamin premix

Source: Courtesy of the University of Minnesota, reprinted from *Feeding the Dairy Herd,* Extension Bulletin 218.

until calving and may require additional nutrients to obtain desired condition at calving. If supplemental feed is needed, the suggested rations in table 5–12 may be used as guidelines.

Feeding Dry Cows and Springing Heifers

Heifers within 2 to 3 months of calving (**springers**) are often kept with the **dry cows** (cows not being milked). Cows are usually dried up (milking stopped) 50 to 60 days before the expected calving date to allow the mammary system and cow to recover from the stress of lactation before starting another milking cycle. Dry cows are separated from the milking herd and grazed on good pasture or fed, free choice, its equivalent in hay and/or silage. Concentrates are generally fed only if required to maintain condition during this last 2 months of gestation. Mineral mix and salt are also provided free choice.

Dairy cows are usually in very good condition during the dry period. High-producing cows cannot eat enough feed to fulfill nutritive requirements in early lactation and rely on stored body fat for milk requirements. This loss of weight must be restored prior to the next calving and lactation, usually by feeding extra in late lactation just before the dry period. The resulting weight gain must be carried through the dry period.

At 2 to 3 weeks prior to expected calving, cows are placed in a maternity barn or small trap depending on the climate of the area. Good pasture or high-quality grass hay is provided free choice along with a gradual increase in concentrate ration (challenge or lead feeding) until cows are eating about 1 1/2 pounds of concentrates per 100 pounds of body weight. This feeding program is maintained until the animal enters the milking herd.

▓ **Springers** heifer within 2 to 3 months of calving

▓ **Dry cows** cows not producing milk; they are usually made to go dry 60 days prior to the next calving and lactation

TABLE 5–12	Rations for 700-pound Heifers That Are Gaining 1.5 Pounds per Day

Ration 1
42 lbs. corn silage (33% DM)
1 lb. grain mix:
    160 lbs. corn and cob meal
    1705 lbs. 44% supplement
    98 lbs. dicalcium phosphate
    12 lbs. limestone
    20 lbs. trace mineral salt
    5 lbs. vitamin premix

Ration 2
50 lbs. sweet corn cannery silage
  (20% DM)
5 lbs. grain mix:
    1274 lbs. corn
    425 lbs. oats
    247 lbs. 44% protein supplement
    10 lbs. dicalcium phosphate
    19 lbs. limestone
    20 lbs. trace mineral salt
    5 lbs. vitamin premix

Ration 3
28 lbs. oat silage
2 lbs. grain mix:
    1960 lbs. corn and cob meal
    5 lbs. limestone
    10 lbs. dicalcium phosphate
    20 lbs. trace mineral salt
    5 lbs. vitamin premix

Ration 4
15 lbs. grass hay
5 lbs. grain mix:
    1975 lbs. corn and cob meal[a]
    20 lbs. trace mineral salt
    5 lbs. vitamin premix

Ration 5
7 lbs. alfalfa hay
20 lbs. corn silage
2 lbs. grain mix:
    1940 lbs. corn and cob meal
    35 lbs. monosodium phosphate
    20 lbs. trace mineral salt
    5 lbs. vitamin premix

Ration 6
7 lbs. grass hay
20 lbs. corn silage
3 lbs. grain mix:
    1314 lbs. shelled corn
    438 lbs. oats
    211 lbs. 44% supplement
    8 lbs. dicalcium phosphate
    4 lbs. limestone
    20 lbs. trace mineral salt
    5 lbs. vitamin premix

Ration 7
15 lbs. alfalfa hay
3 lbs. grain mix[b]:
    955 lbs. barley
    1000 lbs. oats
    20 lbs. trace mineral salt
    20 lbs. monosodium phosphate
    5 lbs. vitamin premix

Ration 8
20 lbs. corn stover (stalkage)
3 lbs. grain mix:
    1100 lbs. corn and cob meal
    865 lbs. 44% supplement
    20 lbs. trace mineral salt
    10 lbs. dicalcium phosphate
    5 lbs. vitamin premix

Source: Courtesy of the University of Minnesota, reprinted from *Feeding The Dairy Herd,* Extension Bulletin 218.
[a]Could substitute barley-oats (50-50 mixture).
[b]Could substitute corn and cob meal

Managing the Milking Herd

Ideally, the normal lactation is 305 days with a 60-day dry period. The typical lactation curve is shown in figure 5–7. In practice, the length of a cow's lactation period may vary from 270 days or less to more than 400 days. Shorter periods normally result when the cow is bred back soon after calving or is dried up due to illness. Longer periods result primarily from problems in getting the cow bred. The average calving interval is about 400 days.

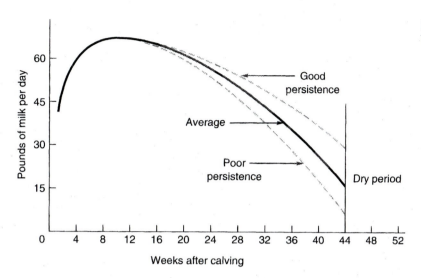

FIGURE 5–7
Typical lactation curve.

Milk production is fairly high immediately after calving. This amount increases for approximately 4 to 6 weeks until the cow reaches maximum production. From this point, there is a gradual decline until the end of lactation. High-producing cows with high 305-day lactation records have good persistence, or the ability to maintain relatively high levels of production throughout the entire lactation period. The average decline in production after peak lactation is about 6 percent per month. Lactation records of 25,000 to 30,000 pounds of milk are not uncommon today.

Production records also vary according to the age of the cow. Cows that freshen at 2 years of age can be expected to increase production by 25 percent to peak production years. Most cows reach maximum production at 6 to 8 years of age (fourth or fifth lactation). Production per year decreases thereafter.

Production records [DHI and DHIR (Dairy Herd Improvement Registry)] usually take into account factors that cause lactations to vary and report a lactation corrected to 6 years of age, 305 milking days, milked twice a day. Other corrections are made for month calved and location of cow. Additional correction factors are used to equalize records for cows milked three times a day instead of two, for cows milked for more or less than 305 days, for cows at an age other than 6 years, or for milk produced with a specific milk fat content. Details can be obtained from breed associations and agriculture extension services. Thus, when cows' records are reported they are usually corrected to reflect the norm for fair comparison.

Feeding the Milking Herd.   Because dairy cows are ruminants, dairy cattle rations normally consist of high-quality legume and nonlegume roughages (e.g., pasture, silage, haylage, and/or hay), with high-quality, palatable concentrates to supplement the roughage for maximum production. Although all nutrients are important, the main concern of dairy farmers is energy. It is difficult, if not impossible, for a high-producing dairy cow to consume enough of any ration to meet the tremendous energy required to produce large quantities of milk.

The highest level of milk production occurs in the first 6 to 12 weeks of lactation, a critical period in feeding. Fresh cows entering the herd should receive all the concentrates they can handle safely (2 to 2 1/2 pounds per 100 pounds of body weight in dry matter). This amount is in addition to the high-quality roughage that cows will eat free choice. Combining forage and concentrate, cows will be able to eat about 3 to 4 pounds of dry ration per 100 pounds of body weight. To supply as much energy as possible, concentrates are used at maximum levels, usually 60 percent of the ration. When concentrates exceed 60 percent, the percentage of milk fat decreases sharply. In addition, increases in rumen parakeratosis, displaced abomasum, and other metabolic disorders are common when feeding cows at excess concentrate levels. Thus the total ration is usually 60 percent concentrate and 40 percent roughage in early lactation.

Body Condition Scoring.   As mentioned earlier, a good dairy cow is expected to lose body weight during peak production. Ideally, 80 percent of the herd should lose no more than 60 pounds between calving and 30 days in milk. At around 50 to 60 days in milk, a typical mature, older cow is out of negative energy balance and will begin to gain 4 to 5 pounds of body weight per week. Cows that lose too much weight during the first 6 weeks postpartum will not be able to regain reproductive efficiency. **Body condition score** is regularly used to evaluate where an individual cow stands relative to the ideals for its stage of lactation. Body condition or fat cover is one indication of a cow's energy reserves. Changes in fat cover reflect shifts in energy balance. Virginia Polytechnic Institute and State University developed a 5-point scoring system on which an extremely thin cow scores a 1 and an obese cow scores a 5. Research has demonstrated a negative relationship between body condition of dairy cows and milk yield per unit of metabolic body weight. Dry cows that score 3.5 produce the most milk yield in the next lactation. As milk production drops, nutritive requirements decline and cows are fed accordingly. Cows are fed less concentrate and more roughage in later lactation, with modifications made according to body condition scores (see figure 5–8).

▨**Body condition score**   system that reflects the amount of subcutaneous fat stored in the body; helpful in determining energy levels needed when balancing rations

Dairy Concentrates.   The major function of a concentrate feed, such as corn or barley, is to supply the additional energy required for maximum milk production that cannot be provided by forages. Protein concentrates, such as cottonseed and soybean meal, adjust the protein level of the feed (see table 5–13). About 50 percent of the crude protein and net energy and 80 to 90 percent of the fiber requirements are typically met in the forage portion of the diet.

The percentage of protein needed varies directly with the protein content of the roughage. The concentrate mixture of energy and protein feeds usually varies between 12 and 18 percent crude protein, with 14 to 16 percent most commonly utilized on a dry matter basis. Increasing the concentrate feeding, however, cannot compensate fully for poor-quality roughages. To meet the requirements for high milk yield, high-quality forage must be fed.

Concentrates are fed in either the milking barn or drylot prior to milking. For those not interested in balancing rations, a good rule of thumb for concentrate feeding is 1 pound of concentrate for each 2 pounds of milk above a 20-pound minimum for grass forage diets and above a 30-pound minimum for legume forage diets. For example, a cow producing 60 pounds of milk would get alfalfa hay plus 15 pounds of grain or corn silage plus 20 pounds of grain.

Forage Quality.   Milk production declines linearly with increasing forage maturity for high-producing dairy cows. The change in alfalfa hay quality with advancing maturity is primarily reflective of the changes in the leaf–stem ratio. There is much higher protein and net energy and greater digestibility in leaves

No Matter How You Look At It...
Body Condition Scoring

Is An Important Part of Modern Dairy Management.

In the dairy cow, body condition is an indicator of the amount of stored energy reserves and changes with different stages of lactation. Fresh cows in peak lactation tend to be in a negative energy balance and therefore lose body condition. Late lactation cows, dry cows and low producers are in a positive energy balance and gain condition. There is no one ideal body condition score. There is a range of desirable scores which change for individual cows over the different stages of each lactation.

Dairy farmers should regularly evaluate the body condition of their cows and heifers so they can fine-tune feeding and management practices. Adequate body reserves are necessary to maintain health, production and reproductive efficiency. Underconditioned cows are prone to reduced milk production and poor persistency of lactation. Overly conditioned cows are predisposed to calving difficulties, fatty liver syndrome, impaired reproduction and metabolic disorders.

Body condition scoring of cattle is an essential management tool for the progressive dairy farmer. It can be mastered with a little training and good observation skills, using both sight and touch to evaluate each cow.

BCS = 3

BCS = 1

BCS = 4

BCS = 2

BCS = 5

Photos by Craig Johnson

FIGURE 5–8 Body condition scoring.
From Elanco Animal Health.

No Matter How You Look At It...
Body Condition Scoring
...Is An Important Part of Modern Dairy Management.

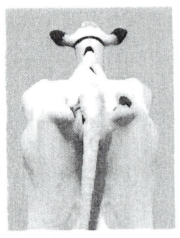

BCS = 1

Deep cavity around tailhead. Bones of pelvis and short ribs sharp and easily felt. No fatty tissue in pelvic or loin area. Deep depression in loin.

BCS = 2

Shallow cavity around tailhead with some fatty tissue lining it and covering pin bones. Pelvis easily felt. Ends of short ribs feel rounded and upper surfaces can be felt with slight pressure. Depression visible in loin area.

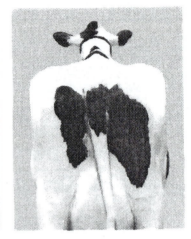

BCS = 3

No cavity around tailhead and fatty tissue easily felt over whole area. Pelvis can be felt with slight pressure. Thick layer of tissue covering top of short ribs which can still be felt with pressure. Slight depression in loin area.

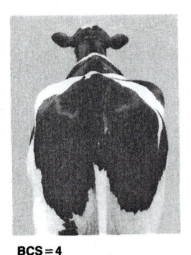

BCS = 4

Folds of fatty tissue are seen around tailhead with patches of fat covering pin bones. Pelvis can be felt with firm pressure. Short ribs can no longer be felt. No depression in loin area.

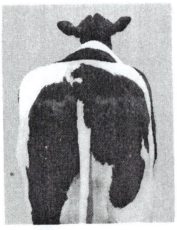

BCS = 5

Tailhead is buried in thick layer of fatty tissue. Pelvic bones cannot be felt even with firm pressure. Short ribs covered with thick layer of fatty tissue.

Elanco Animal Health
A Division of Eli Lilly and Company
Lilly Corporate Center
Indianapolis, Indiana 46285, U.S.A.

AI 7140

Photos by Craig Johnson

FIGURE 5–8   Body condition scoring, continued.
From Elanco Animal Health.

118

TABLE 5–13	Annual Requirements for Dairy Cow Producing 20,000 Pounds of Milk	
Nutrient	**Yearly Total**	**Typical Percentage from Forage**
Dry matter (lb.)	16,380	55
Crude protein (lb.)	2,800	49
NE_L[a] (Meal)	12,000	52
NDF[b] (lb.)	4,800	79
Chewing time (hr.)	4,700	90
Calcium (lb.)	155	51
Potassium (lb.)	150	97

Source: Adapted from M. E. McCullough, *Hoard's Dairyman,* 1992.
[a]NE_L, net energy for lactation.
[b]NDF, neutral detergent fiber.

than in stems. For high-producing cows, the recommended stage of maturity for harvest of alfalfa is bud stage for the first cutting and prebloom for all other cuttings. Harvest is preferred in the early heading-out stage for most grass hays and at the boot stage for small grains.

Dairy Feeding Systems.   Recently, feeding systems other than those previously mentioned, including challenge feeding and the complete ration system, or total mixed ration (TMR), have proven successful. **Challenge feeding,** or lead feeding, refers to the gradual increase in the level of concentrates that both challenges the cow to reach its genetic potential in milk production and prevents the metabolic disorder called **ketosis.** The cow is fed at a rate of 4 to 6 pounds per day, starting 2 to 3 weeks prepartum. This ration is increased at the rate of 1 pound per day until the cow is consuming 1.5 percent of its body weight in concentrates (dry matter basis). For a 1,500-pound cow this would be 15 to 22.5 pounds per day. After calving, the grain ration is increased until the cow reaches maximum milk production or maximum feed intake, whichever comes first. This concept is aimed at finding the potential of cows to secrete milk. Following the first production-level test, the cow's concentrate level is adjusted accordingly. A minimum of 1.5 percent of the cow's body weight should be fed as roughage (dry matter basis).

A complete ration, or TMR, is prepared by mixing the concentrates and forages into one single feed. The ration is balanced for energy and protein and fed free choice to the milking herd. Milk production levels usually separate the milking herd, and a complete ration is formulated for each group. Thus a separate complete ration is fed to dry cows and cows producing less than 50 pounds of milk per day, 51 to 80 pounds of milk, and more than 90 pounds of milk per day. With this system, no additional grain is fed in the milking barn. Large commercial and family-run dairies have found this system labor saving because it utilizes feeding wagons with scales and automated delivery systems. Complete ration examples are given in table 5–14.

It costs more to feed high producers, but the milk income is much higher and the net return above the cost of feed is also increased. Rather than cheapen the quality of the diet, many dairy farmers prefer to mechanize the feeding

▓ Challenge feeding gradual increase in the level of concentrates fed to challenge cows to reach their maximum genetic potential for milk yield and to prevent ketosis; also known as lead feeding

▓ Ketosis metabolic disorder caused by sudden need for energy and the excessive formation of ketone bodies in the blood; also known as acetonemia, pregnancy disease, and twin-lamb disease

Type of Forage	Period 1	Period 2	Period 3	Period 4

TABLE 5–14 Sample Rations for Cows in Various Feeding Periods[a] with Various Forage Types and Combinations

Type of Forage	1	2	3	4
Legume Forage				
Alfalfa hay (lbs.)	22	29	29	22
Grain mix (lbs.)	41	33	19	6
Oats (lbs.)	580	630	660	630
Shelled corn (lbs.)	1180	1280	1300	1300
44% Supplement (lbs.)	200	60	0	0
Dicalcium (lbs.)	20	10	20	50
Trace mineral salt and vitamins (lbs.)	20	20	20	20
Corn Silage, Limited Hay				
Alfalfa hay (lbs.)	6	6	6	6
Corn silage (lbs.)	45	63	63	45
Grain mix (lbs.)	36	27	14	2
Oats (lbs.)	465	460	460	650
Shelled corn (lbs.)	1000	935	900	1300
44% Supplement (lbs.)	480	550	585	0
Dicalcium (lbs.)	20	25	30	30
Limestone (lbs.)	15	10	5	0
Trace mineral salt and vitamins (lbs.)	20	20	20	20
Legume (1/2), Corn Silage (1/2)				
Alfalfa hay (lbs.)	11	15	.15	11
Corn silage (lbs.)	30	40	40	30
Grain mix (lbs.)	38	29	16	4
Oats (lbs.)	520	540	580	660
Shelled corn (lbs.)	1075	1095	1200	1300
44% Supplement (lbs.)	360	325	185	0
Dicalcium (lbs.)	15	20	0	0
Monosodium phosphate (lbs.)	0	0	15	20
Limestone (lbs.)	10	0	0	0
Trace mineral salt and vitamins (lbs.)	20	20	20	20
Grass				
Grass hay (lbs.)	22	29	29	22
Grain mix (lbs.)	42	34	21	7
Oates (lbs.)	550	575	635	655
Shelled corn (lbs.)	1113	1170	1265	1300
44% Supplement (lbs.)	290	210	60	0
Dicalcium (lbs.)	10	10	5	0
Limestone (lbs.)	17	15	15	25
Trace mineral salt and vitamins (lbs.)	20	20	20	20

Production: period 1 = 90 lbs., period 2 = 80 lbs., period 3 = 50 lbs.; fat test = 3.8% all periods and 1300-lb. cow.

Dry matter levels: hay = 90%; corn silage = 33%; and grain = 88%.

Forage content (100% DM); alfalfa = 16% CP and 33% CF; corn silage = 8% CP and 26% CF; and grass = 12 CP and 37% CF.

Source: Courtesy of the University of Minnesota, reprinted from *Feeding the Dairy Herd,* Extension Bulletin 218.

[a]The amount of feed indicated meets the cow's needs. Cows may not be able to consume the indicated amounts in period 1 and 2.

TABLE 5–15	Income over Feed Costs Improves as Production per Cow Increases		
Milk Production	**Feed Cost**	**Income Above Feed Cost**	**Value of Product**
13,086	741	991	1,732
15,354	814	1,213	2,027
16,766	848	1,366	2,214
18,645	908	1,553	2,461

Source: Modified from M. E. Ensminger et al., *Feeds and Nutrition,* 2nd ed. (1990), 808.

TABLE 5–16	Production's Impact on Financial Measures				
	5 Different Herds				
	1	**2**	**3**	**4**	**5**
Rolling Herd Average (RHA) milk (lb.)	16,670	18,860	20,067	21,283	23,363
Number of cows	61	81	67	72	72
Milk sold per worker (lb.)	489,247	553,411	532,171	600,479	594,077
Feed cost per hundredweight ($)	3.79	3.61	3.69	3.84	4.07
Debt per cow ($)	2,568	2,715	2,556	2,074	2,278
Net income per cow ($)	229	334	473	553	645
Operating profit, margin ratio (%)	8.6	11.7	15.7	17.7	18.5
Return on assets (%)	3.3	4.9	6.2	7.4	8.7
Return on equity (%)	5.2	8.0	10.4	10.7	13.0

Source: Wisconsin Center for Dairy Profitability, 257 Holstein herds, 1994.

process and thus save on labor. Table 5–15 demonstrates the increased net value of feeding for higher milk yield.

Table 5–16 demonstrates the increase in feed costs per hundredweight of milk for higher-producing cows. However, when the milk sold per worker increases, it more than compensates for the extra expense in feed. Thus, higher-producing herds are generally more profitable even with the increase in feed costs.

Balancing a Dairy Ration. The first step in dairy feeding is to determine the nutritive needs for maintenance, growth, pregnancy, and lactation. National Research Council (NRC) requirements (minimal values) will provide you with a good range. These minimum values do not allow for any margin of safety, animal differences, feed differences, losses of certain nutrients in storage, or stress. The following are rules of thumb while using NRC tables:

- Use an $NE_{(l)}$ minimum of 0.76 megacalories per pound of dry matter (DM) in early lactation and 0.69 megacalories per pound in late lactation.
- Use a crude protein minimum of 19 percent (DM basis) in early lactation, decreasing to 13 percent in late lactation.
- Limit urea to a maximum of 0.4 pound per day or 1 percent of grain mix.

- Use a minimum of forage DM of 1.5 percent of body weight. Feed high-quality legume forage during early lactation.
- Feed a minimum of 17 percent acid detergent fiber (ADF) in the DM during early lactation, increasing to 22 percent during the dry period. Feed a minimum of 28 percent neutral detergent fiber (NDF) during early lactation, increasing to 35 percent during the dry period.
- Feed 0.5 to 1.0 percent trace mineralized salt in the grain mix and 1 percent Ca-P in the grain mix.
- Incorporate into the ration sufficient vitamins A, D, and E to meet requirements.
- Do not chop or grind the ration too fine.
- Feed total mixed ration and balance to meet all requirements.
- Do not exceed 12 to 16 milligrams per deciliter MUN. Higher levels will reduce milk production and are usually caused by a more costly protein diet.

Negative Energy Balance.   As previously mentioned, dry matter intake (DMI) peaks at 10 to 14 weeks postpartum. Milk yield peaks much earlier, at 4 to 8 weeks postpartum. Because of this difference in timing there is a negative energy balance in high-yielding cows during early lactation. Cows mobilize their body tissue reserves (fat deposits), which results in weight loss. As DMI increases and milk production declines, weight stabilizes.

To minimize negative energy balance, remember that NDF in a ration is negatively correlated with DMI. Minimum fiber levels for normal rumen function have been estimated to be a minimum of 21 percent ADF and 28 percent NDF during the first 3 weeks of lactation and reduced to 19 and 25 percent during lactation peak to meet dietary energy needs.

DMI decreases with consumption of fermented feeds and is 20 percent less with grasses than with legumes. With high-producing cows, it is common to feed sources high in fat to provide a concentrated energy form to the cow. Sources high in unprotected polyunsaturated fatty acids (soybean meal [SBM] or corn oil) have a greater negative effect on rumen microbes and depress fiber digestibility more than saturated fatty acid sources (tallow). Scientists have created rumen-protected fats that are more resistant to microbial action in the rumen, increasing milk fat percentage.

For many years, nutritionists thought that protein leaving the rumen and entering the small intestine was of good quality and met the essential amino acid requirements of the cow. E. Jordan and L. Swanson (Oregon State University), among other research scientists, pioneered new protein terminology for the ruminant animal. They showed that protein quality could actually be lowered through rumen microbial degradation and that the ruminal ammonia produced could have a negative effect on cow fertility. The following is only a partial list of some of the new protein terminology that evolved from this research:

- *Degraded intake protein (DIP):* Intake crude protein that is broken down (degraded) by microorganisms in the rumen.
- *Undegraded intake protein (UIP):* Crude protein that is not broken down in the rumen; instead, it is swept out of the rumen to the abomasum and small intestine for breakdown there and absorbed in the form of peptides and amino acids.

About 60 percent of the crude protein in a typical dairy cow ration is DIP, which breaks down into ammonia in the rumen. Microbes must convert the ammonia to microbial protein in their own cells if the dairy animal is to receive any benefit. If rumen ammonia levels are excessively high, the ammonia is absorbed into the blood and either recycled or excreted in the urine as urea. High rumen ammonia levels can lead to infertility.

- *Milk urea nitrogen (MUN):* About 12 to 16 milligrams per deciliter are normal. Higher levels mean there is either too much protein or not enough rumen-available energy in the ration. If urea levels in the milk are too low, usually DIP, UIP, and/or energy levels are insufficient. High MUN levels are due to feeding carbohydrates that do not digest readily enough to match the protein source.
- *Dried poultry litter (DPL):* Fed as a nitrogen source to growing heifers and dry cows; a commercial heating process destroys all pathogens. Feed 1 to 5 pounds per head per day.[8]

COMMON DAIRY METABOLIC DISORDERS

Bloat (Ruminal Tympany)

Although not considered a true disease, bloat causes widespread discomfort and death to ruminants totaling annual losses in excess of $100 million. The sign of bloat is a swelling to abnormal proportions of the left side of the animal. Severe cases cause pressure on the diaphragm and lungs by the entrapment of the normal gases of fermentation; the pressure on the respiratory organs induces gasping for breath. It is thought that toxic substances produced during this period are inhaled or absorbed through the lungs, causing death.

The exact cause is not clear, but it is well-known that legume pastures, alfalfa hay, and high-concentrate feeds are most likely to produce bloat. Two types are recognized, a gaseous form and a gas bubble form. The gaseous type may be relieved by encouraging the animal to walk, passing a tube down the esophagus, or, as a last resort, puncturing the rumen using a trocar and cannula. The bubble type must be burst like a balloon. A *surfactant* (something to prevent or break the tension needed to produce a bubble) is usually quite effective. Vegetable oils and even detergent soap powders have been used in emergency situations. However, a prepared surfactant such as poloxalene is recommended. It may be used as a drench or in critical cases injected through the wall into the rumen.

Prevention of bloat is preferred over treatment. Filling cattle on dry hay prior to grazing legume pastures, using a heavy stocking rate to prevent rank growth, and using surfactants are common procedures that have been very effective. Soluble leaf proteins, saponins, and hemicelluloses are the primary foaming agents that cause the bubble or foamy bloat. Salivary mucin is an antifoaming agent, but saliva production is reduced with succulent forages. Legume forages (alfalfa and clover) have a higher percentage of protein and are digested more quickly, thus contributing to an increased likelihood of bloat.

---

[8]Dr. Joan Jeffrey, Extension Poultry Veterinarian, University of California–Davis.

Calf Scours (White Scours)

Scours, or diarrhea, is one of the least understood diseases of cattle. About 10 to 12 percent of all calves die from calf scours in the first 30 days of their lives. Death rates commonly exceed 50 percent, and animals that do recover often are stunted for the rest of their lives.

Scours is generally thought to be caused by a bacterial or viral invasion. However, research has revealed that the condition is much more complicated. It may be caused by bacteria, viruses, and environmental conditions, such as high concentrations of cattle, lack of colostrum, overfeeding, vitamin A deficiency, and parasitism. Because of such a wide variety of causes, it is extremely important to determine the specific reason for each outbreak. A veterinarian or diagnostic laboratory scientist can isolate the specific organism or other causes, which is the key to successful treatment.

The principal damage caused by scours is loss of water, bicarbonate, sodium, and potassium from the blood and body fluids. Irritation to the intestines by invasion of the microorganisms produces a body reaction to try to flush out the invading organisms. The calf thus passes watery feces and loses weight. If 15 percent of the body weight is lost as a result of dehydration, the calf goes into a coma and dies.

■**Electrolyte** electrically charged dissolved substance

The key to treating dehydration is to replace fluids lost from the body through **electrolyte** therapy. The calf can be orally drenched at recommended rates while the diagnostic laboratory is determining the specific organism responsible for the diarrhea. These ingredients work somewhat like sports drinks, which professional athletes drink to prevent excessive dehydration and loss of electrolytes (ions).

Once the specific cause has been diagnosed, a recommended antibiotic may be given orally or injected to treat the viral or bacterial invasion. If calves respond to antibiotics, treatment should not be stopped too quickly. Twelve hours is about the maximum effective time for injectable antibiotics. The normal recommendation is to continue treatment for 1 to 2 days after the scours has cleared up.

Other preventive measures include giving adequate amounts of vitamins A and D to cows before calving, keeping calves in clean environments, disinfecting stalls, isolating infected calves, dipping calves' navels in iodine immediately after birth, and making sure newborn calves receive a full feed (about 2 quarts) of colostrum milk within the first hour after birth. Colostrum should be checked for high antibody (immunoglobulin) levels. Newborn calves have no antibodies to provide natural protection against disease until they take in the colostrum. Calves' ability to absorb immunoglobulins is substantially reduced after 24 hours. Calves should receive 10 to 12 percent of birth weight as first milking colostrum (one-half within the first 4 hours); however, calves should be monitored to ensure they do not gorge themselves, thus creating a food-induced diarrhea.

Displaced Abomasum

A left-displaced abomasum commonly occurs in cattle fed large amounts of concentrates, corn silage, or finely chopped roughages. It is thought that the aforementioned feeding causes abomasal atony and produces gas. Because the abomasum is loosely suspended in the abdominal cavity, it can be moved from its normal position and become trapped under the heavy rumen.

Medical management can include treatment without surgery. Depending on the severity and economic value of the cow, one of three common treatments can be used:

1. Parenteral therapy corrects the problem through nutritional management, and rolling (casting) moves the cow from one side to the other, causing the abomasum to float back into place (25 percent recovery rate, $30).
2. In closed surgical management (blind surgery), the rolled and closed suture technique locates the abomasum when the cow is rolled on its back. A suture holds the abomasum in place (88 percent recovery rate, $100).
3. Abdominal surgical management may consist of standing abomasopexy, standing omentopexy, pylormentopexy, and recumbent abomasopexy open abdominal surgeries (94 percent recovery rate, $200).[9]

Goiter

Goiter is an enlargement of the thyroid gland in the neck, which causes a swollen, lumpy appearance on both sides of the neck. A deficiency of iodine in the diet brings on goiter. The thyroid gland needs iodine to manufacture the hormone thyroxin. When the supply is reduced, the thyroid gland is stimulated to grow larger to produce what the body demands. The feeding of iodized salt is a simple cure in the early stages and an easy preventive measure. There is no effective cure for advanced cases.

Grass Tetany (Hypomagnesemic Tetany)

Grass tetany is indicated by a nervousness and tetany (twitching) of the muscles. The animal may breathe rapidly, stagger, or fall when grazing, giving rise to the common name, *grass staggers.* The cause is a low blood magnesium level that usually originates from soil or forage deficiencies. Treatment consists of intravenous injection of a magnesium salt. Preventive measures include the feeding of a free-choice supplement containing magnesium. Grass tetany causes a reduction of Mg levels in the cerebrospinal fluid, which leads to hyperexcitability, muscular spasms, convulsions, and death.

▩ Grass tetany metabolic disorder caused by deficiency of magnesium in the diet; also known as grass staggers and hypomagnesemic tetany

Ketosis (Acetonemia, Hypoglycemia, Pregnancy Disease)

The name ketosis comes from the odor of ketones that affected cattle give off on their breath, urine, and milk. The odor is very offensive and probably best described as something similar to but much stronger than that of nail polish remover. This sweet odor characterizes ketosis, although a definite diagnosis should come from a veterinarian because of other conditions that create the same signs. Besides a thorough physical examination, a Rothera's test for ketone bodies (usually in the urine) will help with diagnosis.

The first form of ketosis is referred to as the wasting form. A decrease in appetite and milk production over a period of 2 to 4 days, rapid body weight loss, and constipation—but normal temperature, pulse, and respiration rates—are clinical signs of the wasting form. Milk production drops severely, animals may

---

[9]R. Franck, "DA Dilemma," *Dairy Herd Management,* Vol. 34(6), June 1994.

die, and those that recover may not regain former production levels until the next lactation. Although ketosis is more common in dairy cattle, it affects all domestic animals. In dairy cattle, it usually occurs as a result of a negative energy balance in the first 6 weeks postpartum. In an effort to correct this condition, the cow mobilizes body fat. During the process of converting this stored energy, ketone bodies are produced.

As the disease progresses, cows will develop additional symptoms, including mild staggering, partial blindness, walking in circles, depressed appetite, excessive chewing on nothing, and salivation. The cow may appear to recover, but signs recur intermittently.

This disorder requires the expertise of an experienced veterinarian. The very complicated metabolic disorder involves the elimination of ketones in the urine and an exhaustion of the glucose stored in the liver. Because cattle have small glucose reserves, the condition can become quite serious.

Prevention consists of keeping cattle neither too fat nor too thin before calving because ketosis often occurs just prior to or just after birth. Cows should be fed so that their body condition score at calving is 3.5 (see earlier section, "Body Condition Scoring"). About 2 to 3 weeks prepartum, cows should be challenge fed to prepare their rumen microbes for larger amounts of concentrates at calving. Secondary ketosis can result from retained fetal membranes, metritis, mastitis, displaced abomasa, fatty livers, and environmental stresses.

Treatment of ketosis includes intravenous administration of 500 cc of a 50 percent glucose solution. Oral administration of propylene glycol for 2 days (100 grams daily) is sometimes recommended. Veterinary consultation is recommended for severe cases for both diagnosis and treatment.

Milk Fever (Parturient Paresis) or Hypocalcemia

▓ **Milk fever** potentially fatal metabolic disorder caused by a sudden need of calcium following birth and initiation of lactation; also known as parturient paresis

Milk fever occurs at or shortly after calving but seldom until the second calf. The cow shows extreme weakness, wild eyes, and a loss of consciousness with the head curled toward the body like that of a sleeping dog. Body temperature is generally subnormal. The cause is a deficiency of calcium brought on by rapid mammary demands for milk production. Treatment usually consists of intravenous injection of calcium gluconate. Untreated animals are likely to die. Prevention methods include reducing the calcium levels in the dry cow ration, thereby promoting the body's release of parathyroid hormone (PTH). PTH release increases absorption of calcium across the intestinal walls and the release of stored calcium from the bones. Newer research involves the use of dietary cation–anion difference (DCAD) to manipulate the ionic charge of minerals fed to control hypocalcemia.

Poisonous Plants

Special problems of grazing livestock on pasture and open range include the possibility of ingesting poisonous plants. It is more common when the pastures are in poor condition from overgrazing, low soil fertility, or other conditions of poor management. In the eastern United States, *tall fescue toxicosis* is the most important natural toxicity problem. Mixing legumes with fescue helps reduce the number of cases of this disorder. Symptoms include swelling of one or both hind hooves and lameness. In the West, *larkspur poisoning* is the major problem. More recently, increases in *fiddleneck* and *groundsel poisoning* have been reported. Fiddleneck contains an alkaloid that causes liver degeneration. Ingesting 5 to 10 percent of body weight (4 to 8 pounds for an 800-pound heifer) per day could be fatal within 2 weeks.[10] Groundsel also contains toxic alkaloids that damage liver function.

Rickets

Rickets is a nutritional disorder caused by a deficiency of calcium, phosphorus, and/or vitamin D. It is characterized by the improper calcification of the organic matrix of bones in growing calves and other animals. Improper calcification results in weakened, soft bones that lack normal density. Bowed limbs and pathological fractures are common symptoms. Treatment includes correction of the diet early on in the course of the disorder.

Traumatic Gastritis (Hardware Disease)

Because cattle do not chew when they first swallow their food, they frequently ingest small pieces of baling wire, nails, screws, and so forth. This ingested hardware usually gets lodged in the honeycomb structure of the reticulum. Puncturing of the reticulum and sometimes the heart, which borders it, causes the disorder known as hardware disease. To prevent hardware disease, cattlemen place a magnet in the reticulum to hold metal, thus preventing puncture of the wall. The disease is cured by surgically removing the metal and administering broad-spectrum antibiotics. Symptoms of hardware disease include an arched back, elevated body temperature, anorexia, massive weight loss, shallow respiration, reluctance to move, and the cow "going off feed."

White Muscle Disease

White muscle disease in young calves or lambs is associated with a deficiency in selenium, vitamin E, or both. Affected animals demonstrate paralysis of the hindlimbs, a dystrophic tongue, and degeneration and necrosis of both cardiac and skeletal muscle.

X Disease (Hyperkeratosis)

X disease is a skin condition resulting in thickened hide, loss of hair, watery eyes, and diarrhea (see figure 5–9). Death losses can exceed 75 percent. The cause is consumption of oil- or grease-type products, to which cattle have a peculiar attraction. The toxin found in lubricants and wood preservatives interferes with the conversion of carotene to vitamin A. No treatment is recognized. Prevention by keeping cattle away from sources appears the best solution. Manufacturers in many areas of the country no longer put chlorinated naphthalene into farm lubricants and wood preservatives, thus lessening the incidence of this disorder.

---

[10]Dr. Art Craigmill, Toxicologist, University of California–Davis.

FIGURE 5–9 An animal showing an advanced case of X disease.
Courtesy of the U.S. Department of Agriculture

BREEDING, SELECTING, AND JUDGING DAIRY CATTLE

The average dairy cow in the United States remains in the milking herd for a little more than 3 years, yielding 1.5 heifer calves. Thus, 33 percent of the milking herd must be replaced each year, through either buying or raising. When heifers are being selected for eventual replacement, pedigree information is very important. With cows, milk production and type information is very important. Reasons for culling include the following:

- Low production (26 percent)
- Chronic health and injuries (16 percent)
- Milking qualities (fast, slow)
- Disposition (quiet, nervous)
- Type (particularly udder conformation)
- Mastitis (12 percent)
- Reproduction (22 percent)
- Dairy purposes (9 percent)
- Died (8 percent)
- No reason given (7 percent)

See table 5–17 for information about the relationship between culling rates and profitability.

Judging attempt to rank or place animals in the order of their excellence in body type

Judging is an attempt to rank or place animals in the order of their excellence in body type. There is considerable debate as to the degree or correlation of type and production and whether they are positively or negatively correlated. Well-attached udders are less subject to injury and mastitis infection, and strong legs hold up longer. Whether judging (comparing several animals) or selecting (comparing one animal with an ideal type), observation must be based on knowledge of animal parts and function. Students must be familiar with the parts of the dairy cow and recognize them by name. Figure 5–10 gives the proper name and location of such parts.

TABLE 5–17 How Culling Rates Affect Dairy Profitability

Forced Cull Rate	Voluntary Cull Rate	Overall Cull Rate	Yearly Return[a]
19.4 (%)	7.8 (%)	27.2 (%)	$42,000
16.5 (%)	8.6 (%)	25.1 (%)	$44,300
13.5 (%)	9.8 (%)	23.3 (%)	$46,500
1.6 (%)	15.2 (%)	16.8 (%)	$55,600

[a]Returns are on labor and management for various culling rates for a 100-cow herd.
Source: Dr. Gary Rogers, Pennsylvania State University.

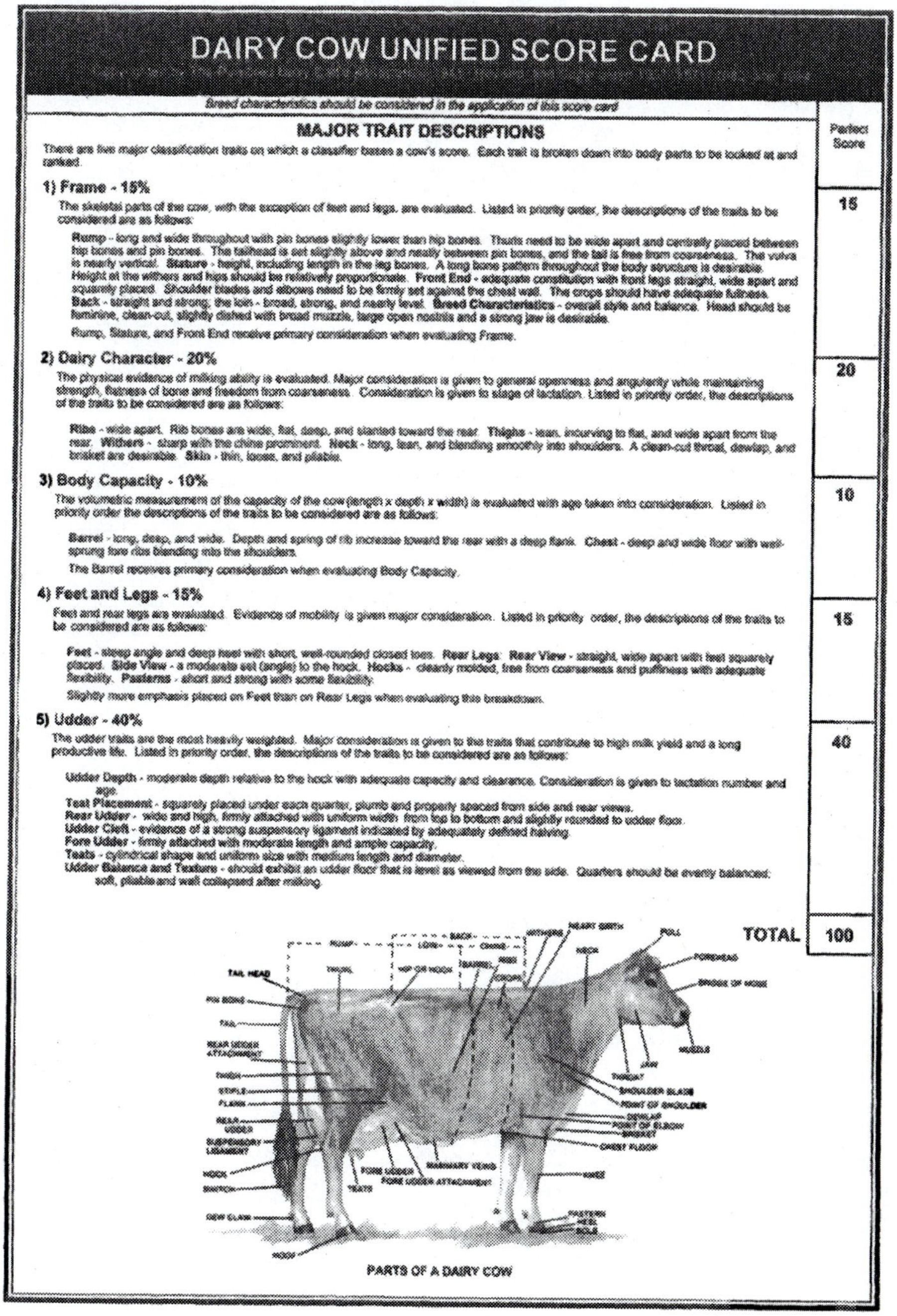

DAIRY COW UNIFIED SCORE CARD

Breed characteristics should be considered in the application of this score card

MAJOR TRAIT DESCRIPTIONS

There are five major classification traits on which a classifier bases a cow's score. Each trait is broken down into body parts to be looked at and ranked.

	Perfect Score
1) Frame - 15% The skeletal parts of the cow, with the exception of feet and legs, are evaluated. Listed in priority order, the descriptions of the traits to be considered are as follows: **Rump** - long and wide throughout with pin bones slightly lower than hip bones. Thurls need to be wide apart and centrally placed between hip bones and pin bones. The tailhead is set slightly above and neatly between pin bones, and the tail is free from coarseness. The vulva is nearly vertical. **Stature** - height, including length in the leg bones. A long bone pattern throughout the body structure is desirable. Height at the withers and hips should be relatively proportionate. **Front End** - adequate constitution with front legs straight, wide apart and squarely placed. Shoulder blades and elbows need to be firmly set against the chest wall. The crops should have adequate fullness. **Back** - straight and strong; the loin - broad, strong, and nearly level. **Breed Characteristics** - overall style and balance. Head should be feminine, clean-cut, slightly dished with broad muzzle, large open nostrils and a strong jaw is desirable. Rump, Stature, and Front End receive primary consideration when evaluating Frame.	15
2) Dairy Character - 20% The physical evidence of milking ability is evaluated. Major consideration is given to general openness and angularity while maintaining strength, flatness of bone and freedom from coarseness. Consideration is given to stage of lactation. Listed in priority order, the descriptions of the traits to be considered are as follows: **Ribs** - wide apart. Rib bones are wide, flat, deep, and slanted toward the rear. **Thighs** - lean, incurving to flat, and wide apart from the rear. **Withers** - sharp with the chine prominent. **Neck** - long, lean, and blending smoothly into shoulders. A clean-cut throat, dewlap, and brisket are desirable. **Skin** - thin, loose, and pliable.	20
3) Body Capacity - 10% The volumetric measurement of the capacity of the cow (length x depth x width) is evaluated with age taken into consideration. Listed in priority order the descriptions of the traits to be considered are as follows: **Barrel** - long, deep, and wide. Depth and spring of rib increase toward the rear with a deep flank. **Chest** - deep and wide floor with well-sprung fore ribs blending into the shoulders. The Barrel receives primary consideration when evaluating Body Capacity.	10
4) Feet and Legs - 15% Feet and rear legs are evaluated. Evidence of mobility is given major consideration. Listed in priority order, the descriptions of the traits to be considered are as follows: **Feet** - steep angle and deep heel with short, well-rounded closed toes. **Rear Legs**: **Rear View** - straight, wide apart with feet squarely placed. **Side View** - a moderate set (angle) to the hock. **Hocks** - cleanly molded, free from coarseness and puffiness with adequate flexibility. **Pasterns** - short and strong with some flexibility. Slightly more emphasis placed on Feet than on Rear Legs when evaluating this breakdown.	15
5) Udder - 40% The udder traits are the most heavily weighted. Major consideration is given to the traits that contribute to high milk yield and a long productive life. Listed in priority order, the descriptions of the traits to be considered are as follows: **Udder Depth** - moderate depth relative to the hock with adequate capacity and clearance. Consideration is given to lactation number and age. **Teat Placement** - squarely placed under each quarter, plumb and properly spaced from side and rear views. **Rear Udder** - wide and high, firmly attached with uniform width from top to bottom and slightly rounded to udder floor. **Udder Cleft** - evidence of a strong suspensory ligament indicated by adequately defined halving. **Fore Udder** - firmly attached with moderate length and ample capacity. **Teats** - cylindrical shape and uniform size with medium length and diameter. **Udder Balance and Texture** - should exhibit an udder floor that is level as viewed from the side. Quarters should be evenly balanced; soft, pliable and well collapsed after milking.	40
TOTAL	100

FIGURE 5–10 Dairy Cow Unified Score Card.
Courtesy of the Purebred Dairy Cattle Association.

| AYRSHIRE | BROWN-SWISS | GUERNSEY |
| HOLSTEIN-FRIESIAN | JERSEY | MILKING SHORTHORN |

BREED CHARACTERISTICS

Except for differences in color, size and head character, all breeds are judged on the same standards as outlined in the Unified Score Card. If an animal is registered by one of the dairy breed associations, no discrimination against color or color pattern is to be made.

AYRSHIRE

Strong and robust, showing constitution and vigor, symmetry, style and balance throughout, and characterized by strongly attached, evenly balanced, well-shaped udder.

HEAD—clean cut, proportionate to body; broad muzzle with large, open nostrils; strong jaw; large, bright eyes; forehead, broad and moderately dished; bridge of nose straight; ears medium size and alertly carried.

COLOR—light to deep cherry red, mahogony, brown, or a combination of any of these colors with white, or white alone, distinctive red and white markings preferred.

SIZE—a mature cow in milk should weigh at least 1200 lbs.

HOLSTEIN

Rugged, feminine qualities in an alert cow possessing Holstein size and vigor.

HEAD—clean cut, proportionate to body; broad muzzle with large, open nostrils; strong jaw; large, bright eyes; forehead, broad and moderately dished; bridge of nose straight; ears medium size and alertly carried.

COLOR—black and white or red and white markings clearly defined.

SIZE—a mature cow in milk should weigh a minimum of 1500 lbs.

MILKING SHORTHORN

Strong and vigorous, but not coarse.

HEAD—clean cut, proportionate to body; broad muzzle with large, open nostrils; strong jaw; large, bright eyes; forehead, broad and moderately dished; bridge of nose straight; ears, medium size and alertly carried.

COLOR—red or white or any combination.

SIZE—a mature cow should weigh 1400 lbs.

BROWN SWISS

Strong and vigorous, but not coarse. Size and ruggedness with quality desired. Extreme refinement undesirable.

HEAD—clean cut, proportionate to body; broad muzzle with large, open nostrils; strong jaw; large, bright eyes; forehead, broad and slightly dished; bridge of nose straight; ears medium size and alertly carried.

COLOR—solid brown varying from very light to dark. Muzzle is black encircled by a mealy colored ring, and the tongue, switch and hooves are black.

SIZE—a mature cow in milk should weigh 1500 lbs.

GUERNSEY

Size and strength, with quality and character desired.

HEAD—clean cut, proportionate to body; broad muzzle with large, open nostrils; strong jaw; large, bright eyes; forehead, broad and slightly dished; bridge of nose straight; ears medium size and alertly carried.

COLOR—a shade of fawn with white markings throughout clearly defined. When other points are equal, clear (buff) muzzle will be favored over a smoky or black muzzle.

SIZE—a mature cow in milk should weigh at least 1150 lbs.

JERSEY

Sharpness with strength indicating productive efficiency.

HEAD—proportionate to stature showing refinement and well chiseled bone structure. Face slightly dished with dark eyes that are well set.

COLOR—some shade of fawn with or without white markings. Muzzle is black encircled by a light colored ring, and the tongue and switch may be either white or black.

SIZE—a mature cow in milk should weigh about 1000 lbs.

FACTORS TO BE EVALUATED

The degree of discrimination assigned to each defect is related to its function and heredity. The evaluation of the defect shall be determined by the breeder, the classifier or the judge, based on the guide for discriminations and disqualifications given below.

HORNS
No discrimination for horns

EYES
1. Blindness in one eye. *Slight discrimination*
2. Cross or bulging eyes. *Slight discrimination*
3. Evidence of blindness. *Slight to serious discrimination*
4. Total blindness. *Disqualification*

WRY FACE
Slight to serious discrimination

CROPPED EARS
Slight discrimination

PARROT JAW
Slight to serious discrimination

SHOULDERS
Winged: *Slight to serious discrimination*

TAIL SETTING
Wry tail or other abnormal tail settings. *Slight to serious discrimination.*

CAPPED HIP
No discrimination unless affects mobility.

LEGS AND FEET
1. Lameness – apparently permanent and interfering with normal function. *Disqualification.*
Lameness – apparently temporary and not affecting normal function. *Slight discrimination.*
2. Evidence of crampy hind legs. *Serious discrimination.*
3. Evidence of fluid in hocks. *Slight discrimination*
4. Weak pastern. *Slight to serious discrimination.*
5. Toe out. *Slight discrimination.*

UDDER
1. Lack of defined having. *Slight to serious discrimination*
2. Udder definitely broken away in attachment. *Serious discrimination.*
3. A weak udder attachment. *Slight to serious discrimination.*
4. Blind quarter. *Disqualification.*
5. One or more light quarters, hard spots in udder, obstruction in teat (spider). *Slight to serious discrimination.*
6. Side leak. *Slight discrimination.*
7. Abnormal milk (bloody, clotted, watery). *Possible discrimination.*

LACK OF SIZE
Slight to serious discrimination.

EVIDENCE OF SHARP PRACTICE (Refer to PDCA Code of Ethics)
1. Animals showing signs of having been tampered with to conceal faults in conformation and to misrepresent the animal's soundness: *Disqualification.*
2. Uncalved heifers showing evidence of having been milked. *Slight to serious discrimination.*

TEMPORARY OR MINOR INJURIES
Blemishes or injuries of a temporary character not affecting animal's usefulness: *Slight discrimination.*

OVERCONDITIONED
Slight to serious discrimination.

FREEMARTIN HEIFERS
Disqualification.

FIGURE 5–11 The Dairy Cow Unified Score Card is used to judge any breed of dairy cattle, either in the show ring or on individual farms.
Courtesy of the Purebred Dairy Cattle Association.

The Dairy Cow Unified Score Card is used by all dairy breed associations for a description of proper dairy type (see figures 5–10, 5–11). The score card is broken down into five main divisions: frame (15 points), dairy character (20 points), udder (40 points), feet and legs (15 points), and body capacity (10 points). A detailed study of the Dairy Cow Unified Score Card indicates the appearance and point values of desired dairy type. Although the function of each division for milk production is briefly given, further explanation is useful in connecting dairy type and milking ability.

Frame includes the rump, stature, front end, and back of the cow. The rump should be long and wide, with the pin bones slightly lower than the hip bones. Good dairy character denotes an angular body with a lack of beefiness (see figure 5–12). This angularity indicates that the cow will convert feed to milk rather than to fat. A dairy cow should carry enough flesh to be in presentable condition and not be too thin (representing unthriftiness) or too fat (representing more meat than milk production).

Body smoothness and openness are evidence of milking ability. A long, wide, and deep barrel strongly supported by well-sprung ribs, along with a large heart girth (see figure 5–12), should exemplify the body capacity.

The udder traits are the most heavily weighted. There should be moderate udder depth relative to the hock, with adequate capacity and clearance. The teats should be squarely placed and hang plumb. The rear udder should be attached high and wide and appear slightly rounded to the udder floor. An udder cleft is evidence of a strong median suspensory ligament. The fore udder should be firmly attached, with moderate length and ample capacity.

Feet and legs are evaluated from both the side and rear view. From the side view a moderate set or angle to the hock and a strong and short flexible pastern are evidence of mobility and longevity. From the rear, straight legs that are wide apart with the feet squarely placed are preferred.

Dairy cattle are judged in show rings, where one cow is compared with other cattle, or on farms, where the cow is compared with the ideal cow

FIGURE 5–12 The dairy character and body capacity essential for milk production: an angular body with openness, a long, wide barrel, and large heart girth.
Reprinted with permission from J. Blakely and D. H. Bade, *The Science of Animal Husbandry*, 6th ed. (Englewood Cliffs, NJ: Prentice-Hall, 1994).

indicated by the Dairy Cow Unified Score Card. This second type of judging is a form of selection called classification and is usually done by a person who represents a breed association. The person assigns a numerical score from the Dairy Cow Unified Score Card, and the cow is classified according to the breed standards. Type classification for breeds is excellent (90 points and over), very good (85 to 89 points), good plus (80 to 84 points), good (75 to 79 points), fair (65 to 74 points), and poor (below 65 points). This scoring system may vary among the different breeds and is given only as an example.

Since 1983, all five major dairy breed associations have adopted a linear means of evaluating cattle. Traits are scored from one biological extreme to the other on a continuous scale. These traits were recommended by a committee from the National Association of Animal Breeders (NAAB) as part of a new uniform functional traits program. One purpose of the linear-type scoring system is to describe a phenotype without assigning merit to a particular score. Individual dairy breeders can then judge or select replacement stock or sires (based on the sire's offspring's phenotype) and still allow for individual preferences of dairy farmers.

Dairy cattle, once classified, may carry their highest classification as a part of their production record. In most breed associations, cows' records reflect the highest classification only. Thus, reclassification only for animals that may go into higher classification scores is normal. Milk production records and cow classification scores are used in selecting cows and cow daughters as replacements. Figure 5–13 illustrates a cow with both production records and classification scores.

Overall, the correlation between milk production and type classification of a cow is not high. Research conducted by J. Honnette et al. at Virginia Polytechnic Institute and State University and L. S. Shapiro and L. Swanson at Oregon State University showed a negative correlation between most type traits and

FIGURE 5–13
Leete Farms Betty's Ida of the Ayrshire breed was classified as excellent and had a 305-day production record of 37,170 pounds of milk containing 4.3 percent fat.
Reprinted with permission from J. Blakely and D. H. Bade, *The Science of Animal Husbandry,* 6th ed. (Englewood Cliffs, NJ: Prentice-Hall, 1994).

FIGURE 5–14 A future veterinarian learning behavior, showing, judging, and handling of a dairy heifer on a college dairy.
Courtesy of L.A. Pierce College Agriculture Department.

milk yield and type traits and fertility, respectively. Still, selection for strong udder support and feet and legs has been shown to increase longevity.

Judging is still important in teaching the behavior, physical conformation, and terminology of dairy cattle. Many agriculture and veterinary programs require some degree of judging and speaking terminology as part of their undergraduate curriculum (see figure 5–14).

Use of Artificial Insemination versus Natural Breeding

After evaluating the dairy farmers' cows and deciding not only which cows to keep and which to cull, the next step is choosing the sire to use to improve on the traits already existing in the herd. There are many advantages in using artificial insemination over a herd sire. Besides the obvious—eliminating the dangers of keeping a bull and venereal disease transmission and having access to sires from across the nation—there is a financial advantage as well. Drs. Jim Smith and Warren Gilson, at the University of Georgia, recently completed a study using 905 southern state herds. They showed the advantages of using artificial insemination over natural breeding. The herds using artificial insemination produced more than 1,900 pounds of milk per year more than the natural-breeding herds. At $12 per hundredweight for milk, artificially inseminated cows netted more than $200 for their owners. In addition, as previously mentioned, disease control is much easier using artificial insemination. Table 5–18 provides statistics for comparison of artificial insemination and natural breeding.

TABLE 5–18	Comparison of Artificial Insemination and Natural Breeding	
	Artificial Insemination	**Natural Breeding**
Number of cows	139.9	161.7
Number of milking cows	121.4	135.9
Percentage of proven sires	80.2	1.1
Percentage of young sires	16.6	0.2
Rolling herd average milk	17,116	15,179
Rolling herd average fat	600	525
Rolling herd average protein	545	487
Mature equivalent (ME) (all lactations)	18,528	16,436
Days open	148.1	150.4
Calving interval	14.1	14.2
Days dry	66.9	69.9
Percentage entering the herd	31.6	30.5
Culling percentage	31.4	30.6
Reproduction culls (percentage of all cows)	8.8	6.4
Reproduction culls (percentage of culls)	27.9	20.6
Predicted transmitting ability for dollars (milking cows)	28.7	−8.9
Predicted transmitting ability for dollars (sires of milking cows)	104.8	63.8

Source: J. Smith and W. Gilson, University of Georgia DHI records from four southern states. Reported June 30, 1998, on Dairy Discussion List, Dairy–L @UMDD.UMD.EDU.

Dairy Genetics

To understand how to choose a dairy sire, two basic principles must be known. The first is the relationship between the various traits being selected, and the second is the amount of a given trait that can be inherited. In Chapter 2, phenotype was described as the physical expression of an animal. This expression is related to the genotype (genetic makeup) and its interaction with the environment ($P = G + E$).

The environment can have a far greater effect on phenotype than does genetics. Environmental parameters include housing, weather, milking equipment, labor, and general management. If we can eliminate as much of the environmental stressors as possible, we then can achieve greater genetic potential for the cow. In addition, when trying to evaluate the genetic worth of a sire or dam, environmental effects must be removed. To do so, a broad geographic range of cows milking under varying conditions, management, and weather must be used.

Genetic correlation measures the extent to which one gene influences the production of other traits (i.e., milk fat and protein). Most of the economically important traits are highly correlated. A high correlation means that when you breed to increase milk protein, as an example, milk fat increases as well. The correlation between the two is estimated at 80 percent. Simply stated, 80 percent of the genes that influence the production of milk fat also influence the production of milk protein.

TABLE 5–19	Amount of Genetic Progress in Primary Selection Trait Possible When Selection Is Also Carried Out on Secondary Traits
Number of Traits Selected	**Progress in Primary Trait**
1	1.00
2	0.71
3	0.58
4	0.50
5	0.45
6	0.43
7	0.35
8	0.32

Source: J. F. Keown, Department of Animal Sciences, University of Nebraska–Lincoln, 1999.

Thousands of genes affect the production of milk and its components. Milk, milk fat, and milk protein are quantitative traits. Quantitative traits generally yield a greater variation in each phenotype.

When selecting a sire, the most economical traits needed in the particular mating should be chosen. The more traits selected, however, the slower the progress in genetic improvement. Table 5–19 shows the decrease in genetic progress as the number of individual traits selected is increased.

The dairy industry has always led the way for artificial insemination. Competition among breeding services has resulted in very thorough testing of dairy sires and their effectiveness. As explained earlier, dairy cattle performance is measured by corrected lactation records. Sire proofs are calculated from this information. The USDA calculates and publishes data on all the cows in such testing programs and their relationship to bulls. A sire with many high-producing daughters is sought after for artificial insemination breeding.

Based on a comparison of the bull's daughters with their contemporaries (herd mates that are in similar lactation and calve in the same season), a **predicted transmitting ability (PTA)** is calculated. This calculation estimates a sire's or dam's potential to transmit yield productivity to its offspring. The genetic base is established every 5 years from production and type data recorded by DHI and breed associations. The USDA publishes PTA values for dollars (PTA$), milk (PTAM), fat (PTAF), protein (PTAP), cheese yield dollars (PTA$ cheese), percentage of fat (PTA%F), and percentage of protein (PTA%P). Some breed associations also calculate a PTA for type. Thus a PTA of +1200 milk means that mature daughters of this bull or cow will produce 1,200 pounds of milk more per lactation than mature daughters of a bull with a PTA of +0 milk. Another term used is percentage of reliability (Rel%), which indicates the confidence in the PTA values. Low Rel% means that the PTA is more likely to change when significantly larger numbers of daughters are added to the summary. This number may range from 1 to 99 percent.

Some breeds have an index used to measure total performance ratings. The Holstein Association uses the type production index (TPI) as a method of ranking bulls on their overall performance. Developed more than 20 years ago, this index includes the following ratio of traits within its new formula: 4 Production: 2 type: 1 health. Within the production component the new formula has a ratio of 5 protein: 2 fat to address consumer demand for products with more protein but less fat. Within the type component there is a ratio of 1 type: .65 udder

Predicted transmitting ability (PTA) estimate of a sire's or dam's potential to transmit yield productivity to its offspring

comp.: .35 feet and legs comp. and within the health component there is a ratio of .9 productive life: .1 somatic cell score. (Source: Holstein Association USA, 2/11/00.)

The udder composite is a compilation of udder traits (e.g., teat placement, udder width, udder cleft, udder depth, and udder height). The feet and legs composite considers foot angle, rear and side views, and so forth. Type index is included to predict an animal's longevity.

Another commonly used index is the net merit index (NM). Geneticists at the USDA's Animal Improvement Programs Laboratory (AIPL) developed the NM in 1994. It incorporates longevity and somatic cell information into an existing production-related index known as milk fat protein dollars (MFP$). The NM currently uses the following ratios: 10(MFP$):4(PTA productive life):1(PTA somatic cell score).

In February 1999, the USDA-AIPL released two new indexes with its sire summaries. The first, the fluid merit index (FM), replaces the milk fat dollars (MF$) values and also includes somatic cell and longevity information. The index has a ratio value of 10MF$:4PTA PL:1PTA SCS. Dairy farmers selling their milk to a cheese plant would not want to use this index, because it does not credit income received for the protein content of the milk. Most cheese plants pay premiums for added protein levels, and thus many dairy farmers are encouraged to increase the genetic potential for protein in their milking herds.

The cheese merit index (CM) is the latest introduction by the USDA-AIPL. The CM replaces the cheese yield dollar (CY$) value and provides longevity and somatic cell score data for producers whose whole milk is made into cheese. The ratio for the CM is 10 CY$:4 PTAPL:1 PTA SCS.

Selected sires should have high PTA and high index values and a reliability of at least 70 percent. Among these sires, the final selection is based on other traits, such as calving ease, udder composite score, foot and leg composite score, and 14 primary linear descriptive traits. Remember, the more traits selected on, the slower the progress in genetic improvement (see table 5–19). Selection should always be made on PTA$ first because the daughters of high-PTA$ bulls will return the most profit to the farm. Dairy bulls are currently tested and identified for various genetic disorders, such as *bovine leukocyte adhesion deficiency* (*BLAD*) (BL, BLAD carrier; TL, tested BLAD normal).

Common Dairy Genetic Disorders

Holsteins.   *Mule foot* (**syndactylism**) is a genetic condition in which one or more of the hooves are solid in structure rather than cloven. The mortality rate is high in calves with mule foot. It is described as a simple autosomal recessive inherited condition.

An animal with BLAD typically dies at 2 to 7 months of age due to an inability to fight infection. BLAD is a lethal, simple autosomal recessive trait found only in Holstein cattle. The white blood cells (WBCs) of animals with BLAD lack the MAC-1 protein. Without it, WBCs are unable to penetrate the walls of blood vessels to fight off infectious disease. The NAAB and the Holstein Association implemented a program to identify carriers to control the spread. Approximately 13 to 14 percent of Holstein sires used for artificial insemination have been identified as carriers. The Holstein Association publishes a list of BLAD-positive sires. The USDA-DHIA sire summaries list carrier bulls with a *BL after their names and those confirmed to be noncarriers as *TL.

▪ Syndactylism
genetic disorder in Holstein cows in which one or more of the hooves are solid in structure rather than cloven; also known as mule foot

Jerseys. *Limber legs* is a genetic condition of Jersey cattle controlled by a simple recessive gene. Calves with the disease that are not born dead are unable to stand. They have incompletely formed muscles, ligaments, tendons, and joints.

Rectovaginal constriction is a simple autosomal recessive trait resulting in severe narrowing or constriction of the anus and/or vestibule. Males have anal stenosis. Vaginal constriction can lead to severe dystocia.

MILK SECRETION AND MILKING MACHINES

For students to understand the milking operation, it is necessary to understand the internal anatomy of the udder and the physiology involved in milk secretion and milk letdown.

Anatomy of the Mammary Gland

The mammary glands of high-producing dairy cows are abnormally developed glands, the result of decades of selective breeding. Through commercial dairying, they are subjected to great physical strain, with little opportunity for rest or repair. Cows with high milk production capacity are believed to be those that have genes that make it possible for their endocrine glands to secrete optimal amounts of hormones required for mammary gland growth, milk secretion, and milk ejection. Prolactin, growth hormone, and thyroxin are all important in the maintenance of lactation. In addition, genetics plays a major part in providing the necessary structure or anatomy of the udder (which is why we emphasize udder composite when we evaluate dairy cows for breeding).

The udder of a cow is divided into four separate quarters. The fore quarters are usually about 20 percent smaller than the rear quarters, and each quarter is independent of the other three (see figure 5–15). To review the basic internal anatomy of one of the quarters of the udder, we will follow the path of milk from the point of synthesis to the end of the teat.

Milk is secreted by individual secretory grapelike units called **alveoli.** These small units range from 0.1 to 0.3 millimeters in diameter and consist of an inner lining of myoepithelial cells that surround a hollow cavity, the *lumen.* The primary milk secretion hormone is prolactin (produced in the anterior pituitary gland). Prolactin initiates and maintains the myoepithelial cells' ability to secrete milk. The myoepithelial cells actually secrete the milk by taking raw materials from the blood supply and synthesizing them into milk. The synthesized milk is secreted into the lumen of the alveolus, which when full contains about one-fifth of a drop of milk. A group of such alveoli in a grapelike cluster is termed a *lobule.* The lumen of each alveolus connects directly with the stem of the cluster of *tertiary* ducts of the lobule. The tertiary ducts drain into the *secondary* ducts, which drain into the large, or *primary,* ducts. A slight constriction at the junction of one duct to another prevents complete drainage of milk. The primary ducts carry the milk to the *gland cistern.* The gland cistern is the collection point of all ducts and holds about 1 kilogram (2.2046 pounds) of milk. The gland cistern drains through the *annular ring* of the upper teat into the *teat cistern,* or cavity inside the teat. Milk is prevented from leaking from the teat cistern by the action of a *sphincter muscle,* which surrounds and closes the *streak canal.* The streak canal is the opening from the teat cistern to the outside of the teat.

■ Alveoli individual grapelike structures of the mammary gland that secrete milk

FIGURE 5–15   The udder is divided into four separate quarters, each independent in its milk-producing function. The figure shows the basic internal anatomy of one of the quarters of the udder showing the site of milk synthesis (alveolus) and milk evacuation (Streak canal).

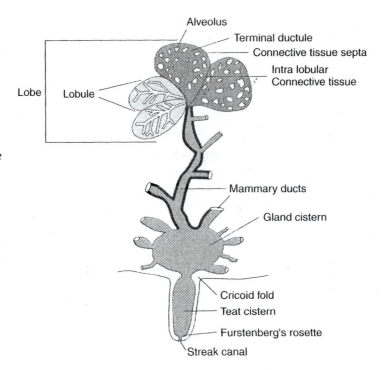

FIGURE 5–15   The udder is divided into four separate quarters, each independent in its milk-producing function. The figure shows the basic internal anatomy of one of the quarters of the udder showing the site of milk synthesis (alveolus) and milk evacuation (Streak canal).

Physiology of Milk Secretion

Because little or no milk is secreted or synthesized during the milking process, all of the milk is present in the udder at the time of milking. Milk is formed or secreted by the cow in the interim between milking. In this interim, milk is synthesized in each functioning epithelial cell of the alveolus and is expelled into the lumen of the alveolus. Because all milk constituents and precursors are transported to the alveolus by the bloodstream, a great amount of blood must pass through the udder in the synthesis of milk. It has been estimated that from 300 to 500 pounds of blood pass through the udder to synthesize 1 pound of milk.

Milk synthesis is most rapid immediately after milking; the first synthesized milk fills the normal storage places in the udder. Neither the size of the udder nor mammary pressure increases significantly in the first hour after milking. The natural storage spaces of the udder hold about 40 percent of the milk present at milking time, and the other 60 percent is accommodated through stretching of the udder. The udder increases about one-third in size during the interim between milking. With the filling of the natural storage spaces of the udder and the initiation of stretching, mammary pressure increases. With the increased pressure, the secretion rate is slowed so that after 6 hours, it is slightly less than it was in the previous hour. The secretion rate continues to slow with increased pressure until an equilibrium is reached, and, if milk is not removed and mammary pressure exceeds 40 millimeters of mercury, milk is reabsorbed.

At the time of milking, the milk has been previously synthesized and stored in the udder. About 40 percent of the milk is stored in the large duct system and cisterns, and 60 percent of the milk is stored in the small duct systems and the alveoli.

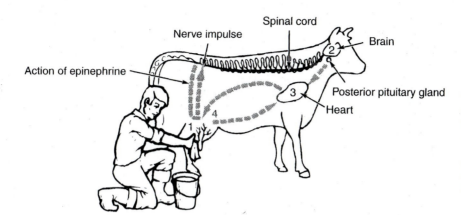

FIGURE 5–16 Milk letdown is caused by various stimuli that force milk from the storage places of the udder.
Reprinted with permission from J. Blakely and D. H. Bade, *The Science of Animal Husbandry*, 6th ed. (Englewood Cliffs, NJ: Prentice-Hall, 1994).

Milk Letdown.   Only a small quantity of the milk present in the udder is immediately available by natural drainage to the milker. This milk is located in the cisterns of the udder. Most of the milk is stored in the small ducts and alveoli, where natural drainage is prevented. Some mechanism is necessary, therefore, to force the milk into the large ducts and cisterns. This expulsion of milk from the alveoli and small ducts is termed **milk letdown** (see figure 5–16). Without milk letdown, only about 2 pounds of milk per quarter could be obtained in milking.

Milk letdown is a nervous reflex produced by various stimuli, including sucking of the teat by a calf, manipulation of the teats in washing or milking, auditory and visual stimuli, and other pleasant sensory stimuli regularly associated with milking. Such stimuli cause the release of the hormone oxytocin from the posterior lobe of the pituitary gland into the bloodstream. Oxytocin is produced by the hypothalamus and stored in the posterior pituitary. It reaches the udder within a few seconds and causes contraction of the tissue of the alveoli and small ducts, forcing milk into the larger duct system. After milk letdown, mammary pressure increases more than 25 percent due to this expulsion mechanism. Because milk letdown lasts only 6 to 8 minutes, milking must be accomplished within this letdown period for maximum production.

Inhibition of milk letdown   When cows are frightened, angered, in pain, or ill treated, they will not let down their milk. This inhibition of milk letdown is caused by the release of epinephrine from the adrenal gland. Epinephrine inhibits the action of oxytocin and lasts for 20 to 30 minutes. Thus rough treatment of dairy cows cannot be tolerated for both animal welfare and simple economic reasons. Scared, nervous, and anxious cows hold up their milk and are more prone to mastitis and other disorders.

Residual milk   At a normal, thorough milking, not all of the milk is removed. Additional milk, termed *residual milk,* may be obtained after normal milking by an injection of oxytocin. The amount of residual milk is variable but is usually about 20 percent of the total milk produced. In most states, if not all, this practice would be illegal because it could adulterate the milk.

■ **Milk letdown**
expulsion of milk from the alveoli and small ducts into the gland cistern of the udder

The Milking Process

After milk letdown, milk is under pressure in the large ducts and cisterns but is prevented from draining through the streak canal by the sphincter muscle (**Furstenberg's rosette**), which does not relax in milk letdown. In the milking process, some method must be used to force open the streak canal and allow milk to flow from the teat.

Hand Milking. In hand milking, the opening between the gland cistern and the teat cistern is closed by squeezing the teat between the index finger and the thumb. Milk trapped in the teat cistern is then forced downward and through the streak canal by compressing the teat against the palm with the fingers.

Machine Milking. Unlike hand milking, in which the milk is forced through the streak canal, machines use a vacuum (or negative pressure) to remove milk and to massage the teat. Machines use two vacuum systems: a continuous vacuum and an alternating vacuum.

These vacuum systems are utilized in a double-chambered teat cup assembly consisting of a teat cup shell that forms the outer wall and a flexible inflation that forms the inner wall. An airtight chamber is formed between the inflation and the teat cup shell. The inner chamber, or inside the inflation, supplies a constant negative pressure of 13 inches mercury to the teat end. (Normally, the vacuum gauge at the pump may read 15 inches mercury. However, this level decreases as the distance to the teat cup increases.) This constant negative pressure causes a pressure differential between the inside of the udder and the inside of the inflation or liner. The higher pressure in the teat forces the teat orifice open, and milk is allowed to flow through the streak canal. Thus milk is not squeezed or forced through the streak canal but is pulled or sucked from the teat by a constant negative pressure.

Unless relieved, the constant vacuum will cause internal hemorrhaging and irritation to the teat tissue. An alternating vacuum and atmospheric air operate in the chamber between the inflation and the shell to alleviate this problem. A negative pressure in the outer chamber counteracts the continuous vacuum inside the inflation, and the inflation remains open. This phase is termed the *milking phase.* The vacuum inside the inflation draws the milk from the teat, as described previously (see figure 5–17).

When atmospheric air is allowed to enter the chamber, the vacuum inside the inflation causes it to collapse. This phase is termed the *resting phase.* The closed inflation relieves the teat of the constant vacuum, thus preventing

FIGURE 5–17 The milking and resting stages of a milking machine are created by alternating vacuum and air between the shell and inflation. Reprinted with permission from J. Blakely and D. H. Bade, *The Science of Animal Husbandry,* 6th ed. (Englewood Cliffs, NJ: Prentice-Hall, 1994).

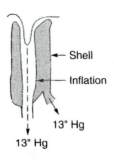

Milking phase

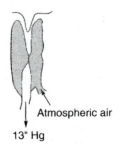

Resting phase

congestion of blood and massaging the teat to maintain stimulation and proper milk letdown.

This vacuum and atmospheric air are alternated to the teat assembly by a pulsator, which changes the vacuum and atmospheric air through use of electrical current and a magnetic valve. The pulsation rate is the number of cycles of alternating vacuum and atmospheric air per minute, 45 to 60 cycles per minute on most machines. The milking or pulsation ratio is the proportion of time spent under vacuum and atmospheric air and is usually approximately 60:40. The inflation is thus open 60 percent (vacuum is on and in milking phase) and closed 40 percent (atmospheric air is present and in resting phase).

Modern milking parlors allow the milking of hundreds of cows/person/day in a clean and comfortable environment (see figures 5–18, 5–19a, b). The newer parlors include computer monitored milking machines which assist in the management of the herd.

Milking machine parts and terms   Following is a list of milking machine parts and their functions:

1. *Milking machine:* The mechanical milking system and all auxiliary equipment.
2. *Vacuum pump:* Either a piston or a rotary pump that produces the vacuum used in milking. Because the vacuum used is only a partial vacuum of 1/2 atmospheric pressure (13 inches mercury), the function of the pump is to remove part of the air as it comes into the air inlets.
3. *Airflow meter:* The metering device that measures cubic feet per minute of air at a given vacuum level.
4. *Vacuum level:* The degree of vacuum in a milking system during operation, expressed as inches of mercury differential measured from atmospheric pressure and indicated by a conventional vacuum gauge.
5. *Vacuum tank:* A vessel or chamber in the vacuum system between the pump and the point of air admission that reduces and stabilizes pressure differentials.
6. *Vacuum controller or regulator:* An automatic air valve in the vacuum system that prevents the vacuum from exceeding a preset level by admitting atmospheric air as needed.

FIGURE 5–18
Modern rotary milking barn.
Courtesy Westfalia-Surge, LLC.

FIGURE 5–19
Modern Herringbone
milking parlor, (a).
Schematic drawing of
a modern Herringbone
milking parlor, (b).
Courtesy Westfalia-
Surge, LLC.

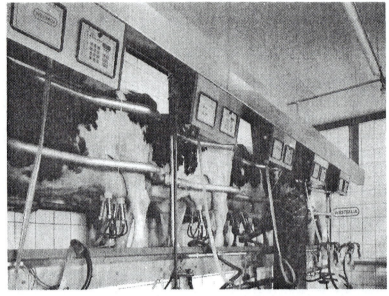

(a)

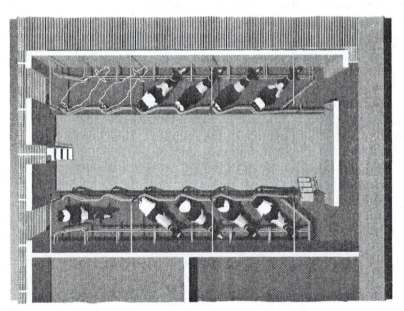

(b)

7. *Vacuum reserve:* The additional air-moving capacity of the vacuum pump
   after the requirements of the milking units, bleeder holes, operating
   accessories, and air leaks have been met; equal to the volume of air
   entering through the controller.

8. *Pulsator:* The mechanism that permits alternate vacuum and atmospheric
   pressure to exist between the rubber teat cup liner and the metal shell.
   This unit creates the massaging action.

9. *Claw:* The sanitary manifold that spaces and connects the four teat cup
   assemblies and the milk hose.

10. *Teat cup assembly:* Made up of the shell, inflation, and air hose.

11. *Liner (inflation):* The rubber part of the milking machine in actual contact with the cow's teat.
12. *Shell:* The cylindrical metal part of the teat cup.
13. *Milk cup:* A milk reservoir adjoining the claw between the milk tubes and the milk hose.
14. *Milk hose:* Hose that connects pipeline or buckets to the claw.
15. *Receiving vessel or jar:* Container that receives milk from pipeline; source of vacuum from the vacuum pump.
16. *Releaser:* Releases milk from under vacuum and discharges it to atmospheric pressure.
17. *Milk pump:* A high-speed pump that moves milk from the releaser to the milk tank.
18. *Milk tank (bulk tank):* A refrigerated storage tank for on-the-farm storage of milk.

Suggested Milking Procedure.   With the physiology of milk letdown and the mechanics of milking machines in mind, the following milking procedures are recommended:

- *Regular time:* Cows are creatures of habit. It is extremely important that milking times are as exact as possible.
- *Same milker:* Cows will sense a change in milking staff. Milk yield is reduced in most herds on days when the relief milker takes over.
- *Cows primed:* Wash the cow's teats and udder with warm water (120 to 130°F).
- *Massage and dry:* Massage and dry off the udder and teats with individual paper towel for cleanliness and proper stimulation.
- *Strip cup:* Strip two or three squirts of foremilk into a strip cup to check for abnormal milk (which would reveal the possibility of mastitis).
- *Predip:* Predip the teat ends in a predip sanitizer solution. Leave the solution on for 30 seconds and then wipe it off. The teat must be completely dry.
- *Attach cup:* Attach teat cup 1 minute after priming to allow maximum use of the milk letdown reflex. Cow milks in 5 to 6 minutes.
- *Remove cup:* Remove when udder is empty. Most large dairies have automatic detachers. Those that do not should recheck the cow within 3 minutes of attaching the machine to avoid overmilking.
- *Postdip:* Postdip teats using an approved teat dip (of I or Cl) to avoid attracting flies, prevent invasion of bacteria, and help drain the remaining milk at the teat end.
- *Sanitize:* Sanitize the teat cups between cows. Many large dairies have automatic backflush units that sanitize the teat cups between each cow.
- *Feed:* Feed cows immediately after they leave the barn to encourage them to stand for 15 to 20 minutes after milking. This position is especially effective at reducing coliform mastitis.

Mastitis.   **Mastitis** is an inflammation of the udder. It decreases dairy industry profits because of reduced milk production and higher costs of medication, withholding milk from treated cows, and replacing animals culled from the herd because of infection and low production. Milk losses due to mastitis exceed $1 billion yearly in the United States. Treatment and involuntary culling

Mastitis inflammation of the mammary gland

costs increase annual losses to the U.S. dairy industry to nearly $2 billion per year.[11] Mastitis occurs in either **clinical** or **subclinical** forms. Clinical forms represent about 25 percent of the cases; affected cows are characterized by fever, depression, weakness, loss of appetite, reduced milk production, abnormal milk, and possibly one or more swollen, hardened, or sensitive quarters of the udder. Abnormalities such as cloudiness, flakes, blood, or lumps occur in the milk. Subclinical mastitis represents 75 percent of the cases, but no signs of it are apparent without special testing. There is, however, a loss in milk production. Subclinical cases can develop into clinical ones unless found and treated.

The causes of mastitis are complex and they vary, but a good control program can minimize outbreaks and the severity of the disorder. Three primary areas of control include environment, cow susceptibility, and microorganisms that invade the teat end.

Various testing techniques help determine the type and cause of mastitis and the infection level within a herd or individual cow. The California mastitis test (CMT) is a simple and economical test that detects mastitis while it is still in the subclinical stage. It can tell the dairy farmer about the status of the mastitis problem and indicate what steps need to be taken in preventing further spread. The CMT causes WBCs produced due to the infection to form a gel-like substance. The amount of gel and consistency is in direct proportion to the number of WBCs in the milk. The reaction is scored as negative (no infection), trace, 1, 2, or 3 (definite mastitis present).

A DHI somatic cell count (SCC) taken once a month can reliably identify cows with subclinical mastitis. There is a strong relationship between SCC and milk yield. As SCC goes up, milk production drops. Somatic cells are body cells, many of which are sloughed off from the lining of the mammary gland during the natural loss of secretory cells. Most somatic cells are WBCs, which move into the udder to fight off bacterial infections.

Although many of our nation's herds exceed SCCs of 700,000 per milliliter, well-managed herds can attain SCCs of less than 100,000 per milliliter. This substantial decrease in SCCs can increase milk yield by more than 3,200 pounds per 305-day lactation per cow.

Linear score, another term used to indicate SCC levels or mastitis severity, is also being used by the DHI associations. The linear scores range from 0 to 10. For each increase in linear score (e.g., from 2 to 3) the number of somatic cells doubles. The USDA now ranks sires based on their daughters' linear scores. There is a 25 percent heritability on linear scores.

Control of mastitis centers around keeping the disease from spreading and treating present infections. Anything that causes injury to the udder, such as improper milking procedures, improper functioning of milking machines, dragging the udder through mud, and so on, or any contact the udder has with unsanitary conditions increases mastitis in the herd. Thus caution during the milking procedure, maintenance of the milking machine, and especially care on the part of the milker are important in mastitis control.

Several companies have recently manufactured vaccines to aid in the control of specific types of bacterially caused mastitis. These vaccines have proven

---

[11]University of Nebraska–Lincoln, NebGuide, 1999. Basic Principles of Mastitis Control NebGuide G95-1253A. Published by Cooperative Extension, Institute of Agriculture and Natural Resources, University of Nebraska-Lincoln.

TABLE 5–20	Common Types of Mastitis-Causing Pathogens	
Type of Bacteria	**Source**	**Means of Spread**
Streptococcus agalactiae	Infected udders	Cow to cow by contaminated udder wash rags, teat cups, hands, and so forth
Staphylococcus aureus	Infected udders, skin	Cow to cow by contaminated udder wash rags, teat cups and hands and by milking wet cows
Coliforms	Manure, bedding	Environment to cow by wet, dirty lots and bedding, milking wet teats, and poor udder preparation
Mycoplasma spp.	Infected cows, udder shedders, nasal and possibly vaginal secretions in calves	Infusion procedures in dry and lactating cows; cow to cow spread via milking machines and hands

Source: Modified from K. Jacobsen, *Bovine Veterinarian*, How to Identify Mastitis Pathogens, p. 6. September 1998.

to be beneficial in reducing the incidence of some forms of *Escherichia coli* (*E. coli* J-5 bacterin) and some *Staphylococcus aureus* infections. The primary mastitis-causing organisms are *S. aureus* and *Streptococcus agalactiae*. Of these two *S. aureus* is much more difficult to eradicate from a herd. *S. agalactiae* can successfully be eradicated by treating all cows as they go dry (stop milking) and by proper sanitation (including teat dipping and sanitizing machines between cows) during the cow's lactation. *S. aureus* occasionally walls itself off within the udder, making it difficult to treat and causing the formation of scar tissue.

Other common mastitis-causing organisms include *Mycoplasma* species (contagious) and *E. coli* (environmental). See table 5–20 for a more complete list. Even though *Mycoplasma mastitis* is not as common as the previously mentioned bacteria, it is highly contagious and very resistant to therapy. Incidence of coliform mastitis can be reduced by feeding cows immediately after milking so that they are forced to stand 15 to 20 minutes while their teat orifices close.

Intramammary Injections.   Of the two general types of organisms causing mastitis, contagious and environment, the environment generally causes the greatest problems in herds averaging below 200,000 SCCs. Environmental mastitis usually takes hold only when there are low numbers of contagious mastitis organisms present. Key points in controlling environmental mastitis include the following[12]:

1. Monitor bulk tank cultures on a regular basis.
2. Keep all lactating and dry cows clean, dry, and comfortable 24 hours per day.
3. Clip the long hairs from the flanks and udders of all lactating cows.
4. Use a minimal amount of water in prepping cows' udders prior to milking. Clean only the teats and dry thoroughly with a single service towel before attaching the machines.
5. Remove the first three to four strips of milk by forestripping each teat at every milking.

[12]Janet Bosch, Milk Quality Specialist, American Breeders Service, DeForest, WI, 53532 Personal Communication, 1998.

6. Apply a properly labeled germicidal predip to each teat and wipe dry after 30 seconds.

7. Avoid liner slips during milking.

8. Dip teats with a proven germicidal barrier dip immediately after machines are removed from the udder.

9. Feed cows immediately after milking to keep them standing while the teat ends close.

10. Keep all equipment in running order; replace rubber parts that crack and harbor bacteria.

11. Treat all quarters of every cow going dry.

12. Control flies.

13. Provide proper ventilation.

Bovine somatotropin and other parameters that affect milk quality and mastitis rates   When dairy cows produce more milk they are susceptible to additional stress. Added stress of any kind can weaken the immune system and increase the incidence of not only mastitis but all other pathogen-causing disorders. High milk yield, in and of itself, however, does not cause mastitis. The lack of proper management needed to handle the increase in milk yield is attributed to increases in udder inflammations.

Bovine somatotropin (BST) increases the growth rate of heifers by 10 percent and stimulates the development of secretory tissue in the mammary gland. Milk cows can produce 10 to 20 percent more milk using only 5 percent more feed. BST-treated cows (according to Michigan State University researchers) give more milk with no more health problems than untreated herd mates.[13] Milk and meat from BST-treated cows present no danger to humans, says a committee report released by the United Nations Food and Agriculture Organization.[14] This international body agrees with statements already made by the U.S. Food and Drug Administration, the World Health Organization, the American Medical Association, and the American Dietetic Association. In the United States 20 to 30 percent of dairy farmers are using BST on 20 to 65 percent of their cows. It is best to use BST on cows recently fresh, healthy, and in a positive energy balance. Profit margins range from 50 to 75 cents per cow per day.[15]

BST works by stimulating the production of IGF-1 by the liver and other organs. IGF-1 causes mammary cells to increase milk production by milk-secreting cells. Milk synthesis increases, in addition to allowing for better feed efficiency.

BST is not the only management tool available to increase milk production. In the early 1980s, researchers at Michigan State and Oregon State Universities reported a 10 percent increase in milk yield when 16-hour daylight was maintained in the milking herd. This increase in yield more than paid for the electricity, power poles, and lights.

Through selective breeding, dairy farmers have been able to not only breed for stronger attached udders, better teat placement, and higher milk yields but also to affect the actual milk components produced by this "new cow." Select-

---

[13] *Dairy Today,* 11(7):16, 1995.

[14] Food and Agriculture Organization of the United Nations Press Release, March 5, 1998.

[15] M. Hutjens, University of Illinois, Dairy Discussion List, June 16, 1997. Dairy–L @ UMDD.UMD.EDU

TABLE 5–21	Variations in Milk Composition over Time		
Components	**1929[a]**	**1992[b]**	**Difference (%)**
Fat (%)	4.5	3.6	−20
Protein (%)	3.8	3.2	−15.8
Lactose (%)	4.9	4.7	−4.1
Ash (%)	0.72	0.72	0.00
Total Solids (%)	13.9	12.2	−12.2

[a]Overman et al., "Studies of the Composition of Milk," *Illinois Exp. Sta. Bul.* 325, 1929.
[b]Bachman, *Large Dairy Herd Management* (1992).

ing for cows that produce more milk in a shorter period of time has the potential of increasing mastitis rates (larger teat orifice is more prone to bacterial contamination). Selecting for higher milk levels, higher protein levels, and lower fat percentages (what consumers want) is difficult because of the high correlation among fat, protein, and milk yield. However, as table 5–21 demonstrates, milk composition has changed significantly during the past century. Most evidence points to a change in dairy cattle breeds, from a very diverse mixture to one dominated primarily by the lower-fat-producing Holstein cow.

HERD HEALTH DISORDERS

Animal caretakers—whether they are herdsmen, milkers, feeders, or laborers— need to recognize basic health parameters of the animals in their care. Although it is the job of the veterinarian to make a diagnosis and to prescribe a treatment, the caretaker must recognize when something is abnormal and be able to work with the veterinarian in preventing, treating, and recognizing signs of illnesses. The normal rectal temperature of cattle should fall within the range of 100.4 to 102.8°F. Most dairy cows maintain a 101.5°F temperature. On very warm days, it is not uncommon for their body temperature to rise a degree and still be considered normal. Temperature, pulse, and respiration (TPR) are some of the first parameters evaluated before proceeding to specific diagnosis. Nonmetabolic dairy cattle diseases are described in the following sections. Students should refer to *The Merck Veterinary Manual* (8th edition), or *Diseases of Dairy Cattle,* by W. C. Rebhun (Baltimore, Md.: Williams & Wilkins), for more detailed information. Diseases that affect beef cattle are also included in this section.

Anaplasmosis

Anaplasmosis is a protozoan disease of cattle produced by a destruction of red blood cells that causes anemia and death. The most common signs of anaplasmosis do not develop quickly. Body temperature rises slowly, there is poor appetite, and weight loss progresses. Anemic victims show a slight jaundice or yellowing of the pink membranes. This anemia is one of the main clinical diagnostic characteristics of anaplasmosis. Progression to recovery or death may take from 2 days to 2 weeks. Only mild discomfort and few deaths occur in calves; yearlings appear sicker but usually recover. The most severely affected are cattle 2 years of age or older, exhibiting 20 to 50 percent death losses. Abortion is the common result of pregnant cows infected with anaplasmosis. All affected cattle appear hyperexcitable just before they die, and many attack attendants. Cattle that recover will experience symptoms of anemia on and off for several months thereafter.

The worst form, acute or peracute, is deadly; it is characterized by high temperature, anemia, difficult calving, and death within 24 hours. Anemia makes the heart beat so hard that a "jugular pulse" can be seen at the jugular vein. A microscopic organism, *Anaplasma marginale,* is the cause of the disorder. These small protozoa attach themselves to red blood cells, causing them to rupture. Oxygen cannot be carried through the bloodstream, and death occurs through internal suffocation.

The disease is transmitted by insect vectors or unsanitary needles and dehorning instruments. These instruments should be disinfected between use on each animal, and insects should be controlled through the use of back rubbers and spray to keep down infestation of ticks, flies, and so on.

A curious and distressing situation sometimes arises the year after an outbreak of anaplasmosis. Some immunized cows that were protected against anaplasmosis have healthy calves that die suddenly within days of birth, yet the cow is not affected. The problem (neonatal isoerythrolysis) has been defined as a genetic rarity. Anaplasmosis vaccine is made from the red blood cells of donor animals. The problem starts if there is a factor in the genetic makeup of the vaccinated cow that causes it to produce antibodies to those red blood cells in the vaccine. At birth, those antibodies are found in the colostrum milk. If the calf is genetically sensitive to those antibodies when it nurses, the antibodies that are ingested destroy the red blood cells, causing jaundice and death. Boosters should not be administered during late gestation.[16]

The problem almost always appears to affect the most vigorous and healthy calves because they nurse so rapidly after birth and consume the antibodies at the most potent level. In beef cattle, when the cow is milked completely of colostrum milk soon after birth and before the calf nurses, the level of antibodies will be sufficiently reduced and will create no reaction.

In vaccinated cattle, a mortality rate as high as 30 percent might be anticipated. Because the condition involves genetics, changing bulls after an outbreak of anaplasmosis and vaccination has been somewhat effective because of dilution of the genetic makeup that apparently lowers sensitivity.

The condition appears worse in Charolais cattle due to closer inbreeding practices that result from there being fewer animals available than in other breeds. However, the situation has occurred in all breeds and in crossbreeds.

Veterinarians have prescribed treatment for anaplasmosis. Broad-spectrum antibiotics such as tetracycline (20 milligrams per kilogram of body weight administered intramuscularly four times at 3-day intervals) can eliminate the infection. For more valuable animals, blood transfusions may be recommended. In beef cattle herds, tetracycline has also been added to the feed for 30 days, an alternative treatment that has proved effective. Acaricides and fly control measures should be used to reduce vector populations.

---

[16]S. D. Lincoln, "Infectious Causes of Hemolytic Anemia—Anaplasmosis," in *Large Animal Internal Medicine*, ed. B. P. Smith (St. Louis: Mosby, 1990).

Anthrax

Anthrax is a disease that affects the entire body (septicemia), most often producing sudden death. Cattle appear to drop dead for no apparent reason. The key sign of anthrax is a carcass that bloats very quickly and discharges blood the color of tar from the rectum, nose, and other body openings. Rigor mortis does not set in as quickly as normally expected, blood fails to clot, and the animal assumes a "sawhorse on its side" appearance because of the characteristic bloat (see figure 5–20). Extreme care should be taken if anthrax is suspected as the cause of death because it is extremely contagious to both people and animals. Blood samples, taken with extreme caution, can be submitted to a qualified laboratory for culture and identification under a microscope to verify whether the disease is present. If anthrax is confirmed or even suspected, the carcass and any contaminated bedding should be quickly buried at least 6 feet deep and topped with adequate quicklime before complete burial. An alternate disposal method is complete cremation to kill all spores.

The peracute (very fast) form of the disease is most common. Only about 1 to 2 hours elapse from infection to death. Someone observing the animal at the point of infection and watching through progression (which is very unlikely) would see muscle tremors, difficult breathing, total collapse, and convulsion ending in death. The previously mentioned discharges at the body openings and bloat begin to develop almost immediately.

The acute form, running its course in about 48 hours, is exhibited by obvious depression, a reluctance to move alternating with periods of hyperexcitability, a temperature of about 107°F, rapid breathing, congested membranes (possibly bleeding), lack of appetite, and abortion by pregnant cows. Milk production declines to almost nothing, and the little milk produced may be blood tinged or deep yellow in color. Swelling in the throat and tongue is common. Death is expected to be 100 percent in the peracute form and about 90 percent in the acute form, even with treatment.

FIGURE 5–20
Typical appearance of the carcass of an animal that has died of anthrax. Note the bloated condition, which occurs soon after death due to rapid decomposition. Courtesy of the U.S. Department of Agriculture.

Anthrax is often confused with less contagious conditions. Amateurs may misdiagnose red water, lightning strikes, several types of poisoning, and acute bloat as anthrax. However, extreme caution should be used in any case because of the possibility of the highly contagious anthrax, which affects both people and animals.

Bacillus anthracis is the microorganism that causes anthrax. On exposure to the air the organism forms a spore called anthrax bacillus. It has been known to live in the soil in a viable condition for more than 60 years. A preventive program of annual vaccination is good protection against the disease; however, it is recommended only in areas where the disease is a problem.

The most common method of transmission is associated with rough, stemmy feeds that have been grown in areas where anthrax is known to exist. By puncturing membranes of the mouth or digestive system, the organism may be eaten or inhaled or gain entrance to the body through these injuries. Absorption of the spore through these minor wounds induces onset of the disease.

Treatment is almost without response, although antibiotics such as penicillin and antiserum are prescribed in the very early stages of the disease. Because the disease is almost impossible to detect in the very early stages, the only real defense is prevention through vaccination.

Blackleg

True blackleg is caused by *Clostridium chauvoei*. Other blackleg-type diseases caused by the Clostridial organism are malignant edema and enterotoxemia.

Signs of blackleg are inflamed muscles, severe toxemia (poisoning), and a death rate approaching 100 percent. Strangely, the youngest, healthiest, fastest-growing animals between 6 months and 2 years of age are usually first affected. The highest rate of outbreaks is in the warmer months of spring and autumn.

The first sign of blackleg is usually a dead calf. High fatality rates within 24 hours of onset of signs are common. Because it occurs so quickly, signs are seldom seen; however, close observation of younger cattle, once an outbreak is known to have occurred, should reveal some telltale signs of infection during the first 24 hours. These signs are obvious lameness, a swelling of upper parts of the leg, and a slight dragging of the hind toe as if the calf were showing signs of exhaustion. Hot, painful swelling is observable; later the same swelling may be cold and painless. Depressed appetite and temperatures of 105 to 106°F are characteristic.

The upper part of the leg often produces discoloration and a gassy swelling under the skin. When touched, these swellings have a dried, crackling, tissue-paper feeling. Occasionally, lesions are seen at the base of the tongue, heart muscle, diaphragm, brisket, or udder. Blackleg is contagious and may be spread by direct contact from one animal to another. The only reliable means of control is by vaccination. All calves 4 to 6 months of age should be vaccinated; this guideline may be modified somewhat on the advice of a local veterinarian. The most common vaccine is a blackleg–malignant edema bacterin that produces a high degree of immunity in 10 to 12 days and lasts for 9 to 12 months or longer. Treatment is usually ineffective once calves are infected, but a few cases may respond to tetracycline or penicillin.

Brucellosis (Bang's Disease)

Abortion after the fifth month of pregnancy is the key sign of brucellosis. The cow may conceive afterward but usually has poor or irregular calving records. Occasionally the cow will carry the fetus to full term, but the calf usually dies and the cow has a retained placenta, along with metritis. In other instances, a cow or heifer may abort once or twice, then calve normally for the rest of its life. Although the cow or heifer is immune to the disease, it remains a spreader of the condition.

Bulls are affected by the same disease through the scrotum, which appears swollen and reddish. Known as orchitis, brucellosis infections can result in sterility in bulls. In both cases, the microscopic bacterial organism *Brucella abortus* is the cause.

The U.S. government recently developed a new vaccine (RB51) that has shown greater efficacy than the previously used Strain 19 in eradicating brucellosis. As of March 31, 1999, only six herds were infected in the entire United States. Forty-five states have now been certified brucellosis free; Florida, Louisiana, Texas, Missouri, and Kansas still struggle with the disease. Even with all of this success, farmers must be very careful in monitoring and vaccinating in the United States. Mexico has a high percentage of human brucellosis cases due to *B. melitensis* (infection in goats) and still has sizable problems with cattle brucellosis as well.[17]

Brucellosis in people is called undulant fever. It is spread through contact with unpasteurized milk. It can be spread to animals through direct contact with infected cattle or contaminated grass, ground, or water. A very common method of transmission is through other cows licking the calf of an abortive cow.

No known treatments exist. Thus government regulations have been enforced to try to eradicate the disease through a search and destroy mission. Strict enforcement of a cleanup program has reduced the incidence of brucellosis from 50 percent of the cattle herds in the United States in the 1930s to only six herds total in 1999. Complete eradication is predicted within 1 to 2 years.

Calf Diphtheria (Necrotic Laryngitis)

Diphtheria is a disease of suckling calves. Drooling and the appearance of infected patches of dead, yellow tissue at the edges of the tongue are the most obvious symptoms. The same soil organism that causes foot rot also causes calf diphtheria (*Fusobacterium necrophorum*). It is thought that the calves' sharp teeth injure the soft tissue in the oral cavity. The infection spreads among calves fed from common utensils or those kept in close contact with each other. Death may occur if treatment is delayed.

Treatment consists of penicillin (22,000 units per kilogram, administered intramuscularly, twice daily) alone or in combination with sulfa drugs (143 milligrams per kilogram the first day, followed by 70 milligrams per kilogram daily) for 7 to 14 days. A tracheostomy may be required for calves that have severe dyspnea.

---

[17]J. L. Alley, Report of the U.S. Animal Health Association Committee on Brucellosis, U.S. Animal Health Association, 1999.

FIGURE 5–21   This steer shows excess saliva flowing from its mouth, one of the symptoms of foot-and-mouth disease. Courtesy of the U.S. Department of Agriculture.

FIGURE 5–22
One of the cows in a Mexican dairy herd dead from foot-and-mouth disease. Courtesy of the U.S. Department of Agriculture.

Foot-and-Mouth Disease

Foot-and-mouth disease is a dreaded disease around the world. The first signs are watery blisters in the mouth (see figure 5–21) and between the claws of the hooves. The udder may also be affected, and high temperatures are common. Mortality rates are low, but the economic losses are catastrophic because of the strict quarantine and eradication programs carried out by almost all governmental agencies of developed countries (see figure 5–22). Foot-and-mouth disease is feared because it is so highly contagious. However, the United States has not had an outbreak since 1929. Caused by a virus, foot-and-mouth disease has no cure. Most governmental agencies rely on the quarantine of imported cattle until they are certain no animal is infected or a carrier. Because of the indistinguishable symptoms of vesicular stomatitis and foot-and-mouth disease, this disorder has resurfaced as a major concern (see the later section "Vesicular Stomatitis").

Foot Rot (Interdigital Phlegmon)

Foot rot is common in cattle in feedlots, corrals, or muddy, confined areas. The skin between the toes becomes red and swollen, sometimes breaking open (see figure 5–23). Hooves may become deformed, and lameness can develop.

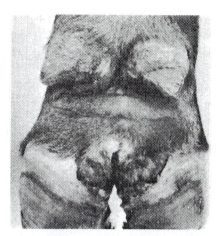

FIGURE 5–23 Foot of a steer showing foot rot. Courtesy of the U.S. Department of Agriculture.

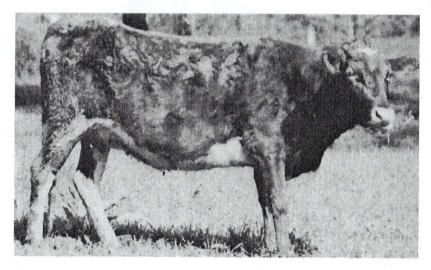

FIGURE 5–24 Dairy heifer with Johne's disease. Courtesy of the U.S. Department of Agriculture.

A soil organism (*Fusobacterium necrophorum*) is the major cause of this inconvenient disease. Other organisms, such as *Bacteroides melaninogenicus, Escherichia coli,* and *Staphylococcus aureus,* may be involved synergistically. Losses are caused by failure of cattle to eat because of pain and slight fever, as well as from lowered milk production and lameness. The major preventive measure is to keep cows out of muddy barnyards, swamps, and stagnant ponds. Disinfectant footbaths have helped in some herds. Treatment by trimming the affected parts of the hooves and walking animals through a suitable disinfectant (such as iodine solution) is also effective. Copper sulfate–sulfanilamide (1:4) powder and Kopertox[18] solutions have also proven effective in treating this disorder.

Johne's Disease (Paratuberculosis, Chronic Specific Enteritis)

Johne's disease is an infectious, incurable disease characterized by reappearing diarrhea and gradual loss of condition (see figure 5–24). Refusal to eat and eventual death may occur. Johne's disease produces one of the most serious forms of

[18]Kopertox, Fort Dodge Laboratories, Inc., Fort Dodge, Iowa.

diarrhea. This infectious intestinal disorder creates emaciation, thickening or folding of the intestinal wall, a condition of edema characterized by bottle jaw, a very offensive odor of the diarrhea but no blood, and a generally weakened condition. The disease may recur at intervals throughout the life of the animal. It is slow to develop; most animals affected are 2 to 6 years of age before they develop signs. The tuberculosis-like disease is caused by the small bacterial organism *Mycobacterium paratuberculosis*. Although vaccines are available, they are not normally recommended and treatment is not very successful. Culling of the affected animal is the most practical recommendation.

Johne's disease affects cattle, sheep, goats, deer, antelope, wild rabbits, and llamas. Approximately 1.4 percent of U.S. beef and 2.6 percent of U.S. dairy cattle are infected with *M. paratuberculosis*.[19]

It is most commonly transmitted via ingestion of feces by young cattle, but it can also be transmitted in semen, milk, and across the placenta. This disorder is best controlled by proper sanitation, testing, and culling. Calfhood vaccination is controversial in that it can reduce incidence but does not really eliminate the infection. Conflicting data hints that *M. paratuberculosis* may be the causative organism in human Crohn's disease. Hence, because of this zoonotic possibility, caution should be used with animals identified with Johne's disease and humans.

Two types of tests have been used to identify cattle suffering from Johne's disease. The *intradermal test* is used to detect infected animals in the herd before an outbreak occurs. The *serological test* is used once an outbreak occurs to identify the types of organisms that caused the problem. Neither of these tests has been able to identify subclinical cases with much accuracy. Fecal culture is much more specific, but it takes weeks to run the test.

Leptospirosis

The kidney is affected in leptospirosis, but in such a variety of ways that the following conditions may occur: abortion (see figure 5–25), mastitis (blood-tinged milk coming from a noninflamed udder), high temperature, jaundice (yellowing of the pink membranes, especially around the eye, vulva, penis, and gum),

FIGURE 5–25 A cow and her aborted fetus, caused by Leptospirosis. Courtesy of the U.S. Department of Agriculture.

[19]M. T. Collins, Johne's Information Center, University of Wisconsin–Madison School of Veterinary Medicine, April 1998.

wine-colored urine (red water disease of cattle), and anemia. Very young animals commonly die; older cattle have a death rate of only 5 percent. Necropsies reveal gray to white lesions on the liver. Recovered animals appear dull and listless. Definite diagnosis is made only through laboratory culture.

The cause of leptospirosis is microscopic bacteria. There are more than 200 serovars or strains, but the three that most commonly infect cattle are *Leptospira hardjo, L. pomona,* and *L. grippotyphosa. L. canicola* and *L. icterohaemorrhagiae* are also found in cattle. Icterus, anemia, and hemoglobinuria are common symptoms of these five serovars. Leptospirosis is a contagious disease that affects all farm and companion animals and is zoonotic. Infection is spread by contact with urine or urine-contaminated feed or water. The microscopic agglutination test (MAT) is the most commonly used serologic test for the diagnosis of leptospirosis. This test measures antibodies to leptospirosis that usually appear within 12 days of initial infection. Leptospirosis is prevented by annual vaccinations of all serovars affecting the herd, confinement rearing, and management methods to reduce the transmission from rats and wildlife.

Lumpy Jaw (Actinomycosis) and Wooden Tongue (Actinobacillosis)

A very hard swelling of the bony tissue about the head, usually the jaw, is characteristic of *actinomycosis.* It may also affect the tissues of the throat area, causing wooden tongue. An infection of the bone develops gradually, and by the time the swelling is noticed, it is difficult to treat. Swelling may break through the skin, discharging pus and a sticky, honeylike fluid containing little white granules. This eruption actually forms a pathway for drainage, spreading the organism into the surrounding area.

Inability to eat, excessive salivation, and a chewing motion of the tongue characterize wooden tongue. The hard, swollen tongue may have ulcers along the edge, and movement of the tongue from side to side by the observer causes obvious pain. The tongue later becomes shrunken and immobile. Lymph nodes may swell, rupture, and exude pus.

Lumpy jaw is caused by the microorganism *Actinomycosis bovis.* Wooden tongue is caused by *Actinobacillus lignieresii.* Prevention is the key to controlling both these disorders. Rough, course feeds or other objects that could injure the lining of cows' mouths when they chew should be eliminated. Infected animals should be removed from the herd.

Transmission of the organism is through contamination and injection into the body caused by eating sharp objects such as rough hay or feeds containing awns, trash, or wire. Treatment consists of the use of iodides for both actinomycosis and actinobacillosis (lumpy jaw and wooden tongue). An intravenous 20 percent solution of sodium iodide was the common treatment in the past but is no longer recommended because of food safety concerns. Treatment is rarely successful in chronic cases. When antibiotics such as sulfa drugs and penicillin stop the progression of the disease the disfigurement usually remains.

Mad Cow Disease (Bovine Spongiform Encephalopathy)

Signs of mad cow disease, a transmissible spongiform encephalopathy, are primarily neurological in nature and end in the death of the cow. The disease was first diagnosed in Kent, Great Britain, in 1986 and is thought to be caused by the scrapie prion or slow virus. There is no known treatment. Public health concerns have concentrated on two human diseases with similar symptoms, Creutzfeldt-Jakob disease (CJD) and kuru, which have also been blamed on a slow virus or prion. There is no evidence, however, that the transmissible spongiform encephalopathies of humans are acquired from animals.

The feeding of meat and bonemeal contaminated with the scrapie virus is thought be a major source of transmission to cattle. In 1988, Great Britain banned the use of ruminant-derived protein (offal) in ruminant rations as a means of prevention. On August 4, 1997, the United States similarly banned the feeding of beef and lamb by-products to other ruminants. In addition, the United States, in 1989, banned the importation of live cattle and zoo ruminants into the United States from the United Kingdom and in 1997 from the Netherlands (where one case was reported). The British government has further banned human consumption of certain bovine by-products as an additional safety precaution. Symptoms of mad cow disease include kicking, aggression, and incoordination leading to excessive salivation, severe incoordination, frenzy, and death.

Diagnosis includes histologic examination of the hindbrain. There is no test during incubation. Brain samples are sent to the USDA's National Veterinary Services Laboratory, Ames, Iowa. Treatment has proven ineffectual. As of December 1994 there were 134,202 confirmed cases in Great Britain affecting 52.1 percent of dairy herds and 13.7 percent of beef herds. Incidence is declining. As of April 1996 Great Britain had already destroyed 165,000 cows as a part of its control method.

Malignant Edema

Malignant edema is also called gas gangrene (a similar wound-derived infection in people is common). Signs are similar to those of blackleg, except this disease affects cattle of any age and it usually occurs after an injury. High fever, swellings that make a crackling sound when touched, and production of a thin, reddish fluid from inflamed areas are major signs of this disorder. Gaseous air bubbles under the skin cause swelling around the area of a wound. Body temperature is normally elevated. If the disease progresses throughout the body, there will be a depression followed by death.

The cause is the bacterium *Clostridium septicum*. Entrance by this organism is gained through cuts and scratches. Difficult births, vaccinations, castrations, navel infections, surgical procedures, and parasitic infestations all allow the organism entrance. Sanitary conditions are a must because the organism is soilborne and tends to build up in environments with heavy concentrations of cattle. Treatment consists of massive doses of tetracycline and penicillin both systemically and around the wound. There is a vaccine (see the earlier section "Blackleg") to effectively prevent the occurrence of malignant edema. Calves should be vaccinated as early as 2 months of age (two doses 2 to 3 weeks apart are recommended in high-risk areas).

Pinkeye (Infectious Keratitis, Infectious Conjunctivitis) or Infectious Bovine Keratoconjunctivitis)

General signs of pinkeye are a reddening of the membranes of the eye, excessive tears, and pain when exposed to light. A milky film may cover the eyeballs.

Temporary or permanent blindness may result. In severe cases, the eyeball may be lost. The course of the infection may run from 4 to 8 weeks or longer.

Pinkeye rarely causes death in cattle but is extremely serious because of weight loss and decreased milk production. This disease appears suddenly, spreads rapidly, and is almost always associated with large numbers of flies.

The most frequently involved causes in cattle are the bacterium *Morexella bovis* plus the irritation brought on by bright sunlight, dust, wind, flies, or weed seed. In addition, the infectious bovine rhinotracheitis (IBR) virus (involved mostly in the fall and winter rather than the customary summer incidents); ultraviolet light; an overdose of phenothiazine, a worming compound that creates sensitivity to light; and marginal amounts of vitamin A are also common causes. Ideal treatment includes frequent applications of topical antibiotics, daily subconjunctival antibiotic injections, and topical atropine (to maintain cycloplegia). Dust bags and insecticide tags can be used to reduce the numbers of face flies, which spread this disorder. Bacterins have also been proven effective at reducing the incidence and severity of *M. bovis* in the herd.

Pneumonia

Signs of pneumonia include wide leg stance, wheezing sounds from chest and lungs, nasal discharge, extension of tongue, and labored breathing. The cause is usually a virus condition brought on by exposure to cold. Water vapors inhaled into the lungs during improper drenching and inhalation of chemicals or dust are also common causes. Drenching with the head held high rather than the normal, level condition should be especially guarded against. Treatment with antibiotics is a standard practice to prevent secondary infections. Broad-spectrum antibiotics, such as penicillin and tetracycline, are often recommended, along with isolation and prevention of undue stress. Steroid drugs may be recommended if the problem was precipitated by chemical or dust inhalation. The death rate is usually 10 to 20 percent of affected animals if treated and as high as 75 percent if untreated.

Red Water Disease (Bacillary Hemoglobinuria)

Red water disease is an acute, highly fatal disease. It usually occurs in springtime and is characterized by high fever and a breakdown of hemoglobin in the blood, which travels through the kidneys to cause port wine–colored, blood-tinged urine (hemoglobinuria). The blood-tinged urine is the key sign to watch for in this disease. In addition, jaundice (a yellow coloring of the pink membranes) is usually seen around the eye and vulva (in females) or prepuce (in males).

This disease is rapidly fatal (usually within 12 hours from onset of signs), with a nearly 100 percent mortality rate. Signs of the disease are labored breathing and grunting, an arched back indicating abdominal pain, a weak and rapid pulse, temperature elevation, and a swollen brisket. The manure of an affected animal is very dark, and the urine, as previously mentioned, is wine colored. Because the disease is so rapidly fatal, signs are usually not observed until one or more animals have died from the disease and closer observation is necessary. The cause is the bacterium *Clostridium hemolytica.*

The disease is most commonly spread by flooding from an infected area to a clean one. It is also thought to be associated with liver flukes, which, along with the disease itself, are known to be spread by snails. Treatment consists of very high doses of tetracycline with very high levels of dextrose and electrolyte solutions given intravenously. Even then, the outcome of the treatment is doubtful. Prevention is the key to eliminating losses from this disease. A red water vaccine is available for herds in areas of high risk. Vaccination is recommended every 6 months; if only one vaccination per year is desired, it should be done in the spring when the disease is most prevalent. Adequate drainage of pasture to control flooding and snails, thought to be spreaders, is another method of control.

Shipping Fever (Hemorrhagic Septicemia)

The first signs of shipping fever usually occur after an animal has undergone the stress of castration, vaccination, dehorning, weaning, working, being chilled or wet, or moving a considerable distance to a new location. A sudden onset of high temperatures (104 to 106°F), depression, going off feed, difficult breathing, coughing, runny nose and eyes, unthrifty appearance, and diarrhea characterize the disease (see figure 5–26). Affected animals may die within 3 weeks or may recover but seldom do well in a feedlot situation.

Although stress is the precipitating factor, the disease is most often caused by the *Pasteurella multocida* organism. This complex disorder may be brought on by a multitude of other organisms. Rough handling and shipping long distances are usually associated with the susceptibility of the animal to bacterial infection.

Treatment consists of high levels of antibiotics such as penicillin, streptomycin, tetracycline, and Tylosin, combined with steroid therapy. Vitamin injections may also be used as a supportive therapy. Isolation of the sick animals in a warm, dry area and treatment as if they had pneumonia are often recommended. Because stress is known to be the major contributing factor to this disease, this part of the complex disorder is reduced by not doing all the jobs

FIGURE 5–26 A calf infected with shipping fever (hemorrhagic septicemia). Courtesy of the U.S. Department of Agriculture.

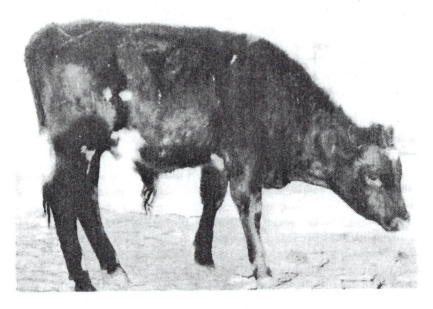

of weaning, dehorning, castration, and so on at one time. Cattle are worked moderately and gently through a series of jobs over a period of time. The shock to the system is apparently reduced in this way, allowing the body's natural systems to resist the organisms that try to gain a foothold. Recently there have been reported outbreaks in Yellowstone's bison, as well as in cattle herds in Asia, Africa, and the Middle East.

Trichomoniasis

Abortion in heifers or cows at very early stages (2 to 4 months) of gestation is the key sign of trichomoniasis, and an infected uterus often follows abortion to keep cows from coming back into heat. This is a true venereal disease of cattle caused by the protozoal microorganism *Trichomonas fetus.* The organism is passed from cow to cow by bulls that are infected with it. An inflamed prepuce on the bull is exhibited in fully developed stages of the disease. Although the disease may be transmitted by artificial insemination, commercial semen is usually treated with antibiotics and poses no threat of infection.

Prevention includes the use of artificial insemination with semen from bulls that are certified by Certified Semen Services (CSS). CSS health requirements eliminate infected bulls from breeding through testing, sanitation, and addition of antibiotics to semen. Treatment consists of eliminating infected and carrier animals. Valuable animals have been treated with three injections of ipronidazole at 24-hour intervals along with long-acting tetracycline.

Tuberculosis

Tuberculosis, although rare in U.S. dairy cattle, is a potential cause of respiratory distress. The causative organism is *Mycobacterium bovis.* This organism also causes tuberculosis in most warm-blooded vertebrates, including people. The introduction of milk pasteurization was a major step in the fight against human tuberculosis. Regulatory control efforts have also wiped out the once real threat of tuberculosis among cattle in the United States, but this disease continues to be a potential threat to humans and animals in many developing nations.

In the United States, the primary means of control is through test and slaughter of reactors to the tuberculin test. Although the chief reservoirs of infection are cattle and people, badgers, bison, opossums, kudu, deer, llamas, elk, and both domestic and feral pigs continue to be a source of recontamination to cattle and humans.

Vesicular Stomatitis

Vesicular stomatitis virus (VSV) is a viral disorder that causes lesions indistinguishable from those of foot-and-mouth disease in cattle and pigs. Horses are also susceptible to VSV. VSV is endemic in parts of South and Central America and in parts of the southern United States. VSV was a suspected foreign animal disease in a horse reported to the USDA on May 1, 1995, in New Mexico. Symptoms include blisters and ulcers on the tongue, gums, and lips; fever; sores on teats and feet; and excessive salivation. Milk production drops by 50 percent. The incubation period is 2 to 8 days. Insect vectors spread infection to cattle, pigs, sheep, goats, bobcats, raccoons, monkeys, rodents, and deer. Economic losses to dairy farmers can be catastrophic. VSV is a reportable disease; state and federal regulatory veterinarians should be contacted for help in diagnosis and in ruling out foot-and-mouth disease. Positive diagnosis is made by enzyme-linked immunosorbent assay (ELISA). There is no specific treatment for this disorder.

Vibriosis (Bovine Genital Campylobacteriosis)

Campylobacteriosis, formerly called vibriosis in cattle, is caused by one of two microaerobic inhabitants of the gastrointestinal tract of poultry, dogs, cats, sheep, and cattle. In humans this microbe causes systemic disease (e.g., fever, gastroenteritis, meningitis, endocarditis, and arthritis). It is most commonly spread by contaminated water or milk.

In cattle, *Campylobacter fetus venerealis* is transmitted by ingestion and spread through the blood to the placenta, where it causes sporadic abortion. *Campylobacter fetus fetus* (formally *C. fetus intestinalis*) is transmitted venereally. Diagnosis can be made by use of the vaginal mucus agglutination test (VMAT). The use of artificial insemination and vaccination of the herd are primary methods of control. The vaccine should be administered 4 weeks before the breeding season.

Warts (Fibropapillomas)

Warts are the most common tumors found in dairy cattle. Animals between the ages of 6 months and 2 years are at the greatest risk of infection. Previous exposure to the virus causing warts seems to provide a degree of immunity. Vaccination can help prevent warts, and administration of two doses of commercial vaccine is recommended at 2 to 3 months of age.[20] In cattle, warts are more commonly found on the head, neck, and shoulders. Occasionally, warts may involve the venereal regions, where they cause great pain and can lead to secondary infection.

Hairy Foot Warts (Papillomatous Digital Dermatitis)

Hairy foot warts, a relatively recent disorder that affects large numbers of dairy herds across the country, is highly contagious. (figures 5–27, 5–28.) The use of oxytetracycline and lincomycin has proven to be effective in treating the disorder. A more recently developed Hygieia's Serpens species bacterin[21] is also available as a preventive measure. Topical sprays, containing copper sulfate, antibiotics, and a surfactant product[22] have been marketed as a preventive and treatment regime in the United States. The causative agent of hairy foot wart is a bacterium and not a virus.

---

[20]Dr. Erich Studer, Carnation Farm, Washington, November 10, 1997.

[21]Jim Wallis, Hygieia Biological Laboratories, P.O. Box 8300, Woodland, CA 95776. Telephone, 530-661-1442; toll free, 888-494-4342 (888-HYGIEIA); fax, 530-661-1633; e-mail, Hygieia @ compuserve.com.

[22]Victory by Surge.

FIGURE 5-27 This cow had extreme pain to direct touch but was still weight bearing. Many cows seem to accommodate the pain by walking more on their toes, as seen here, resulting in clubfootedness. Courtesy of Hygieia Biological Laboratories.

FIGURE 5-28 It is easy to see from where the name *hairy footwart* came. This lesion is quite mature, and because the disease is being recognized more quickly it is likely that few warts will progress to this late stage. Lesions are usually described by their appearance. Early lesions are red, moist, and have little or no raised area to the margins. Courtesy of Hygieia Biological Laboratories.

Infectious Bovine Rhinotracheitis, Bovine Herpesvirus I, (BHV-1), or Red Nose

Infectious bovine rhinotracheitis (IBR) is a major cause of viral abortion in U.S. dairy herds. It is described as an infection of the upper airway and trachea that may also cause infectious pustular vulvovaginitis, resulting in endemic abortions and neonatal septicemia. Immunity from previous exposure and from vaccination is short-lived; hence, annual vaccinations are required to protect the herd. The clinical signs of IBR include elevated temperatures (105 to 108°F), depression, anorexia, rapid respiration (40 to 80 breaths per minute), heavy serous nasal discharge, a grunting cough, and coarse tracheal rales. Treatment is similar to that for pneumonia and shipping fever in cattle.

Bovine Respiratory Syncytial Virus

A more recent respiratory virus affecting cattle is that identified as bovine respiratory syncytial virus (BRSV). Respiratory distress caused by BRSV was first reported in Europe during the 1970s and has more recently been diagnosed in the United States (1980s) in both dairy and beef cattle.[23] Clinical symptoms include elevated temperature (104 to 108°F), depression, anorexia, cough, nasal and lacrimal discharge, and increased respiratory rate. Treatment primarily consists of antibiotics used to control secondary infections. Reducing stress, correcting dehydration, and other supportive therapy may be necessary.

Bovine Viral Diarrhea

Bovine viral diarrhea virus (BVDV) is one of the most commonly diagnosed viruses in bovine abortion cases. It is also one of the most commonly reported viruses associated with multiple viral infections of the respiratory tracts of calves. Prevention of BVDV includes improved sanitation and the use of inactivated and modified live vaccines. The modified live vaccines should not be used on pregnant or highly stressed cattle. Treatment includes supportive care and the use of antibiotics to prevent or treat secondary infections (including bacterial pneumonia).

Bovine Neospora Abortion

The first documented case of bovine neospora abortion in California was in 1985. By 1997 it had cost the California dairy industry $35 million. The disease is transmitted directly from cow to fetus. In 1995 a new ELISA test was developed for diagnosis of bovine neospora abortion from blood samples. The causative organism has been identified as *Neospora caninum,* a coccidial protozoa that causes abortion in cattle. The same organism infects dogs, goats, sheep, and cats. Symptoms in dairy cattle include abortion, stillbirth, calves born weak or paralyzed, and calves born with nonsuppurative encephalitis. Abortion can occur throughout gestation. As of June 1999, no vaccine had been approved for its prevention.

Cryptosporidiosis

Cryptosporidiosis is an enterocolitis caused by the coccidial parasite *Cryptosporidium parvum.* It is not host specific and thus is capable of infecting not only young calves but also pigs, dogs, cats, horses, and people. It is estimated that more than 90 percent of dairy farms are infested with this parasite. It has been implicated in outbreaks of diarrhea in many mammals, including humans. For example, in April 1993, the city of Milwaukee blamed *C. parvum* for causing thousands to become ill from the ingestion of contaminated city water. *Cryptosporidium* species are found in wildlife feces and infect people who come in contact with them (e.g., by consumption of contaminated water).

---

[23]W. C. Rebhun, *Diseases of Dairy Cattle* (Media, PA: Williams & Wilkins, 1995).

FIGURE 5–29 As previously discussed in chapter 4, fistulated cows serve as a reservoir of large numbers of micro-organisms that can be easily extracted and placed in cows' "off-feed." This Holstein fistulated heifer is one that is typically found in many large dairy and beef operations. Photo by author.

With dairy calves, housing in isolation hutches (rather than group housing) significantly decreases incidence levels. Calves that are 1 to 4 weeks of age are most susceptible. Symptoms include anorexia, diarrhea, and tenesmus. There is no effective specific treatment, just supportive therapy such as rehydration, correction of acidosis, and maintenance of energy requirements.

Disease and parasite prevention is a constant concern throughout the life of most animals. Farmers vaccinate for the more important diseases (previously mentioned) by the use of syringes and nebulizers. Periodic drenching using a drenching gun, spraying, dipping (see figure 5–29) and anthelmintic products helps fight both internal and external parasites. Parasites cost the livestock industry more than $100 million per year.[24]

Six major categories of parasites affect domestic animals: protozoa, nematodes, trematodes, cestodes, acanthocephalans, and arthropods. We do not have space in this text to include all of the parasites making up the aforementioned list. We have attempted to include the most common and economically important ones to cattle producers.

Dairy and beef ranchers generally divide parasites into one of two major groups: those affecting the outside of the body, external parasites (e.g., lice, mange mites, and heel fly), and those primarily affecting the inside of the body, internal parasites (e.g., nematodes, or roundworms, and coccidia [a major protozoa]).

Cattle Grubs and Warbles (Hypodermosis of Cattle)

Heel flies attach their eggs on the hair of the lower legs of cattle. In 3 to 7 days, the eggs hatch and the larvae penetrate the skin and travel through the muscles and connective tissue and along the nerves. They secrete enzymes to help with their movement. They eventually break through the back of the host through cysts, or warbles. Pupae emerge from this stage, and 1 to 3 months later flies emerge to start the cycle all over again. It takes a full year for the cycle to

---

[24]J. E. Strickland, University of Georgia Cooperative Extension Service, December 3, 1997.

complete itself. Treatment has to start with control of the heel fly population. Dust bags help to minimize irritation of the flies. Systemic insecticides have been approved for beef cattle and dairy heifers but should not be given to lactating dairy cows.

Lice

Lice are wingless, flattened insects with legs that are adapted for grasping hair shafts. They are generally host specific. Cattle are most commonly infested with biting lice. Biting lice feed by scraping matter from the dermis and base of the hairs. Immature lice are called *nymphs.* Sucking lice also infect cattle. They have narrow, pointed heads and feed on blood. Both biting and sucking lice are associated with unthrifty appearance, rough hair coat, lowered milk production, pruritus, and reduced weight gains. Heavily infested animals lose clumps of hair as they attempt to rub and relieve the discomfort of louse feeding.

The preferred feeding areas of lice include the neck, back, face, dewlap, and base of the tail. Whole-animal sprays control most adult lice but do not kill nymphs. Many systemic parasiticides provide long-term louse control, but not all are approved for use on lactating dairy cattle. Ivermectin can be used in dairy calves and heifers. Pyrethrin or permethrin preparations (sprays and powders) and coumaphos products are available for use in dairy cattle.

Mange (Barn Itch)

Mange reportable contagious disease caused by mites; also called scabies

Mange is a very contagious disorder caused by mites. Chorioptic mange is the most common to infect dairy cattle and is seen more frequently in winter months. The primary signs are pruritus, violent swishing of the tail, and rubbing of the tail and perineum against stationary objects. Coumaphos is an effective spray against this type of mite. All cattle in the herd, not just those exhibiting signs, should be treated at the same time. Demodectic mites are normal inhabitants of the hair follicles and sebaceous glands of cattle but seldom cause any clinical disease.

Sarcoptic mange causes the disorder known as *scabies.* Scabies is a reportable disease in cattle because it can be transmitted to people. It primarily infects the head, neck, and shoulders. Symptoms include pruritus, papules, alopecia, loss of weight, and a dramatic drop in milk production. Treatment requires a call to a regulatory veterinarian. For lactating cows, heated 2 percent lime sulfur is used as a spray or dip. For nonlactating cows, a 0.3 percent coumaphos dip repeated in 14 days is very effective.

Psoroptic mange is another reportable disease of dairy cattle. This mange has similar symptoms to those of sarcoptic mange. Nontreated cows become weak and emaciated and can die from secondary infections. Treatment involves calling a regulatory veterinarian; the protocols are similar to those of sarcoptic mange.

Ringworm (Dermatophytosis)

Dermatophytosis is very common in dairy calves (ages 2 to 12 months) and occasionally infects adult cattle as well. Grouping of young cattle, especially during winter months, increases the incidence. Ringworm affects the keratinized layers of the skin, causing alopecia and crusting. The fungus can live for extended periods on inanimate objects, increasing the likelihood of spreading the disorder to other animals. The most common causes include the fungi *Trichophyton mentagrophytes* and *T. verrucosum.*

Treatment includes topical sprays or dips, as well as systemic products such as griseofulvin (7.5 to 60 milligrams per kilogram administered orally for 7 or more days). Disinfecting the premises and all fomites (with hypochlorite solution) coming in contact with the cattle (e.g., posts and stanchions) is required to prevent further outbreaks. Attenuated fungal vaccines have shown some promising results in European research. As of June 1999, these products had not been approved in the United States. Ringworm can be spread to people and to cats, so precautionary measures such as wearing gloves and keeping cats away from infected cattle should be practiced.

Cestodiasis

Tapeworms (*Moniezia benedeni*) in cattle are believed to be nonpathogenic; hence, they will not be discussed here.

Coccidiosis

Coccidiosis is of major concern in raising dairy calves, especially in group housing. The major pathological coccidia affecting cattle are *Eimeria bovis* and *E. zuernii*. *E. bovis* costs U.S. dairy farmers $62 million per year. Losses affect 77 million young cattle in United States, killing 80,000 cattle annually.[25] Fecal contamination of water and feed (conditions more common with group housing of calves) allows the ingestion of infective oocysts. Clinical signs of coccidiosis in calves include bloody diarrhea, depression, and anorexia. Tenesmus is also a common symptom. Treatment includes the oral administration of coccidiostats. Amprolium, monensin, lasalocid, and decoquinate are drugs most commonly used to treat at-risk calves.

Nematodes

Nematodes are commonly referred to as *roundworms*. They are a common concern of pastured cattle. The major nematode parasite of cattle is *Ostertagia ostertagi* (also known as the *brown stomach worm*). Clinical signs include variable amounts of diarrhea, weight loss, poor hair coats, anorexia, and anemia. Treatment includes minimizing pasture contamination and worming heifers before placing them on pasture and then again 3 to 8 weeks after turnout. Common anthelmintics approved for nonlactating dairy animals include albendazole, fenbendazole, Ivermectin, Levamisole, oxfendazole, Thiabendazole, and Clorsulon.

Trematodes

Trematodes are flatworms such as liver flukes. Liver flukes infest beef cattle more than dairy and primarily in areas along the Gulf Coast and western United States. Treatment includes albendazole and Clorsulon. Control measures should include reducing exposure of cattle to snails, which act as intermediate hosts.

---

[25]J. U. T. Quigley, *Calf Notes,* December 3, 1997.

FIGURE 5–30
Automated sprinklers
and crowd gates facili-
tate the movement
and preparation
before milking in this
California dairy.
Photo by author.

MISCELLANEOUS DAIRY CATTLE HOUSING

To accommodate dairy animals in all types of weather newer, more flexible housing has been sought by those in the industry. One such example is *Polydome* plastic hutches. Calves stay warmer in the winter and are protected to a greater degree from the winds and rains than in the traditional wooden models.

Automated sprinkler systems and pneumatically controlled crowd gates lessen the labor and time required to wash, dry, and bring in cattle during the milking process (see figure 5–30).

SUMMARY

At the birth of the United States almost every family had a milk cow to provide needed dairy supplies for the family. Today, the dairy industry is highly mechanized and relies on the increasingly scientific innovations of dairy engineering, veterinary medicine, and animal nutrition. Dairies are generally located near dense urban populations, providing consumers with a fresh supply of milk daily. Since 1950, milk production per cow has more than tripled; the industry thus requires fewer cows and less feed and provides more product to the consumer.

Milk's wholesome image has partially been tarnished by the use of BST in many U.S. dairy herds. Although BST use has been proven to be safe to both the cow and consumer the general public's lack of understanding of basic science has contributed to this negative image. Dairy farmers and the industry will have to do a better job in educating the public if they hope to continue to use BST and gain its acceptance.

We have included a fairly complete listing of common bovine disorders and their symptoms, prevention, and treatment in this chapter. Beginning students are not expected to memorize this compilation, but familiarity with common disorders in students' general geographic locations will better enable them to prevent and/or treat the disorders early so as to minimize economic disaster on the farm. A healthy animal is a requirement for both economic viability and humane considerations.

EVALUATION QUESTIONS

1. Four of the leading states in milk production are _____ , _____ , _____ , and _____ .
2. The Ayrshire breed is colored _____ and white.
3. The dairy breed used mostly for veal and milk is _____ .
4. The dairy breed with the highest fat percentage is _____ .
5. The dairy breed with a golden yellow color to its milk is _____ .
6. The oldest breed of dairy cattle is _____ .
7. Milking cows are usually milked _____ days and dried up _____ days before the next lactation.
8. Cows are usually rebred at _____ days after calving.
9. The _____ is used by all dairy breed associations as the standard for judging and classifying cattle.
10. The main points of a dairy cow and their respective points (in judging) are

    a. _____ _____

    b. _____ _____

    c. _____ _____

    d. _____ _____

    e. _____ _____

11. The mammary system of a cow is divided into _____ independent parts.
12. Milk is actually secreted in grapelike structures called _____ .
13. The period of nonlactation between two periods of lactation is called a _____ period.
14. _____ is an association that dairy farmers join to participate in its record-keeping and management plans and is operated jointly by the USDA and state colleges of agriculture of land grant universities.
15. _____ is adjustment of milk with different fat percentages to equivalent amounts on an energy basis.
16. _____ is an infertile female calf born cotwin to a bull.
17. A metabolic disease characterized by excessive ketone body formation is _____ .
18. _____ means naturally hornless.
19. A count of _____ cells helps determine inflammation of the mammary gland.
20. _____ means diarrhea.
21. _____ is the physical conformation of an animal.
22. _____ causes undulant fever in people.
23. _____ is a method of feeding cows aimed at finding their potential to secrete milk and used to prevent metabolic shortage in energy.

24. _____ is the circumference of the body just back of the shoulders of an animal; used to estimate body weight.
25. _____ are age conversion formulas applied to milk production records of young cows to compare their milk yield with that of more mature cows.
26. _____ is called parturient paresis.
27. _____ describes an animal that has a crooked hock, which causes the lower part of the leg to be bent forward out of a normal perpendicular straight line.
28. _____ is the most common treatment for milk fever.
29. _____ is the most common limiting nutrient for high milk yield.
30. _____ is the number of dairy cows in the United States.
31. _____ is the rolling herd average for U.S. dairy herds.
32. _____ is the nation with the highest-producing dairy cows.
33. _____ is the world's record per cow for annual milk yield (in pounds).
34. _____ is the most important type trait in a dairy cow.
35. The hormone _____ stimulates milk secretion.
36. The hormone _____ stimulates mammary duct development.
37. The optimum number of daylight hours for maximum milk yield is _____ .
38. _____ parts blood must pass through the udder to produce one part milk.
39. _____ is also known as syndactylism.
40. PTAT is the abbreviation for _____ .
41. Casting is used as a temporary treatment of _____ .
42. _____ is also known as barn itch.
43. _____ is a cause of cattle grubs.
44. _____ is a zoonotic disorder that affects as many as 90 percent of dairy farms by this parasitic organism.
45. _____ is a test for determination of BF% in milk.
46. _____ was the year question 45 was developed.
47. _____ is the state that produces the most milk (total pounds).
48. _____ is the state with the largest number of dairy cows.
49. _____ is also called Illawara.
50. _____ is caused by a deficiency of vitamin E or Se in dairy calves.
51. Another name for BSE is _____ .
52. _____ are molds that develop on feed that produce toxins.
53. The hormone _____ stimulates milk letdown.
54. The hormone _____ counteracts milk letdown.
55. The hormone _____ stimulates an increase of up to 20 percent more milk with only 5 percent more feed.
56. _____ is another name for a carousel-type milk barn.
57. _____ is the recommended breeding method for dairy cattle.
58. An agent that destroys worms in the digestive tract is _____ .
59. The tendency of two or more traits that vary in the same direction or in opposite directions due to common forces or influences is known as _____ .

60. _____ is a chamber that connects the four teat cups to the milk line.
61. _____ is the standardization of lactation records to the level of yield that would have been attained by each cow if it had been a mature cow and calved in the month of the year of highest calving frequency for its breed.
62. _____ is a disease characterized by pneumonia or septicemia. The highest incidence occurs in animals subjected to stress.
63. _____ is also known as acetonemia.
64. _____is a disease that causes chronic diarrhea and weight loss resulting from infection with the bacterium *Mycobacterium paratuberculosis*.
65. _____ is also known as bacillary hemoglobinuria.
66. CSS stands for _____ .
67. _____ is caused by *Mycobacterium bovis*.
68. _____ is a viral disorder that causes lesions indistinguishable from those of foot-and-mouth disease.
69. _____ is also known as bovine genital campylobacteriosis.
70. _____ is the common name for infectious bovine keratoconjunctivitis.
71. _____ is also called gas gangrene.
72. _____ is the common name for bovine spongiform encephalopathy.
73. _____ is caused by *Actinomycosis bovis*.
74. _____ is the most commonly used test for diagnosis of leptospirosis in cattle.
75. _____ is the common name for interdigital phlegmon.
76. The primary milk secretion hormone is _____ .
77. The spores from the fatal disease _____ can live in the soil for more than 60 years.
78. _____ is a fatal protozoan disease of cattle that destroys the red blood cells, causing anemia and death.
79. _____ is the normal rectal temperature of a dairy cow.
80. _____ is the mechanism that permits alternating vacuum and atmospheric pressure to exist between the teat cup liner and shell.
81. _____ is also known as traumatic gastritis.
82. _____ is used to regularly evaluate where an individual cow stands in body condition relative to the ideals for its stage of lactation.
83. Established colleges of agriculture in each state are called _____ .
84. Intake crude protein that is broken down by microorganisms in the rumen is called _____ .
85. MUN is an abbreviation for _____ .
86. _____ is the common name for ruminal tympany.
87. _____ should only be used as a last resort for the treatment of ruminal tympany.
88. _____ is caused by a deficiency of iodine in the diet.
89. _____ is caused by a deficiency of magnesium in the diet.
90. The disorder _____ is prevented by the use of DCADs.
91. _____ is also known as hyperkeratosis.

92. The Holstein Association uses _____ as a method of ranking bulls on their overall performance.
93. _____ is the common name for papillomatous digital dermatitis.
94. _____ is also known as red nose.
95. _____ is one of the most commonly diagnosed viruses in bovine abortion cases and is easily prevented with vaccination and improved sanitation.
96. _____ is caused by the heel fly.
97. _____ are wingless, flattened insects.
98. _____ and _____ are two reportable external parasites common to dairy and beef cattle.
99. _____ is the common name for nematodes.
100. _____ is an example of a trematode.

DISCUSSION QUESTIONS

1. What are some chief feed sources for a dairy animal of the following nutrients?
   a. Energy (Carbohydrates & Fats)
   b. Protein
   c. Vitamins A, D, and E
   d. Minerals Ca and P
2. List the five major dairy breeds with their native home, coat color, and distinguishing characteristics.
3. Be able to fill in the major parts of the dairy cow.
4. Draw and label the major parts of the bovine mammary gland.
5. Write a practical and workable calf-feeding program.
6. Describe common dairy buildings, barns, and designs (herringbone, side-open, rotary barns, parallel barns, flat barns, parlors, and freestalls).
7. Describe how milk is removed from the cow—by machine, by hand milking, and by the calf. Discuss mechanism, method, and time.
8. Why should you never give more than two gram-negative vaccines at one time?
9. Why should you *not* vaccinate cattle on days when temperatures are >85°F and humidity >40 percent?
10. What are PTA, PTAT, and calving ease?
11. Describe negative energy balance. Explain the difference between dry matter intake peaks and milk yield peaks (timing).
12. Review all major diseases and parasites and their causes, symptoms, preventive measures, and cures for dairy cattle.
13. What is mature equivalent? What is the standard lactation of a cow?
14. What are three determining factors in when to breed a dairy heifer?

CHAPTER 6

Dairy and Meat Goat Industry

OBJECTIVES

After completing the study of this chapter you should know:

- The importance of goats and why they are found in certain regions of the world in relatively large numbers.
- The distinguishing characteristics of the most common breeds of goats found in the United States.
- The similarities and differences in management of dairy goats and dairy cattle, including similarities and differences between milk production records and reproductive characteristics.
- The feeding techniques, feed, and nutritional management of goats.
- The common health disorders of goats, including the symptoms, prevention, diagnosis, and treatment of each common disorder.

KEY TERMS

Billy chevon	Disbudded	Mohair
Browse	Forage	Older goat
Brucellosis	Forbs	Pleiotropic effect
Cabrito	Hermaphroditism	Psoroptic mange
Chevon	Kemp	Wattles
Demodectic mange	Malta fever	

INTRODUCTION

Goats have been domesticated for more than 12,000 years; they have served humankind longer than cattle and sheep. In many areas of the world, people consume more milk and milk products from goats than from any other animal.[1] There are nearly 60 recognizable breeds numbering more than 639 million head

---

[1]L. S. Shapiro, *Principles of Animal Science* (Simi Valley, CA: Ari Farms, 1997).

TABLE 6–1	Goat Numbers around the World

Area or Country	Number of Goats[a] (in Millions)
Africa	175
Asia	401
Europe	16
North and Central America	16
South America	25
World	639

[a]Food and Agriculture Organization Production Yearbook, 1995. Rome, Italy.

(see table 6–1). Most goats are found in dry climates where temperature extremes are great. In Africa, the Middle East, Asia, South America, and the Caribbean islands goats are favored meat animals, and their hides are valuable export items. Goats are considered to be nature's best herbicide. Recently, researchers in Nebraska demonstrated using goats to select out spurge heads while walking through high bromegrass.

In the United States, goats were among the animals brought along with the Virginia settlers. These goats were used for milk and meat. Depending on the source, there are somewhere between 2 and 4 million head of goats in the United States; Texas leads in Angora, meat, and bush goats, and California leads in dairy goats. In 1998, there were 1,781 dairy goats in 45 herds on official Dairy Herd Improvement test in California. They averaged 1,945 pounds of milk, 3.50 percent fat, and 3.16 percent protein. The national average is 1,775 pounds of milk, 3.64 percent fat, and 3.17 percent protein in 483 goat herds and with 12,256 goats on official DHI test.

The first purebred dairy goats were imported to the United States in 1893. Today, dairy goats are scattered throughout the United States; most are kept close to large metropolitan areas where goat milk is marketed. The leading dairy goat states, in order of numbers of lactations per year, are California, Oregon, Wisconsin, Washington, Arizona, New York , and Pennsylvania.

▓ **Mohair** fleece of the Angora goat

Goats used in the United States are grouped as dairy, **mohair,** and meat breeds, although both dairy and mohair (Angora) breeds are also used for meat. In the United States, eight breeds, six of which are considered dairy, are represented in substantial numbers. They are the Alpine, Angora, LaMancha, Nubian, Oberhasli, Pygmy, Saanen, and Toggenburg breeds. Some additional minor breeds will be briefly mentioned in this chapter. The dairy goat industry consists of six major purebred breeds:

1. *Nubians* are the most numerous of U.S. and Canadian dairy goat breeds. The (Anglo) Nubian breed was developed in England by crossing British dairy stock with Indian and Egyptian types. They are shorthaired and heat tolerant and have meatier carcasses than the Swiss breeds. Nubians are noted for their long, pendulant ears and Roman noses. They produce milk higher in butterfat content but lower in quantity than the Swiss breeds. Nubians come in many colors and color patterns. Mature does weigh about 130 to 135 pounds, and bucks average 175 to 180 pounds. Nubians tend to be meatier than the other dairy breeds (see figure 6–1).

FIGURE 6–1
Nubian doe.
Courtesy of the
American Dairy Goat
Association.

FIGURE 6–2
Toggenburg doe.
Courtesy of the
American Dairy Goat
Association.

2. *Toggenburgs* (Toggs) originated in the Toggenburg Valley of Switzerland. This breed is small, with short and compact bodies. The does weigh about 100 to 135 pounds at maturity, and the bucks weigh closer to 150 to 175 pounds. Toggs have a shade of brown to the body, with characteristic erect white ears, white facial markings, and white lower legs (see figure 6–2). They can be long or medium haired, and they have beards. Toggs are usually polled, although some Toggs have horns. Many Toggs have *wattles* hanging from their necks. Toggs are good milkers; they average close to 2,000 pounds of milk per 305-day lactation and have records close to 8,000 pounds.

3. *French Alpine* goats originated in France from Swiss foundation stock. They have a wide variation in color, including white, gray, brown, and black in characteristically named color patterns (see figure 6–3). They are large, rugged animals with alert eyes and erect ears. Bucks weigh 170 to 180 pounds, and does weigh about 125 to 135 pounds.

4. The *Saanen* breed originated in the Saanen Valley of Switzerland. They are the largest dairy goat breed and the most popular breed in Israel, Australia, New Zealand, and many European countries. Saanens are white or cream colored, with long or short hair and erect ears (see figure 6–4).

FIGURE 6–3
Alpine doe.
Courtesy of the
American Dairy Goat
Association.

FIGURE 6–4
Saanen doe.
Courtesy of the
American Dairy Goat
Association.

Saanens are polled, although occasionally one is born with horns. Both bucks and does are bearded, and bucks may have a tuft of hair over the forehead. They are called the Holsteins of dairy goat breeds because of the large amounts of low-fat milk they produce and their large bodies. Saanen bucks weigh 185 pounds, and does weigh up to 140 pounds. An Australian Saanen doe held the world record for milk production of 7,714 pounds in 365 days until 1997, when a U.S. Toggenburg doe milked 7,965 pounds of milk, 312 pounds of fat, and 240 pounds of protein in 305 days.

5. *American LaMancha* is a relatively new breed of dairy goat. It was developed in 1958 when breeders in California crossed goats of Spanish origin with purebreds of Swiss origin. Small rudimentary ears, like gopher or elf ears (see figure 6–5), are the main distinguishing characteristic of this breed (responsible gene is genetically dominant). All colors and patterns are found in LaMancha goats. The percentage of milk fat from LaMancha goats is higher than that of the Swiss breeds.

6. *Oberhasli* goats are medium sized and chamois in color (bay)—usually solid red or black with a few white hairs. Markings include two black stripes down the face from above each eye to a black muzzle, nearly all black forehead, and black stripes from the base of each ear coming to a point just back of the poll and continuing along the neck and back as a dorsal stripe to the tail (see figure 6–6). This breed used to be known as Swiss Alpine and also originated in Switzerland. They are well adapted for high-altitude mountain grazing and are moderate milk producers.

7. *Sable* goats are also from Switzerland (from Saanen parents). They have the same characteristics as Saanen goats with the exception of color. They are dark, or sable, colored.

FIGURE 6–5
LaMancha doe.
Courtesy of the
American Dairy Goat
Association.

FIGURE 6–6
Oberhasli doe.
Courtesy of the
American Dairy Goat
Association.

FIGURE 6–7   An
Angora goat showing
a full fleece of mohair.
Courtesy of L.A. Pierce
College Agriculture
Department.

Nondairy goats include the following:

8. *Angora* goats originated in Turkey. Both sexes are horned and open faced with pure white (occasionally black) hair over the rest of the body. The hair of the Angora goat is called mohair. The outer coat consists of long locks or strands of mohair (see figure 6–7). The Angora produces 1 or more inches of growth each month. Selection should be made against the undesirable large, chalky white hairs called **kemp** and for the more valuable finer, smooth, and straight (uncrimped) hair. The mohair should be soft to the touch. Goats are clipped twice a year and yield 6.5 pounds per clipping. Angora goats have thin, long, pendulous ears. Mature bucks

Kemp   large, chalky white weak, heavy medullated fibers found in wool. They are highly objectionable as they do not take a dye and are weak in strength

FIGURE 6–8
Pygmy doe and recently arrived kid nursing. Note after-birth hanging from the vulva.
Photo by author.

weigh from 125 to 175 pounds, and mature does weigh 80 to 90 pounds. Texas is the home of 95 percent of Angora goats in the United States, and 60 percent of the world's Mohair is produced in Texas.[2]

9. *Pygmy* goats originated in the western part of Africa (Nigeria and the Cameroons). They are white to gray and black in color and appear small, cobby (the Pygmy is an achondroplastic dwarf), and compact (16 to 21 inches at withers) (see figure 6–8). Pygmy goats are precocious breeders and usually have twins and sometimes triplets. They are adaptable to humid and hot climates and are resistant to the tsetse fly. Full-grown males stand at 20 inches, making them excellent animals for petting zoos and for young children. Their high butterfat content (6.5 percent) of their milk enables them to produce rich-tasting milk for families who want their own milking animal.

10. *Boer goats* originated in the Eastern Cape Province of South Africa. They are nonseasonal breeders selected primarily for their meat. Because of their extended breeding season, Boer goats can have three kiddings every 2 years and can average 200 percent kidding rates (twins). Mature does weigh 200 to 265 pounds, and mature bucks weigh as much as 240 to 380 pounds (see figure 6–9). Boer goats were introduced into the United States in 1993.[3] Besides being used as meat goats, Boer goats are valued for their skins. Only the shorthaired goats have acceptable skins. Breeds of goats that compete for quality skins include the Maradi, small East African, Somali, Black Bengal, Moxoto, Marota, and Sahil breeds.

---

[2]J. R. Gillespie, *Animal Science* (Albany, NY: Delmar, 1998).
[3]American Boer Goat Association, P.O. Box 140615, Austin, TX 78714.

FIGURE 6–9 Boer goats can be used to weed orchards and vineyards, as seen here on the 240-acre teaching farm laboratory at L.A. Pierce College.
Photo by author.

11. *Spanish goats* are primarily raised for brush clearance and for meat production. They do better on poor forage than other meat breeds such as the Boer goat. These goats are highly prolific, are relatively small (75 pounds at maturity), and can survive with minimal care.

12. *Cashmere goats* are a type, not a breed. Most goat breeds, except the Angora, can produce cashmere (the fine underdown). The majority of the world supply of cashmere has come from Afghanistan, Iran, Outer Mongolia, India, and China; the Kashmir goat, found primarily in the province of Kashmir, was thus developed. This goat is white in color and has erect ears and long, twisted horns. Recently, Australia and New Zealand began producing cashmere fiber from their goats. Cashmere fibers must be separated, either by combing out the down or by using a commercial dehairer on sheared fibers. The longer and finer down is used in knitted garments, and the shorter fibers are used in woven fabrics.

Cashmere is valued for its soft down or winter undercoat of fiber. It is an extremely fine fiber. Estimated world cashmere production is 3,000 to 4,000 tons per year. Most is made into the most luxurious, expensive cloth and knitwear found in Scotland. The demand for cashmere has always exceeded the supply.

DAIRY GOAT MANAGEMENT

The dairy goat is often thought of as a miniature dairy cow. Although there are many similarities, distinct differences exist that make goat dairying an important part of today's agriculture. In the United States, dairy goats are second to cattle in milk production for human consumption, yet throughout the world dairy goats supply the majority of milk for humans in many countries.

Like the dairy cow, the dairy goat has been bred and selected throughout history to produce large amounts of milk. Desirable body conformation of the

FIGURE 6-10
Modern dairy goat barn. Note the two-teat-cup automated pipeline system. Photo by author.

dairy cow and goat is similar. The individual glandular structure of the goat's udder, with the alveoli, milk ducts, gland cistern, and teat anatomy and function in milk production, is like that of cattle. Conversion of feed nutrients to milk is very similar in both. Lactation of 305 days with a 60-day dry period is also the norm for both species.

Characteristics unique to the goat for milk production make it a different enterprise. The most obvious is the udder of the goat. Instead of the four teats and udder divisions of the cow, the dairy goat's udder has only two divisions and teats (see figure 6-10). The dairy goat is very efficient in milk production. In general, 7 goats would produce as much milk as 1 cow, yet 10 goats can be fed on 1 cow's ration. It is not uncommon for a 125-pound doe to produce more than 4,000 pounds of milk in one 305-day lactation.

The dairy goat is only 1/10 the size of a cow, so it is much easier to maintain. As we shall see, nutrient requirements are less, and goats eat a wide variety of feeds, converting them into milk. It is easy to see why goat dairies range from a few goats in the backyard to large commercial dairies with hundreds of milking does.

CHARACTERISTICS OF GOAT MILK

Goat milk has been known for its nutritional and medicinal value since biblical times. Compared with cow milk, goat milk has the following characteristics:

1. Goat milk is whiter in color.
2. The milk fat globules are smaller and stay in emulsion with the milk in goat milk. Fat must be separated by a mechanical separator because it will not rise to the surface over a reasonable time.
3. Goat milk casein and milk fat are easier to digest than those in cow milk. There are more shorter-chain fatty acids in goat milk than in cow milk.

4. The protein curd is softer in goat milk, which allows the production of special cheeses.
5. Goat milk is higher in the minerals calcium and phosphorus and A, E, and various B vitamins.
6. Goat milk is used for people who are allergic to cow milk or who have various digestive disorders.

FEEDING GOATS

There are three main products of goats: milk, meat, and mohair. The end product dictates the nutritional needs and management of the goat. Mature goats have digestive tracts similar to those of other ruminants. Their gastrointestinal tracts consist of the rumen (12 to 24 liters), the reticulum (1 to 2 liters), the omasum (1 liter), and the abomasum (3.5 liters), followed by a 100-foot-long (11-liter) intestinal canal. Goats, however, are more efficient ruminants than sheep or cattle. They are able to consume more dry matter for their body size: 5 to 7 percent of body weight as compared with only 3 to 3.5 percent of the body weight of a cow in dry matter intake. Goats are also more efficient than cattle or sheep in digesting coarse, fibrous feeds. Hence goats will consume and use roughages that cattle or sheep will not touch. Dairy goats in high production require well-balanced rations high in protein and energy. Available **browse** (brushy plants), **forbs** (broadleaf plants), and other **forage** (grass) will satisfy many of the nutritive needs of goats that are raised on ranges.

As with other milking animals, however, the nutritive requirements of lactating does are much more substantial than those of dry does or growing animals. Thus the diet for does in milk should be based on high-quality roughages plus excellent pasture or a concentrate feed. About 2 to 3 pounds of leafy, immature hay per day per goat or good-quality pasture is required, plus 0.5 pound of a 16 percent concentrate ration per quart of milk produced. Excellent pasture can replace up to half of the concentrate required per day. An example of a concentrate mixture for goats is 40 percent corn, 20 percent oats or barley, 23 to 25 percent wheat bran, 15 percent soybean or cottonseed meal, 1 percent salt, and 1 percent calcium–phosphorus supplement. The concentrate mixture should be coarsely ground because goats dislike finely ground, dusty feeds. Plenty of clean water and a salt–mineral supplement should be given free choice to all goats.

When pasturing goats, care must be taken to keep them away from volatile feeds, which will taint their milk with odors. Withdrawal from pasture 2 to 4 hours prior to milking may be required to produce good-tasting milk. In addition, major scent glands, located around the horn base, produce a goat odor that can be picked up in the milk. The purpose of this scent is to stimulate estrus in both male and female goats.

Feeding does during the dry period is important for development of the unborn kids and obtaining proper body condition of the does for maximum milk production. Seventy percent of the birth weight of the unborn kids is developed during the dry period. During this time does are fed good-quality pastures alone or 1 to 2 pounds of concentrates per day on poor- to fair-quality pastures. The concentrate mixture fed to milking does will also suffice for dry does.

Goats are unique grazers. They are able to produce mohair (Angora goat's hair) and meat from otherwise nonproductive lands. Goats are excellent

Browse leaves, shoots, and twigs of brush plants, trees, or shrubs fit for grazing especially by sheep, goats, and deer

Forbs nongrass broadleaf nonwoody plants, including sage, shinoak, and saltbush, commonly found on range and eaten by livestock

Forage term used synonymously with roughage which describes the total plant material consumed by an animal. Forages have greater than 18 percent crude fiber

climbers, will travel longer distances in search of preferred forages than cattle or sheep, and are thus well suited to mountainous or hilly, rough ranges. Angora and Spanish or meat-type goats are raised primarily on rangelands. Goats eat a wide range of plants and prefer brushy browse. They are used to help control brushy and weedy plant species in many ranges. Due to their different diet, they are very compatible with sheep and cattle, and combining the two species will increase total range production. Goats are very hardy and require little or no protection from the climate, except after shearing, when sheds may be required in cold, wet weather. Angora and Spanish goats do not need extra feed if good pasture is provided. Supplemental feeding is advised only when pastures are short due to drought or cold weather.

Angora goats and Spanish goats are raised for goat meat production. Spanish goats differ from other goats in that they are not seasonal breeders. Hence they are bred year-round to supply a constant supply of meat. A goat that weighs 100 pounds may yield a carcass of 50 pounds. Goat meat has similar cuts to those of lamb but is much leaner and is unique in flavor, palatability, and tenderness. The loin is the most valuable and most tender cut of meat. **Cabrito** (Spanish for little goat) is the meat from kids of Spanish goats (first few days of life). **Chevon** is meat from goats slaughtered at weaning or older. Chevon is classified by age. *Billy chevon* is an uncastrated male slaughtered between 9 and 18 months of age. *Older goat,* meat taken from mature goats, is used primarily in processed meats such as goat sausage, frankfurters, bologna, and chili con carne.

▓ **Cabrito**   meat from young goats

▓ **Chevon**   meat from goats slaughtered at weaning or older

LACTATION

The standard lactation of a goat is the same as that for a dairy cow—305 days and twice-a-day milking. The lactation period for a goat is 7 to 10 months, with a 2-month dry period. Milk yield per day will increase after kidding to a maximum at about 2 to 3 months postpartum. Milk production then declines gradually during the rest of the lactation period. As with dairy cows, maximum milk yield per year is obtained from goats with good persistence (slower decline in milk production after the peak).

In 1977, Mackness Myna (British hybrid) broke the goat milk record with 5,963 pounds at 3.75 percent fat. Since 1977, this record has been broken many times (see table 6–2). One U.S. Toggenburg bred by Katrina Kay of western Texas produced nearly 8,000 pounds in only 305 days as a 3-year-old.

REPRODUCTION IN THE GOAT

For general information on reproduction of the goat, please review chapter 3 and table 6–3 in this chapter.

As mentioned earlier, goats are seasonal breeders and will cycle August through January, with the months of September, October, and November the prime months for breeding. During the breeding season, does will come into heat every 21 days for a 1- to 2-day heat period. Heat signs in goats are similar to those in cattle (e.g., restlessness, frequent urination, constant bleating, swollen vulva, riding other does, and standing for other does to ride). Does bred on the second day of heat realize high pregnancy rates. A healthy buck can serve at least 30 does. Mating is accomplished by taking the doe to the buck for a

TABLE 6–2	Milk Production **Records by Breed**						
				Average Production			
	Record Milk[a]	California Milk[b]	National Milk[c]	Fat		Protein	
	(lb.)	(lb.)	(lb.)	(lb.)	(%)	(lb.)	(%)
Toggenburg	7,965[d]	2,171	1,710 (1968)	63	3.2	58	3.0
Saanen	6,571[e]	2,139	1,998 (1899)	67	3.5	58	3.1
Alpine	6,416[f]	1,974	1,979 (2031)	72	3.5	63	3.1
Nubian	5,940[g]	1,893	1,618 (1567)	69	4.4	56	3.6
LaMancha	5,400[h]	1,767	1,771 (1699)	67	3.9	56	3.3
Oberhasli	4,665[i]	2,346	1,663 (1605)	57	3.6	48	3.0

[a]American Dairy Goat Association (ADGA), Spindale, NC: June 2, 1999.
[b]Tri-County Goat Newsletter, Tulare-Kings-Fresno Counties, California, March 1998.
[c]ADGA–USDA Statistics, ADGA, Spindale, NC: June 1999.
[d]Western-Acres Zephry Rosemary, 1997.
[e]JC-Reed's Cloverhoof Haley, 1997.
[f]Donne's Pride Lois, 1982.
[g]Skyhill's Elisha, 1996.
[h]Tyler Mt. May's Priscilla, 1991.
[i]Catoico Summer Storn, 1997.

TABLE 6–3	**Summary of Goat Reproductive Characteristics**
Gestation length	144 to 155 days
Estrous cycle	21 days
Estrus	1 to 2 days
Age at puberty	120 to 365 days
Size of kids at birth	1.5 to 11 pounds (average 5.5 pounds)

single service. Bucks are kept separate from milking does because the milk can take up their strong odor. Artificial insemination can be used, and it is becoming more popular. Frozen semen may be purchased and stored, as for dairy cattle. An added advantage to artificial insemination in dairy goats is the elimination of the buck's odor, which is so readily picked up in the doe's milk.

Breeding for polled condition (to eliminate the need for dehorning) was a major cause of infertility in goats. It caused a marked increase in *hermaphroditism*—having reproductive organs of both sexes. In goats, polled is dominant over horns. A **pleiotropic effect** results when one gene controls or contributes to the phenotype of more than one trait. Linkage occurs when two genes for different traits are located close together on the same chromosome. When selecting for the polled condition, both polled and hermaphroditic goats result.

▌**Pleiotropic effect** results when one gene controls or contributes to the phenotype of more than one trait

PARTURITION (KIDDING)

Prior to parturition, the does are put into a clean, dry, quiet, draft-free quarter of about 30 square feet per doe. Goats usually have easy kidding with few complications. Normal kidding occurs in about 30 minutes. Multiple births are normal for goats; twins and triplets are common. Multiple births can be encouraged through selection and proper nutrition during gestation.

After kidding, does should expel the afterbirth in about 4 hours. If they do not, a veterinarian should be consulted. Hay is fed free choice to does, and grain is fed the day after kidding. They should start with only 1 to 2 pounds of concentrate per day and increase gradually until they are consuming 0.5 pound of concentrate per quart of milk produced.

At kidding, the mucus must be cleaned from the noses of the young and their navels dipped in iodine to prevent infection. Newborn kids need colostrum, either by nursing or through a bottle.

MANAGEMENT OF KIDS FROM BIRTH TO PUBERTY

Dairy goats give colostrum milk for 3 to 5 days after kidding. The colostrum milk must be given to the young kids four to five times a day for the first 3 to 5 days by nursing or by feeding from a bottle, depending on the management of the owner. Under natural suckling, kids consume small amounts of milk at very frequent intervals. Bottle-feeding is more labor intensive, but kids receive more individual attention and are much easier to handle after weaning. Kids raised on bottles should receive about 8 percent of their body weight in milk per day, fed in two to three portions. For large goat herds, self-feeder units such as a lamb bar can successfully reduce labor. If not properly managed, lamb bar feeding increases nutritional scours and death rates among young kids.

As with dairy calves, kids can be successfully raised on milk replacer. Lamb milk replacer is generally used with young kids 2 weeks and older. During the first 2 weeks, it is generally recommended that caretakers feed only goat milk or at least a mixture of goat milk with replacer. Kids are weaned from milk or milk replacer at 6 to 12 weeks of age, depending on the amount of concentrate mix eaten per day. Kids should be encouraged to eat a good 20 percent protein starter ration (such as the calf starter mentioned in chapter 5). High-quality hay and fresh water should be made available to kids. Early consumption of hay and grains promotes rumen development. When kids are eating hay and grain well, milk can be discontinued. Kids should weight 2 to 2.5 times their birth weight when they are weaned.

Kids are *disbudded,* not dehorned. (Disbudding is the practice of removing the horn buds at a very early age.) Disbudding should be done between 3 and 14 days of age. The hair should be clipped, a local anesthetic such as lidocaine given to decrease pain, and a hot electric disbudder held over the area for approximately 15 seconds. A topical spray should be applied to reduce problems with flies on the resulting wound.

Buck kids to be slaughtered under 2 months of age do not need to be castrated. Bucks not kept for breeding and not slaughtered before 2 months are castrated at 2 to 5 days of age. In castrating bucks, the lower one-third of the scrotal sac is cut with a knife and the testicles are squeezed through the opening. The cords are then cut by scraping with a sterilized knife. Iodine or topical spray is applied. Some managers also use a Burdizzo clamp (bloodless castrator) or the elastrator (elastic bands) method. Use of the elastrator has been criticized, however, because of the potential development of gangrene. Buck kids kept for replacements should be removed to separate pens at 2 to 3 months of age to prevent premature pregnancies.

Replacement kids should be fed so that they reach breeding size at 9 to 10 months of age. Managers can ensure such a growth rate by providing good-quality pastures or hay plus 1 pound of concentrate mixture per day. The concentrate mixture can be eliminated when high-quality pastures are available. The goal should be a 110-pound doe (Nubian, Alpine, and Saanen) freshening at 12 months of age and 90 pounds for the lighter breeds (LaMancha and Toggenburg).

HERD HEALTH

Herd health is achieved through nutritional management, disease control, reproductive management, parasite control, environmental management, and overall stress reduction. Goats are subject to many of the diseases and parasites discussed in the chapters on diseases of cattle and sheep. Some of the major differences are mentioned in brief in the following section. For a more detailed description, please see chapters 5 and 8.

Bluetongue

Goats are generally highly resistant (unlike sheep) to bluetongue. See chapter 8.

Brucellosis

Brucellosis is an abortion disease of goats. In goats it is caused by *Brucella melitensis,* a different organism from that which causes brucellosis in cattle. It is, however, still zoonotic. In people it is called *Malta fever*. In Mexico there is a very high incidence of both brucellosis in goats and Malta fever in people. In the United States, control is by test and slaughter. Vaccination with the Rev. 1 strain is a preventative method used in countries where *B. melitensis* is endemic.

Caprine Arthritis Encephalitis

Caprine arthritis encephalitis (CAE) is caused by a virus that infects the central nervous system. Symptoms include polyarthritis, ataxia, partial paralysis, hard udder, and chronic wasting. CAE is host specific. It is widespread in the dairy goat breeds and less so in the meat and fiber breeds. There is no specific treatment, but regular foot trimming and the use of nonsteroidal anti-inflammatory drugs improve the condition of animals. Quarantine and blood testing of new herd animals is a recommended control method.

Caprine Pleuropneumonia

Caprine pleuropneumonia is a highly fatal, highly contagious disease that affects goats in the Middle East, Eastern Europe, Africa, and Asia. Its causative organism is one of a few *Mycoplasma* species. Vaccines are available in some countries. Quarantine of affected herds is also advisable.

Contagious Ecthyma (Sore Mouth/Orf)

Contagious ecthyma is an infectious dermatitis contagious to goats, sheep, and humans. It is most severe, however, in goats. Symptoms include lesions on the skin of the lips and mucosa of the mouth. Occasionally, sores are also found on the feet. Because it is caused by a virus, antibiotics are used primarily to control secondary infections. A live viral vaccine is available but should be used only if all the animals in the herd are vaccinated.

Chlamydial Abortion (Enzootic Abortion)

Chlamydial abortion is the most common type of infectious abortion in goats in the United States. It is caused by *Chlamydia psittaci.* Abortions usually occur during the last month of pregnancy. Tetracycline is a traditional treatment. No chlamydial vaccine has been approved for goats, but there is one for sheep.

Enterotoxemia

Enterotoxemia is caused by *Clostridium perfringens* types C and D. It is also known as overeating disease. When goats are fed diets high in carbohydrates, the bacteria multiply rapidly and release a toxin. Symptoms include diarrhea, depression, ataxia, coma, and death. Prevention includes feeding more frequent and smaller meals. Toxoid vaccines are available for young kids as well.

Foot Rot

Foot rot incidence can be reduced by trimming hooves at least four times a year and by keeping goats out of wet, marshy areas.

Goat Pox

Goat pox is a serious and often fatal disease that causes widespread lesions on the mucous membranes and the skin. The disease is confined to southeastern Europe, Africa, and Asia. It is not found in the United States.

Johne's Disease

Johne's disease (paratuberculosis) is a serious disorder in goats that causes them to appear unthrifty, emaciated, and unproductive. Diarrhea is not as common in goats as it is in cattle (see related section in chapter 5).

Ketosis (Pregnancy Disease or Pregnancy Toxemia)

Overconditioning a pregnant doe predisposes her to ketosis. It is much more rare in goats than in sheep. Keeping does in good body condition is the best preventive measure.

Listeriosis

Listeriosis is a bacterial disorder caused by *Listeria monocytogenes.* It is a zoonotic disorder and affects a wide range of farm and companion animals. Transmission is by the fecal–oral route. In goats its symptoms include encephalitis, facial paralysis, and circling. In pregnant does, abortion, stillbirths, and neonatal deaths occur. Penicillin is the drug of choice in treating listeriosis.

Mastitis

Pregnant does should have a 45- to 60-day dry period to allow the mammary gland a period of rest and for does to get ready for their next kidding. At drying off, the udder should be treated with a dry cow mastitis antibiotic. The same rules of milking and mastitis prevention listed for dairy cows should be followed for dairy goats (see chapter 5).

Milk Fever

Milk fever is very rare in dairy goats. When it occurs, the same precautions and treatment outlined for dairy cows should be followed (see chapter 5).

Tetanus

Tetanus is a very highly fatal disorder caused by *Clostridium tetani*. In most cases, the organism is introduced into the tissues through wounds (e.g., disbudding, castrating, ear tagging, or deep puncture wounds from wire and nails). Symptoms include tonic spasms, difficulty in chewing of food, bending of the extremities when startled, and, for sheep, goats, and pigs, falling to the ground and exhibiting a tetanic spasm of the spine. Control is achieved by active immunization. A tetanus antitoxin should be administered to young animals not previously immunized at time of castration, disbudding, or ear tagging.

External Parasites

Lice (Pediculosis).   Goats are susceptible to both biting and sucking lice. Treatment includes the spraying or dipping of all goats in the herd with coumaphos.

Mites.   *Chorioptes caprae* is fairly common in goats. It causes mange, with symptoms of papules and crusts on the feet and legs. It is a reportable disease but has been eradicated from the United States. It is still common in Europe, New Zealand, and Australia.

In the United States, it is more common to see goats with *Psoroptic mange* caused by *Psoroptes cuniculi*. These mites affect the ears but can also spread to the head, neck, and body. With Angora goats, severe damage to the mohair occurs. Lactating dairy goats should only be treated with a lime–sulfur solution. Meat and fiber goats can be dipped with coumaphos or injected with Ivermectin. Psoroptic mange is a reportable disease.

Demodectic mange also occurs in goats (*Demodex caprae*). Localized lesions can be infused with Lugol's iodine or rotenone in alcohol.

Ringworm.   Ringworm is uncommon in production flocks of sheep and goats. When it occurs, daily topical treatment of equal parts iodine and glycerin to infected areas is helpful.

Internal Parasites

Goats tend to have more internal parasites than dairy cows, especially in confined management. Internal parasites increase in numbers during the preparturient period; thus, the dry period is an ideal time to deworm. Stomach worms, coccidia, and liver flukes (depending on the geographical area) are of major concern.

SUMMARY

Goats have been domesticated for more than 12,000 years. In the United States, goats have been restricted geographically; most goat herds are found in California and Texas. The feeding and management of milk goats is very similar to that of milk cows. Slight differences in reproductive, disease, and milking management exist in this seasonally bred animal. The number of milk, meat, and pet goats will most likely continue to increase into the twenty-first century.

EVALUATION QUESTIONS

1. The two major goat-producing areas in the world are _____ and _____ .

2. In the United States, Angora goats are located mostly in the state of _____ .

3. The _____ and _____ dairy goat breeds yield milk high in milk fat content.

4. The _____ breed is popular because it produces a large quantity of low-fat milk (Holstein of the dairy goat breeds).

5. The goat's udder has _____ divisions, each with its own glandular structure and teat.

6. The goat has a _____ digestive system.

7. Goats can consume _____ percent of their body weight in dry matter per day.

8. Goats are bred primarily during the months of _____ through _____ .

9. Gestation in the goat is about _____ months long.

10. Kids are weaned from milk or milk replacer at _____ to _____ weeks of age.

11. Most mohair is produced in the countries of _____ and _____ .

12. _____ is a type but not a breed of goats that produce a fine underdown.

13. A goat will produce about _____ pounds of mohair per clipping.

14. The standard lactation of dairy goats is _____ days.

15. _____ is meat from goats slaughtered at weaning or older.

16. _____ is the dairy goat breed that has long, pendulant ears and produces a high-butterfat milk.

17. _____ is the dairy goat breed with no ears.

18. _____ is the number of goats worldwide.

19. _____ is the largest dairy goat breed and most popular in Israel, Europe, Australia, and New Zealand.

20. The breed of goat that produces mohair is _____ .

21. _____ is the origin (area of the world) of cashmere.

22. Pygmy goats originated in _____ .

23. _____ are the shoots, twigs, and leaves of brush plants found growing on rangeland.

24. The condition in which the goat contains reproductive organs of both sexes is called _____ ; it is typically found when trying to breed for the polled condition.

25. Kids raised on bottles will consume _____ percent of their body weight per day.
26. _____ is the breed of goat that originated in Turkey.
27. _____ is the meat goat breed from southern Africa.
28. _____ is a small, flat, wingless insect with sucking mouthparts that is parasitic on the skin of animals.
29. _____ is a disease that affects the central nervous system of young kids and induces head tremors, loss of coordination, and partial paralysis. In adult goats, arthritis affects the feet and other limb joints; a hard udder is also common.
30. _____ is also called sore mouth.
31. _____ are broadleaf plants.
32. A 100-pound goat will normally yield a _____-pound carcass.
33. Meat taken from mature goats is called _____ .
34. The world goat record for milk production is from the _____ breed.
35. _____ is a genetic term describing when one gene controls the phenotype of more than one trait.
36. Buck kids that are not kept for breeding should be castrated by age _____ .
37. _____ is a disease in goats that causes Malta fever in humans.
38. _____ is the most common type of infectious abortion in goats.
39. _____ is also known as overeating disease.
40. _____ is a highly fatal goat disorder caused by a soil organism penetrating a deep puncture wound.
41. _____ is a common U.S. reportable type of mange in goats.

DISCUSSION QUESTIONS

1. What is caprine pleuropneumonia? Why is it important?
2. What is the importance of Johne's disease?
3. Why are goats so prized around the world?

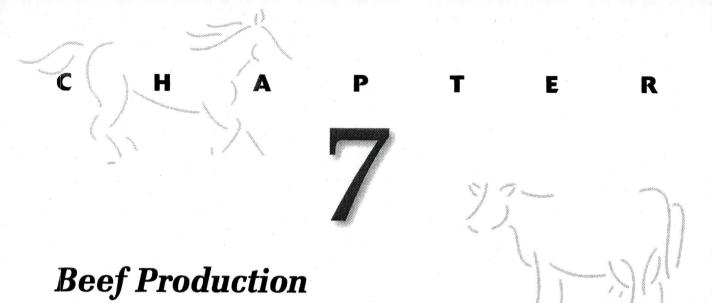

CHAPTER 7

Beef Production

OBJECTIVES

After completing the study of this chapter you should know:

- The importance of the beef industry to U.S. agriculture, including its history and future (according to the trends).
- Distinguishing characteristics of the various beef cattle breeds commonly found in the United States.
- The importance of the F_1 cross, its meaning, and what this cross is generally used for in commercial cattle raising.
- The slaughtering process commonly used in the United States and its impact on animal welfare and animal rights discussions.
- The meaning of Ahimsa and Shehitah and their relationship to animal slaughter.
- The meaning of dressing percentage and animal grading and comparison with other commonly slaughtered livestock species.
- The importance of the fifth quarter.
- The importance of the use of federally owned land in producing beef economically in the United States.
- Changes in the use of growth-promoting substances in producing U.S. beef.
- Beef cattle facilities, equipment, and management in modern-day beef cattle husbandry.
- Common parasites that affect the profitability of a beef enterprise and the overall health of cattle.
- The importance and impact of the Animal Medicinal Drug Usage Clarification Act (AMDICA).
- Methods of selecting, breeding, and judging beef cattle.

KEY TERMS

Ahimsa	β-adrenergic agents	Casting
Animal rights	Breeder cattle production	Castration
Animal unit	Burdizzo	Chemical castration
Animal welfare	Canner	Chemotherapeutic agents

Choice	Hormones	Select
Commercial	Ionophores	*Shehitah*
Commercial (program)	Marbling	Slaughter cattle
Cornual nerve block	Maturity	production
Creep feeding	Methane inhibitors	Standard
Cutter	Performance testing	Stocker
Dehorning	Prime	Termination cross
Elastrator	Probiotics	Texture
F_1 cattle (program)	Progeny testing	Utility
Feeder cattle production	Quality grades	Wintered
Good	Registered program	Yield grades

INTRODUCTION

Beef production is still the largest segment of agricultural industries in the United States, accounting for more than 17 percent of all farm marketing and more than 38 percent of the total cash receipts from all livestock and poultry products.[1] Texas is the leading state for beef production. The per capita beef consumption in the United States was 67.9 pounds in 1998[2] compared with 94.4 pounds in 1976. This decline in per capita consumption can be attributed to consumers' concern about cholesterol and a switch to poultry and fish products (see table 7–1). For the past three decades, cattle numbers in developing countries have increased due to a greater demand for beef. Uruguay and Argentina have the highest per capita consumption of beef at 133 and 152 pounds, respectively. As long as the United States has areas that are unable to sustain growth of crops but that will maintain adequate supplies of rough forages, the beef cattle industry will continue to develop and thrive.

U.S. REGIONS OF BEEF PRODUCTION

The U.S. beef industry is divided into cattle-raising regions based on similarities of conditions in each of the regions. Total numbers of cattle and cattle loading states are summarized in tables 7–2, 7–3, and 7–4.

TABLE 7–1	Average U.S. Per Capita Consumption of Meat (Boneless Weight in Pounds)								
Year	Beef	Pork	Veal	Lamb	Total Red Meat	Chicken	Turkey	Total Poultry	Total Red Meat and Poultry	Fish
1960	59.8	48.9	4.1	3.1	115.9	16.2	4.9	24.4	140.3	10.3
1970	79.8	48.6	2.0	2.1	132.5	25.2	6.4	34.3	166.8	11.7
1980	72.2	52.6	1.2	1.0	127.0	33.1	8.1	41.2	168.2	12.4
1990	64.1	46.7	0.9	1.1	112.9	41.5	13.9	56.7	169.3	15.0
1999	62.1	51.1	0.7	0.7	114.6	53.2	14.1	67.7	182.3	15.0

Source: U.S. Department of Agriculture (USDA), April 1999.

---

[1]National Cattlemen's Beef Association, 1999.

[2]*Insight to the Beef Industry,* 12(1):4, 1999.

TABLE 7–2	U.S. Cattle Inventory as of April 1, 1999

98,522,000 total cattle
42,615,000 cows
33,472,000 beef cows
9,143,000 dairy cows
19,604,000 heifers (> 500 pounds)
555,000 beef replacement heifers
406,000 dairy replacement heifers
999,400 other heifers
16,836,000 steers (> 500 pounds)
2,276,000 bulls (> 500 pounds)
17,190,000 calves

Source: U.S. Department of Agriculture (USDA), April 1999.

TABLE 7–3	Top Five Cattle Production States (as of January 1, 1998)	
Texas	14,300,000	
Kansas	6,860,000	
Nebraska	6,650,000	
California	4,800,000	
Missouri	4,300,000	

Source: Livestock Marketing Information Center, 1998.

TABLE 7–4	Top Five Feedlot States (as of January 1, 1998)	
Texas	2,860,000	
Kansas	2,370,000	
Nebraska	2,300,000	
Colorado	1,140,000	
Iowa	1,000,000	

Source: Livestock Marketing Information Center, 1998.
Note: California ranked seventh, with 375,000 head.

Northeast Region

The Northeast is of minor importance to commercial beef production. Beef herds are small and are often a supplementary enterprise by part-time farmers. The farmers in this area are predominantly dairy farmers; thus, beef production will remain stable and is not expected to increase in the future. Approximately 1,779,800 cattle are being raised in this region.

Corn Belt and Lake States

An increase in beef production in the Corn Belt and lake states has resulted from more animals being home-finished with locally grown grains. Beef herds are small, and increases in calf production stem from efficient pasture management on old cropland and the conversion of some dairy herds to beef herds. Future expansion in these two regions is limited by the inability of beef to compete for land with crops that yield a higher profit. Any increase will come from better management of existing herds. The lake states are currently home to 4,272,000 head of cattle, and the Corn Belt is home to 11,570,000 head.

Southeast

Beef production has increased most dramatically in the Southeast. The transition of cropland (especially cotton) to improved pastures has led to a rapidly expanding cow-calf industry. The trend has been toward specialization of farms, leading to increasing numbers and higher quality of existing beef herds. Beef production will continue to increase but at a slower rate, with increases in grass-fattened calves and cow-calf operations. The Southeast has 17,590,000 head of cattle.

Northern Plains

Beef cow numbers have doubled in the northern plains in the past 20 years. Current estimates are 18,218,000 head of cattle from Kansas north to North Dakota. Beef herds are becoming a major enterprise rather than a supplementary, part-time venture. Future growth will depend on how well cattle can compete for land with grain crops. The trend is toward forage production.

Southwest

The Southwest has always been the largest cattle-producing area in the United States. Much of the land is rangeland, timberland, or rough terrain that can be used most efficiently for grazing. A dramatic increase is noted in finishing feedlot operations, particularly in the Texas Panhandle. A ready source of sorghum grains and a supply of feeder calves have developed the feedlot industry in this region. Future increases will be small because of the maximum use of land suited only for grazing. Some increase may result from a switch from sheep and goat to beef production, more efficient management of existing herds, and further increases in the feedlot industry. Currently there are 21,933,000 head of cattle in this region.

Mountain and Pacific Regions

The mountain region has doubled beef production in the past 20 years, mostly because of expanding feedlot operations, especially in Colorado and California. Beef herds are kept in the mountainous regions and are quite large. Future expansion will come from further increases in feedlot operations and better management of existing herds. Competition for land from crops and urban centers will limit expansion in the Pacific region. The mountain region has 9,774,000 head of cattle, and the Pacific region has 5,165,100 head.

The primary purpose of the beef cattle industry is the production of meat for human consumption, although hides and other by-products are of economic importance. By-products, including hides, variety meats, tallow, and numerous other products, account for more than 10 percent of the value of a steer. Hides continue to be the single most valuable by-product, accounting for 57 percent of all by-product value in 1999 (see table 7–5).

TABLE 7–5	Value of Cattle By-Products per 1,000-Pound Steer		
Year	Hides	Total	Hides as Percentage of Total
1980	$27.77	$57.22	49
1990	$60.29	$91.50	66
1999	$45.53	$80.46	57

The beef industry includes three systems of production:

1. *Feeder cattle production:* Based on finishing cattle quickly (180 days or less, gaining an average of 2 pounds per day or more); feed conversion averages 7 pounds to 1 pound.
2. *Slaughter cattle production:* Involves selecting fattened cattle (by packer buyers), judging their live weight and conformation, estimating carcass traits and yield, slaughtering the cattle, and postmortem inspection.
3. *Breeder cattle production:* Production of cattle (breeding), both purebred and commercial herds.

The oldest and most established form of animal husbandry in the United States might well be the cow-calf program. The cow-calf business is normally separated into three basic categories:

1. *Registered program:* Based on selection of a breed to fit the environment. It is usually on small acreages; the market must be in a condition to absorb or demand the type of purebred animals being produced. These ranchers are selling genetics and not beef.
2. *F_1 Cattle:* Refers to a crossing of two pure breeds of cattle. Hybrid vigor or heterosis results from this crossbreeding. Heterosis increases weaning weights, disease resistance, mothering ability, and production of milk. Because heterosis lasts only one generation, this production technique is known as *termination cross,* in which no offspring is kept. Rotational crosses—breeding crossbred heifers to a bull of a breed that is not used in the original cross (three-way crosses)—have become common as well. The F_1 cross is usually sold as feeder cattle (to be fattened in the feedlot). Some of the large feedlots handle hundreds of thousands of cattle at a time. They are mechanized and use sophisticated computerized management to feed, provide medical treatment, and inventory their animals.
3. *Commercial:* Usually takes advantage of registered bulls, either to maintain the same bloodlines or to crossbreed to grade cows. Crossbred bulls are not recommended for breeding. They look better (because of their hybrid vigor) than they perform. Therefore, unless a definite system is planned to keep a heterogeneous mixture of blood in offspring, the crossbred bull is much less important than the crossbred cow. Commercial cattle breeders can use a system of confinement on improved pasture (controlling diseases and other problems that come from concentrations in small quarters) or the open range method. Stocking rates vary from 250 square feet per cow in drylot systems and 1 to 3 acres per cow on well-managed, improved pastures to 640 acres or more per cow in open range. Most commercial herds have spring calvings, although many herds split their calvings into a spring and fall herd.

TABLE 7–6	**Common Breeds of Beef Cattle**	
Standard Breeds	**Exotic Breeds**	**Breeds of American Origin**
Angus	Blond d'Aquitaine	Barzona
Brahman	Chianina	Beefmaster
Brown Swiss (beef)	Gelbvieh	Brangus
Charolais	Hays Converter	Charbray
Devon	Limousin	Polled Hereford
Galloway	Lincoln Red	Polled Shorthorn
Hereford	Maine-Anjou	Red Brangus
Milking Shorthorn	Marchigiana	Santa Gertrudis
Red Angus	Murray Grey	
Red Poll	Piedmontese	
Scotch Highland	Pinzgauer	
Shorthorn	Simmental	
	Sussex	
	Welsh Black	

■ Creep feeding feeding young animals extra feed in an area separate from their dams; the opening to the area is too narrow or too low for the dams to enter, but the young animals can enter freely

Most commercial cows run on pasture during the grass season until weaning time (at approximately 6 months of age). Some farm managers use **creep feeding** to supplement calf growth while nursing. In some areas, the calves are held after weaning and **wintered** on winter forages or hay and supplement feed and then sold as yearlings. Although 97 percent of the beef cattle in the United States are of commercial breeding, most commercial calves are sired by purebred bulls.

■ Wintered   how an animal is fed through the winter months usually consists of cheaper home-grown roughages for dry pregnant cows

Competition among purebred cattle breeders is fierce. Because of this competition, breeders have been forced to use the most modern scientific methods for breed improvement, including embryo transfer, embryo splitting, and, possibly in the future, cloning. Breeders of purebred stock have the responsibility of furnishing foundation or replacement stock to other breeders of purebred stock as well as providing bulls for commercial producers. Hybrid vigor derived from sound crossbreeding programs has placed an even greater demand on breeder producers to provide F_1 females for use in commercial breeding programs. Hundreds of beef breeds are available. They are generally divided into three groups: (1) standard breeds, (2) exotic breeds, and (3) breeds of American origin (see table 7–6).

BEEF BREEDS

No one breed excels above all others in all points of beef production. Hence, some choice must be made based on production goals, environment, and availability of replacements and breeding stock and so forth. See insert of Plates A through D.

Amerifax

Amerifax is a cross between American Beef Friesian and Angus (see figure 7–1). The cattle are noted for good calving and mothering ease. Both sexes reach puberty at an early age, so they can be successfully used for breeding as yearlings. Amerifax grow rapidly for the first 2 years and then level out before they are too large to process. Both sexes display a gentle disposition. They are polled and solid red or black in color. Cows weigh 1,000 pounds on average.

FIGURE 7–1 One of the newest American breeds of cattle, Amerifax are genetically five-eighths Angus and one-eighth Beef Friesian.
Courtesy of Amerifax Cattle Association.

Angus

The American Angus that we know of today started with four bulls from Scotland in 1873. Originally they were thought of as freaks of nature because of their naturally polled heads and solid black color. The Angus has a smooth hair coat and is comparatively small in size; bulls weigh up to 1,900 pounds, and cows weigh 1,500 pounds. The Angus is small at birth and small at weaning. Desirable traits attributed to the Angus include cold tolerance, mothering and milking ability, early maturity, little calving difficulty, good rustling ability, high fertility, and, perhaps the most outstanding quality, a reputation for producing an excellent-quality carcass with small bones, high muscling, and a low percentage of fat covering. Common criticisms of the Angus are lack of size, overprominent shoulders in some bulls, and bulls not trailing cows in sparse open range as well as bulls from other breeds. However, when crossed with the native Texas Longhorn cows the Angus produced animals that survived well over the winter and actually had increased size and weight. There are 12,570,230 total registered Angus in the United States, making them the second most popular breed.

Ankina

Ankina Breeders, Inc., founded in 1975, establishes minimum performance standards of yearling weight ratios to determine eligibility for registration. Ankina are black or dark brown and polled; scurs are acceptable (see figure 7–2). Any mating plan that results in five-eighths registered Angus and three-eighths Chianina with proof of purity from the registration certificates is acceptable, provided the foundation animals are free from genetic defects.

Ankole-Watusi

In spite of their unusual look, Ankole-Watusi cattle possess many traits that are sought after in the cattle industry today and are used in many crossbreeding programs. Their horns are their most striking feature: they may reach 8 feet in length (largest of any breed in the world) and 9 inches in diameter at the base. This unique feature allows them to throw off excess heat; the horns offer

FIGURE 7–2 The result of selectively mating the maternal traits of a registered Angus with the terminal sire merits of a full-blooded Chianina is the Ankina.
Courtesy of Ankina Breeders, Inc.

FIGURE 7–3 Ankole-Watusi, the longest and largest-horned cattle in the world, originated in Africa.
Courtesy of Dr. James Blakely.

a radiator-like effect for cooling so that the cattle can survive on the hot plains of Africa (see figure 7–3).

Barzona

Barzona is another breed that originated in the United States. The Bard Kirkland ranch in Arizona during the 1940s and 1950s first produced this newly recognized breed. Mixing one-fourth each of Africander, Hereford, Angus, and Santa Gertrudis cattle produced the Barzona. It is medium size with medium to heavy weaning weights. The Barzona's desirable traits include the ability to thrive under harsh conditions, heavy weaning weights, superior feed conversion, fast growth, heat tolerance, and mothering ability. It is widely adaptable to conditions that include mountainous ranges, dry regions, and other locations with sparse vegetation. The Barzona has been criticized for lack of uniformity compared with the more established breeds. The Barzona is red in color and has large horns. The cattle are well adapted to the deserts of Arizona (see figure 7–4).

FIGURE 7–4 The Barzona breed was developed for the Arizona range, but it performs well in most other areas as well. Courtesy of Dr. James Blakely.

Beefmaster

Beefmaster cattle were developed in Southern Texas by Tom Lasater in 1931. They were created to perform more efficiently in the harsh environment of the southwestern United States. Lasater began a systematic crossing of Hereford, Brahman, and Shorthorn cattle. The result was a new breed of beef cattle that were 50 percent *Bos taurus* (English) and 50 percent *B. indicus* (humped cattle). Beefmaster are predominantly red in color and most are horned.

Belted Galloway

Belted Galloway may be one of the oldest breeds known to humans according to written records that date back to 1530. They were brought to the mountainous and wooded southwestern region of Scotland by the Viking hordes that overran what is now Great Britain in the ninth century. They named this region Gaul after their king, and it later became known as Galloway. These small and hardy animals retain that name today. Typically the cattle are black or dun in color with a white belt engirding their bodies. Protected by a fine, mossy undercoat and rough outer hair, they can survive the harshest of environments.

Blonde d'Aquitaine

The Blonde d'Aquitaine has been bred since the sixth century in Europe, where they were used as draft animals. This use explains their muscle development, hardiness, and docile temperament. Some are polled; those that are not have horns that are thick at the base and light in color, graduating to darker tips. This breed is established and catalogued mainly in France.

Braford

The Braford was developed in the United States by crossing Herefords and Brahmans, with a characteristic five-eighths Hereford and three-eighths Brahman standard. This heritage gives the characteristic medium size with unusual color patterns such as brindle, mottled, and various combinations that one might expect using the Hereford and Brahman stock available (see figure 7–5). The Braford association, however, does not believe that a color is economically

FIGURE 7–5
Crossing the Brahman and Hereford breeds gave rise to the Braford breed.
(a), Braford bull,
(b), Braford cow-calf pair.
Courtesy of the International Braford Assoc. Inc.

(a)

(b)

relevant so no color standard exists. The Braford produces medium-sized calves with heavy weaning weights. The desirable traits are ability to thrive under little management; heat, insect, and disease tolerance; good mothering ability; and efficient conversion of feed, especially forages, to beef. The Braford lacks the cold tolerance necessary for production in extreme northern states.

Brahman

Brahmans have a very diverse history in many countries of the world. The American Brahman, however, was developed from four predominant Indian breeds with a beef standard in mind. With all of this at hand, a new bovine species emerged that would thrive in a region and climate where insects, diseases, and temperature extremes are common. Brahmans still contribute to the economies of many countries of the world today. Brahmans are the sacred cattle of India. They are usually steel gray in color, but solid red is also common. They are easily identified by their drooping ears and prominent hump over the withers. Bulls weigh 1,800 to 2,000 pounds, and cows weigh approximately 1,200 pounds.

Brahmousin

Brahmousin are classified as five-eighths Limousin and three-eighths Brahman. This breed is ideal for southern parts of the United States where heat and insect tolerance are important. Brahmousin cattle are primarily polled and are noted for good mothering and optimum milk production. They are red, light red, or blonde in color.

Brangus

Oklahoma and Texas shared in the development of the Brangus, which is three-eighths Brahman and five-eighths Angus, in the 1930s. The color is black with the characteristic hump from the Brahman side of the breeding and the characteristic polled nature from the Angus side. The Brangus is heavy and produces medium-sized calves with heavy weaning weights. The desirable characteristics include thick, beefy conformation; good growth rate; heat tolerance; insect and disease resistance; and good mothering ability. The Brangus is criticized for having a bad disposition and lack of thickness in the hindquarters.

Charbray

Charbray is a U.S.-made breed that originated in the Rio Grande Valley of Texas. The original cross involved five-eighths Charolais and three-eighths Brahman. Today the standards are more relaxed, allowing three-fourths Charolais and one-fourth Brahman to seven-eighths Charolais to one-eighth Brahman. Light tan at birth, they usually change to a creamy white before weaning. The Charbray is large, with cows weighing up to 2,200 pounds and bulls up to 3,200 pounds. They are heat and insect tolerant (see figure 7–6).

FIGURE 7–6
Crossing the Brahman and Charolais breeds gave rise to the Charbray breed.
Photo by author.

Charolais

One of the oldest of the French cattle breeds, the Charolais are considered to be of Jurassic origin and have been developed in the district around Charolles in central France. There is historical evidence that these white cattle were being noticed as early as 878 C.E. The first Charolais came into the United States from Mexico in 1934 and grew rapidly in popularity. By 1997, more than 2 million head of Charolais had been registered and recorded in the United States. In France, Charolais cattle are used for milk, meat, and draft. Charolais are white or creamy white, with pink skin. Bulls weigh from 2,000 to more than 2,500 pounds; cows weigh 1,250 to more than 2,000 pounds.

Chianina

The name *Chianina* (kee-a-nee-na) comes from the Chianina Valley in the province of Tuscany in central Italy. It is one of the oldest breeds of cattle in Italy and dates back to before the Roman Empire. Full-blooded Chianina range in color from white to steel gray and have black pigmented skin. A major study at Texas Tech University that compared Chianina with the typical British crosses showed that Chianina beef has 36 percent less fat and calories. The first test of its kind, it showed this breed to have set the standards for high-quality lite beef in the United States. Chianina is the largest breed of cattle in the world. Mature bulls weigh up to 4,000 pounds and stand 6 feet high at the withers. Cows weigh up to 2,400 pounds.

Devon

Devon is a dual-purpose breed that originated in southwestern England in the counties of Devon and Somerset. The Devon was introduced into the United States about 1623. It was one of the first breeds introduced into the New World. The color is deep ruby red, ranging to chestnut. Because of their color pattern, Devons have been nicknamed "rubies." Devons have horns that are creamy white with dark, generally black tips, a characteristic that makes them easy to distinguish from other breeds. They are medium size at maturity and produce medium calves with medium weaning weights. Their desirable traits are beefiness, dual use, and good milking ability. Their adaptability to the United States would appear to be in breeding up and crosses on existing breeds. Devons are also noted for their docility and ability to adapt to temperature extremes. However, they are criticized for slow growth and slow finishing in feedlots.

Dexter

Dexter is one of the smallest purebred bovine in the world. Bulls should not exceed 44 inches in height or 1,000 pounds. Cows average no more than 42 inches tall and 750 pounds. There are two types: a smaller dwarf animal that carries a lethal gene and a proportionate nondwarf type that does not carry the lethal gene. There are several degrees of dwarfism in Dexter cattle; the key is to produce nondwarf offspring. This practice is risky at best because the lethal gene can show up anyway.

Dutch Belted

Mainly known as a dairy breed, the Dutch Belted breed produces medium-sized cows and bulls that range from 1,200 pounds to 1,700 pounds. Their distinctive color pattern is usually black but can be red with a white belt around the middle. They originated in the Netherlands, where they are known as Lakenvelders. The term *laken* means a blanket around the body. Although always a rare breed, they neared global extinction by the 1970s. A small handful of American breeders have attempted to preserve this breed.

Galloway

Galloways originated in Scotland and were first imported into the United States in 1866. Color characteristics are black, dun, belted, white, or red. Galloways are naturally polled.

Gelbvieh

Gelbviehs originated in Bavaria (southern Germany) and are a dual-purpose breed (milk and meat). Gelbviehs are solid golden red to rust colored and are horned. Mature bulls weigh 2,000 to 2,500 pounds; cows weigh from 1,200 to 1,500 pounds.

Gelbray

The Gelbray breed was developed on the Yeager ranch in Ardmore, Oklahoma. It is a three-way cross of Angus, Gelbvieh, and Brahman. Gelbray are hardy and can withstand harsh climates. They exhibit good mothering qualities and ease of calving.

Hereford

The first Herefords, from Herefordshire, England, came to the United States in 1817. The first breeding herd was established in 1840 by William Sotham and Erastus Croning in Albany, New York. Herefords in the United States date from the Sotham-Corning origin. These cattle were tough and had the bred-in ability to survive rough ranching conditions and provide improved beef quality in the process. The trademarks of Herefords are white faces and distinctive red bodies. They can be horned or polled. The breed has adapted and flourished in every region of the United States. There are 19,600,538 registered Herefords in the United States, making them the most popular breed of beef cattle in the nation.

Limousin

The golden red Limousins are one of the oldest breeds on the European continent. Pictures of the cattle found in cave drawings near Montignac, France, date back some 20,000 years. Limousins are highly known as work animals in addition to their beef qualities. At the end of their work lives they are fattened for slaughter. Limousins were first noticed by cattlemen in the United States in the 1960s and were imported in 1971. Bulls weigh about 2,400 pounds, and cows weigh about 1,300 pounds.

Maine-Anjou

The Maine-Anjou is actually a cross between the French Mancelle breed and the English Durham. The breed association was formed in 1909 and changed the name to Maine-Anjou, taking the names of the Maine and Anjou River Valleys. It is one of the largest breeds developed in France. The coloring is very dark red with white markings on the head, belly, rear legs, and tail. The Maine-Anjou was imported into Canada in 1969. Bulls weigh about 2,500 pounds, and cows weigh 1,900 pounds.

Marchigiana

Marchigiana (mar-key-jah-nah) is the most popular breed in Italy. It makes up approximately 45 percent of the total cattle population in Italy and can be found throughout the world. Because the skin of the cattle is black under the white hair, they have a natural resistance to cancer of the eye, pinkeye, sunburn, and other skin problems. They also have a tremendous ability to thrive in extreme heat or cold. Mature bulls weigh 2,650 to 3,100 pounds, and cows weigh from 1,400 to 1,800 pounds.

Murray Grey

The Murray Grey was developed along the Murray River in Australia in the early 1900s. A light roan Shorthorn cow was bred to various Aberdeen Angus bulls and dropped only gray calves. There were only 12 cows by 1917. These 12 naturally polled cows became the nucleus of the breed. A genetic phenomenon accounts for the gray color. Murray Greys were first introduced into the United States in 1969 via imported semen. Since then the breed has expanded to almost every state in the union. Bulls weigh about 2,000 pounds, and cows average 1,200 pounds; they are considered very docile.

Piedmontese (Piedmont)

The Piedmontese were allowed to evolve over the centuries in the Alpine regions of northern Italy. These unique cattle are genetically designed to produce low-fat and yet tender beef. Originally this breed was a dual-purpose breed.

Pinzgauer

Refinement of early native red Bavarian cattle gave way to the Pinzgauer breed. Its distinctive markings include a dark red coat and a white topline that is narrow at the shoulders and becomes wider toward the rump. The tail and underline are also white. Even when bred to other breeds of cattle this rigid color pattern surfaces. Pinzgauers rarely have eye problems and are resistant to most external parasites. Cows mature at 1,300 to 1,650 pounds, and bulls top 2,900 pounds. They were imported into the United States in 1974. There are approximately 30,000 head in the United States today.

Red Angus

The Red Angus is the same breed as the Black Angus except for one unique aspect. During ancient crossbreeding programs, breeders were trying for polled animals. The result was a recessive gene that would produce red offspring, one in four to be exact. Even today, 1 in every 500 Angus in black herds is red. This characteristic is recessive so that it breeds true.

Red Brangus

Red Brangus is a cross between Red Angus and Brahman. This combination provides the carcass quality and calving ease of the Angus bloodlines with the hardiness and hustle of the Brahman.

Salers

Salers (Sa-lair) are native to the Auvergne region of south-central France. It is a very old breed known for its beef and milk production (for cheese) until recently. The cattle are deep cherry red, with white switch and spots under the belly. They were one of the last Europeans breeds to be imported into North America (1975). On average, bulls weigh 2,530 pounds, and cows weigh 1,540 pounds.

Santa Gertrudis

Touted as America's first breed, the Santa Gertrudis took the best of three worlds and produced an outstanding animal that could thrive on otherwise unproductive rangeland (see figure 7–7). The breed was named after the Spanish land grant where Captain Richard King first established the King Ranch (southern Texas). His idea to produce a rugged breed started with systematically crossbreeding Shorthorn and Hereford cattle with Brahman. The ultimate cross of Brahman and Shorthorn proved to be the optimum blend, and in 1940 the cross received official recognition from the U.S. Department of Agriculture (USDA) for three-eighths Brahman and five-eighths Shorthorn as a pure breed.

Shorthorn

The Shorthorn breed is recognized as the genetic foundation of more than 30 different breeds of cattle known today (which means the breeds can trace a portion of their parentage back to the Shorthorn), documenting the breed's influence and adaptability. Shorthorn cattle are also known for their ability to produce large amounts of milk and are used as draft animals in addition to producing quality beef. This breed also was the first to be registered (1822). The cattle can be any color but are usually red, white, or any combination of the two. Their horns are rather short and incurve. Bulls weigh 2,400 pounds, and cows weigh 1,700 pounds.

Simmental

Simmentals are from the Simme valley of Switzerland and are a triple-purpose breed (milk, meat, and draft). They were first imported into the United States in 1890. They are usually red and white spotted with white on the face, underline, and switch. Mature bulls weigh 2,500 pounds, and cows weigh 1,600 pounds.

FIGURE 7–7   (a), King 55, King Ranch herd sire. He was the first bull of the Santa Gertrudis breed to be awarded all 4 stars in the Superior Sire Program. Steps include Superior performance in fertility, growth, performance of offspring, beef quality of offspring, and absence of genetic defects. (b), Santa Gertrudis cow and calf. Developed by the King Ranch, the Santa Gertrudis was recognized as the first beef breed ever produced in America and the first world wide to be so recognized in over a century.
Courtesy of the King Ranch Archives, King Ranch, Inc.

(a)

(b)

Sussex

Sussex originated in England. They were first imported into the United States in 1883. Sussex are deep mahogany red with white switch and ivory horns. They are noted for lean meat with a high dressing percentage.

Tarentaise

Surprisingly, the Tarentaise (TAIR-en-taze) are known for their milk production in France. In France they are not dual-purpose cattle but are used solely for the production of a Gruyere-type cheese called Beaufort. However, in America, the Tarentaise has been a great asset to the beef cattle industry. They were first imported into the United States in 1973. Tarentaise have an inherent high resistance to common respiratory and shipping diseases. Mature bulls weigh about 1,800 pounds, and cows weigh about 1,150 pounds.

Texas Longhorn

Brought to the New World by Christopher Columbus from Spain (1493, Santo Domingo), the Texas Longhorn endured for 500 years as one of the hardiest and most well-known American breeds. Synonymous with the Old West, nearly 300 years after setting foot in America, millions of these remarkable animals inhabited the southwestern United States. About 10 million were driven to railheads to feed populations in the North, East, and West in the mid-1800's. However, by the turn of the twentieth century their popularity began to fade and interest turned to the new English breeds. In just 40 years of fenced-in land and plows this breed came closer to extinction than the buffalo. Texas longhorns were known for their endurance and resistance to disease and hardship. They could withstand long trail drives and survive an incredible distance without water.

White Park

White Park cattle were brought to what is now Great Britain in 55 B.C.E. by the Romans and left to run wild after the descent of Rome. This distinctive breed was eventually domesticated and kept in large enclosures similar to game preserves—hence the name White Park cattle. Remarkably, this breed was able to naturally purify itself by self-eliminating poor mothers, hard calvers, and those that lacked vigor and hardiness to survive in the wild. They are white cattle with a natural dark pigmentation of the ears, eyes, nose, front legs, and udder. They are naturally polled and are comparable in size to the Hereford. White Park cattle were first brought to the United States during World War II to preserve seedstock in case of German invasion, and most ended up in Illinois.

THE BEEF CARCASS

Although indication of quality can be noted in the live animal, the final test of a beef animal is slaughtering to yield the product demanded by consumers. Packer buyers use selection techniques to determine the price they are willing to pay for an animal. The price reflects what buyers consider to be a reasonable gamble that will allow their companies to make a profit. If the buying decisions are right most of the time, the buyers remain buyers.

Slaughtering Process

Veterinarians provide premortem (before-death) inspection. If an animal is determined for any reason to be unfit for human consumption, it is tagged and must be used for nonedible products. Examiners who find animals questionable tag them as suspect; the final decision rests with the veterinarians doing the postmortem (after-death) inspection, which all carcasses must pass. Animals passing the premortem inspection proceed directly to the kill floor.

Stunning. For humane reasons, in both the United States and in Europe animals to be slaughtered are rendered unconscious as quickly and painlessly as possible without causing the heart to stop pumping. Because of safety factors, most stunning is done by a captive bolt (see figure 7–8). Except for kosher-slaughtered animals (which are not stunned), this procedure is used almost nationwide in the United States. Which method is most humane in the slaughter of animals is controversial, but both attempt to minimize the pain to the animal.

FIGURE 7–8
Animals are rendered unconscious or stunned by a device known as a captive bolt.
Courtesy of Michael De La Zerda, Texas A&M University.

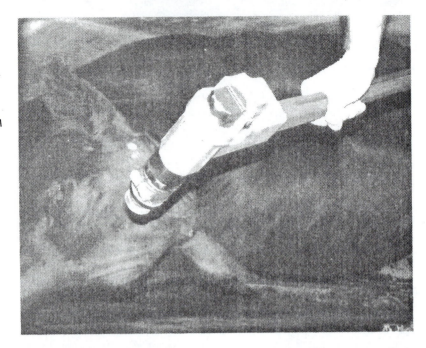

FIGURE 7–9 The stunned body is hoisted, the jugular veins and carotid arteries are severed by sticking with a sharp knife, and blood is drained with the aid of the still-beating heart.
Courtesy of Michael De La Zerda, Texas A&M University.

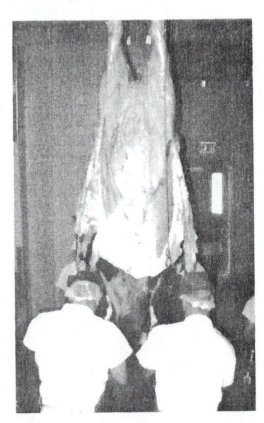

FIGURE 7–10 Skinning involves removal of the hide, feet, and head.
Courtesy of Michael De La Zerda, Texas A&M University.

Sticking.   The stunned body is hoisted by the rear feet using special shackles (see figure 7–9). Because the heart is still beating, the blood is actively pumped from the body after a sharp knife is used to sever the jugular veins and carotid arteries in the neck. After the heart ceases to beat, the carcass continues to drain blood because of the hanging position. Good drainage is important to the appearance of the retail cuts.

Skinning.   After drainage is complete, the animal is lowered on its back so that the hide may be partially removed. The carcass is rehoisted (see figure 7–10), and the skinning process is completed by removal of the feet, hide, and head. The carcass is then eviscerated (intestinal organs removed), as shown in figure 7–11. The kidneys are left in the body. The carcass and viscera are inspected, and parts (such as the liver) may be rejected for human consumption. In some cases, the entire carcass is condemned.

Halving.   The tail is removed and the backbone split (see figure 7–12) with a power-driven saw to create two halves—a tight side and a loose side. The left half is always the tight side, so called because the fat adheres closely to the kidneys and backbone, giving a more desirable appearance to such retail cuts as T-bone steaks.

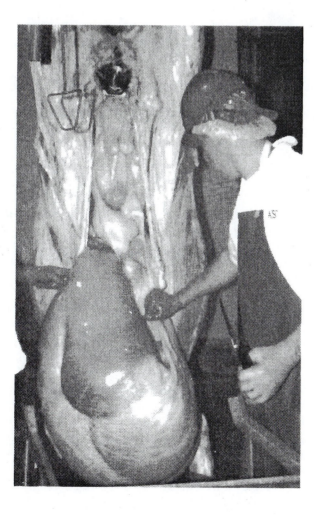

FIGURE 7–11
Evisceration allows inspection of the carcass and viscera to determine suitability for human consumption.
Courtesy of Michael De La Zerda, Texas A&M University.

FIGURE 7–12   The carcass is halved by removing the tail and splitting the backbone with a power-driven saw. Courtesy of Michael De La Zerda, Texas A&M University.

FIGURE 7–13   The carcasses are washed, cooled, and cut between the 12th and 13th ribs to expose the loin eye. Courtesy of Michael De La Zerda, Texas A&M University.

Cooling.   Before further inspection, the hot carcass halves are washed; some may be shrouded and then sent to the coolers, where the temperature is held at 34°F, for 24 hours before ribbing. USDA inspectors inspect and pass on wholesome carcasses. An inspection stamp on the meat signifies its acceptance as fit for human consumption. These inspections are mandatory in the United States on the state and national levels.

Grading.   A knife is used to cut between the 12th and 13th ribs (ribbing) to the backbone (see figure 7–13). This cut exposes the loin-eye area and condition that, along with other observations, are used to determine the quality of beef represented by a value given by a government grader (discussed later in this chapter).

Dressing Percentage.   Yield, as it is often called, is the percentage of the product left in the cooler as compared with live weight. Conformation and finish influence dressing percentage the most, provided a normal fill is allowed. The more muscular, fatter animals are expected, on the average, to grade higher and dress (yield) higher (see figure 7–14). Because the viscera are part of the weight subtracted from the live weight to arrive at dressing percentage, it is easy

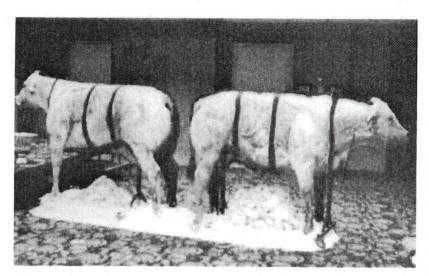

FIGURE 7–14
Carcasses may differ in muscling, grade, and other factors. In this famous frozen steer exhibit from Colorado State University, the steer on the right shows an excessive fat trim (note the trim under the belly) compared with the steer on the left. Reprinted with permission from J. Blakely and D.H. Bade, *The Science of Animal Husbandry*, 6th ed. (Englewood Cliffs, NJ: Prentice-Hall, 1994).

TABLE 7–7	Comparison of the Dressing Percentages of Various Livestock Species
Beef	1,000 pounds = hide (80 pounds) + offal (300 pounds) + carcass (620 pounds) Dressing percentage = 52 (canner) to 63 (choice and prime)
Lamb	100 pounds = pelt (15 pounds) + offal (35 pounds) + carcass (50 pounds) Dressing percentage = 50
Pork	230 pounds = offal (50 pounds) + carcass (172 pounds) Dressing percentage = 70 to 74
Chicken broiler	4 pounds = feathers (0.3 pound) + offal (0.8 pound) + carcass (2.9 pounds) Dressing percentage = 73

to see how an extra fill of water or feed can add to selling weight but lower yield. Packer buyers take the possibility of fill into account when bidding. If there is any question, buyers bid very conservatively, so it is to sellers' advantage not to allow the cattle undue fill just before sale.

Dressing percentage varies with each animal, but some average figures for the grades involved are prime and choice, 63 percent; standard and good, 60 percent; commercial, 56 percent; and utility, cutter, and canner, 52 percent. See table 7–7 for a comparison of dressing percentages with various livestock species.

Aging.   For optimum tenderness, beef carcasses that are fat enough to seal out air can be aged for 2 to 5 weeks at temperatures between 34 and 38°F. This temperature is slightly above freezing and allows the natural enzymes to break down the connective tissue (collagen) surrounding the cells, tenderizing the meat.

Slaughtering versus Animal Rights

Opinions vary widely as to ethical methods of slaughtering cattle or whether slaughter of cattle is ethical at all. On the one hand, the Hindu belief of *Ahimsa* (noninjury) emphasizes the respect of and consideration for human and animal life. The greatest care must be taken to prevent the injury of any living being, engendering vegetarianism.

In the Jewish faith, both the taking of animal life and the eating of flesh are governed by laws written in the Bible to show respect for the animal to be killed and at the same time to prevent callousness or indifference to killing. A special process known as *Shehitah* must be followed for the killing and eating of animals. *Shehitah* laws are designed to secure the maximum freedom from pain. The slaughter method includes the quick cutting of the trachea, esophagus, and carotid and jugular blood vessels without any interruption in the cutting and with a nontearing knife. Animals killed by hunting or by stunning first and animals that show any organic blemishes are not considered kosher, even if they fall into the category of a kosher animal.

For most Americans, animal protein is a necessity for both good diet and pleasurable cuisine. Most people in the United States ask only a reasonable assurance of noncruelty to animals.

Kosher Slaughter

Beef used for the Jewish market must be slaughtered according to religious customs. A special double-edged, razor-sharp knife is used to sever the jugular veins and carotid arteries. Stunning is not permitted, and only one stroke of the blade is allowed. The animal must be killed in such a way as to minimize any pain or anxiety of death. Jewish law forbids the consumption of the meat of animals injured (stunned or otherwise) before slaughter. In addition, Jewish law forbids the consumption of undrained beef. Because the hindquarters are considered to have small veins, they are not utilized in the Jewish trade; only the forequarter is marketed. In some areas with large Jewish populations, all cattle are kosher slaughtered to cater to the local population.

Animal rights
philosophy that animals have the same basic rights as human beings; against speciesism

We previously talked about *Ahimsa*, the Hindu belief in noninjury of other life forms. Those involved in the **animal rights** movement in the Western world believe that animals have moral rights equal to those of humans and are totally opposed to not only biomedical research using animals, sporting and other entertainment events using animals, and product testing using animals but also the use of animals for clothing and food. They claim that animals should have the same basic rights granted to human beings. The term *speciesism* denotes having a preference in treatment of one species over another. To do so, animal rights activists claim, is no better than showing preference for one race over another. This debate will most likely intensify if legislation is passed prohibiting the eating and slaughtering (by certain methods) of animals within the United States.[3] It is important that the livestock and veterinary industry become proactive in this regard prior to such legislation taking hold. One such researcher, Dr. Temple Grandin of Fort Collins, Colorado, has designed countless facilities, methods of slaughtering, and methods of handling livestock all with these goals in mind: reducing the anxiety, pain, and suffering of animals we are going to eat. Dr. Grandin is highly regarded by both the livestock and animal rights communities.

Animal welfare
philosophy that humankind has dominion over animals and, as such, has responsibility for their well-being

The majority of Americans are not part of the animal rights community. Most fall into the category of the **animal welfare** movement. Those involved in

---

[3]L. S. Shapiro, *Applied Animal Ethics* (Albany, NY: Delmar, 2000).

Shaggy Brown Highland. Photo by John Seralin.

Beefmaster. Courtesy of Beefmaster Breeders Universal.

Angus. Courtesy of Christy Collins. Used with permission.

Plate A

Red Brangus. Courtesy of Christy Collins. Used with permission.

Brahman. Courtesy of Christy Collins. Used with permission.

Charolais. Courtesy of Christy Collins. Used with permission.

Plate B

Polled Hereford. Courtesy of Christy Collins. Used with permission.

Limousin. Courtesy of Christy Collins. Used with permission.

Maine Anjou (Black 3/4 Blood). Courtesy of Christy Collins. Used with permission.

Plate C

Texas Longhorn. Photo by Francois Gohier, courtesy of Photo Researchers, Inc.

Simmental. Courtesy of Christy Collins. Used with permission.

the animal welfare movement believe that a reduced and minimal number of animals should be used in research and those that are used should be treated as humanely as possible. Humane treatment includes proper housing, disease prevention, proper nutrition, and humane euthanasia or slaughter. The concept of animal welfare implies that humankind has dominion (power or right) over animals and, as such, has responsibility for animal well-being.

From an animal science perspective there is a greater productive response of animals when they are given proper care. Dairy animals, in particular, give much more milk, demonstrate greater reproductive efficiency, and show less mastitis when they are handled in a more gentle manner. Animal science personnel need to communicate to the nonfarming public, however, the difference between what an animal wants and what an animal needs.

Carcass Grades and Grading

Beef carcass grades are based on the quality and palatability of the meat (USDA *quality grades*) and the quantity or cutout yield of trimmed, boneless, major retail cuts (USDA *yield grades*). Since 1976, both quality and yield grades have been assigned to a beef carcass. Grading is not mandatory but is a service provided by the USDA and paid for by the processor. Because consumers depend mainly on USDA quality and yield grades in purchasing, most processors find the service advantageous.

U.S. Quality Grades.   One result of consumer demand for beef is the need for a systematic way to express meat quality and carcass value. The Federal Meat Grading Service was established by an act of Congress in 1925. Standards for dressed beef were soon established, and grading of beef carcasses began in May 1927. USDA quality grades for beef have been revised since 1927 to ensure that they serve as a reliable guide to the eating quality of beef. USDA quality grades are *prime, choice, select* (*good* for calves and lamb), *standard, commercial, utility, cutter,* and *canner.* They are based on carcass characteristics such as maturity, marbling, texture of the lean meat, and color of the lean meat. Because marbling and maturity cannot be directly determined on live cattle, live quality grade is largely based on an evaluation of fatness. Thus, the amount of finish of a steer is used to determine or predict marbling. The actual grade must be assigned (determined) after slaughter.

Maturity is determined by the size, shape, and ossification of the bones and the color and texture of the lean meat; more mature animals have darker and tougher lean meat. *Marbling* refers to the intermingled fat in the lean muscle as evaluated between the 12th and 13th ribs (place of ribbing). *Texture* or firmness of the lean meat and color of the lean meat are also evaluated at the 12th and 13th ribs. Carcasses are grouped according to sex classes. Male classes consist of steers, bullocks (young bulls), and bulls. Female classes are heifers and cows.

Five different maturity groups (A, B, C, D, and E) are recognized for the approximate age at slaughter: 9 to 30, 30 to 48, 48 to 60, 60 to 96, and 96 months or older. Marbling is broken down into slightly abundant, moderate, modest, small, slight, traces, and practically devoid. The relationship between marbling and maturity with quality grades is shown in figure 7–15. As seen, the higher USDA quality grades (prime, choice, good, and standard) are reserved for animals in early maturity groups (A and B) with adequate marbling. In 1976

Relationship between marbling, maturity, and carcass quality grade*

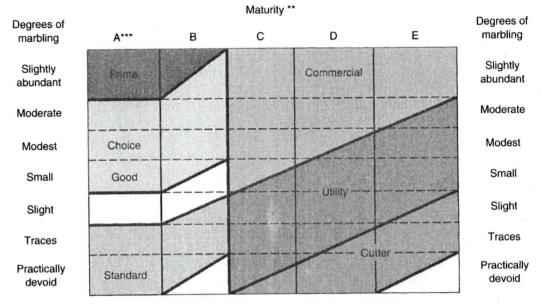

* Assumes that firmness of lean is comparably developed with the degree of marbling and that the carcass is not a "dark cutter."

** Maturity increases from left to right (A through E).

*** The A maturity portion of the Figure is the only portion applicable to bullock carcasses.

FIGURE 7–15 Relationship between marbling, maturity, and carcass quality grade. Reprinted with permission from J. Blakely and D.H. Bade, *The Science of Animal Husbandry*, 6th ed. (Englewood Cliffs, NJ: Prentice-Hall, 1994).

marbling standards were reduced and now allow more animals in the prime, choice, and good grades. This change reflects today's consumer preference for lean meat and the cattle feeders' practice of feeding animals for early marketing. Overfeeding or overfinishing animals is inefficient and not practiced in today's feedlots. Examples of U.S. quality grades for slaughter steers are shown in figure 7–16.

U.S. Feeder Cattle Grades. In 1979, feeder cattle grades were revised to reflect the expected weight of the animal when it grades low choice, yield grade 3. Thus feeder cattle grades reflect the feeding value of the animal. Feeder cattle grades are applied to cattle less than 36 months of age and are based on three general characteristics: frame size, muscle thickness, and thriftiness.

Frame size Frame size refers to the animal's skeletal size in relation to its age. Small-framed animals finish out faster than large-framed animals. There are three frame sizes. Large-framed (L) animals are tall and long bodied for their age. They will be expected to produce a U.S. choice carcass (0.5 inch fat at the 12th rib) at 1,000 pounds for heifers and 1,200 pounds for steers. Medium-framed (M) animals have slightly larger frames for their age and would be expected to produce a U.S. choice carcass at 850 to 1,000 pounds for heifers and 1,000 to 1,200 pounds for steers. Small-framed (S) animals would be expected

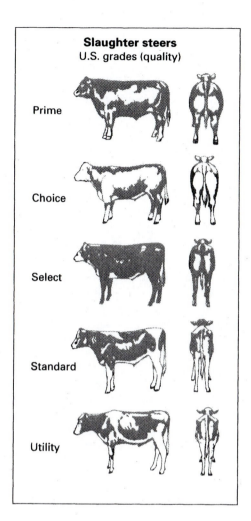

Slaughter steers
U.S. grades (quality)

Prime

Choice

Select

Standard

Utility

FIGURE 7–16
Slaughter steers.
USDA quality grades.
Courtesy of the U.S.
Department of Agriculture.

to produce a U.S. choice carcass at under 850 pounds for heifers and under 1,000 pounds for steers.

Muscle thickness   Muscle thickness refers to the relationship between muscle development and skeletal size. Three standards for thickness are designated:

Number 1: Thick muscling
Number 2: Moderate muscling
Number 3: Slight muscling

Thriftiness   Thriftiness refers to the apparent health of an animal and to its ability to grow and fatten normally. Unthrifty animals would not be expected to perform normally due to such conditions as disease, parasitism, emaciation, or double muscling. Unthrifty cattle are graded U.S. inferior regardless of size or muscling.

The resulting grades for feeder cattle are as follows:

Large frame number 1	Medium frame number 1	Small frame number 1
Large frame number 2	Medium frame number 2	Small frame number 2
Large frame number 3	Medium frame number 3	Small frame number 3

U.S. Yield Grades. Beef of the same quality grades varies widely in fatness, which directly affects the yield or cutability of beef carcasses for the retail markets. In the 1960s, retailers stressed the importance of controlling cutability or yield in carcasses as well as quality. The USDA in 1965 provided yield grades to identify the quantity or amount of salable meat from beef carcasses. There are five USDA yield grades numbering 1 through 5; yield grade 1 has the highest yield of retail cuts and yield grade 5 the lowest. Carcasses in each yield grade are expected to yield about 4.6 percent more retail cuts than carcasses in the next lower yield grade. The expected yields of boneless retail cuts for carcasses of yield grades 1, 2, 3, 4, and 5 are 52.6 percent or above, 52.6 to 50.3 percent, 50.3 to 48.0 percent, 48.0 to 45.7 percent, and 45.7 to 43.4 percent, respectively.

Yield grades, expressed as a whole number or as a percentage, are determined by an equation based on four factors: hot carcass weight, ribeye area at the 12th rib, fat thickness at the 12th rib, and estimated percentage of kidney, pelvic, and heart fat. The equation for determining yield grade is as follows: Percentage closely trimmed, boneless retail cuts from the round, loin, rib, and chuck = 51.34 to 57.84 (inches fat thickness over ribeye muscle) − 0.462 (percentage kidney, pelvic, and heart fat) + 0.740 (square inches of ribeye area) − 0.0093 (pounds hot carcass weight). Figure 7–17 gives examples of yield grades.

FIGURE 7–17
USDA yield grades for slaughter cattle.
Courtesy of the U.S. Department of Agriculture.

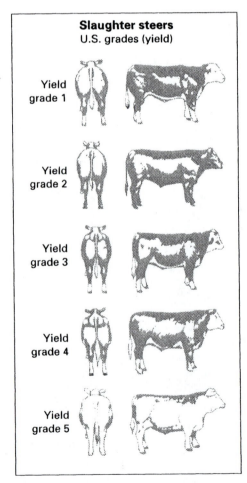

Slaughter steers
U.S. grades (yield)

Yield grade 1

Yield grade 2

Yield grade 3

Yield grade 4

Yield grade 5

Wholesale and Retail Cuts

The beef halves are usually divided into fore and hind quarters between the 12th and 13th ribs after inspection and grading and further divided into major wholesale and retail cuts as illustrated in figure 7–18. Most beef carcasses are distributed from packing houses to markets or restaurants in the form of halves, quarters, or wholesale cuts, depending on customer demands. Wholesale cuts are then broken down into the familiar retail cuts found along supermarket meat counters.

The very-low-quality grades such as cutter and canner are usually not sold in halves or cuts. The meat is boned out and sold as boneless cuts or used in a

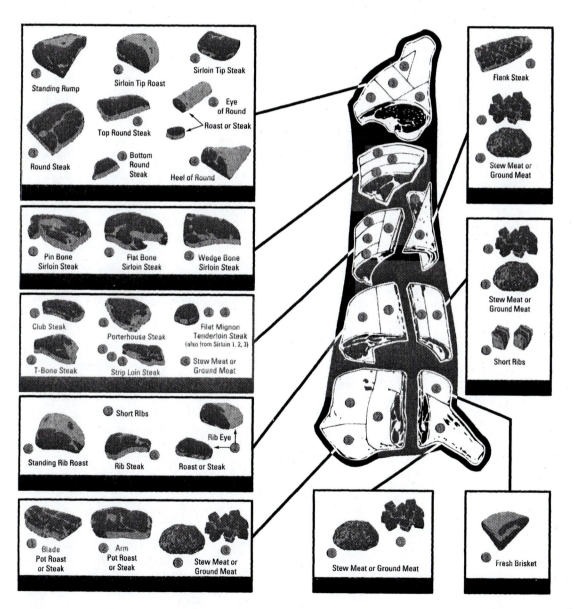

FIGURE 7–18   Major wholesale and retail cuts of beef.
Courtesy of the U.S. Department of Agriculture.

variety of prepared meats such as sausage, wieners, canned meat products, and so on. This meat is completely acceptable, edible, and nutritious. The reason for boning out is simply to make a better appearance in comparison to higher grades.

Beef By-Products

The fifth quarter   In slaughtering and marketing beef, only about 60 percent of the steer is ultimately converted to a beef carcass and less than that is converted to retail cuts. For example, a 1,000-pound market steer is divided into four quarters, which weigh about 600 pounds. From this 600 pounds (unless it is choice, then it may be as much as 633 pounds) fat and bone will be trimmed at the retail counter. Thus out of a 1,000-pound steer one will receive only 440 pounds of beef + 400 pounds of the fifth quarter (by-products). The by-products include everything of value that is produced on the killing floor other than the dressed carcass. The by-products are divided into edible and inedible products. As previously mentioned, by-products contribute over 10 percent of the packer's profit margin. Utilization of the entire animal is the goal of every packer.

1,000-pound steer = 440 pounds of beef + 400 pounds fifth quarter (by-products)

Hides make up more than half of the dollar value of all by-product sales. Shoes, belts, purses, furniture, clothing, athletic equipment, and even such musical instruments as drumheads can be made from the hide. About 60 to 70 percent of U.S. hides are bound for world trade (mostly to Japan).

Hair hides are purchased complete with hair, which is trimmed away. Upholstery stuffed with animal hair was once popular, but synthetics have wiped out this market. Artists' paintbrushes (camel hair brushes) are also made from animal hair. However, the best brushes can only use the fine hair of the animal's ear.

Hides are soaked in a salty brine to cure for a minimum of 24 hours before tanning. Most tanners are specialists and buy hides from the slaughter facilities. Heavy hides are soaked for several weeks in tanning solutions made from wood bark to give maximum pliability; chromic salts are used for lighter, less particular hides.

Feed additive for livestock   Tallow and greases are second to hides in cash value for by-products. They average 60 pounds per animal. Fats are rendered to be used as edible products for humans such as oleomargarine or in animal feeds to raise the energy content. Vegetable margarine and shortening have replaced much of their use in cooking. Edible tallow's major domestic use is as an additive in livestock and pet foods because it is a cheaper source of protein than meat. Domestic sales of pet food is in the billions of dollars per year. Fats are also used in chemicals, plastics, lubricating oils, detergents, antifreeze, paints, cellophane, and a host of other items.

Soap is made from inedible tallow. Synthetics replaced a great deal of natural soap, but environmentalists have rekindled interest in natural soap products.

Variety meats from the fifth quarter have a high nutritional content but have low appeal for most American consumers. These meats include the following:

Hearts	Spleens (milt)	Kidneys
Stomachs (tripe)	Livers	Brains
Tongues	Sweetbreads (thyroids)	Ox joints

Gelatin is made from collagen or cartilage and is used in the following edible foods:

Salads   Chewing gum   Desserts
Marshmallows   Dissolvable medicine capsules

Inedible gelatin is also obtained from by-products and is sold for the following uses:

Photographic film   Plasterboard   Printer rollers
Matches   Window shades   Wallpaper
Glue and adhesives   Sandpaper

The pharmaceutical industry has manufactured more than 140 medications derived from cattle glands. However, the total weight of glands is less than a pound per animal. Thus, processing is very expensive. Many medicines now rely on biotechnology. For example, insulin from the pancreas is now made in large silos from genetically engineered *Escherichia coli* bacteria; it took 1,500 animals to produce 1 ounce of insulin, 10,000 pituitary glands to produce 1 pound of adrenocorticotropic hormone (ACTH) used to treat human adrenal gland problems, and so forth.

STOCKER SYSTEMS

Stocker cattle are heifers used to replenish the cow herd or cattle that remain on the farm or ranch for additional gain before entering the normal American system of fattening cattle (see figure 7–19). **Stocker** is a growing phase between weaning and feeder. Cheap gains (e.g., by using public land such as that managed by the Bureau of Land Management [BLM] or the U.S. Forest Service; costs in recent years have ranged from $1.35 to $2.25 per animal unit month) are necessary during this phase. The feeder phase is a finishing phase in the feedlot (ideally in 3 to 5 months) prior to slaughter. Feed conversion averages 6 to 8 pounds per pound of gain. It is much more expensive to put on gains during the feeder phase than during the stocker phase of production.

▌**Stocker** beef animal in the phase between being weaned and ready for finishing in the feedlot. Stockers are fed for growth rather than finish

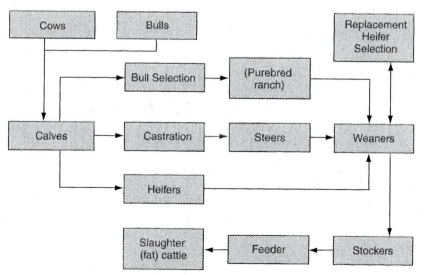

FIGURE 7–19 Flow sheet for beef production.

A 250- to 600-pound stocker can be handled by roughing through the winter, grazing on winter pasture, grazing on summer pasture, or a combination of these methods. The main objective here is to produce either replacement cattle for the herd or a mature animal for feedlot, resale, or slaughter that has been grown solely on roughage feeds. This system is the most common worldwide system of producing beef. Although it is unusual in the United States to actively produce grass-fed cattle, it is the only method in most parts of the world. Humans more efficiently utilize grains; however, roughages that the human population cannot utilize are converted through ruminants into a desirable, edible product—beef.

Animal Unit

Most cheap gains are put on cattle that are pastured on government or publicly owned land. The federal government owns 320 million acres on which grazing is allowed in the 11 western states. In 1988 alone, more than 5.6 million head of livestock grazed on this land. The management of the publicly owned land is generally delegated to either the BLM or the U.S. Forest Service. Roughly 12 million ranchers and farmers pay the U.S. government grazing fees to feed livestock on this range. The BLM has to manage not only domestic livestock but also an estimated 1.9 million big game animals (of which 1.1 million are deer). The feeding requirements of wildlife must be balanced out with the grazing capacity of the land. When additional forage is available the BLM issues grazing rights to farmers and ranchers for their livestock. A fee is charged per animal for grazing based on the expected consumption of forage. In 1996, the basic fee was $1.35 per animal unit month (AUM). An AUM is a common animal denominator based on feed consumption (see table 7–8). It is assumed that one mature cow represents an **animal unit.** Then, the comparative (to a mature cow) feed consumption of other age groups or classes of animals determines the proportion of an animal unit that the animals represent.

Ration of one mature cow = ration of five mature ewes = 1 animal unit

1,500-pound cow will consume 50 percent more feed than a 1,000-pound cow

Also, the period of time to be grazed affects the total carrying capacity. If an animal is carried for 1 month, it will take 1/12 of the total feed required to carry the same animal for 1 year. For this reason, the term *animal unit months* is used.

▓ **Animal unit**
generalized unit for describing the stocking density or carrying capacity of feed on a given area of land; one animal unit is the amount of feed a 1,000-pound cow with calf will consume

FATTENING CATTLE

The feedlot business is based on finishing cattle quickly (see figure 7–20). Utilizing a full-feed program with high-concentrate rations, thin calves weighing 350 to 750 pounds can be finished in 180 days or less. The feed conversion rate under this system can be expected to be about 7 pounds of feed per pound of gain, with an average daily gain of 2 pounds or better. Younger calves are more efficient but somewhat slower gainers than older cattle. Some feeders invariably seek thin calves, especially if they are partially mature, because of rapid weight gains from sufficient nutrition.

Another feedlot method is to use fleshy calves that have been creep fed and are conditioned to eating. These calves continue to gain at a steady rate once full feed is reinstated. This better-quality calf may range in weight from 500 to

TABLE 7–8	Animal Units
Type of Livestock	**Animal Units**
Cattle	
Cow, with or without unweaned calf at side	1.0
Bull, 2 years old or older	1.3
Young cattle, 1 to 2 years old	0.8
Weaned calves to yearlings	0.6
Horses	
Horse, mature	1.3
Horse, yearling	1.0
Weanling colt or filly	0.75
Sheep	
5 mature ewes, with or without unweaned lambs	1.0
5 rams, 2 years old or older	1.3
5 yearlings	0.8
5 weaned lambs to yearlings	0.6
Swine	
Sow	0.4
Boar	0.5
Pigs to 200 pounds	0.2
Chickens	
75 layers or breeders	1.0
325 replacement pullets to 6 months of age	1.0
650 8-week-old broilers	1.0
Turkeys	
35 breeders	1.0
40 turkeys raised to maturity	1.0
75 turkeys to 6 months of age	1.0

FIGURE 7–20
Feedlots handle thousands of cattle for fattening.
Photo by author.

800 pounds and be finished in 170 days or less. The average daily gain and feed required should be quite similar to those of thin calves. The shorter feeding period may offer a distinct advantage when the goal is a specific marketing time.

Older cattle, generally termed medium-grade yearlings, are often selected when a ration containing maximum roughage is necessary. At certain times of the year some cattle feeders have coarse roughages that must or should be utilized. Because older cattle have more fully functioning rumina, they are able to convert poor-quality roughage to beef. Research indicates that roughage fed the first half of the period followed by full-feed concentrate provides better gains and efficiency than a 50-50 mixture. Mature yearlings weighing only 700 pounds often gain up to 2 pounds per day on 10 to 12 pounds of these coarse feeds per pound of gain.

Conclusions on finishing cattle immediately indicate that large amounts of concentrated feeds are needed with small amounts of roughage and considerable skill. Weanling calves are good gainers because they are growing as well as fattening. Thus grains are economically marketed through cattle. Some roughage produced on farms and ranches that might otherwise go unused can be marketed through the use of older cattle with more highly developed digestive systems.

Growth-Promoting Substances

Growth-promoting substances are compounds that either occur naturally or mimic naturally occurring compounds. Most commonly, they include estrogen or testosterone. Growth promotants have been used successfully in the United States to increase lean production and feed efficiency since the 1960s. Growth-promoting products in the United States are manufactured in the form of implants (size of the lead in a number 2 pencil). The implant is placed beneath the skin on the backside of the animal's ear. About 63 percent of all cattle and 90 percent of all fed cattle are implanted in the United States.[4] Implanted cattle show an 8 to 12 percent increase in carcass weight per unit of feed.

Some currently used substances that promote growth and/or improve feed efficiency include the following:

■ **Ionophores** class of antibiotics that are used extensively as feed additives for cattle

1. **Ionophores:** Class of antibiotics that are extensively used as feed additives for cattle. Major ionophores include strains of *Streptomyces fungi* and monensin (*Rumensin*), *lasalocid, salinomycin, lysocellin,* and *narasin.* These compounds are also used as coccidiostats in poultry. Feeding ionophores to cattle consistently improves feed conversion efficiency and often improves daily gain. The ionophores work by changing the rumen fermentation, resulting in an increased proportion of propionic acid. The use of ionophores also inhibits the growth of gram-positive bacteria.

2. *Methane inhibitors:* Methane production reduces the efficiency of rumen fermentation. Inhibition of methane production would increase the efficiency of ruminant production as well as reduce the methane emissions (linked to global warming). Examples include chloroform, iodoform (cannot feed them), and ionophores.

■ **Probiotics** microbes used as feed additives

3. **Probiotics:** Microbes used as feed additives. Live microbial feed supplements, which benefit the host animal by improving its gastrointestinal microbial balance (yogurt—*Lactobacillus acidophilus* and *Streptococcus faecium*—and yeast), must be resistant to bile and to stomach acids. They inhibit the growth of pathogens, provide digestive enzymes, and so forth.

---

[4]Beef Facts Index, *Growth Promotants in Cattle Production,* National Cattlemen's Beef Association, Denver, CO, May 1995.

4. *β-adrenergic agents:* Repartitioning agents (norepinephrine analogs that stimulate β-adrenergic receptors) that repartition nutrients from fat to protein synthesis, causing increased muscle mass and decreased body fat (e.g., clenbuterol, cimaterol, and ractopamine).

5. *Chemotherapeutic agents:* Arsenicals, nitrofurans, and sulfonamides that promote growth activity but are not antibiotics. Some of these agents show bacteriostatic and coccidiostatic action in poultry, swine, and other animals as well.

6. *Hormones:* Estrogenic hormones implanted in the ear (see figure 7–21) are widely used to improve feed efficiency and average daily gain in cattle. *Melengestrol acetate (MGA),* a synthetic progesterone, is a common feed additive used in feedlot heifers' diets (0.25 to 0.50 milligrams per head per day) to suppress estrus and thus improve feed efficiency (5 percent) and growth rate (5 to 11 percent). *Compudose* is an ear implant (estradiol) that demonstrates a 10 to 15 percent increase in average daily gain (ADG) and a 5 to 10 percent increase in feed efficiency (FE). *Ralgro* implant (resorcyclic acid lactone from corn mold) produces a 10 percent increase in ADG and a 5 to 10 percent increase in FE. *Synovex S* implant (progesterone and estradiol) produces a 10 to 15 percent increase in ADG and a 5 to 10 percent increase in FE. *Synovex H* implant (testosterone and estradiol) produces a 10 percent increase in ADG and a 5 to 10 percent increase in FE.

Hormones have been added to beef cattle feeds since 1952, when Iowa State University announced the addition of diethylstilbestrol (DES) to cattle feed to increase weight gain and feed efficiency. In 1973, DES was pulled from the animal feed market because of claims that large amounts of the feed could cause cancer. Since the 1970s, improved growth promotants have been developed to reduce the cost of producing beef. One of the greatest challenges is marketing this beef to a public unaware of its safety.

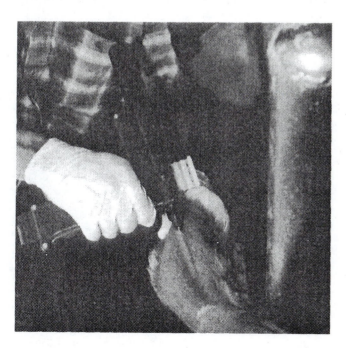

FIGURE 7–21
Ear implant used to administer growth promotants in beef cattle.
Courtesy VetLife.

BEEF CATTLE HUSBANDRY, FACILITIES, AND EQUIPMENT

Beef cattle are capable of adapting to a wide range of environmental, feeding, and management situations. This adaptability allows for a variety of managerial skills needed to process cattle, feed them, and move them to market. Still, there are some basic considerations of providing food, water, shelter, and comfort that are universal, regardless of the topography, weather, and other environmental conditions.

Shelter for Cattle

Beef cattle on pastures with some natural windbreaks may not need buildings for shelter. Cattle use windbreaks to decrease wind chill and avoid exposure to extreme cold. In regions of the United States that experience extreme temperatures, cattle need protection from wind, rain, cold, and heat. Extremes in temperatures can cause diseases and a general inability to function properly. Often a cold, rainy condition is the most dangerous; some shelter is required to block the wind and rain. Loose, open shelters (see figure 7–22) are adequate for range cattle in most areas.

Shade should be provided in areas where heat stress is common. Shade trees are preferable to man-made sources of shade. Cattle in the feedlot are often under more stress than range cattle and thus require more attention to their shelter needs. Shaded areas of various kinds have been used in feedlots. Open sheds should allow 30 to 50 square feet for each animal unit. Housing should be simply constructed, easy to clean, and reasonably priced.

The water requirements of heat-stressed cattle are much higher than normal. Increased access to water should thus be provided in extreme heat conditions. Cattle raised in hot, dry climates should have to travel no farther than 1/2 mile in rough country and no more than 2 miles on flat terrain to obtain clean, fresh water.[5]

FIGURE 7–22
Loose, open shelters such as the one shown provide enough protection for range cattle in most areas. Reprinted with permission from J. Blakely and D.H. Bade, *The Science of Animal Husbandry,* 6th ed. (Englewood Cliffs, NJ: Prentice-Hall, 1994).

[5]N. Merchen, *Guidelines for beef cattle husbandry,* Agricultural Animal Guide Workshop, Purdue University, Indianapolis, Indiana, June 12–14, 1997.

Fences for Pastures and Corrals

Fence construction throughout the world ranges from canals in Holland to thickly planted hedges or vines in France to the split rails and stones used in the United States before the invention of barbed wire. All fences serve the same function, restricting cattle to a specific area. Modern fences for range use consist of either wood or steel posts with usually two to four strands of barbed wire. Post spacing varies according to cattle concentration; common spacing is 10 to 20 feet in concentrated grazing areas or 50 to 120 feet for suspension-type fences in sparsely stocked areas. Figure 7–23 illustrates some different types of fences. Woven wire is often used instead of barbed wire in areas of concentrated grazing or where sheep also utilize pastures.

Corral fences need to be stronger and more durable than range fences. The materials used for construction of corral fences are wood, metal cables, pipe

FIGURE 7–23
Smooth, high-tensile wire is used in this state-of-the-art fence. Note that posts are center bored for wires with one inside wire offset and electrified. The crossover walk allows access by foot. Courtesy of Dr. James Blakely.

FIGURE 7–24
Osage corner posts
are quite effective in
areas of the world
where rock is readily
available.
Reprinted with
permission from
J. Blakely and D.H. Bade,
*The Science of Animal
Husbandry,* 6th ed.
(Englewood Cliffs, NJ:
Prentice-Hall, 1994).

or sucker rods, and heavy woven wire. Fences are usually 5 feet or higher with post spacing from 4 to 10 feet. The materials used vary with availability and cost.

One of the most creative forms of fencing utilizes an Osage corner post (see figure 7–24). First reported in areas of Oklahoma where large, flat, heavy rocks dot the pastureland, this type of corner post is reported to be immovable. Simplicity is the key to the Osage corner post. A large wire ring of hog netting, several feet across, is held in place by one or two metal fence posts driven into the ground to give stability to the ring in the initial stages of filling. Large rocks are then placed inside the ring without mortar and with only minimal effort in arrangement. When completed, wire strands are attached by completely encircling the stack of rocks. With so much weight and load-bearing surfaces, the pile of rocks takes on the characteristics and resistance to being moved of a boulder several times its size.

A relatively new innovation in fencing for all classes of livestock is the high-tensile wire fence. In 1973, a New Zealand sheep rancher, John R. Wall, wrote to U.S. Steel in the United States for a few thousand feet of type III galvanized wire for an experimental fence to replace barbed wire. The next year, a huge tree fell across the high-tensile wires and pinned one of the fences to the ground. When the tree was removed, the wires sprang back into place and the only repair needed was to replace staples on the posts.

Manufacturers' tests show that high-tensile wire fencing has nearly twice the breaking strength of two-ply barbed wire. Each wire stretched to the recommended 250 pounds of tensile will withstand at least 1,200 pounds of livestock pressure or extremes in temperature without losing its elasticity. This type of fencing needs to be properly grounded to protect it from lightning or the possibility of overhead electrical lines falling on it. However, high-tensile fencing will stop charging cattle, sheep cannot squeeze between the wires, horses

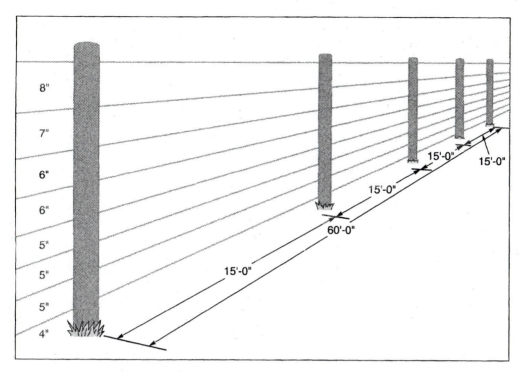

FIGURE 7–25 An eight-wire fence with line posts 60 feet apart and droppers every 15 feet has replaced barbed wire in many parts of New Zealand and Australia.
Reprinted with permission from J. Blakely and D.H. Bade, *The Science of Animal Husbandry,* 6th ed. (Englewood Cliffs, NJ: Prentice-Hall, 1994).

cannot weigh it down, and the cost is about the same as a five-strand, two-ply barbed wire fence.

The estimated life is 35 to 50 years, and high-tensile wire requires a different type of installation, corner posts installed to manufacturers' specifications, and specific tools for the fence to perform as it should. Figure 7–25 shows the recommended spacing of line posts and setting of wires.[6]

Pens and Corrals

The arrangement and sizing of pens is important in feedlots and confined beef operations. Cattle pens should be easily accessible by livestock, and vehicles afford ease in feeding, cleaning, and caring for the animals, provide comfort and adequate room for all animals, and allow for future expansion. Figure 7–26 illustrates a well-designed pen layout for a modern feedlot. Pens should be well drained and even cemented, if possible, in wet areas with heavy rainfalls.

Pens usually should allow 40 to 50 square feet of space for each mature animal. The number of animals per pen varies from 50 to 200, depending on the management plan of the operation. All pens should have adequate feed and water facilities for the cattle. Feed bunks should be 30 to 36 inches wide and about 20 inches from the ground. Each animal should be allowed 20 to 24

---

[6]England Distributing, Route 3, Bloomfield, IA 52537.

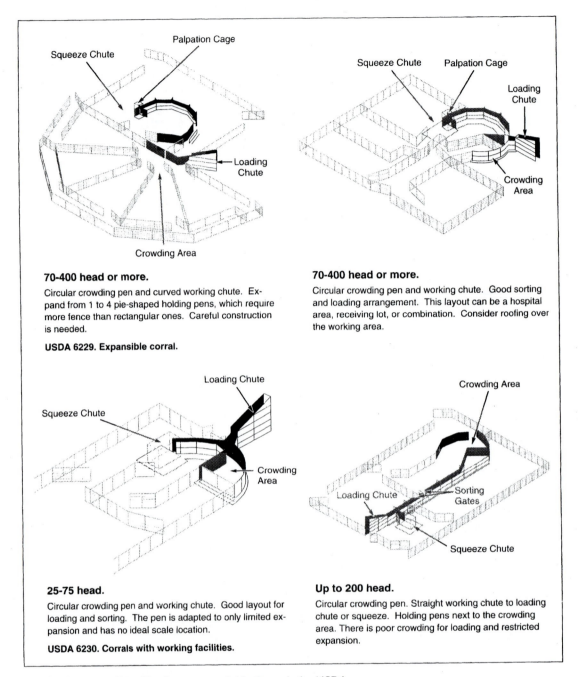

70-400 head or more.

Circular crowding pen and curved working chute. Expand from 1 to 4 pie-shaped holding pens, which require more fence than rectangular ones. Careful construction is needed.

USDA 6229. Expansible corral.

70-400 head or more.

Circular crowding pen and working chute. Good sorting and loading arrangement. This layout can be a hospital area, receiving lot, or combination. Consider roofing over the working area.

25-75 head.

Circular crowding pen and working chute. Good layout for loading and sorting. The pen is adapted to only limited expansion and has no ideal scale location.

USDA 6230. Corrals with working facilities.

Up to 200 head.

Circular crowding pen. Straight working chute to loading chute or squeeze. Holding pens next to the crowding area. There is poor crowding for loading and restricted expansion.

FIGURE 7–26 Plans like these are available through the USDA.
Reprinted with permission from J. Blakely and D.H. Bade, *The Science of Animal Husbandry,* 6th ed.
(Englewood Cliffs, NJ: Prentice-Hall, 1994).

inches at the feed bunk. A clean water source is a must. Water facilities vary in size, shape, and design. Concrete and metal troughs are the most commonly used water troughs. Drinking fountains have been used successfully to supply fresh water at all times. Water troughs should be heated in cold regions to prevent freezing and shaded in hot regions to keep water cool. Troughs

FIGURE 7–27
Various types of
watering systems can
be used for beef
cattle. The water
trough at top supplies
water to three
different pastures.
Courtesy of the U.S.
Department of Agriculture.

(a)

(b)

should be cleaned regularly for sanitation and disease prevention. Figure 7–27 illustrates feasible examples of watering systems from which a breeder may choose.

The layout of a working corral is one of the most important points in feedlot or beef cattle equipment design. Corrals are highly specialized facilities built to assist in the handling, sorting, and restraining of both cows and calves. Good corrals consist of a crowding area or holding area that leads into a working chute. The chute allows for sorting cows and calves into desired groups, into separate pens, or into other chutes. Sorted cattle can then be moved through scales, a restraining chute, loading chute, or holding pens for future care. Figure 7–28 provides only one example of a well-designed corral layout. Note both the rounded design of the crowding pen to avoid cornering and injuring cattle and the working facilities (squeeze chute and scale that are covered and well protected).

FIGURE 7–28 An example of a well-designed corral and working area. Reprinted with permission from J. Blakely and D.H. Bade, *The Science of Animal Husbandry*, 6th ed. (Englewood Cliffs, NJ: Prentice-Hall, 1994).

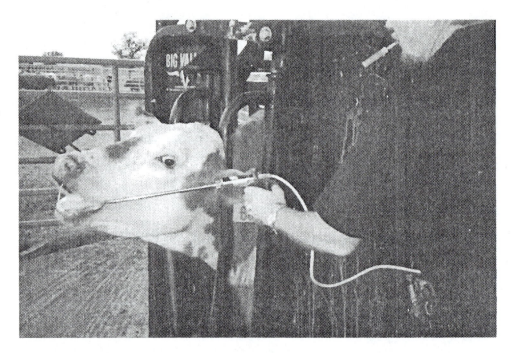

Restraining Equipment

Most animal care procedures, such as dehorning, vaccination, castration, weighing, identifying, treating for sickness or disease, and artificial insemination, require restraining of individual animals. Early restraining was accomplished by using ropes (roping the animal in a pen using tow ropes pulling in opposite ways). The popular rodeo event of steer roping, restraining a steer by roping the head and hind legs, began in this way. Another method was to throw the animal by a rope, as shown in figure 7–29. The principle is to make a loop over the neck, a half hitch around the heart girth, and another half hitch around the flank-loin area. Another rope or halter holds the animal to a post or fence. By pulling straight back on the rope, the half hitch is tightened, resulting in a paralysis of the hind legs due to pressure on the nerves. A very large animal can be thrown to the ground in this way with surprising efficiency and can be worked on with ease. This method is commonly called *casting*.

Today's facilities for restraining can be a chute with a blocking gate, a chute with a head clamp, a squeeze chute, or the modern chute that performs the functions of head clamp, squeeze chute, tilting table, and scale all in one. One person operates hydraulically controlled levers. All heavy-use chutes should be concrete, just wide enough to allow one animal to move through at a time, and designed to keep slippage at a minimum.

Equipment for Specific Functions

Castration.   *Castration* is performed to reduce animal aggressiveness, to prevent physical danger to other animals in the herd and to handlers, to enhance reproductive control, to manage genetic selection, and to satisfy consumer preferences regarding the taste, texture, and tenderness of beef. Although intact

FIGURE 7–29 Even very large animals can be thrown and restrained by a rope.
Reprinted with permission from J. Blakely and D.H. Bade, *The Science of Animal Husbandry,* 6th ed. (Englewood Cliffs, NJ: Prentice-Hall, 1994).

cattle (noncastrated) gain more weight on less feed,[7] the aforementioned safety and palatability considerations usually outweigh this slight economic consideration.

Castration should be performed as early as possible (2 to 2 1/2 months of age) to minimize stress to the animal. Local anesthetic should be considered for animals over 4 months of age. Many seed stock kept for possible use as replacement bulls are later culled because they fail to meet minimal breeding expectations. For these and any other older animals being castrated, anesthesia and techniques and procedures to control bleeding must be used.[8] Bulls castrated after 8 months of age develop a staglike appearance (which is a very objectionable characteristic for animals leaving the feedlot), and they tend to bleed more and have greater weight loss than those castrated earlier.

Cattle breeders have long practiced castration by use of a sharp knife. In this method, the scrotum is washed with an antiseptic solution and the bottom third of the scrotum is removed by the knife. The testicles are pushed out of the scrotum, and the cords are either severed or crushed using an emasculator (one edge has teeth and one edge cuts). The emasculator can be used to crush the spermatic cord versus cutting through it with a knife. The wound is dusted with a powder to prevent infection and to aid in healing. This system allows drainage and fairly fast healing. An alternate method is to slit the scrotum vertically on both sides and remove the testicles through the slits. Care should be taken to slit the scrotum low enough to allow fluids to drain during healing. These methods should only be performed at times of the year when flies are not a problem.

---

[7]D. G. Addis, Research Reports, University of California.

[8]S. E. Curtis, *Guide for the Care and Use of Agricultural Animals in Agricultural Research and Teaching,* 1999.

FIGURE 7–30
Castration equipment.
Reprinted with
permission from J. Blakely
and D.H. Bade, *The
Science of Animal
Husbandry,* 6th ed.
(Englewood Cliffs, NJ:
Prentice-Hall, 1994).

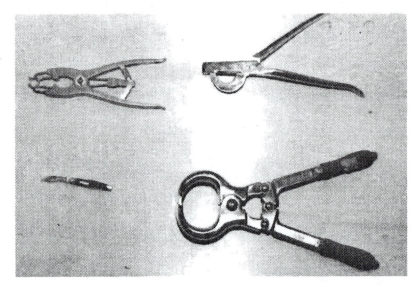

▓ **Elastrator**
bloodless means of
docking tails or
castrating using rubber
bands to cut off the
circulation, causing that
part of the body to
wither and then slough
off

Another method of castration is use of either an **elastrator** or a *Burdizzo* pincer (see figure 7–30). The elastrator stretches a rubber ring that is released around the scrotum right above the testicles. Blood circulation is stopped below the ring, and the testicles and that part of the scrotum actually die and slough off. This method works best on calves under 2 months of age. Bands without special elastrators should not be used for castration of calves older than 1 week of age. When the elastrator is used, tetanus antitoxin should be administered simultaneously.

The Burdizzo, a bloodless castrator, works much like the elastrator, with the cords crushed by the pincer, cutting off circulation to the testicles only. The lower part of the scrotum is not affected. Cords from each testicle are pushed to the side of the scrotum, where they are clamped about 1 1/2 to 2 inches above the testicle. There is no advantage to using anesthesia with this method.[9]

Chemical castration[10] involves placing an injectable castrating agent directly into the testicles of young bulls weighing up to 150 pounds. The solution also desensitizes the treated area once it is injected, reducing or eliminating shock or stress from irritation or swelling. The product creates some swelling, however. Within 60 to 90 days the testicles and spermatic cords are destroyed.

▓ **Dehorning**
removal of horns or
their developing cells
from cattle, sheep, and
goats

Dehorning.   **Dehorning** of cattle is necessary to prevent bruises, damaged hides, and other injuries to other animals, especially during their transport and handling. Hornless cattle require less space at the feed bunk and water trough and are less hazardous for humans to work with. Cattle should be dehorned while they are still young calves. Young animals can be dehorned with less stress or shock to the animal. When it is necessary to dehorn cattle older than 1 month of age, local anesthesia (*cornual nerve block*) should be used to control pain. Cattle should be observed for shock, infection, and hemorrhage after the dehorning procedure. Many methods are used to dehorn cattle (see figure 7–31);

---

[9]C. C. Chase Jr., R.E., Larsen, R.D. Randel et al. *Journal of Animal Science* 73:975–980, 1995.
[10]Chem-Cast.

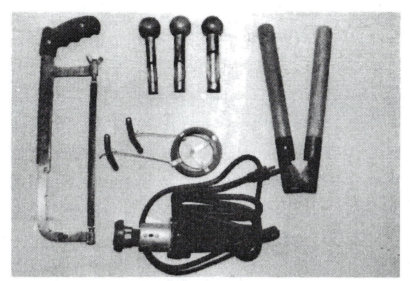

FIGURE 7–31
Dehorning equipment.
Reprinted with
permission from
J. Blakely and D.H. Bade,
*The Science of Animal
Husbandry,* 6th ed.
(Englewood Cliffs, NJ:
Prentice-Hall, 1994).

the one used depends on the age of the animal to be dehorned and the experience of the person doing the dehorning. Common methods of dehorning include the following:

1. *Dehorning saw:* Only used on mature animals and the horn is only tipped 2 inches or so. It is used primarily in the range country for cattle over 4 months of age.
2. *Tube dehorner:* Used when a calf is very young (it scoops out the horn button). Bleeding is stopped by grabbing the artery with hemostats and pulling it; heat, blood stopper, or the like may also be used.)
3. *Barnes dehorner:* Used when the calf is 5 to 6 months old.
4. *Keystone:* Used for an older calf.
5. *Superior:* Used for 3- to 4-month-old calf.
6. *Electric dehorner:* Consists of applying a specially designed hot iron to the horn bud of young cattle. This system is bloodless and should only be used on calves with horn buttons less than 3/4 inch in length.
7. *Mechanical:* Long-handled keystone.
8. *Chemical dehorning:* Involves an alkali paste (potassium or sodium hydroxide) placed around the horn button during the first week of life. The paste destroys the growth or development of the horn. Care must be taken so that the paste does not drip into or get rubbed into the eye of the calf or (for nursing calves) get rubbed onto the udders of cows.

Branding and Identification.   Many options are available for identification of cattle, including ear tags, ear tattoos, photographs of color markings, and, the most popular, branding (see figure 7–32). Branding allows for identification of ownership as well as a permanent number for individual identification. Two methods of branding are commonly used: the traditional hot iron brand and the freeze brand. For the hot iron brand, the iron should be hot but not quite red hot and be held firmly on the animal for 5 seconds; electric brands can also be used in this method. For freeze brand, a newer method, copper irons kept in dry ice or liquid nitrogen are held firmly on the animal for about 30 seconds. White hair grows on the branded area about 3 months after branding. Although it can be

FIGURE 7–32
Equipment used in identification of cattle, (a). Freeze branding, (b). White hair grows on the branded area. Photo (a) reprinted with permission from J. Blakely and D.H. Bade, *The Science of Animal Husbandry,* 6th ed. (Englewood Cliffs, NJ: Prentice-Hall, 1994). Photo (b) by author.

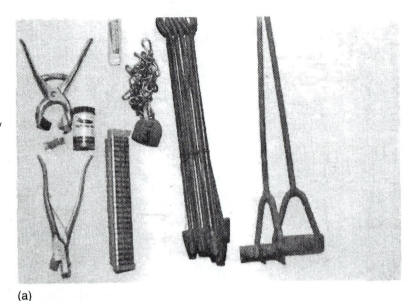

(a)

(b)

used on white breeds because of the different shade of white that grows in the brand, this method is especially suited for dark-colored cattle because of the ease of reading the white brand. Individual cow identification is more common on large ranches. The percentage of beef cows using individual cow identification is summarized in table 7–9.

Vaccination and Drenching.   Disease and parasite prevention is necessary throughout the lives of most animals. Vaccination for common diseases in the area is easily done at branding and castration time by use of various syringes. Periodic drenching using a drenching gun (see figure 7–33) helps fight parasites and other intestinal disturbances.

Vaccination schedule for beef cattle   Cattle are vaccinated for infectious bovine rhinotracheitis (IBR), bovine viral diarrhea (BVD), bovine respiratory

TABLE 7–9	Percentage of Beef Cows Using Individual Cow Identification and Their Methods (as of January 1, 1997)
Methods	**Cows (%)**
Hot iron brand	14
Freeze brand	2.7
Ear notch	6.2
Microchip transponder	0.0
Brucellosis ear tag	13.1
Other metal ear tag	1.9
Plastic ear tag	56.8
Ear tattoo	9.6
Other method	2.2
None	30.2

Source: USDA, *Reference of 1997 Beef Cow-Calf Management Practices,* Part I. June 1997.

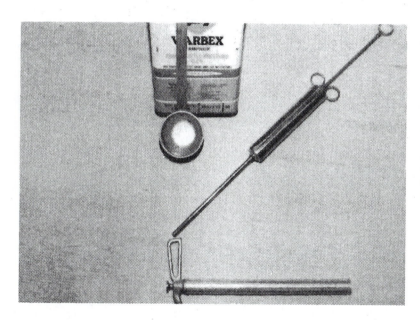

FIGURE 7–33
Drenching guns and systemic pour-on insecticides help in combating parasites and diseases.
Reprinted with permission from J. Blakely and D.H. Bade, *The Science of Animal Husbandry,* 6th ed. (Englewood Cliffs, NJ: Prentice-Hall, 1994).

syncytial virsu (BRSV), Parainfluenza-3 virus (PI_3), *Haemophilus somnus* Pasteurella, blackleg, malignant edema, and leptospirosis at 4 to 6 months of age; boosters are administered 2 to 4 weeks later. They are vaccinated for brucellosis in calfhood. Each year before breeding, cattle are vaccinated for IBR, BVD, leptospirosis, trichomoniasis, and campylobacteriosis.

Spraying and Dipping.   External parasites are controlled by use of insecticides sprayed with a high-pressure pump (see figure 7–34) or by use of a dipping pit (see figure 7–35). Either method will effectively control most external parasites (e.g., lice and summer flies). A routine procedure for range and feeder cattle of the Southwest is the use of a systemic pour-on insecticide to prevent the development, under the hide, of the heel fly larvae (grub or ox warble). Dr. Temple Grandin has developed a dipping vat with a curved chute, round

FIGURE 7–34
Modern spraying devices can be calibrated to produce a mist automatically over a large, specified area.
Reprinted with permission from J. Blakely and D.H. Bade, *The Science of Animal Husbandry,* 6th ed. (Englewood Cliffs, NJ: Prentice-Hall, 1994).

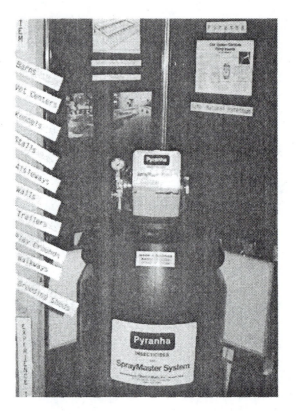

crowd pen, and wide curved alley.[11] These features lessen the anxiety among the herd, thus lowering stress and depression.

Parasites continue to be an economic problem to the cattle rancher or farmer; 72.8 percent of all producers dewormed some of their cattle in 1996.[12] Most dewormed based on tradition rather than an assessment of animal condition or laboratory testing. Parasites cost the cattle industry $200 million a year in lost production.[13] Common relief for beef external parasites include chemical pesticides or treated ear tags for flies and deworming medicaments for various internal parasites. The most common internal parasite of cattle is the brown stomach worm. Calves from 7 to 15 months of age are primarily affected. Symptoms include scours, anorexia, and weight loss; death is occasionally seen in severe infestations. Table 7–10 summarizes common cattle parasites in the United States.

New Regulations for Animal Drugs. In 1998, the Animal Medicinal Drug Usage Clarification Act (AMDICA), a federal law enforced by the U.S. Food and Drug Administration (FDA), and AB 611 (a California statute) took effect. Both laws concern the use of over-the-counter (OTC) drugs. An OTC drug can be purchased without prescription. However, the label dose, species, indications for use, and withdrawal times must be followed. A prescription is required from

---

[11]Dr. Temple Grandin's web page, http//www.grandin.com, Feb. 20, 2000.

[12]USDA, *Reference of 1997 Beef Cow-Calf Management Practices,* Part II. July 1997.

[13]J. Gerke, "Parasite Control," *Drovers Journal* 125(3):24, 1997.

FIGURE 7–35 The use of dipping vats ensures good coverage of insecticide over the animal's body.
Reprinted with permission from J. Blakely and D.H. Bade, *The Science of Animal Husbandry,* 6th ed. (Englewood Cliffs, NJ: Prentice-Hall, 1994).

TABLE 7–10	Common Cattle Parasites in the United States	
External Parasites	**Cause and Types**	**Control**
Flies	Horn fly, face fly, heel fly, housefly, horsefly, deerfly, stable fly, and screw worm fly	Aerosol, bait, biological, dusting powder, ear tags, feed additive, spray pour-on, and traps
Gnats	Buffalo gnats *(Simulidae)*	Environmental aerosol, dips, ear tags, sprays, and pour-ons
Ticks	Rocky Mountain wood ticks, blacklegged, American dog, and Boophilus	
Lice	*Damalinia bovis*	Dip, dusting powder, ear tags, injectable, and spray
	Haematopinus eurysternus	
	Linognathus vituli	
Mange mites	*Sarcoptes scabiei*	Injectable, pour-on, powder, spray, and dip
	Psoroptic ovis	
	Chorioptes bovis	
	Demodex bovis	
	Psorergates bos	

Internal Parasites	**Cause**	**Control**
Brown stomach worm	*Ostertagia ostertagi*	Block, drench, feed additive, injectable, and intraruminal
Cooperia	*Cooperia punctata*	
Hookworm	*Bunostomum phlebotonum*	
Large stomach worm	*Haemonchus placei*	
Nodular worm	*Oesophagostomum radiatum*	Paste, pour-on, and soluble powder
Tapeworm	*Moniezia expansa*	
Thread-necked worm	*Nematodirus helvetianus*	
Threadworm	*Strongyloides papillosus*	
Trichostrongylus	*Trichostrongylus axei*	
Whipworm	*Trichuris*	
Lungworm	*Dictyocalus viviparus*	
Liver flukes	*Fasciola hepatica*	

Source: J. Gerke "Parasite Control Guide." *Drovers Journal* 125(3):52, 1997.

a licensed veterinarian anytime an OTC drug is used in an extralabel manner. The prescription is required to ensure the quality and safety of the beef being produced. The veterinarian's label must contain information for extended withdrawal times, dosage, route of administration, species, indicated conditions or diseases, and precautions for use.

Grooming Equipment

Equipment used to groom cattle from head to tail is important to prevent diseases and infections (e.g., foot rot), to control external parasites, and to add beauty to the animal. Various equipment used in grooming is illustrated in figure 7–36.

SELECTION OF BEEF CATTLE

The showring used to be the main arena for cattle producers selecting their breeding stock. It involved eye appraisal or judging an animal in comparison to others in the ring. Breed characteristics played a major part in show winnings. Functional or performance based traits were also evaluated. These traits were based on the animal's conformation and helped to determine muscling, freedom from waste, structural soundness and reproductive efficiency. An animal's physical appearance, however, is only partially based on genetics (traits capable of being passed on to offspring). Thus herd and individual records (including pedigree) are extremely important in evaluating future performance. These records summarize the performance of both the animal and its ancestors and should include production testing.

Cattle Terminology

To intelligently discuss observations of the superiority of one animal over the other, it is necessary to know cattle terminology. It is also extremely important

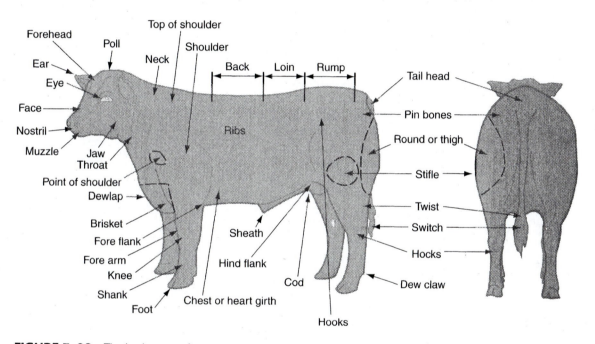

FIGURE 7–38   The body parts of a steer.

for veterinary students to learn these terms so that they can discuss structural conformation with ranchers and farmers. Preparing animals for show (see figure 7–37) and for judging contests is part of the process used to help students learn animal conformation, terminology, behavioral traits, and breed characteristics. Figure 7–38 shows the major parts of a steer that should be known before attempting an evaluation of cattle. It is more descriptive to say that an animal is thicker through the quarter with a wider, more bulging round that lets down into a deeper twist than to say that an animal is broader across the rear. Familiarity with parts of an animal indicates interest, knowledge of the field of animal science, and especially a critical eye in livestock judging. A college course in animal judging is strongly recommended for all preveterinary and all animal science students.

Judging the Steer

The steer or a class of steers should always be judged in a logical manner. Judges generally view all animals first from the side, then from the rear, and lastly from the front. In the showring, animals are usually led around in a circle so that the judge can get an overall view of them. Then the judge starts the logical sequence of evaluating all animals against each other.

The judge looks for size for age from the side view (see figure 7–39). Perhaps the most important overall observation here is length. The recent trend has been toward cattle that are showing more stretchiness because the more expensive cuts are in the area of the backbone. An animal should be straight across the topline, long from its hooks to its pins (lots of round steak), fairly deep in the flank (more round steak), and trim about the middle (less waste, more meat). A tight middle means a higher dressing percentage of 60 to 65 percent in high-quality steers. The steer should stand on sound bone (an indication of muscling) and should have uniformity in appearance, which indicates a good-quality animal.

Figure 7–40 indicates the influence of economics on selection, based on appearance. Although 49 percent of the weight of the average steer is in the rear portion illustrated, 70 percent of the value is represented by it. Animals with much length, strength in the topline, and heaviness in the quarter are highly valued.

After judging from the side and making comparisons among them, most judges have the animals turn for a rear view. A steer that is wide and muscular through the center of the quarter is illustrated in figure 7–41. The top of the round should be straight and full in the tail head area. Adequate depth from the top of the tail head to the twist and a wide, bulging round is desired, indicating a large volume of choice steaks. Also from the rear, judges observe the placement of the rear feet. An animal that is heavy boned (large circumference) and tracks wide has a muscular loin eye (back strap), as in figure 7–41. This muscle is often measured in carcass shows. The production goal is 1 square inch or more per hundred pounds of live body weight. Heavy bones in the live steer

FIGURE 7–39 The side view shows length, trimness, muscle, and bone. Reprinted with permission from J. Blakely and D.H. Bade, *The Science of Animal Husbandry*, 6th ed. (Englewood Cliffs, NJ: Prentice-Hall, 1994).

indicate heavy muscling in the carcass. Moving close enough to look over the top from the rear, the uniformity of width from pins to hooks to crops is noted.

From the front, one should observe the thickness through the forequarter and the fullness in the dewlap and brisket. A full, pendulous brisket indicates excessive fattening. A thin, rather loose dewlap indicates an animal that is carrying less finish, which is becoming more important in today's market as the consumer demands lean, muscular cuts. Bone is also observed in the front legs much the way it is from behind; that is, the desirable animals are thick

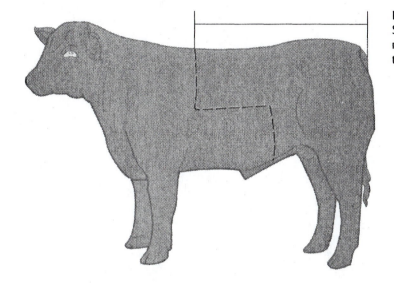

FIGURE 7–40
Seventy percent of the retail value is found in the top rear.

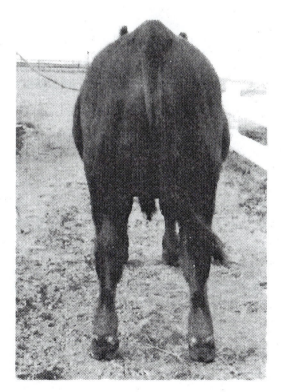

FIGURE 7–41
Width, muscle, and bone from the rear view.
Reprinted with permission from J. Blakely and D.H. Bade, *The Science of Animal Husbandry,* 6th ed. (Englewood Cliffs, NJ: Prentice-Hall, 1994).

boned and stand on straight, heavy legs, indicating thickness of muscling in other areas that are correlated to this observation.

Handling the steer is the final step in estimating carcass traits and grade. A thin, firm, mellow covering of fat (finish) is expected to cover the ribs, loin, and back of superior steers. A loose, pliable skin is desired because it indicates carcass quality.

Judging is never more than the comparison of two animals, regardless of the size of the class. The judge always compares one animal with another; then, once a decision has been made between them, the winner of these preliminary comparisons is compared with the winner of other comparisons. Judging is a very simple process if taken in this systematic manner. To illustrate the thought process involved in a typical livestock judge's selecting a grand champion for a show, the following reasons might be given for placing a particular animal number one:

> I placed this steer grand champion, because he was the longest steer in the class, being neater and trimmer about the middle, indicating a higher dressing percentage than any other steer in the class. He was straight in his topline and firm in his quarter, showing thickness of round, great length from his hooks to his pins, deep in the flank, and standing on straight, firm bone. He is thicker through the quarter, shows greater depth in his twist, and shows more uniformity in width throughout, carrying the broadness from the loin up through the crops and over his back than any of the steers standing below him. He is carrying a desirable amount of finish over his back, loin, and ribs but does not indicate an excessive degree of fat, due to the uniformity of his brisket and the thinness of his dewlap. This steer will hang up the higher percentage of primal cuts and the highest grading carcass of any steer in his class.

Judging is the process of seeing through the live animal and evaluating the final product, the carcass. Figure 7–42 shows a muscled, well-finished carcass with large loin eye. This is what any judge really sees when the final slap on the rump is made.

FIGURE 7–42
Carcass measurements include loin-eye development and backfat thickness. Courtesy of the U.S. Department of Agriculture.

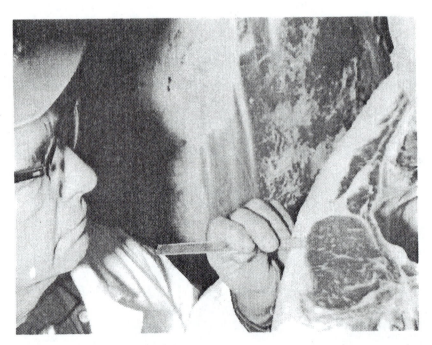

Pedigree

Cattle have been selected based on the supposed merit of their ancestors for hundreds of years. This system is still used heavily in some purebred herds, not always with happy endings. A bloodline, or being a descendant of an outstanding individual, does not always mean that outstanding traits will result from selection and mating to that descendant.

Pedigrees are most useful when the most recent individuals are weighted most heavily; each generation may reduce the potential by one-half because half the genes of an offspring come from each parent. Generally, the parents and grandparents are about as far back as it is recommended to go for genotypic material.

Fads and promotional advertising may encourage mistakes in pedigree selection. Breeders must be careful that names are backed by the right kind of performance desired. Some undesirable inherited traits (such as dwarfism) have been disadvantages that have accompanied seemingly progressive selections based on pedigree.

The best use of pedigree information is as the basis for selection of young animals before their performance or that of their progeny is known. Traits expressed late in life, such as longevity or continued soundness, are arguments for pedigree surveillance. Also, traits expressed in only one sex, such as milk production, have been effectively selected for in numerous instances. For example, a sire that consistently produces daughters and granddaughters with high milk production and mothering ability is more likely to contribute strength to a pedigree than a bull with descendants weak in those traits. Young bulls selected from the strong pedigree will more likely produce daughters with improved milk production and mothering ability.

Pedigrees serve as a useful tool once their limitations are realized. However, once an individual animal's performance and that of its progeny are measured, the pedigree should receive less weight in an individual's evaluation.

Animal Performance

Measuring the individual animal's ability to perform efficiently and economically involves both simple and complex techniques. Prolificacy is often measured in females by observation of regular heat periods and the ability to settle quickly; both are noted as positive for this trait. Males are observed for masculinity, aggressiveness, and good sperm count. A long life means longer production and is a trait observed and selected for by memory in many commercial herds and by pedigree in registered herds. Efficient growth from birth to weaning to production is measured in terms of rate of gain, amount and type of feed consumed, and quality (which determines price) of the end product. Data on an individual bull may be helpful in selection of potential breeding stock. The information commonly recorded and evaluated includes the following:

1. *Birth weight:* Taken within 24 hours of birth.
2. *Adjusted 205-day weight:* The weaning weight adjusted to 205 days by computing from less than or more than this number of days by use of factors obtained from testing of thousands of cattle.
3. *Adjusted 365-day weight:* Yearling weight.
4. *Ratio:* Bull's relative performance compared with that of its contemporaries for adjusted 205- and 365-day weights.

5. *Scrotal circumference.*
6. *Calving ease:* Indicates which bulls can be used on heifers.
7. *Expected progeny difference (EPD) and accuracy (ACC):* Important terms used in understanding beef sire summaries. This prediction is similar to that of Predicted Transmitting Ability (PTA) for dairy bulls. An EPD combines into one figure (a reproductive index), a measurement of genetic potential based on the individual's performance and the performance of related animals (i.e., sire, dam, and other relatives). EPD is expressed as a positive or negative value, reflecting the genetic-transmitting ability of a sire. The most common EPDs reported for bulls, heifers, and cows are as follows:
   a. *Birth weight EPD:* Expected birth weight deviation of calves from this individual, excluding maternal influence. This is the relative birth weight, with equal uterine conditions assumed.
   b. *Milk EPD, weaning EPD:* Weight at 205 days of calves sired by, or out of, this individual, excluding maternal influence.
   c. *Yearling weight EPD:* Weight at 365 days, excluding maternal influence.
   d. *Maternal breeding value (MBV) ratio:* Evaluates the transmitting ability of sires for maternal performance of their daughters. An MBV ratio of 102.3 would indicate a sire whose daughters average 2.3 percent above the average of the breed.

An example of a bull's EPD: A bull with a birth weight EPD of +2.5 with an accuracy of 0.70 would have offspring 2 1/2 pounds heavier than the average breed offspring. The accuracy of this estimate would be 70 percent. In choosing sires, a comparison of varying EPDs and ACCs can be used. A bull with a +25 EPD for yearling weight would expect to have progeny that average 25 pounds above the average of all sires that have yearling progeny records.

Simple Measuring Techniques.

Pencil, pad, and laptop   A pocket notepad is helpful to record data such as date of birth, birth weight, date of first heat, breeding dates, and so on. With the price reduction of laptop computers, many ranchers and farmers are using this modern tool and then later transferring the records to their larger office computers.

Perhaps the most widely used tool of selection in the United States is the livestock scale. Large or small, stationary or portable, it has become an indispensable part of measuring techniques. Calving weight, weaning weight, and weight at calving are the three most accepted, useful observations (see figure 7–43).

After the initial weight at birth, growth to weaning is largely a reflection of the dam's milk production and the calf's inherited ability to gain. Calves that are heavier at weaning are usually grouped together for selection of the best for breeding replacements or keepers. For comparison purposes, weights should be adjusted for sex and age variations. At weaning, bulls are about 15 pounds heavier than steers, and steers are about 20 pounds heavier than heifers.

Yearling weights indicate growing ability, mostly a function of inherited qualities provided that feed, health, and other factors are normal. Because half the genes come from each parent, this weight measurement also indicates the sire and dam's potential for contributing further genetic material.

Weight of the calf at first calving completes the cycle of weight observations from birth to reproduction. Cows with lighter calves are culled first even

FIGURE 7–43 The scale has become an indispensable tool in livestock selection techniques. Reprinted with permission from J. Blakely and D.H. Bade, *The Science of Animal Husbandry,* 6th ed. (Englewood Cliffs, NJ: Prentice-Hall, 1994).

though their past records may have been good. In a few years, by culling out the normal 20 percent, a herd of cows may be selected that were superior calves, grew well after weaning, and are reproducing the heavier, higher-grading calves.

Bull-testing stations   Most stations now have one or more centers designed to measure economically important traits such as rate of gain and feed efficiency. Other measurements may be added, but these two are common to all. The purpose is to gain information on young sire prospects after weaning and before breeding. A controlled, identical environment over a specified period (usually 140 days minimum) makes more accurate comparisons possible. The higher-gaining, more efficient bulls are more likely to produce calves with these characteristics than the low gainers and poor converters.

Testing of yearling bulls offers the newest and best genetics. It is not uncommon today to find yearling bulls weighing 1,300 pounds. Buying tested yearling bulls gives producers an excellent way of finding the genetic traits needed to improve their herds. Yearling bulls give 1 more year of service than the traditional 2-year-old bulls purchased 10 to 15 years ago. Many producers buy a mixture of unproven but tested yearling bulls and the more traditional tested and proven 2-year-old bulls.

Complex Measuring Techniques.

Ultrasonic device (sonoray)   A device similar to sonar, which is used by submarines to detect objects under water, has been produced on a small, portable scale for use in livestock research. The device emits high-frequency sound waves and records the time it takes for an echo to bounce back from the junction between two tissues (e.g., fat and lean). These echoes are displayed on a readout instrument that allows for plotting of the shapes and sizes of some

FIGURE 7–44
Ultrasound is used to measure the loin-eye area in live animals. Courtesy of the U.S. Department of Agriculture.

muscles. The loin-eye muscle (longissimus muscle) is plotted with accuracy using this method and is sometimes used to supplement bull-testing data. The obvious advantage is that important data with relatively accurate measurements of backfat and longissimus muscle area of potential seed stock can be obtained without surgery or slaughter (see figure 7–44). This information helps eliminate the costly testing of littermates and half siblings.

The more recently developed real-time ultrasound equipment allows for cross-sectional imagery of the longissimus muscle. Thus both backfat thickness and the loin-eye muscle can be accurately measured in a live animal.

Cutout data   Some bull-testing stations equipped with slaughter facilities use cutout data to add feed efficiency and even ultrasonic measurements to rate of gain. A steer, preferably a full brother to the bull being tested, completes the same program. The steer is then sacrificed to obtain actual data. The most useful guides are percentage of lean cuts and square inches in area of the loin-eye muscle. Additional data often include grade, dressing percentage, and marbling scores. This information, coupled with the performance of the bull, gives a more valid projection of the potential improvement based on this line of selected genetic material.

Production Testing

Production testing encompasses *performance testing* (individual merit) and *progeny testing* (merit of offspring). It is used most effectively when based mainly on characteristics of economic importance that are highly hereditary. For example, the following traits are often considered of highest economic value:

1. Reproductive efficiency
2. Mothering ability

3. Rate of gain
4. Economy of gain
5. Longevity
6. Carcass merit

Production testing is not a new idea; it was used by the Romans more than 2,000 years ago. Of course, techniques have advanced significantly. The basis, in the beginning as now, is that individuals will differ in their performance and that traits are inherited. Thus production testing as a selection technique is a systematic measurement of economically important traits, the recording of these traits, and the use of the recorded information for selecting superior and culling inferior producers. Production testing is the most infallible of all the methods of selection, particularly for selecting young herd sires. To be effective, the superior characteristics to be selected for must be of economic importance and highly heritable. Objective measurements such as pounds, inches, and so on should be used. Minimum standards are set forth, and animals that fall below these standards are culled.

Improvement in an economically important trait may be attributed to heredity and environment. If the traits selected for are influenced mainly by environment, the heritability of those characteristics will be low and little progress can be expected. Therefore, traits that are on the upper end of the scale of probability of heritability should be used (see table 7–11).

Developing a Production-Testing Program.   Generally speaking, five basic steps are necessary to embark on a production-testing project. Both performance testing (merit of the individual) and progeny testing (merit of the offspring) make up the broader procedure of production testing.

Make permanent markings   Each cow and sire should be marked by branding, tattooing, and so on. Furthermore, for ease of identification from a distance, these marks should be easily read on foot, on horseback, or from the cab of a pickup. Brands work well in this regard. However, some breeders prefer to supplement brands or tattoos with large plastic ear tags that are easily read and may be color coded to quickly differentiate among groups or bloodlines (see figure 7–45).

TABLE 7–11	Heritability of Some Economically Important Characteristics in Beef Cattle
Characteristic	**Heritability**
Birth Weight	Medium to high
Weaning weight	Low to medium
Daily rate of gain	Medium to high
Feed efficiency	Medium
Live and carcass grade	Medium
Area of loin eye	Very high
Milking and mothering ability	Medium
Reproductive efficiency	Low

FIGURE 7–45
Cows and sires are permanently marked and made easy to identify for production-testing purposes.
Reprinted with permission from J. Blakely and D.H. Bade, *The Science of Animal Husbandry,* 6th ed. (Englewood Cliffs, NJ: Prentice-Hall, 1994).

Record ages   Ages should be recorded for all cows, sires, and calves in a herd. If records do not exist for cows, they should be mouthed by a qualified person for an estimate. Aging cattle is necessary because comparisons among families must be adjusted for age influence. Fair comparisons of the young, the old, and cows at their peak of performance are not possible unless age influence is considered. Weaning weights of the calves are the adjusting factor because very young and old cows wean lighter calves than cows in their prime.

Observe from birth   Each calf should be weighed at birth and records made concerning the date, weight, sex, dam, sire (if known), and any other information that may be important to the breeder. A permanent identification (ear tattoo or brand is most common) should link the pair. Again, for ease of identification, ear tags are suggested to save needless rechecking.

Record information   Weaning weights and grades at 6 to 8 months should be recorded and information updated for each calf, each dam, and each sire. An example of one form of headings is shown in figure 7–46. Data may need to be adjusted. These forms can be added to for further rate of gain, feed efficiency, and other measurements.

Make decisions based on records   Production testing really pays off when the records are put to work and sentiment is retired. The cows that produce the lightest, least desirable calf crop should be culled regardless of personal preference for such minor things as markings, provided that they do not weigh heavily on the economic market. The sires that produce the lightest calves should also be culled.

a. CALF & YEARLING WEIGHTS & GRADES RECORD

Sire	Dam			Calving Record				Weaning Record						Yearling Record			
Herd Identity Number	Herd Identity Number	Age of Dam	Calf Number	Sex	Date Calved	Birth Weight	Weaning Weight	Age in days	*Adjusted 205 Day Weight	Weight Per Day of Age	Weight Ratio or Grade	Yearling Weight	Age in Days	**Adjusted 365 Day Weight	Weight Per Day of Age	Weight Ratio or Grade

b. SIRE'S ANNUAL PRODUCTION RECORD

Sire's Herd Identity Number	Calf Number	Sex	205 Day Adjusted Weight	Weight Ratio	Weaning Weight	Age in days	365 Day Adjusted Weight	Grade	Weight Ratio	Grade	Replacements	Remarks

c. LIFETIME PRODUCTION RECORD OF DAM

Herd Identity Number	Year of Birth	Average Total Production at Weaning				Production Index	Replacements	Remarks
		Number of Calves	Adjusted Weaning wt.	Weight Ratio	Grade			

* To compute the adjusted 205 day weaning weight, apply the formula: $\dfrac{\text{actual weight} - 70}{\text{age in days}} \times 205 + 70 = 205$ day weight

then correct the adjusted weight for the age of dam according to the following table:

Percent to be added to calf weights after adjusting for age

Age of dam	Percent to be added
2	15
3	10
4	5
5-10	none
11 and older	5

** To compute the adjusted 365 day weight, apply the formula:

$\dfrac{\text{final weight} - \text{actual weaning weight}}{\text{number of days between weights}} \times 160 + 205$ day adjusted weaning weight = adjusted 365 day weight

FIGURE 7–46 Example of a useful production record form.
Reprinted with permission from J. Blakely and D.H. Bade, *The Science of Animal Husbandry*, 6th ed. (Englewood Cliffs, NJ: Prentice-Hall, 1994).

The heaviest, fastest-growing, most efficient heifers and bulls should be kept for replacements. For best results, the heifers are fed out and yearling weights and grades are taken. The worst (20 percent, for example) are culled out and the superior heifers bred to the superior sires. After first calving, these heifers are again culled, based on the weaning weights of their first calves.

In this manner, the best animals are tested for individual performance (performance testing) and offspring performance (progeny testing), and a strict culling program endorses selection of the best to be bred to the best. Thus production testing removes the grandeur of reputation, the magnificence of ancestors, and the glory of visual excellence to pace the contest with documented results.

A System of Selection

Finally, using a combination of all of the techniques (eye appraisal, pedigree, animal performance, and production testing) discussed, the breeder needs to follow a system of selection that will result in maximum progress (see figure 7–47). Three common systems are tandem selection, independent culling level, and selection index.

Tandem Selection.   One trait is selected for at a time until maximum progress has been achieved; then another trait becomes the target for improvement. *Tandem* means one after another. The process sounds simple. For instance, once cattle that are selected for the polled character breed true for this trait, the same group is selected for weaning weight and so on.

Independent Culling Level.   Production testing fits well with this systematic procedure. Minimum standards are set for several traits. Failure by any animal to meet the minimum standard for any one trait results in removal from the herd to be sold or slaughtered. The obvious disadvantage is that if standards are too high and too many traits are involved, the level culled could be too high to leave

FIGURE 7–47
Success of selection of stockmen through the years can be seen when you compare the ideal animal of the nineteenth century with modern grand champion steers.

sufficient animals to work with. With common sense and some adjustments in the standards from time to time, this system becomes very effective.

Selection Index. Generally accepted as the most efficient of the systems, an index evaluates all important traits and combines them into one figure or score. Higher scores mean a more valuable animal for breeding purposes. The weight assigned to each trait to be included in the index depends on its economic importance, its heritability, and its genetic linkage to other traits. Many breeders accept this method because slightly substandard performance in one trait can be offset by excellence in another trait. The only disadvantage appears when too many traits are included in an index or progress is attempted on characteristics of low heritability or little economic importance. When used with a sensible number of traits of relatively good heritability, the selection index gives perhaps the fairest appraisal of merit.

SUMMARY

The beef industry is still the largest segment of U.S. agriculture. The per capita consumption of beef exceeds 67 pounds per year in the United States and is as high as 152 pounds per year in Argentina. Very few beef ranchers incorporate all phases of the industry. Distinct production systems and regional locations for each exist in modern beef production. The beef industry of the twenty-first century will need to compete not only with the poultry, fish, and pork industries but also with consumers' animal welfare and environmental views. Education of the public, politics, and marketing will be major forces in the industry's future.

EVALUATION QUESTIONS

1. The leading state in beef production is _____ .
2. _____ is the supplementation of young (calves) that are nursing their dams.
3. _____ means heterosis.
4. _____ is the common animal denominator based on feed consumption.
5. The dressing percent is _____ for a broiler, _____ for pork, _____ for lamb, and _____ for beef.
6. _____ is a growing phase or that period in the life of a calf from weaning to around 800 pounds when it is prepared to go on a high-energy finishing ration.
7. _____ is an estimate of palatability based primarily on marbling and maturity and generally to a lesser extent on color, texture, and firmness of lean meat.
8. A young animal that carries sufficient height but insufficient finish for slaughter purposes and will make good gains if placed on feed is called _____ .
9. Calves and yearlings that are intended for eventual finishing and slaughter but are being fed for growth rather than finishing are called _____ .

10. _____ is meat from ruminant animals (with split hooves) that has been slaughtered by a *Shochet* according to ritual laws.

11. _____ means without injury.

12. A feed additive that changes the metabolism within the rumen by altering the rumen microbes to favor propionic acid over acetic acid is _____ .

13. _____ is a synthetic progesterone given to feedlot heifers to prevent them from cycling and thus increasing feed intake and weight gain.

14. Three commonly used growth implants in finishing steers and heifers are _____ , _____ , and _____ .

15. _____ are microbial cultures or ingredients that stimulate cultures capable of modifying the gastrointestinal environment to favor healthy tissue development.

16. _____ is an inorganic or organic compound that inhibits the growth of organisms but is not produced by a living organism. It is commonly used to improve the rate of gain and feed efficiency.

17. The humped cattle are the _____ species.

18. _____ was the first breed developed in the United States.

19. The largest breed of cattle in the world is _____ .

20. The cattle breed with the largest horns is _____ .

21. _____ , _____ , and _____ are generally referred to as the three predominant English breeds.

22. A 1,000-pound choice steer may be expected to yield a carcass that weighs _____ pounds.

23. _____ , _____ , and _____ are the three most widely used weights in a beef animal's record.

24. _____ evaluates all important traits and combines them into one figure or score.

25. Production testing encompasses _____ (individual merit) and _____ (merit of offspring).

26. An animal is stunned rather than killed to keep the _____ functioning.

27. _____ is the muscle exposed by cutting between the 12th and 13th ribs.

28. Removal of intestines is called _____ .

29. Enzymes break down _____ through age.

30. The intermingled fat in muscle tissue is called _____ .

31. The lowest quality grade is _____ .

32. Yield grades range from _____ through _____ .

33. The short plate comes from the _____ quarter.

34. T-bones come from the _____ .

35. Most stunning in the United States is done by a _____ .

36. Cooling temperatures of hot carcasses are typically _____ .

37. Round steak and sirloin come from the _____ .

38. In the _____ selection system, one trait at a time is selected until maximum progress has been made.

39. Most Charolais cattle in the United States were imported from _____ .

40. _____ and _____ are the two most popular beef breeds in the United States.

41. _____ is the most commonly used beef breed for calving ease (crosses).

42. _____ refers to all waste organs or tissues removed from the carcass in slaughtering.

43. The device for measuring the thickness of fat over the animal's back (ribeye area) is called _____ .

44. _____ is an estimate of the percentage of product from a carcass, compared with liveweight.

45. _____ identify the quantity or amount of salable meat from beef carcasses.

46. _____ is a device used in bloodless castration and in the docking of tails of livestock, accomplished by applying a strong rubber band.

47. The percentage yield of hot carcass in relation to the weight of a choice steer is _____ %.

48. _____ is a naturally hornless animal.

49. An evaluation of an animal on the basis of the performance of its offspring is known as _____ .

50. _____ is a measurement of the degree of difference between the progeny of a bull and those of average bulls of the same breed in the trait being measured.

51. Calves in feedlots should gain _____ or more pounds in 180 days.

52. The feed required to produce 300 pounds of gain on light, thin calves is about _____ pounds.

53. The Burdizzo is used to _____ .

54. Anaplasmosis is caused by a minute parasite that affects the _____ .

55. An unexpected death, a bloated carcass with bloody discharge from body openings, and all four feet pointing up is a sign of _____ .

56. Blackleg has been prevented by _____ when cattle are 3 to 4 months of age.

57. Foot-and-mouth disease is best controlled by _____ .

58. _____ is a disease that is often contracted from swampy, poorly drained areas and that is carried by snails.

59. The symptoms of _____ are deformed hooves, lameness, and red and swollen skin between the toes.

60. _____ is the cause of Johne's disease.

61. Symptoms of _____ are labored breathing, extension of tongue, and nasal discharge after cold weather.

62. _____ is a complex disease that occurs after stressful conditions.

63. Consumption of _____ products can cause X disease.

64. Ten million of the _____ breed of cattle were driven to railheads during the 1800s.

65. A full, pendulous brisket means too much _____ .

66. A young bull, typically less than 20 months of age, is called a _____ .

67. All weights except birth weights should be adjusted for _____ and _____ .

68. The _____ is the single most valuable by-product of beef cattle.

69. The _____ breed is the smallest purebred bovine in the world.

70. The most popular breed of beef cattle in the United States is _____ .

71. The left half of the carcass is always the _____ side.
72. Chemical castration should only be performed on bulls that weigh less than _____ pounds.
73. Freeze branding typically involves the use of _____ , which has a temperature of −196°C.
74. _____ corner posts are made with large, flat, heavy rocks surrounded by woven wire.
75. Local anesthesia should be administered to all calves castrated over _____ months of age.
76. _____ is a method of administering local anesthesia when dehorning calves.
77. The most common method of branding cattle in the United States is _____ .
78. _____ and _____ are the most common methods of controlling external parasites in beef cattle.
79. AMDICA is an abbreviation for _____ .
80. _____ testing is of the merit of offspring.

DISCUSSION QUESTIONS

1. Describe the common beef breeds in the United States, including color patterns, size, advantages of their use, those with humps and horns, and so forth.
2. Why are the 205- and 365-day weights so important?
3. What is the difference between yield and quality grading of meat? What are some of the common cuts of beef?
4. What is the importance of scrotal circumference?
5. What makes up the fifth quarter?
6. Give three examples of dual-purpose cattle.
7. Give three examples of humped cattle.
8. What part of the steer does pot roast come from? Short plate? Brisket? Chuck?
9. What is the importance of dehorning? What are the most common methods used?
10. What is the importance of castration? What are the most common methods used?
11. What are ionophores? Of what importance are they in livestock feeding and management?

C H A P T E R

8

Sheep Production

OBJECTIVES

After completing the study of this chapter you should know:

- The historical, current, and future trends of sheep in U.S. animal agriculture.
- Common sheep by-products used in the United States.
- Distinguishing characteristics of common U.S. sheep breeds.
- The differences between ewe, ram, and dual-purpose breeds of sheep.
- The various nutritional and reproductive management principles involved with sheep production.
- How photoperiodism affects the profitability of sheep raising.
- The use of the lambing jug, crutching, tagging, and shearing and the importance of each to profitable sheep management.
- The most common sheep diseases and their symptoms (in brief), diagnosis, treatment, and prevention.
- How predators pose a serious threat to sheep and how some (if not most) shepherds cope with the problem.
- The selection, care, grading, preparation, and marketing of wool.
- The differences between lamb and mutton, slaughter classes, and the grading of lamb.
- The importance and methods of selection, production testing, and breeding of sheep.

KEY TERMS

Broad tail	Dwarfism	Ivermectin
Bummer	Entropion	Ked
Caracul	Ewe breeds	Lambing jug
Carbonizing	Facing	Lamb joint
Carding	Fell	Lanolin
Cockle	Fisking	Libido
Crutching	Fleece	Lice
Cryptorchidism	Grease wool	Lopi
Dual-purpose breeds	Hothouse lambs	Maggots

Offal
Orf
Parrot mouth
Persian lamb
Phytoestrogen
Pizzle rot
Polyestrus
Posthitis
Psoroptic scab
Ram breeds
Ram epididymitis

Rectal prolapse
Salmonellosis
Sarcoptic mange
Scabies
Scouring
Sheath rot
Sheep nose botfly
Sheep scab
Shrinkage
Spider syndrome

Spinose ear tick
Spool joint
Suint
Uremic poisoning
Urine scald
Wether
Wool blind
Woolen
Worsted
Yolk

INTRODUCTION

Sheep were first domesticated from wild animals that existed in Asia some 10,000 to 20,000 years ago. Sheep were one of the first domesticated animals and to this day have no natural defenses against predators other than the farmers and ranchers who care for them. Wool fibers have been found in remains of primitive villages in Switzerland that date back an estimated 20,000 years. Egyptian sculpture from 4000 to 5000 B.C.E. portrays the importance of this species to people. Much mention is made in the Bible of flocks, shepherds, sacrificial lambs, and garments made of wool. The Romans introduced sheep into Western Europe. They prized sheep, anointed them with special oils, and combed their fleece to produce fine-quality fibers that were woven into fabric for the togas of the elite.

Little is known about the original selection and domestication of sheep, but they are thought to have descended from wild types such as the Moufflon, a short-tailed sheep. Wild varieties in Europe and Asia probably served as foundation stock to produce wool, meat, skins, and milk. It appears that selection practices not only removed most of the wild instincts, leaving the species completely dependent on people for management and protection, but also resulted in a lengthened tail. Nearly all domestic breeds today have long tails before docking.

As weaving and felting began to develop as an important element in the advancement of civilization, more definite types and breeds of sheep began to emerge to produce quality fibers at the expense of other traits. The Merino breed of Spain developed into one of the first recognizable fine-wool breeds. It was so prized that the king of Spain made it a crime punishable by death to send any out of the country without his permission.

The English also developed many breeds very early that would adapt to their varying climate. Thus between the years 1000 and 1500 C.E. the two great powers in wool production were Spain and England, and most of the breeds in the United States today can be traced back to breeds or stock exported from those countries.

Domestic sheep were foreign to the New World and were first introduced by Columbus on his second voyage in 1493 to the West Indies. Cortez brought sheep into Mexico in 1519, and Spanish missionaries contributed to their popularity through the teaching of weaving arts to the Indians. By 1890, sheep were found in virtually every state and territory in the United States. Some 100 to 200 years ago, Europeans took the Merino and British breeds of sheep to South America, Australia, South Africa, and New Zealand.

Sheep are docile animals and naturally gregarious, meaning that they stay together in groups. They have an excellent sense of smell and hear well but have poor eyesight. There are approximately 80,000 ranchers[1] raising 8.4 million sheep in the United States. Major sheep-producing states, in order of numbers, are Texas, California, Wyoming, Colorado, and South Dakota (see table 8–1).

Sheep are used for a variety of purposes. Besides producing wool and lamb, sheep are also used for controlling weeds and brush (thus avoiding the need to use herbicides and other possible carcinogens to do the same). In addition, sheep are used to graze under power lines (where mechanical means are greatly reduced), and in California sheep have been used to graze firebreaks.

High-quality wool is used for making clothing and carpeting for consumers. More than 1.7 billion pounds of wool per year (in the United States) are utilized in durable wool blend carpet. A good ewe will produce 8 to 10 pounds of wool per year. In addition, wool can also be used for tasks such as cleaning up oil spills or insulating houses. In the Western world, synthetics have replaced wool in this regard. In third world countries, synthetics are too expensive and are not considered as environmentally friendly. Table 8–2 summarizes additional sheep by-products used by American consumers.

TABLE 8–1	Sheep Numbers by State
Texas	1,650,000
California	1,000,000
Wyoming	680,000
Colorado	535,000
South Dakota	500,000
Total United States	8,457,100

Source: U.S. Department of Agriculture (USDA), *Sheep and Goats*, 1996.

TABLE 8–2	Common Sheep By-Products Used in the United States		
From Hide and Wool			
Lanolin	Rouge base	Felt	Drumheads
Clothing	Insulation	Carpet	Rug pads
Footwear	Asphalt binder	Yarns	Woolen goods
Fabrics	Artists' brushes	Hide glue	Paint and plaster binder
Tennis balls	Ointment base	Baseballs	Pelt products
Sports equipment	Worsted fabric	Upholstery	Textiles
From Fats and Fatty Acids			
Explosives	Mink oil	Floor wax	Solvents
Oleomargarine	Chewing gum	Ceramics	Tallow for tanning
Paints	Shoe crème	Rubber products	Oleo shortening
Makeup	Tires	Rennet for cheese	Chicken feed
Industrial oils	Paints	Dish soap	Biodegradable detergents
Stearic acid	Antifreeze	Crayons	Industrial lubricants
Dog food	Cosmetics	Creams and lotions	Protein dog food
Protein hair conditioner	Protein hair shampoo	Shaving cream	Herbicides

(continued)

---

[1]American Sheep Industry (ASI) Association, Englewood, Co., (303)771-3500, 1999.

| TABLE 8–2 | Common Sheep By-Products Used in the United States (continued) | | |

Retail Meats

| Leg of lamb | Lamb chops | Rack of lamb | Lambecue |
| Pot roasts | Round steaks | Ground lamb | BBQ ribs |

From Intestines

| Sausage casings | Surgical sutures | Tennis racquet strings | Instrument strings |

From Manure

| Nitrogen fertilizer | Phosphorus | Minor minerals | Potash |

From the Bones, Horns, and Hooves

Syringes	Neat's-foot oil	Fertilizer	Gelatin desserts
Bonemeal	Marshmallows	Rose food	Emery boards and cloth
Piano keys	Plywood	Ice cream	Bone charcoal for
Potted meats	Pet food ingredients	Dice	high-grade steel
Gelatin capsules	Collagen cold cream	Laminated wood	Paneling
Bandage strips	Collagen and bone for	Crochet needles	Horn and bone handles
Bone charcoal pencils	plastic surgery	Glycerine	Photographic film
Adhesive tape	Bone china	Wallpaper	Wallpaper paste
Combs and toothbrushes	Buttons	Phonographic records	Cellophane wrap and tape
Bone jewelry	Abrasives	Malts and shakes	Steel ball bearings

Source: The Sheepman's Production Handbook (Englewood, CO: American Sheep Industry (ASI) Association, 1997).

BREEDS OF SHEEP

Because of the vast size of the United States, with its great variety of terrain, elevations, and climate, it is not surprising to find many of the 914 or more sheep breeds estimated to be scattered throughout the world in that nation. There are 47 breeds and types of sheep in the United States; several breeds originated there (figures 8–1 and 8–2). U.S. sheep breeds provide a diverse range of performance for milk, wool, carcass characteristics, reproduction, and growth rate. This genetic variability can be used, especially in crossbreeding programs, to maximize production under varying environmental and management conditions.

All sheep, however, have several characteristics in common. As members of the animal kingdom, they are of the phylum Chordata (backbone), class Mammalia (suckle their young), order Artiodactyla (hooved, even toed), family Bovidae (ruminants), genus *Ovis* (domestic and wild sheep), and species *Ovis aries.*

Within this species many different breeds developed of necessity. Both purebred and crossbred breeds have their place. Successful breeders choose their stock carefully. Some breeds and types available for utilization are described in this chapter. It is important to note that economic considerations such as the price of wool or lamb may cause a shift from one type or breed to another. The leading sheep breeds in the United States based on total numbers registered are summarized in table 8–3.

FIGURE 8–1
Targhee ewe.
Courtesy American
Sheep Industry
Association.

FIGURE 8–2
Polypay, Dubois, Idaho.
Courtesy American Sheep
Industry Association.

TABLE 8–3	Major U.S. Sheep Breeds Based on 1995 Registration Numbers
Breed	**Number**
Suffolk	29,800
Dorset	12,600
Rambouillet	12,400
Hampshire	11,200
Columbia	5,000
Southdown	5,800
Corriedale	3,100
Polypay	2,900
Montadale	2,900
Shropshire	2,600
Cheviot	2,400

Source: American Sheep Industry (ASI) Association, 1999.

TABLE 8–4	Fine-Wool Breed Comparisons						
	Mature Body Weight		**Average Fiber Diameter**		**Grease Fleece Weight (lb.)**	**Yield (%)**	**Staple Length (in.)**
	Ram (lb.)	**Ewe (lb.)**	**(Micron)**	**(Spinning Count)**			
American Cormo	160–200	120–160	17–23	46–56	5–8	50–65	2.5–4
Debouillet	220–275	125–150	18–22	64–80	9.5–14	45–55	3–5
Delaine-Merino	190–240	125–160	17–22	64–80	9–14	45–54	2.5–4
Rambouillet	200–300	140–180	19–24	60–70	10–15	45–55	2.5–4

The most common classification of breeds is by type of wool produced. Other factors such as meat type, color, horned or polled, and adaptability characteristics are considered within each type; the broad classifications are fine-wool breeds, medium-wool or meat breeds, long-wool breeds, dual-purpose breeds, hair and double-coated breeds, and minor breeds.

Fine-Wool Breeds

All the fine-wool breeds in the United States can be traced back to the Spanish Merino, a breed selected primarily for high-quality wool production. They produce a fine-wool fiber that has a heavy *yolk* or oil content. Fine-wool breeds can breed out of season and thus are capable of producing lambs in the fall months. See table 8–4 for summary statistics of fine-wool breeds.

American Cormo.   Cormo sheep originated in Tasmania, Australia, from one-fourth Lincoln crossed to one-fourth Australian Merino and one-half Superfine Saxon Merino. They were first brought to the United States in 1976. Cormo sheep have open faces and are a hardy breed adaptable to harsh climatic conditions. They produce a white, long-stapled, high-yielding, fine-wool fleece with a high degree of fiber uniformity.

FIGURE 8–3
The Debouillet is adapted for rugged, sparse grazing areas. Reprinted with permission from J. Blakely and D.H. Bade, *The Science of Animal Husbandry,* 6th ed. (Englewood Cliffs, NJ: Prentice-Hall, 1994).

Debouillet.   The Debouillet was developed by Roswell and Tatum in New Mexico in 1920. The original cross of Rambouillet and Delaine-Merino on the Ames Dee Jones Ranch produced this well-adapted range sheep. The Debouillet is medium sized, white-faced with wool on the legs, and hardy and can lamb unassisted under range conditions (see figure 8–3). The sheep produce a high-quality, long-stapled, fine-wool fleece.

Delaine-Merino.   The Merino sheep originated in Spain and has had an unbroken line of breeding for more than 1,200 years. The three types of Merino are types A, B, and C. Type C is also called Delaine-Merino. It has very little wrinkle to the skin and is the primary Merino found in the United States. Type A Merino is very wrinkled; type B is slightly wrinkled. Merino ewes breed year-round. White is the characteristic color; the Merino is usually horned (see figure 8–4a-c).

Rambouillet.   France is considered the origin of the Rambouillet breed even though it was developed from Merino stock imported from Spain about 1850. The name is derived from the agricultural experiment station in Rambouillet, France, where the sheep were perfected through selective breeding. The Rambouillet is the foundation of most western U.S. range flocks. The sheep are white faced, with wool on the legs (see figure 8–5). Rambouillet sheep are the largest and most popular of the fine-wool breeds; they are rugged and adaptable to a wide variety of arid range conditions. The breed is also known for having an extended breeding season.

Medium-Wool or Meat Breeds

Many breeds in the medium-wool or meat breed category are selected primarily for meat type; wool production is sacrificed when necessary. The fleece is medium in fineness and in length. These breeds are used primarily for meat production. Medium-wool breeds include Cheviot, Dorset, Hampshire,

FIGURE 8–4

Merino lamb face to face with a Coopworth ewe. Merino ewe is watching from the side, (a). Yearling Merino rams, Bliss Ranch Merinos, Loveland, Colorado, (b). Merino ewe and twin lambs. Bliss Ranch Merinos, Loveland, Colorado, (c).

Photo (a) courtesy of the L.A. Pierce College Teaching Farm Laboratory; photos (b) and (c) courtesy of Bliss Ranch Merinos, Loveland, Colorado.

(a)

(b)

(c)

FIGURE 8–5
The Rambouillet is the largest of the fine-wool breeds.
Courtesy American Sheep Industry Association.

| TABLE 8–5 | Medium-Wool Breed Comparisons |

	Mature Body Weight		Average Fiber Diameter		Grease Fleece Weight (lb.)	Yield (%)	Staple Length (in.)
	Ram (lb.)	Ewe (lb.)	(Micron)	(Spinning Count)			
Cheviot	160–200	120–160	27–33	46–56	5–8	50–65	2.5–4
Dorset	225–275	150–200	26–32	48–58	5–8	50–65	3–4.5
Hampshire	250–350	175–250	25–33	46–58	6–10	50–60	2.5–4
Montadale	200–275	160–180	25–30	50–58	7–11	50–60	3–5
North Country Cheviot	200–300	130–180	27–33	46–56	5–10	50–65	4–6
Oxford	225–325	150–200	28–34	46–54	7–10	50–60	3–5
Shropshire	225–290	170–200	25–33	46–58	6–10	50–60	3–4
Southdown	180–230	120–180	24–29	54–60	5–8	40–55	2–3
Suffolk	275–400	200–300	26–33	46–58	4–8	50–60	2.5–3.5
Texel	190–240	140–185	28–33	46–54	7–10	60–70	3–4
Tunis	175–225	130–160	26–31	50–58	8–12	—	—

Montadale, North Country Cheviot, Oxford, Shropshire, Southdown, Suffolk, Texel, and Tunis (see table 8–5).

Cheviot.   The Cheviot Hills of Scotland is the origin of the Cheviot, a relatively small-statured breed that was imported into the northern United States in 1838. The face and lower legs are white and free of wool (see figure 8–6). The nose is black and both sexes are polled. The head is carried high, and the ears are erect. Outstanding characteristics include vigor, good milking and mothering ability, quality carcasses, and adaptability to rugged grazing conditions.

FIGURE 8–6
The face and lower legs of the Cheviot are free of wool, and both rams and ewes are polled.
Courtesy American Sheep Industry Association.

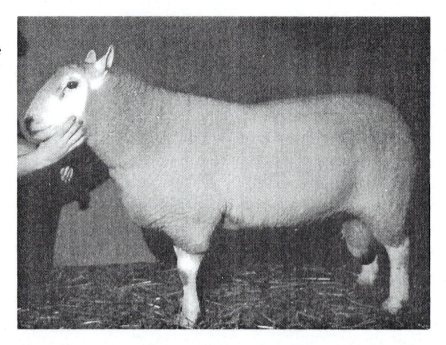

FIGURE 8–7
The Dorset will breed out of season and is very prolific.
Courtesy American Sheep Industry Association.

Dorset.   In about 1880 Dorset, Somerset, and Wiltshire Counties in England first began exporting to the United States a medium-sized, medium-wool breed called the Dorset. The Dorset is white faced. Both horned and polled strains (see figure 8–7) are registered in the Continental Dorset Club. This breed is very prolific, lambs early, and has good milking ability. Both sexes can breed year-round; the rams are among the most active of any breed in hot weather. Because of this characteristic, the Dorset is widely known for its use

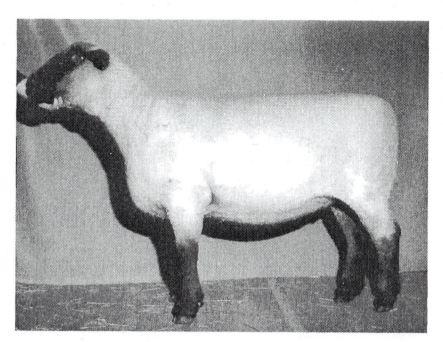

FIGURE 8–8
The black, polled
Hampshire is large
and does well on
rangelands.
Courtesy American
Sheep Industry
Association.

in the production of hothouse lambs (lambs produced during a time of the year that most sheep do not normally reproduce).

Hampshire. Exported from Hampshire County, England, about 1840, the Hampshire breed has received wide acceptance and popularity in the United States as a source of terminal sires for commercial lamb production. The black, woolless ears and nose and lack of horns distinguish the rugged Hampshire (see figure 8–8). This large, robust breed is known for its vigor and strength. It does well on the range, lambs easily, and has heavy, vigorous lambs.

Montadale. In 1931 E. H. Mattingly of St. Louis, Missouri, started crossing sheep to develop a bloodline composed of 40 percent Cheviot and 60 percent Columbia. He eventually (1945) came up with a breed that has both a bare face and bare legs below the knee (see figure 8–9). White hair is required, although black (not brown) spots are acceptable. The Montadale is a polled, medium-sized, mutton-type sheep with good-quality wool.

North Country Cheviot. As might be expected, the North Country Cheviot breed comes from Scotland, and it is white, usually polled, and similar to the Cheviot discussed earlier. The outstanding characteristic is high-quality wool. The breed was first imported into North America in 1944. The sheep are polled and medium to large sized, have white faces and bare heads and legs, and produce medium-wool fleece with good staple length.

Oxford. About 1850, the Oxford was imported into the United States from Oxford County, England. Figure 8–10 shows a representative of the breed. It is polled and varies in color. A very large, mutton-type breed that produces good feeder lambs, the Oxford is best adapted to areas with good feed sources. Oxford sheep originated from a Hampshire and Cotswold cross.

FIGURE 8–9
The Montadale is a cross of the Columbia and Cheviot breeds and is a more recent breed, originating in Missouri in 1945. *Courtesy American Sheep Industry Association.*

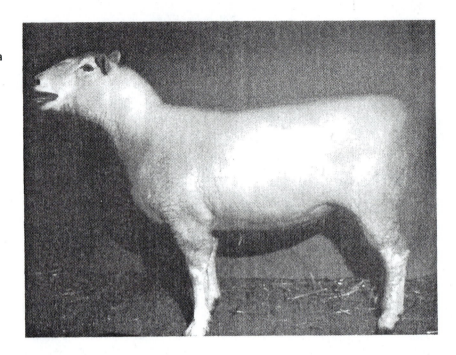

FIGURE 8–10
The Oxford, a mutton-type sheep, was formed by crossing Hampshire and Cotswold strains of sheep in 1833. *Courtesy American Sheep Industry Association.*

Shropshire.   Shropshire and Stafford Counties in England are considered to be the birthplace of the "Shrop." The breed was imported into the United States in 1855. Both sexes are polled and the face is black, but the wool covering is complete from the tip of the nose to the hooves (see figure 8–11). Considered a dual-purpose breed (both wool and lamb), this medium-sized sheep is noted as a prolific producer with good mothering ability in the ewe.

FIGURE 8–11
The Shropshire is considered dual purpose; it is used for both wool and lamb production.
Courtesy American Sheep Industry Association.

FIGURE 8–12
The small- to medium-sized Southdown is noted for producing a high-quality carcass.
Courtesy American Sheep Industry Association.

Southdown. A cluster of hills in southeastern England, the South Downs, lends its name to the breed imported into the United States in 1803. It is one of the oldest sheep breeds. Figure 8–12 shows a typical Southdown. It is similar in appearance to the Shropshire, although normally it has a white or light-colored nose. Its face and legs are covered with wool, but there is seldom much wool below the knees. The Southdown is small to medium in size. The breed is very docile and is ideally suited for intensive management.

FIGURE 8–13
The Suffolk is a popular meat breed and has short wool with some pigmented fibers.
Courtesy L.A. Pierce College Teaching Farm Laboratory.

It is the breed most associated with carcass quality in lambs; it has won the International Grand Champion Fat Lamb Award in Chicago more than 30 times. Carload lot winnings have been just as impressive. The sheep are unexcelled in mutton type, conformation, and carcass quality.

Suffolk.   The Suffolk had its origin in England and was first imported into the United States in 1888. Black, bare face and legs mark the distinctive Suffolk (see figure 8–13). A very large breed (the largest breed in the United States), the Suffolk is heat tolerant and very hardy. It is a good rustler and adapts well to the western range conditions of the United States because it is prolific and a good feeder. The ewe is an excellent milker and mother. It is one of the few breeds that have any aggressive characteristic at all; the Suffolk sometimes "play rough." They are commonly used in the production of crossbred slaughter lambs.

Texel.   The Texel breed comes from the Netherlands, where it has been bred for more than 160 years. It was first imported into the United States in 1990. Texels are hardy, adaptable, medium-sized sheep that are selected for high muscle–bone and lean–fat ratios. They have a white face with no wool on the head and legs and produce a medium-wool fleece. They have extreme muscling and produce high-cutability carcasses.

Tunis.   The Tunis breed originated in Tunisia in North Africa. It was imported into the United States in 1799. The breed was popular in the southern United States until it was almost completely eliminated during the Civil War. The breed is polled, has pendulous ears, has no wool on the head and legs, and has a medium-sized fat tail. The color of the face is reddish brown to bright tan. Tunis sheep are prolific and good milkers and mothers; they produce a medium-wool fleece.

Long-Wool Breeds

Bred chiefly for mutton and size, the long-wool breeds are slow to mature and produce fatter carcasses, conditions that have led to a decline in their importance in the United States. These breeds were developed in England. They

TABLE 8–6	Long-Wool Breed Comparisons						
	Mature Body Weight		Average Fiber Diameter		Grease Fleece Weight (lb.)	Yield (%)	Staple Length (in.)
	Ram (lb.)	Ewe (lb.)	(Micron)	(Spinning Count)			
Border Leicester	225–300	150–225	30–38	40–50	8–12	60–70	5–10
Coopworth	225–275	140–170	30–36	44–50	12–18	—	5–8
Cotswold	250–290	190–220	33–40	36–46	12–15	60	10–12
Lincoln	240–300	200–250	34–41	36–46	12–16	50–70	8–15
Perendale	220–260	120–150	29–35	44–54	8	60–70	4–6
Romney	200–275	150–200	32–39	36–48	10–18	55–70	5–8

FIGURE 8–14
Imported from England, the Leicester has achieved little popularity except in Canada.
Courtesy of Dr. James Blakely.

produce a long, coarse-fiber wool, are hardy and prolific, and are used in many crossbreeding programs. Their wool is used in making heavy clothing, upholstering, and rugs. Long-wool breeds include Border Leicester, Coopworth, Cotswold, Lincoln, Perendale, and Romney. Table 8–6 summarizes physical attributes of the long-wool breeds.

Border Leicester. Developed by the famous English breeder Robert Bakewell, the Leicester was imported into the United States before the Revolutionary War (see figure 8–14). The Border Leicester originated from Leicester and Cheviot crosses. The breed never attracted much attention in the United States but gained some popularity in Canada. The chief use of the Border Leicester appears to have been in improving other breeds. It is well muscled and has excellent-quality fleece for long wool. The Leicester ewe is a good mother and has a white face and a bare head and legs. The sheep yield a long-stapled, lustrous, coarse wool that is much in demand by hand spinners.

Coopworth. Developed in the 1960s from Border Leicester and Romney breeds, the Coopworth breed has increased to more than 11 million head in New Zealand, Australia, eastern Europe, and the United States. Coopworth

FIGURE 8–15
A Coopworth ewe. The Coopworth is a dual-purpose breed with equal emphasis on meat and wool; it has coarse, long, lustrous wool.
Courtesy of the L.A. Pierce College Teaching Farm Laboratory.

sheep were first imported into the United States in the 1970s. Ian Coop of Lincoln College in Canterbury, New Zealand, developed the breed. The Coopworth has replaced the Romney on wet lowlands and easy hill country. Performance data emphasize high lambing percentages, heavy fleece weights, physical soundness, rapid weight gains, easy care, and good mothering abilities. This medium-sized to large sheep has a white face and legs clear of wool (see figure 8–15). There is usually some wool on the poll. Equal emphasis is placed on meat and wool. The wool is used in heavy apparel and carpet.

Cotswold.   Originally from England, the Cotswold was first imported into the United States about 1830. The Cotswold is polled and white, with long, ropy wool that usually falls over the eyes (see figure 8–16). The clean face and forelegs are white. This large breed is used mostly for crossbreeding to improve size, fleshing, and length of fleece.

Lincoln.   Lincoln County, England, is the Lincoln's place of origin. The breed has not gained much popularity in the United States but is widely used in New Zealand, Australia, and South America. Lincoln sheep were first imported into the United States in 1825. The Lincoln is polled, has a bluish white face, and has fleece covering most of the face and legs (see figure 8–17). Lincolns have pointed ears, are polled, and have a prominent forelock of wool. This breed holds the double distinction of being the largest breed of sheep and the heaviest fleeced of the mutton breeds. Because of heavy muscling it is used chiefly for crossbreeding purposes, particularly in South America and especially in crosses with Merinos.

FIGURE 8–16
The Cotswold is a large, polled breed with long white wool. Courtesy of the L.A. Pierce College Teaching Farm Laboratory.

FIGURE 8–17
The Lincoln is a dual-purpose breed used mainly for creating crossbred ewes; it has long, coarse, lustrous wool.
Courtesy American Sheep Industry Association.

Perendale.   Geoffrey Peren of Massey University in New Zealand developed the Perendale by crossbreeding the Cheviot and Romney (see figure 8–18). A dual meat and wool breed, it is an easy-care breed and a classic hearty hill country forager. The Perendale is a small- to medium-sized sheep with pointed ears, white face, and legs clear of wool. There is some wool on the poll, and the nose is black. The wool has exceptional spring, which gives good shape retention to knitted apparel and a high insulation factor in garments. The 11 million Perendale in the world are found mostly in New Zealand, Victoria, and New South Wales in Australia.

FIGURE 8–18
The dual-purpose
Perendale breed, with
equal emphasis on
meat and wool, was
developed for the hill
country.
Courtesy of Kamahi
Perendales, Wyndham,
New Zealand.

FIGURE 8–19   The
large Romney is noted
for high-quality
carcass and fine
fleece.
Courtesy American
Sheep Industry
Association.

Romney.   An English breed referred to as the Kent or Romney Marsh breed
originated in Kent County, once a marshlike area that was drained ages ago. The
Romney is white and open faced, carries a foretop above the eyes, and has little or no wool below the knees (see figure 8–19). It is very popular in New

Zealand, a country that made perhaps the greatest improvement in the breed. The first U.S. imports came from England in 1904, and later imports came from New Zealand. In addition to large size, the Romney is noted for a rapid growth rate, a shorter, finer fleece than other long-wool breeds, and a high-quality carcass. The Romney ewe is a good milker.

Dual-Purpose Breeds

Dual-purpose breeds possess general adaptability and quality production of meat, milk, and wool. Breeds in this category include Columbia, Corriedale, East Friesian, Finnsheep, Panama, Polypay, and Targhee. Table 8–7 summarizes the attributes of the dual-purpose breeds.

Columbia.   The Columbia breed originated in Wyoming and Idaho in 1912, by crossing Lincoln rams and Rambouillet ewes. A white, open-faced, polled head characterizes the Columbia (see figure 8–20). It produces a high-quality fleece, adapts to range areas, and is considered to have the qualities desired by the

TABLE 8–7	**Dual-Purpose Breed Comparisons**						
	Mature Body Weight		Average Fiber Diameter		Grease Fleece Weight (lb.)	Yield (%)	Staple Length (in.)
	Ram (lb.)	Ewe (lb.)	(Micron)	(Spinning Count)			
Columbia	250–350	160–240	23–29	54–62	12–16	45–55	4–6
Corriedale	220–275	150–200	25–31	50–58	10–15	50–60	3.5–6
East Friesian	225–270	160–180	29–33	46–54	9–12	55–65	4–6
Finnsheep	160–220	120–160	24–31	48–60	4–8	50–70	3–6
Panama	250–280	180–210	25–30	50–58	13–15	45–55	3–5
Polypay	190–250	140–180	24–31	48–60	7–10	48–60	3–5
Targhee	200–300	140–200	21–25	58–64	10–14	45–55	3–5

FIGURE 8–20 The white, open-faced, polled Columbia is a favorite of many commercial sheep managers and breeders.
Courtesy American Sheep Industry Association.

FIGURE 8–21 The Corriedale is a good dual-purpose breed. Courtesy of Dr. James Blakely.

commercial sheep manager or breeder. Fancy details for show purposes receive little consideration. Columbia sheep are one of the largest breeds in the United States, yield medium-wool fleeces with good staple length, and are increasingly used as terminal sires to produce fast-growing, lean market lambs.

Corriedale.   The Corriedale Estate in New Zealand was the origin of this respected producer of wool and mutton. Lincoln rams and Merino ewes served as the basic cross, although some Romney and Leicester blood may have been infused in the early stages. The first Corriedale were imported into the United States (Wyoming) in 1914. A polled head and white, open face characterize the blocky Corriedale (see figure 8–21). Corriedale sheep yield heavy, medium-wool fleeces with good staple length and luster.

East Friesian.   The East Friesian breed originated in northern Germany and the province of Friesland in the Netherlands. The breed was imported into the United States from Canada in 1994. East Friesian sheep have the highest milk production of the improved dairy sheep breeds. They are a large but docile, highly prolific, open- and white-faced, polled breed. They produce a coarse-grade, long-stapled wool.

Finnsheep (Finnish Landrace).   Finnsheep originated in Finland, where they are known as Finnish Landrace. They were brought to Canada in 1966 and to the United States in 1968 (see figure 8–22). They are widely used in the United States today in crossbreeding programs because of their unusual prolificacy. Finnsheep are often referred to as the sheep that lamb in litters. It is not unusual for Finnsheep to have triplets, quadruplets, quintuplets, and even sextuplets. They are good mothers and have among the highest ratings for ease of lambing. When a ewe has quadruplets or more, two or three are usually left on the ewe; the rest are given colostrum and raised on lamb milk replacer.

Commercial sheep producers in the United States have used Finnsheep mostly for crossbreeding purposes to increase the percentage of twins and triplets in the original stock. Up to one-eighth Finnsheep in a cross is reported to generate a statistically significant increase in total lamb crop.

Purebred Finnsheep should weigh 100 pounds before breeding and should be at least 125 pounds at lambing. They can often be bred at 7 months of age and lamb before 1 year of age, although special feed and care during gestation will be needed to meet the nutritional requirements of multiple lambs.

FIGURE 8–22 The Finnsheep or Finnish Landrace breed is known for multiple births.
Courtesy American Sheep Industry Association.

The purebred Finnsheep has a large number of black lambs, a characteristic not discriminated against by the association. In spite of this negative feature, the high incidence of multiple lambs makes it an increasingly popular breed for crossbreeding purposes in the United States. Finnsheep have a naturally short tail that often does not need docking.

Panama. Panama sheep originated in Idaho in the early 1900s. They began as a cross between Rambouillet rams and Lincoln ewes and are polled. Because the same two breeds were used to produce the Columbia, the characteristics and appearance are about the same for the Panama and Columbia breeds. However, Panama sheep are more intermediate in size and produce a heavy, dense, medium-grade fleece with a long staple length.

Polypay. The Polypay medium-wool breed of sheep was developed by Clarence Hulet at the USDA sheep experiment station, Dubois, Idaho, and at Nicholas Farms in Sonoma, California, in the late 1960s (see figure 8–2). The breed was established for the purpose of superior reproduction through twins and triplets.

Bloodlines from Rambouillet, Targhee, and Polled Dorset were combined for the characteristics of size, long breeding season, milking ability, herding instincts, and fleece characteristics. To this mix was added the early puberty and high multiple lambing characteristics of the Finnsheep. The result is a breed that produces impressive 200 to 300 percent lamb crops in well-managed flocks. Breeders are now found in several states and lines have been adapted to several areas, including hot, semiarid climates. A breed association was formed in 1980.

Targhee. Another Idaho innovation is the Targhee (see figure 8–1). Developed at the U.S. sheep experiment station at Dubois, Idaho, in 1926, the breed is three-fourths Rambouillet and one-fourth long-wool breeding (from Lincoln–Leicester–Corriedale–Rambouillet crosses).

The Targhee is named after Targhee National Forest, Idaho, where experimental animals were summer pastured. The sheep are white, polled, open faced, rugged, adapted to range areas, very prolific, and intermediate in size. Perhaps the most outstanding characteristic of the Targhee is a heavy fleece yield of long-stapled wool.

Hair and Double-Coated Breeds

Hair and double-coated breeds are noted for being woolless or double coated and are especially adapted to specific climatic conditions. Breeds included in this category are Barbados Blackbelly, California Reds, Dorper, Katahdin, Romanov, and St. Croix. Table 8–8 summarizes comparisons of hair and double-coated breeds.

Barbados Blackbelly. The Barbados Blackbelly originated in West Africa. It was first introduced into the Caribbean region in the 1500s. Early in the twentieth century, this breed was imported into the United States and later crossed with the Mouflon, Rambouillet, and other European breeds. This sheep is a hair breed that is known to resist parasites, tolerates low-grade forage, and breeds out of season.

California Reds. Researchers at the University of California at Davis started the California *Reds* breed in 1970 by crossing Tunis and Barbados sheep. California Reds are deerlike in appearance and respond well to quiet handling. They are born all red but turn a light tan as they mature, except their legs and head, which remain red. Rams possess a full mane of hair growing down the chest and are polled.

Dorper. The Dorper breed was developed in the early 1940s in South Africa. Dorper sheep are a cross between the Blackhead Persian and Dorset Horn breeds. Their color is white or white with black heads. They are very fertile and have an extended breeding season. Dorpers have a mixture of hair and wool but do not require shearing. They are nonselective grazers and yield muscular, high-quality carcasses.

TABLE 8–8	Hair and Double-Coated Breed Comparisons						
	Mature Body Weight		Average Fiber Diameter		Grease Fleece Weight (lb.)	Yield (%)	Staple Length (in.)
	Ram (lb.)	Ewe (lb.)	(Micron)	(Spinning Count)			
Barbados Blackbelly	90–150	70–130	—	—	—	—	—
California Reds	175–220	120–150	28–31	50–54	5–7	50	3–5
Dorper	220–250	170–200	—	—	—	—	—
Katahdin	175–250	120–160	—	—	—	—	—
Romanov	175–240	135–160	28–33	46–54	6–13	65–80	4–5
St. Croix	163	119	—	—	—	—	—

Katahdin.   The Katahdin breed originated in Maine from crosses of St. Croix, Suffolk, and Wiltshire Horn sheep. They are a woolless, easy-care breed and are good milkers and prolific breeders. They possess unusual tolerance of heat and humidity as well as cold environments. Katahdins are the largest of the hair breeds and produce a lean and well-muscled carcass with excellent meat flavor.

Romanov.   Romanov sheep originated in the former Soviet Union (Volga Valley Northeast of Moscow) during the eighteenth century. They were imported into the United States in 1986. They are considered to be one of the most prolific breeds in the world, are hardy, and breed out of season. Romanov sheep are born black and lighten to a soft gray as they mature and develop their secondary fleece coat. The wool color is almost always lost on the first cross with white wool breeds.

St. Croix (Virgin Island White).   St. Croix sheep are found in the U.S. and British Virgin Islands in the Caribbean. They are descended from the hair sheep of West Africa and perhaps a cross including Wiltshire Horn blood and the native Criollo. They are white with some solid tan, brown, and black spots. Both sexes are polled. Dr. W. C. Foote, of Utah State University, first imported St. Croix sheep into the United States in 1975. Lambing rates average 150 to 200 percent, with two lambings a year not uncommon. St. Croix sheep are resistant to parasites and highly adaptable to high heat and humidity.

Minor Breeds

Minor breeds are breeds with smaller populations in the United States. These breeds include Black Welsh Mountain, Blueface Leicester, California Variegated Mutant, Clun Forest, Gulf Coast Native, Icelandic, Jacob, Karakul, Navajo-Churro, Scottish Blackface, Shetland, and Wiltshire Horn. Table 8–9 summarizes features of the minor breeds.

TABLE 8–9   Minor Breed Comparisons

	Mature Body Weight		Average Fiber Diameter		Grease Fleece Weight (lb.)	Yield (%)	Staple Length (in.)
	Ram (lb.)	Ewe (lb.)	(Micron)	(Spinning Count)			
Black Welsh Mountain	100–125	75–100	29–36	44–54	3–4	50–65	3–4
Blueface Leicester	230–270	160–220	29–31	48–54	6–8	60–70	5–6
California Variegated Mutant	150–200	125–150	22–25	58–62	6–12	60	3–5
Clun Forest	175–200	130–160	28–33	46–54	5–9	—	—
Gulf Coast Native	145–180	85–115	26–32	48–58	4–6	—	2.5–4
Icelandic	180–220	130–150	19–30	50–56	4–5	—	1.5–3
Jacob	140–190	90–130	17–35	44–56	3–6	—	4–7
Karakul	175–225	100–150	25–36	44–58	5–10	80–85	6–12
Navajo-Churro	120–175	85–120	22–47	36–62	4–8	60–65	5.5–14
Scottish Blackface	150–175	115–130	28–38	40–54	5–6	—	10–14
Shetland	90–125	75–100	19–29	54–70	2–4	65–80	—
Wiltshire Horn	250	125–150	—	—	—	—	—

Black Welsh Mountain.   Sheep of the Black Welsh Mountain breed were first introduced into the United States in 1972. They are a small breed, standing only 20 to 24 inches tall. They are hardy, and have a long, wool-covered tail and blue skin. Their black wool is short, dense, and without kempy fibers.

Blueface Leicester.   Blueface Leicesters originated in northern England. They were developed by crossing white-faced Leicesters with Border Cheviots. They were bred primarily for production of highly productive crossbred ewes and quality slaughter lambs. They possess good milking ability, are highly prolific, and mature early. Blueface Leicesters are medium to large sized with dark blue skin on their heads that shows through their white hair. They produce a dense, coarse-fiber fleece.

California Variegated Mutant.   The California Variegated Mutant (CVM) breed originated from the all-white Romeldale breed that was developed by crossing the Romney and Rambouillet breeds. The typical color pattern is the badger face with body wool colored cream, dark gray, or silver and the belly, britch, and neck a darker color. The sheep grow a soft, high-yielding, long-stapled, uniform fleece.

Clun Forest.   The Clun Forest breed originated in England and is considered a moderately prolific, easy-care breed of medium size. The sheep are brown or black faced with no wool on the face and legs and produce a medium wool fleece. They are good milkers and foragers and are known for longevity.

Gulf Coast Native.   The Gulf Coast Native breed is the oldest type of sheep in the United States and has existed in the southern regions for several centuries. The origin is unknown. The sheep are open faced with white to brown color, small in size with refined skeletal structure, and without wool on the legs and underline; they produce lightweight, medium-grade fleeces. They are also very tolerant to gastrointestinal parasites and will breed during the summer months.

Icelandic.   Icelandic sheep originated in Iceland and are thus adapted to a harsh, changeable climate and marginal pasture and browse conditions. They are prolific, good milkers and possess exceptional longevity. They are noted as both a meat breed and a source of wool for *lopi* yarn. Fleece colors can be white, tan, brown, gray, or black. The breed is both polled and horned.

Jacob.   A minor breed, the Jacob has been raised in England for more than 350 years. Its origin, however, is unknown. The Jacob is small and multihorned with black spots randomly distributed over its body. It has distinctive black facial markings over each eye and on the nose. The sheep produce a medium-grade wool with some kempy fibers that create a hair effect characteristic of tweed clothing.

Karakul.   The Karakul breed was named for a village in central Asia where the breed originated (see figure 8–23). Karakul sheep are one of the world's oldest breeds. They are black or brown, have long pendulous ears, and are naturally

FIGURE 8-23
Karakul ewe and lambs. The Karakul is a fur sheep kept for production of lamb pelts.
Courtesy of the L.A. Pierce College Teaching Farm Laboratory.

short tailed or broad tailed; only the rams are horned. Few are produced in the United States. This breed is kept for the production of lamb pelts for fur production and produces a rather poor quality of mutton. Pelts vary in quality and are classified as *broad tail* (lambs killed within a few hours of birth), *Persian lamb* (lambs killed at 3 to 10 days of age), and *Caracul* (lambs killed at up to 2 weeks of age).

Navajo-Churro. The Navajo-Churro sheep was developed in the United States by the Navajo Indians in Arizona, New Mexico, and Utah. The breed originated from Spanish Churro, the first type of domestic sheep in North America. Some rams have four fully developed horns, and some ewes also have horns. The ewes cycle naturally out of season, lamb easily, and usually have multiple births. They have a long-hair outer coat and a fine-wool inner fleece that may be white, black, gray, or brown. Their wool is used for hand spinning, specialty garments, and carpets.

Scottish Blackface. The Scottish Blackface originated in Scotland. The sheep are medium sized and black faced with little wool on the head and legs. The ewes are good milkers and very protective of their lambs. Their wool is very coarse in quality and has a long staple length.

Shetland. The Shetland breed originated more than 1,000 years ago in northern Europe. The sheep are noted for longevity and their ability to survive under harsh conditions. They are one of the smallest breeds and have naturally short tails. Shetlands are bred for their production of colorful wool.

Wiltshire Horn. The Wiltshire Horn sheep is an ancient British breed from the Chalk Downs region. The breed reached large numbers during the seventeenth and eighteenth centuries and then almost became extinct prior to

World War I. The Piel Farm in Canada imported Wiltshire Horn sheep to create the Katahdin breed. All the Wiltshire Horn sheep in the United States came from the Piel flock. Both sexes are white, with occasional dime-sized black spots in the undercoat. They grow a heavy coat of coarse hair for the winter. Wiltshire Horn sheep are medium sized; mature rams weigh 250 pounds, and ewes weigh 125 to 150 pounds. Their tails do not need to be docked.

Sheep breeds can also be divided into categories based on their production goals and abilities:

1. *Ewe breeds* are generally white faced, with fine or medium wool, and are noted for longevity, high milk yield, wool production, size, and reproductive performance. Common ewe breeds include Delaine-Merino, Rambouillet, Debouillet, Corriedale, Targhee, Finnsheep, and Border Leicester.
2. *Ram breeds* are the meat-type sheep breeds raised primarily to cross with ewes of the ewe breed category. These sheep are noted for growth rate and carcass characteristics. Common ram breeds include Suffolk, Hampshire, Shropshire, Oxford, Southdown, Montadale, and Cheviot.
3. *Dual-purpose breeds* are used as either ewe or ram breeds. Common dual-purpose breeds include Dorset, Columbia, Romney, Lincoln, and Texel.

SHEEP NUTRITION

Sheep are ruminants. Therefore, many of the principles learned in the previous chapters on dairy and beef cattle and on dairy and meat goats can be applied. There are some major differences, however. Sheep are unique in that they readily consume plants that cattle and goats find toxic or unpalatable. We previously pointed out the dangers of tall Larkspur (*Delphinium* spp.), a native herbaceous forb that is the leading cause of cattle deaths on mountain rangeland. Montana State University researchers report that sheep successfully graze larkspur among other noxious weeds. In the West, sheep provide 85 percent control of spotted knapweed (*Centaurea maculosa*), another invasive and poisonous plant species. More than 810,000 acres in Montana alone are infested with knapweed. Sheep eat leafy spurge (*Euphorbia esula*), an evasive, indestructible weed that infests more than 3 million acres of farm and public lands in 26 northern states.

A variety of rangeland, pasture, and forage crop management practices can enhance forage production and utilization while meeting the nutritional needs of grazing sheep. Sheep are selection grazers. A significant portion of U.S. rangeland is publicly owned and has multiple uses. Unmanaged grazing by any large herbivore can lead to ecological disaster. Consequently, range scientists manage public land in the United States.

Whereas cattle are classified as bulk grazers, usually preferring grasses to other plant types; goats and deer are more selective for quality and tend to prefer foliage and mast of shrubs and trees. Sheep are somewhat intermediate to these extremes but strongly prefer short, tender forbs and grass regrowth. Knowing these differences, range scientists employ combinations of animal species in a grazing plan that can benefit the rangeland as well as the grazing animals.

The potential benefits of distinct selection tendencies of the ruminant species are often underestimated and overlooked.

By combining two species the stocking rate can be increased 10 to 20 percent without an increased burden on the rangeland. By combining three species (cattle, sheep, and goats) the stocking rate can be increased by as much as 50 percent, provided that the rangeland has plant species relished by the added third animal species. The main disadvantage to multispecies grazing is the increased requirements for fencing, handling facilities, and managerial skills.

Feeding during the Breeding Period and during Gestation and Lactation

Feeding extra grain or lush pasture 2 to 3 weeks prior to the breeding season, flushing, is recommended by all managers to increase the number of ova shed from the ovary and thus increase the incidence of twinning. The normal flushing ration is 1/2 pound of oats or corn per head per day. An increase in the lamb crop of 10 to 20 percent is often the result. Legume pastures (such as red clover) should not be used for flushing because of the potential for *phytoestrogen-induced infertility.*[1]

Ewes should be fed to gain 20 to 25 pounds before lambing, which may be accomplished by feeding free-choice legume or mixed legume, hay, silage, or root crops such as turnips. When sheep graze turnips, the planting rate should be kept low (2 to 4 pounds per acre). High planting rates cause smaller tubers. The large tubers reduce the incidence of choking and permit extended grazing days. A protein supplement and a mineral supplement may also be needed. The roughage rations and supplements previously suggested for cattle may be used. Consumption will be about 4 pounds of hay or 12 pounds of silage for the average 150-pound ewe.

Because of the rapidly developing fetus and the extra energy needed for strong lambs at birth, 1/2 pound of grain per ewe per day 30 to 45 days before parturition is suggested. This grain allowance should be increased to 1 pound after birth to supply the energy needed for maximum lactation, particularly important because of the high incidence of twinning. If a legume hay is used, no protein supplement is generally needed. A salt–mineral mixture and clean water should always be available. As soon as pastures become available, the ewes and lambs are gradually switched to them. Lambs may be creep fed but usually are not because, unlike calves, they attain sufficient finish on pasture to meet market demands. Complete nutrient requirements for sheep are provided in tables 8–10 and 8–11.

Feeding Orphan Lambs (Bummers)

Excess lambs in multiple births have to be raised artificially. Milk replacers for lambs should be as similar to ewe's milk as possible. Calf milk replacer does not contain sufficient fat for lambs. Lamb milk replacer should contain 24 to 30 percent fat, 20 to 25 percent milk protein, and 30 to 35 percent lactose. Soy products should not be fed to newborn lambs because the abomasum is not capable of digesting soy proteins and carbohydrates.

---

[1]P. R. Cheeke, *Applied Animal Nutrition,* 2nd ed. (Upper Saddle River, NJ: Prentice-Hall, 1999).

TABLE 8–10 Daily Nutrient Requirements of Sheep

Body Weight		Weight Change/Day		Dry Matter per Animal[a]			Energy[b]					Crude Protein		Ca, g	P, g	Vitamin A Activity IU	Vitamin E Activity IU
							TDN		DE,	ME,							
kg	lb	g	lb	kg	lb	% body weight	kg	lb	Mcal	Mcal		g	lb				
Ewes[c]																	
Maintenance																	
50	110	10	0.02	1.0	2.2	2.0	0.55	1.2	2.4	2.0		95	0.21	2.0	1.8	2,350	15
60	132	10	0.02	1.1	2.4	1.8	0.61	1.3	2.7	2.2		104	0.23	2.3	2.1	2,820	16
70	154	10	0.02	1.2	2.6	1.7	0.66	1.5	2.9	2.4		113	0.25	2.5	2.4	3,290	18
80	176	10	0.02	1.3	2.9	1.6	0.72	1.6	3.2	2.6		122	0.27	2.7	2.8	3,760	20
90	198	10	0.02	1.4	3.1	1.5	0.78	1.7	3.4	2.8		131	0.29	2.9	3.1	4,230	21
Flushing—2 weeks prebreeding and first 3 weeks of breeding																	
50	110	100	0.22	1.6	3.5	3.2	0.94	2.1	4.1	3.4		150	0.33	5.3	2.6	2,350	24
60	132	100	0.22	1.7	3.7	2.8	1.00	2.2	4.4	3.6		157	0.34	5.5	2.9	2,820	26
70	154	100	0.22	1.8	4.0	2.6	1.06	2.3	4.7	3.8		164	0.36	5.7	3.2	3,290	27
80	176	100	0.22	1.9	4.2	2.4	1.12	2.5	4.9	4.0		171	0.38	5.9	3.6	3,760	28
90	198	100	0.22	2.0	4.4	2.2	1.18	2.6	5.1	4.2		177	0.39	6.1	3.9	4,230	30
Nonlactating—First 15 weeks gestation																	
50	110	30	0.07	1.2	2.6	2.4	0.67	1.5	3.0	2.4		112	0.25	2.9	2.1	2,350	18
60	132	30	0.07	1.3	2.9	2.2	0.72	1.6	3.2	2.6		121	0.27	3.2	2.5	2,820	20
70	154	30	0.07	1.4	3.1	2.0	0.77	1.7	3.4	2.8		130	0.29	3.5	2.9	3,290	21
80	176	30	0.07	1.5	3.3	1.9	0.82	1.8	3.6	3.0		139	0.31	3.8	3.3	3,760	22
90	198	30	0.07	1.6	3.5	1.8	0.87	1.9	3.8	3.2		148	0.33	4.1	3.6	4,230	24
Last 4 weeks gestation—Lambing rate expected: 130–150% or last 4–6 weeks lactation suckling single[d]																	
50	110	180(45)	0.40(0.10)	1.6	3.5	3.2	0.94	2.1	4.1	3.4		175	0.38	5.9	4.8	4,250	24
60	132	180(45)	0.40(0.10)	1.7	3.7	2.8	1.00	2.2	4.4	3.6		184	0.40	6.0	5.2	5,100	26
70	154	180(45)	0.40(0.10)	1.8	4.0	2.6	1.06	2.3	4.7	3.8		193	0.42	6.2	5.6	5,950	27
80	176	180(45)	0.40(0.10)	1.9	4.2	2.4	1.12	2.4	4.9	4.0		202	0.44	6.3	6.1	6,800	28
90	198	180(45)	0.40(0.10)	2.0	4.4	2.2	1.18	2.5	5.1	4.2		212	0.47	6.4	6.5	7,650	30
Last 4 weeks gestation—Lambing rate expected: 180–225%																	
50	110	225	0.50	1.7	3.7	3.4	1.10	2.4	4.8	4.0		196	0.43	6.2	3.4	4,250	26
60	132	225	0.50	1.8	4.0	3.0	1.17	2.6	5.1	4.2		205	0.45	6.9	4.0	5,100	27
70	154	225	0.50	1.9	4.2	2.7	1.24	2.8	5.4	4.4		214	0.47	7.6	4.5	5,950	28
80	176	225	0.50	2.0	4.4	2.5	1.30	2.9	5.7	4.7		223	0.49	8.3	5.1	6,800	30
90	198	225	0.50	2.1	4.6	2.3	1.37	3.0	6.0	5.0		232	0.51	8.9	5.7	7,650	32

Nutrients per Animal

Ewes[c]

First 6–8 weeks lactation suckling singles or last 4–6 weeks lactation suckling twins[d]

50	110	−25 (90)	−0.06 (0.20)	2.1	4.6	4.2	1.36	3.0	6.0	4.9	304	0.67	8.9	6.1	4,250	32
60	132	−25 (90)	−0.06 (0.20)	2.3	5.1	3.8	1.50	3.3	6.6	5.4	319	0.70	9.1	6.6	5,100	34
70	154	−25 (90)	−0.06 (0.20)	2.5	5.5	3.6	1.63	3.6	7.2	5.9	334	0.73	9.3	7.0	5,950	38
80	176	−25 (90)	−0.06 (0.20)	2.6	5.7	3.2	1.69	3.7	7.4	6.1	344	0.76	9.5	7.4	6,800	39
90	198	−25 (90)	−0.06 (0.20)	2.7	5.9	3.0	1.75	3.8	7.6	6.3	33	0.78	9.6	7.8	7,650	40

First 6–8 weeks lactation suckling twins

50	110	−60	−0.13	2.4	5.3	4.8	1.56	3.4	6.9	5.6	389	0.86	10.5	7.3	5,000	36
60	132	−60	−0.13	2.6	5.7	4.3	1.69	3.7	7.4	6.1	405	0.89	10.7	7.7	6,000	39
70	154	−60	−0.13	2.8	6.2	4.0	1.82	4.0	8.0	6.6	420	0.92	11.0	8.1	7,000	42
80	176	−60	−0.13	3.0	6.6	3.8	1.95	4.3	8.6	7.0	435	0.96	11.2	8.6	8,000	45
90	198	−60	−0.13	3.2	7.0	3.6	2.08	4.6	9.2	7.5	450	0.99	11.4	9.0	9,000	48

Ewe Lambs

Nonlactating—First 15 weeks gestation

40	88	160	0.35	1.4	3.1	3.5	0.83	1.8	3.6	3.0	156	0.34	5.5	3.0	1,880	21
50	110	135	0.30	1.5	3.3	3.0	0.88	1.9	3.9	3.2	159	0.35	5.2	3.1	2,350	22
60	132	135	0.30	1.6	3.5	2.7	0.94	2.0	4.1	3.4	161	0.35	5.5	3.4	2,800	24
70	154	125	0.28	1.7	3.7	2.4	1.00	2.2	4.4	3.6	164	0.36	5.5	3.7	3,290	26

Last 4 weeks gestation—Lambing rate expected: 100–120%

40	88	180	0.40	1.5	3.3	3.8	0.94	2.1	4.1	3.4	187	0.41	6.4	3.1	3,400	22
50	110	160	0.35	1.6	3.5	3.2	1.00	2.2	4.4	3.6	189	0.42	6.3	3.4	4,250	24
60	132	160	0.35	1.7	3.7	2.8	1.07	2.4	4.7	3.9	192	0.42	6.6	3.8	5,100	26
70	154	150	0.33	1.8	4.0	2.6	1.14	2.5	5.0	4.1	194	0.43	6.8	4.2	5,950	27

Last 4 weeks gestation—Lambing rate expected: 130–175%

40	88	225	0.50	1.5	3.3	3.8	0.99	2.2	4.4	3.6	202	0.44	7.4	3.5	3,400	22
50	110	225	0.50	1.6	3.5	3.2	1.06	2.3	4.7	3.8	204	0.45	7.8	3.9	4,250	24
60	132	225	0.50	1.7	3.7	2.8	1.12	2.5	4.9	4.0	207	0.46	8.1	4.3	5,100	26
70	154	215	0.47	1.8	4.0	2.6	1.14	2.5	5.0	4.1	210	0.46	8.2	4.7	5,950	27

First 6–8 weeks lactation suckling singles (wean by 8 weeks)

40	88	−50	−0.11	1.7	3.7	4.2	1.12	2.5	4.9	4.0	257	0.56	6.0	4.3	3,400	26
50	110	−50	−0.11	2.1	4.6	4.2	1.39	3.1	6.1	5.0	282	0.62	6.5	4.7	4,250	32
60	132	−50	−0.11	2.3	5.1	3.8	1.52	3.4	6.7	5.5	295	0.65	6.8	5.1	5,100	34
70	154	−50	−0.11	2.5	5.5	3.6	1.65	3.6	7.3	6.0	301	0.68	7.1	5.6	5,450	38

(continued)

TABLE 8-10   Daily Nutrient Requirements of Sheep (continued)

Nutrients per Animal

| Body Weight | | Weight Change/Day | | Dry Matter per Animal[a] | | | Energy[b] | | | | Crude Protein | | Ca, | P, | Vitamin A Activity | Vitamin E Activity |
kg	lb	g	lb	kg	lb	% body weight	TDN kg	TDN lb	DE, Mcal	ME, Mcal	g	lb	g	g	IU	IU
Ewe Lambs																
First 6–8 weeks lactation suckling twins (wean by 8 weeks)																
40	88	-100	-0.22	2.1	4.6	5.2	1.45	3.2	6.4	5.2	306	0.67	8.4	5.6	4,000	32
50	110	-100	-0.22	2.3	5.1	4.6	1.59	3.5	7.0	5.7	321	0.71	8.7	6.0	5,000	34
60	132	-100	-0.22	2.5	5.5	4.2	1.72	3.8	7.6	6.2	336	0.74	9.0	6.4	6,000	38
70	154	-100	-0.22	2.7	6.0	3.9	1.85	4.1	8.1	6.6	351	0.77	9.3	6.9	7,000	40
Replacement Ewe Lambs[e]																
30	66	227	0.50	1.2	2.6	4.0	0.78	1.7	3.4	2.8	185	0.41	6.4	2.6	1,410	18
40	88	182	0.40	1.4	3.1	3.5	0.91	2.0	4.0	3.3	176	0.39	5.9	2.6	1,880	21
50	110	120	0.26	1.5	3.3	3.0	0.88	1.9	3.9	3.2	136	0.30	4.8	2.4	2,350	22
60	132	100	0.22	1.5	3.3	2.5	0.88	1.9	3.9	3.2	134	0.30	4.5	2.5	2,820	22
70	154	100	0.22	1.5	3.3	2.1	0.88	1.9	3.9	3.2	132	0.29	4.6	2.8	3,290	22
Replacement Ram Lambs[e]																
40	88	330	0.73	1.8	4.0	4.5	1.1	2.5	5.0	4.1	243	0.54	7.8	3.7	1,880	24
60	132	320	0.70	2.4	5.3	4.0	1.5	3.4	6.7	5.5	263	0.58	8.4	4.2	2,820	26
80	176	290	0.64	2.8	6.2	3.5	1.8	3.9	7.8	6.4	268	0.59	8.5	4.6	3,760	28
100	220	250	0.55	3.0	6.6	3.0	1.9	4.2	8.4	6.9	264	0.58	8.2	4.8	4,700	30
Lambs Finishing—4 to 7 Months Old[f]																
30	66	295	0.65	1.3	2.9	4.3	0.94	2.1	4.1	3.4	191	0.42	6.6	3.2	1,410	20
40	88	275	0.60	1.6	3.5	4.0	1.22	2.7	5.4	4.4	185	0.41	6.6	3.3	1,880	24
50	110	205	0.45	1.6	3.5	3.2	1.23	2.7	5.4	4.4	160	0.35	5.6	3.0	2,350	24
Early Weaned Lambs—Moderate Growth Potential[f]																
10	22	200	0.44	0.5	1.1	5.0	0.40	0.9	1.8	1.4	127	0.38	4.0	1.9	470	10
20	44	250	0.55	1.0	2.2	5.0	0.80	1.8	3.5	2.9	167	0.37	5.4	2.5	940	20
30	66	300	0.66	1.3	2.9	4.3	1.00	2.2	4.4	3.6	191	0.42	6.7	3.2	1,410	20
40	88	345	0.76	1.5	3.3	3.8	1.16	2.6	5.1	4.2	202	0.44	7.7	3.9	1,880	22
50	110	300	0.66	1.5	3.3	3.0	1.16	2.6	5.1	4.2	181	0.40	7.0	3.8	2,350	22

Early Weaned Lambs—Rapid Growth Potential[f]

10	22	250	0.55	0.6	1.3	6.0	0.48	1.1	2.1	1.7	157	0.35	4.9	2.2	470	12
20	44	300	0.66	1.2	2.6	6.0	0.92	2.0	4.0	3.3	205	0.45	6.5	2.9	940	24
30	66	325	0.72	1.4	3.1	4.7	1.10	2.4	4.8	4.0	216	0.48	7.2	3.4	1,410	21
40	88	400	0.88	1.5	3.3	3.8	1.14	2.5	5.0	4.1	234	0.51	8.6	4.3	1,880	22
50	110	425	0.94	1.7	3.7	3.4	1.29	2.8	5.7	4.7	240	0.53	9.4	4.8	2,350	25
60	132	350	0.77	1.7	3.7	2.8	1.29	2.8	5.7	4.7	240	0.53	8.2	4.5	2,820	25

Source: Reprinted with permission from P. R. Cheeke, *Applied Animal Nutrition*, 2nd ed. (Upper Saddle River, NJ: Prentice-Hall, 1999).

[a]To convert dry matter to an as-fed basis, divide dry matter values by the percentage of dry matter in the particular feed.

[b]One kilogram TDN (total digestible nutrients) = 4.4 Mcal DE (digestible energy), ME (metabolizable energy) = 82% of DE.

[c]Values are applicable for ewes in moderate condition. Fat ewes should be fed according to the next lower weight category and thin ewes at the next higher weight category. Once desired or moderate weight condition is attained, use that weight category through all production stages.

[d]Values in parentheses are for ewes suckling lambs the last 4–6 weeks of lactation.

[e]Lambs intended for breeding, thus, maximum weight gains and finish are of secondary importance.

[f]Maximum weight gains expected.

TABLE 8–11 Nutrient Concentration in Diets for Sheep (Expressed on 100 Percent Dry Matter Basis[a])

Body Weight		Weight Change/Day		Energy[b]			Example Diet Proportions		Crude Protein, %	Ca, %	Ph, %	Vitamin A Activity, IU/kg	Vitamin E Activity, IU/kg
kg	lb	g	lb	TDN,[c] %	DE, Mcal/kg	ME, Mcal/kg	Concentrate, %	Forage, %					
Ewes[d]													
Maintenance													
70	154	10	0.02	55	2.4	2.0	0	100	9.4	0.20	0.20	2,742	15
Flushing—2 weeks prebreeding and first 3 weeks of breeding													
70	154	100	0.22	59	2.6	2.1	15	85	9.1	0.32	0.18	1,828	15
Nonlactating—First 15 weeks gestation													
70	154	30	0.07	55	2.4	2.0	0	100	9.3	0.25	0.20	2,350	15
Last 4 weeks gestation—Lambing rate expected (130–150% or Last 4–6 weeks lactation suckling singles[e])													
70	154	180 (0.45)	0.40 (0.10)	59	2.6	2.1	15	85	10.7	0.35	0.23	3,306	15
Last 4 weeks gestation—Lambing rate expected (180–225%)													
70	154	225	0.50	65	2.9	2.3	35	65	11.3	0.40	0.24	3,132	15
First 6–8 weeks lactation suckling singles or last 4–6 weeks, lactation suckling twins[e]													
70	154	−25 (90)	−0.06 (0.20)		2.9	2.4	35	65	13.4	0.32	0.26	2,380	15
First 6–8 weeks lactation suckling twins													
70	154	−60	−0.13	65	2.9	2.4	35	65	15.0	0.39	0.29	2,500	15
Ewe Lambs													
Nonlactating—First 15 weeks gestation													
55	121	135	0.30	59	2.6	2.1	15	85	10.6	0.35	0.22	1,668	15
Last 4 weeks gestation—Lambing rate expected (100–120%)													
55	121	160	0.35	63	2.8	2.3	30	70	11.8	0.39	0.22	2,833	15
Last 4 weeks gestation—Lambing rate expected (130–175%)													
55	121	225	0.50	66	2.9	2.4	40	60	12.8	0.48	0.25	2,833	15
First 6–8 weeks lactation suckling singles (wean by 8 weeks)													
55	121	−50	−0.22	66	2.9	2.4	40	60	13.1	0.30	0.22	2,125	15
First 6–8 weeks lactation twins (wean by 8 weeks)													
55	121	−100	−0.22	69	3.0	2.5	50	50	13.7	0.37	0.26	2,292	15

Replacement Ewe Lambs[f]

30	66	227	0.50	2.9	2.4	65	35	12.8	0.53	0.22	1,175	15
40	88	182	0.40	2.9	2.4	65	35	10.2	0.42	0.18	1,343	15
50–70	110–154	115	0.25	2.6	2.1	59	15	9.1	0.31	0.17	1,567	15

Replacement Ram Lambs[f]

40	88	330	0.73	2.8	2.3	63	30	13.5	0.43	0.21	1,175	15
60	132	320	0.70	2.8	2.3	63	30	11.0	0.35	0.18	1,659	15
80-100	176–220	270	0.60	2.8	2.3	63	30	9.6	0.30	0.16	1,979	15

Lambs Finishing—4 to 7 Months Old[g]

30	66	295	0.65	3.2	2.5	72	60	14.7	0.51	0.24	1,085	15
40	88	275	0.60	3.3	2.7	76	75	11.6	0.42	0.21	1,175	15
50	110	205	0.45	3.4	2.8	77	80	10.0	0.35	0.19	1,469	15

Early Weaned Lambs—Moderate and Rapid Growth Potential[g]

10	22	250	0.55	3.5	2.9	80	90	26.2	0.82	0.38	940	20
20	44	300	0.66	3.4	2.8	78	85	16.9	0.54	0.24	940	20
30	66	325	0.72	3.3	2.7	78	85	15.1	0.51	0.24	1,085	15
40–60	88–32	400	0.88	3.3	2.7	78	85	14.5	0.55	0.28	1,253	15

Source: Reprinted with permission from P. R. Cheeke, *Applied Animal Nutrition*, 2nd ed. (Upper Saddle River, NJ: Prentice-Hall, 1999).

[a] Values are calculated from daily requirements divided by DM intake. The exception, vitamin E daily requirements/head, are calculated from vitamin E/kg diet × DM intake.

[b] One kilogram TDN = 4.4 Mcal DE (digestible energy); ME (metabolizable energy) = 82% of DE.

[c] TDN calculated on following basis: hay DM, 55% TDN and on as-fed basis 50% TDN; grain DM, 83% TDN and on-as-fed basis 75% TDN.

[d] Values are for ewes in moderate condition. Fat ewes should be fed according to the next lower weight category and thin ewes at the next higher weight category. Once desired or moderate weight condition is attained, use that weight category through all production stages.

[e] Values in parentheses are for ewes suckling lambs the last 4–6 weeks of lactation.

[f] Lambs intended for breeding; thus, maximum weight gains and finish are of secondary importance.

[g] Maximum weight gains expected.

SHEEP REPRODUCTION

Sheep are seasonal, **polyestrus** (*poly-*, many; *estrus-*, heat) breeders. In other words, sheep can have numerous cycles during which conception can occur but only during a certain season, the fall. Neither rams nor ewes are fertile during other seasons. A few breeds (e.g., Dorset) may produce lambs out of season.

Ewes have a high incidence of twinning; a 110 to 150 percent lamb crop in a flock is not unusual. Unlike cattle, which can be managed to calve in the spring, fall, or year-round, sheep normally lamb in the spring. Management and feeding practices prior to this time affect the efficiency of reproduction. Ewes that are 3 to 7 years old are more fertile and raise a higher percentage of lambs than do younger or older ewes.

Age

Female lambs in most mutton-type breeds reach puberty at 5 to 7 months of age, normally breeding the first fall at age 8 months or older. Larger, more slowly maturing wool breeds may first breed at age 16 months or older. Rams reach puberty about a month earlier than ewes in each case. The age at puberty is greatly influenced by breed, nutrition, and date of birth. Because estrus is seasonal, lambs born late in the spring may not reach puberty until the second fall season. Crossbred ewes are bred to lamb at 1 year of age.

Photoperiodism and Temperature

Sheep respond to decreased day lengths by showing a greater proportion of ewes in estrus and higher conception rates. High temperatures cause heat sterility in rams and greater early embryonic death in ewes. Careful environmental management before, during, and after breeding is essential to maintain high lambing crops.

This seasonally polyestrus mammal begins the estrous cycle triggered by cool temperatures in the fall. The cycle is repeated every 16 days on the average until conception occurs. The length of heat averages 30 hours. Unlike females of other species, ewes show few external indications of estrus other than standing to be mounted. As previously stated, the length of season varies by breed. Southdown, Cheviot, and Shropshire start cycling early to late fall; Suffolk, Hampshire, Columbia, and Corriedale cycle from August through early winter; and Rambouillet, Merino, and Dorset cycle from midsummer to midwinter.

The gestation length ranges from 144 to 152 days (average of 148 days). The medium-wool and meat breeds have shorter gestation periods, and the fine-wool breeds have longer gestation periods.

Estrus Synchronization and Artificial Insemination

In some intensively managed sheep operations, progesterone is given to synchronize estrus, followed by the use of artificial insemination. One method of synchronization of estrus is to feed progesterone for 12 to 14 days. All the ewes will start to cycle together after removal of progesterone from the feed. The use of implants and intravaginal devices (pessary) has produced mixed results at synchronization. Injection of progesterone followed by injections of Pregnant Mare Serum Gonadotropin (PMSG) the day progesterone is discontinued has shown better synchronization and fertility rates. Two injections of PMSG, 15 to 17 days apart, improve fertility rates in ewes being synchronized. Conception rates are generally low at first estrus. Fresh ram semen is used in breeding sheep. There have been very poor results with frozen semen. In recent years, classes in transcervical insemination of sheep have been offered across the country to increase the genetics available within a given flock.[2] Ram semen can be frozen with a milk, yolk-lactose, or yolk-tris extender.[3] However, fertility is considerably lower than when using fresh semen (less than 12 hours old).[4]

Natural Mating

Tagging the ewes (shearing the locks of wool and dirt from the dock) makes service by the ram more certain. The ram is also trimmed around the sheath. The feet should be inspected at this opportune time and trimmed if necessary. Most range flocks average 3 to 3.5 rams per 100 ewes. The stocking rate for pasture mating is 15 to 25 ewes per ram lamb, 25 to 35 ewes per yearling ram, and 35 to 60 ewes per mature ram. Ram lambs should not be kept in the same breeding pen as mature rams. Males are vigorously productive to 6 or 8 years of age.

To identify progress of mating, the rams may be marked by painting the breast with a thick, colored paste. A ewe that accepts a ram is therefore marked by a smear on the rump. The paste may be made by mixing lubricating oil and yellow ocher for the first application; reapplication may be necessary every 1 to 2 days. After 16 days, the color is changed by mixing the oil with Venetian red and maintained for the next 16 days. The last 16 days, lamp black is mixed with the oil. This method shows which ewes settled (conceived) at the first, second, or third heat period. A ewe that has only a yellow mark was bred early. A red mark over yellow means two services. A black mark over the other colors could mean that the ewe is not going to conceive and may need to be culled from the flock. The colors should always progress from light to dark.

Where ewes are separated into flocks served by only one ram, this marking method can detect a sterile male easily because all ewes are repeat breeders. The problem can be detected in time to bring in a fertile male. Culling poor breeders prevents them from interfering with the good, fertile rams. Without the simple marking system, an entire lamb crop can be skipped and a year of production wasted.

Several diseases or disorders are attributed to poor fertility in rams. They range from reproductive system defects to poor nutrition to microbial infections. All of these problems should be addressed well in advance of the breeding season. Rams should be sheared 6 to 8 weeks before the breeding season for maximum performance. Rams exhibiting signs of pneumonia or other health problems should not be used for breeding. See the later section "Diseases of Sheep" (epididymitis, vibriosis, and sheath rot) for other causes of infertility in rams.

---

[2]Pipestone Artificial Breeders, Pipestone, Minn.

[3]E. S. E. Hafez, *Reproduction in Farm Animals*, 6th ed. (Philadelphia: Lea & Febiger, 1993).

[4]H. J. Bearden and J. W. Fuquay, *Applied Animal Reproduction*, 4th ed. (Upper Saddle River, NJ: Prentice-Hall, 1999).

Gestation and Parturition Care

Sheep are relatively hardy but should have shelter to protect them from extended periods of soaking rain; a plain, open shed facing away from the wind will do. Exercise is very important to good blood circulation. If bad weather forces ewes to spend several days without activity near the hay bunks and feed troughs, they should be moderately walked or encouraged to move about by scattering roughage outside.

Ewes should be tagged again around the dock, flank, and udder as lambing time approaches. Wool with manure and/or burrs on it should be removed to make lambing more sanitary and nursing easier. The practice of clipping the wool from the rear quarters of a ewe prior to lambing is called **crutching.** Tagging is also done at this time. The ewe is crutched by placing it on its rump and taking several strokes with the clippers across the belly just ahead of the udder. Then the wool on the inner side of each leg is sheared off. A free hand is next placed over the two teats, making several strokes from the attachment of the udder to the dock. Crutching is a desirable practice; it ensures cleaner lambing and enables the new lamb to find the udder more readily. It is particularly desirable when the herdsman may not be on hand during lambing because, with uncrutched ewes, the lamb may start sucking on a tag and not find a teat before it is too weak to nurse.

Another procedure worthy of mention is **facing,** the practice of cutting the wool above and below the eyes of sheep to prevent them from becoming **wool blind.** If this procedure is not done in some breeds with heavy face covering, they are apt to be poor foragers and show poor mothering ability.

In large flocks in the West, lambing occurs on the open range with few problems (except for predators) and little assistance, although good shepherds watch closely and are on hand when needed. Smaller flocks are often confined, and ewes "making bag" are moved to a lambing pen (*lambing jug*) that consists of four portable panels, each about 4 feet long. Fresh, clean straw can be used for bedding. Normal presentation is the same as for cattle, and assistance is offered only when obviously needed. When born in confinement, the lambs' navels should be painted with iodine as a precaution against tetanus and other clostridial disorders, which are quite common among confined sheep. Newborn lambs are susceptible to hypothermia. If the weather is very cold, it may be advisable to provide heat lamps to keep the lambs warm. Rubbing with old towels can also be effective in drying off the lambs and stimulating blood circulation.

If all goes well after a few days in the lambing pen, the ewes and their young are turned out. Occasionally ewes disown one or more lambs, which requires the attention and patience of the shepherd. Some methods employed to overcome this rejection are

- Milking the ewe and smearing the milk on the rump of the lamb and nose of the ewe
- Smearing mucous membranes from the lamb's nose on the ewe's nose
- Blindfolding the ewe
- Tying a dog near the lambing pen

The dog sometimes brings out protective maternal instincts, and the other methods heighten the sense of smell to overcome rejection. The same method may

■ Crutching   practice of clipping the wool from the rear quarters of a ewe before it lambs

■ Facing   practice of clipping the wool above and below the eyes of a sheep to prevent wool blindness

■ Wool blind   growth of wool on the face and around the eyes in a manner that obstructs the vision of the sheep

be used to switch an orphan to a ewe that has only one lamb. A ewe that loses a lamb may be persuaded to accept an orphan by rubbing the orphan's back with the dead lamb. A common practice is to remove the skin of the dead lamb and tie it to the back of the adopted lamb. This practice works surprisingly well. The skin may later be removed bit by bit.

One danger of this method is possible infection from the dead lamb if it died of disease. Also, in spite of all efforts, it is sometimes too time-consuming or impossible to foster a **bummer** (orphan lamb) with another ewe. In large operations where there may be many bummers or a triplet or smaller twin that is not getting enough milk, it may be advantageous to utilize an automatic lamb self-feeder using cow's milk or milk replacer. For the first few days, lambs should receive colostrum milked from a fresh ewe, goat, or cow.

▪ **Bummer** orphan lamb

Bottle-fed lambs (bummer lambs) are weaned at about 6 weeks of age. Some shepherds recommend weaning as early as 3 weeks of age. Once the lambs are eating solid foods, it is safe to wean them. They will usually lose weight initially, but in the long run early weaning is cost-effective.

Lambs nursing their dams are weaned at 45 days to 6 months, allowing ewes a short rest (95 to 160 days) before beginning the next breeding season in the fall. The exact age for weaning depends on how the lambs are to be marketed. Lambs may be marketed as grass-fat spring lambs, enter the feedlot for further fattening and be classified simply as lambs when marketed, or kept as ewe lambs and ram lambs for breeding stock.

Intensive Systems for Reproductive Management

Accelerated lambing refers to ewes lambing more frequently than once a year. Ewes that lamb in the fall may breed back as soon as 40 days afterward, enabling a ewe to lamb every 190 days instead of every 365 days. One advantage to this system is the marketing of lambs when their supply is low and the prices are generally high. Disadvantages include low fertility and prolificacy rates and small birth weights.

Some breeds, including Rambouillet, Dorset, Polypay, and Barbados Blackbelly, are better suited for accelerated lambing because they have long breeding seasons. Crosses of these breeds with Finnsheep have also shown good results. The use of special light treatment (16 hours of dark and 8 hours of light for 12 weeks) improves **libido** and fertility during the normal anestrous season.

▪ **Libido** sex drive

Hothouse Lambs. Production of **hothouse lambs** is a highly specialized business. Lambs are dropped in the early fall or winter and sold at the light weight of 30 to 60 pounds live weight. Boston and New York are the markets that demand hothouse lambs, so called because they are kept in protected shelters and pushed to be ready for sale within 6 to 12 weeks of birth. The lambs are usually castrated but not docked because some buyers associate the docked condition with older lambs.

▪ **Hothouse lambs** lambs born in the fall or early winter and finished for market between 6 and 12 weeks of age and weighing from 30 to 60 pounds

Easter Lambs. There is some demand for light (20 to 30 pounds live weight) lambs each year at Easter, principally in the East. Heavier lambs at more variable weights are also accepted, which provides a market for early lambs if the shepherd decides to take advantage of a favorable market.

TAIL DOCKING AND CASTRATION

Tails are docked primarily for sanitation purposes. A long tail soaked in urine and feces attracts flies, increasing the likelihood of fly strike, a potentially fatal disorder. Tail docking should be done as early as possible, but it is best to wait 3 to 4 days after birth to give the lamb a recovery period from the stress of being born. Removal of the tail after 2 months of age should be performed with the animal under local anesthesia and with special care taken to prevent hemorrhaging. The most common methods of tail docking lambs include the elastrator (rubber rings), hot iron cautery, and surgical removal following the application of an emasculator. In all cases, the lamb should be given a tetanus antitoxin to prevent that disorder at the time of docking. Care should be taken to leave sufficient length of tail (at least 1 to 1 1/2 inches) to prevent prolapse of either the rectum or vagina (common when docking too short).

Castration should also be performed 3 to 4 days after birth. Castration is needed to prevent indiscriminate breeding as well as to lessen the aggressive behavior in maturing males and the injuries that normally accompany that behavior. The most appropriate method of castration in sheep depends on the environmental conditions, the experience of the shepherd, and the amount of assistance he or she may have. The elastrator, Burdizzo, and surgical removal are all commonly used. Tetanus antitoxin should be given at the time of castration. Any animals over the age of 2 months should be given anesthesia and special care to minimize hemorrhage and infection.

DISEASES OF SHEEP

The sheep is a very gentle species of animal, almost devoid of self-protective instincts because of its dependence on people. Thousands of years ago, shepherds watched over their flocks to keep them from harm. At that time, management meant little more than frightening off predators. Even if an attacking wolf or wild dog were driven off before seriously wounding an animal, a sheep could still die of shock. Shepherds must thus recognize problems early and take preventive measures. More than other species, sheep have a tendency to lose their desire to live once they are stressed. Diseases, of course, cause stress, and treatment should include not only modern methods and drugs but also the gentle touch of a compassionate handler. As with other classes of livestock, body temperature is often the first signal of developing disorders. The normal temperature range should be 100.9 to 103.8°F.

Genetic or Inherited Abnormalities

Cryptorchidism. *Cryptorchidism* is caused by a simple recessive gene contributed from each parent. Rams with only one testis descended into the scrotum should never be used for breeding even though they are fertile. Ewes that produce such ram lambs and rams that sire them must be culled to stop the progression of this disorder within a flock.

Dwarfism. *Dwarfism* is also a recessive but semilethal genetic disorder. Ewes and rams producing dwarf offspring should be culled. Parrot-mouth dwarfs have been observed in a strain of Southdown sheep. All lambs that were affected died within a month of birth.[5]

Entropion.   Entropion lambs should be culled. The mode of inheritance has not been determined, but it is known to be genetic. If not corrected, the lambs will go blind.

Parrot Mouth.   The lower jaw is too short in lambs with *parrot mouth*. One researcher found 1.4 percent of 7,000 Rambouillet lambs had this disorder. Among those from parents that were affected, the incidence was greater than 16 percent.[6]

Rectal Prolapse.   *Rectal prolapse* is common in black-faced sheep. This serious defect can also be caused by short tail docking, using estrogens as a growth promotant, and feeding young lambs high-concentrate rations.[7]

Spider Syndrome.   *Spider syndrome* is a disorder related to abnormal transformation of cartilage into bone. The front legs are usually bent out from the knees, and the hind legs typically show deformities. It is a recessive genetic abnormality that is more common in Suffolk and Hampshire sheep and considered semilethal. It is thought to have developed as a genetic mutation in the late 1960s, but it was not reported as a widespread problem until the mid-1980s. Tested-free animals can be purchased today. These animals have passed a progeny test such that there is less than a 1 percent chance that they carry the recessive gene in their genotype.[8]

Wool Blindness.   Open-faced ewes raise more lambs and wean more pounds of lamb than do ewes with extremely wooly faces that cause *wool blindness*. *Facing* (practice of cutting the wool above and below the eyes of sheep) helps to prevent wool blindness.

Microbial and Metabolic Disorders

Blackleg.   Although sheep are affected by blackleg to a lesser extent than cattle, signs, treatment, and prevention are the same. See the section "Blackleg" in chapter 5.

Bloat.   Signs, treatment, and prevention are the same as for cattle. See the section "Bloat" in chapter 5.

Bluebag (Mastitis).   Known as bluebag in sheep, this is the same disease as mastitis in cattle. See the section "Mastitis" in chapter 5.

---

[5]R. Bogart, *Improvement of Livestock* (New York: The MacMillan Company, 1959).

[6]F. B. Hutt and B. A. Rasmusen, *Animal Genetics,* 2nd ed. (New York: John Wiley & Sons, 1982).

[7]*The Sheepman's Production Handbook* (Denver, CO: American Sheep Producers Council, 1975).

[8]Spider Syndrome Task Force, Sheep Industry Development Program, Inc., (Primal Communication, 1999).

Bluetongue.   Depressed appetite, inflammation of the inside of the mouth and nose (which may turn blue), frothing at the mouth, and labored breathing are major signs of bluetongue. A red band at the top of the hoof may appear. Although bluetongue is a noncontagious viral disease and fatalities are not high (10 to 30 percent), the disease is dreaded in the southwestern part of the United States where it is found. Insects transmit the virus that causes the disease. Besides sheep, bluetongue is found in wild ruminants and rarely in cattle, goats, and carnivores. Because treatment has largely been ineffective, prevention through the use of a vaccine is recommended.

Caseous Lymphadenitis.   Caseous lymphadenitis (CLA) is a chronic, infectious, and contagious disease of both sheep and goats. It is caused by the bacterium *Corynebacterium pseudotuberculosis* and results in caseous (thick, cheeselike accumulation) abscesses of the lymph nodes and internal organs. Symptoms include reduced weight gain, abscess scars (decreased pelt value), decreased wool growth, reduced milk production, and reduced reproductive efficiency. It is also reported to cause regional lymphadenitis in people. Transmission of the disorder is by mechanical inoculation through the skin; shearing, castration, docking, and working around fencing materials are common means of gaining entrance. Feed bunks are easily contaminated, and the bacteria easily spread as animals rub themselves on soiled surfaces. Incidence of CLA increases with age.

There is no effective treatment for CLA due to the way the bacteria encapsulate themselves within the lymph node. Abscesses can be lanced and flushed daily with a solution of nitrofurazone and hydrogen peroxide. Opened abscesses can spread the disease. All infected animals should be culled and isolated from noninfected sheep immediately. Strict sanitation and immediate treatment of wounds with iodine solutions will lessen the incidence.

Circling Disease (Listerellosis, Encephalitis).   The principal signs of circling disease are awkward staggering, walking in circles, and paralysis. Usually fatal, this disease is caused by a bacterial infection that affects the brain and reduces coordination. In ewes, the disease is sometimes confused with pregnancy disease (ketosis) but may be distinguished by moving the head, which is usually turned to one side in circling disease. The head will always return to the same position if the ewe has circling disease, but it will remain straightened out in sheep with pregnancy disease. There is no prevention other than proper sanitation. Treatment is not very effective, although some response has been noted if antibiotics (such as penicillin G) and supportive therapy (fluids and electrolytes) are given early enough. Listeria organisms can be transmitted to humans through milk and by handling of aborted fetuses.

Enterotoxemia (Overeating Disease, Pulpy Kidney Disease, Apoplexy).
Very common in feedlots, enterotoxemia is responsible for the largest death losses in feedlot lambs. In addition, lambs under 6 weeks of age that are nursing heavy-milking ewes are highly susceptible. Signs include sudden loss of appetite, staggering, convulsions, and death. The cause is a bacterial (*Clostridium perfringens* type D and sometimes type C) development brought on by a high scale of feeding, whether in feedlot or lush pasture. The bacteria are normally present in the digestive tract of most sheep. However, only under heavy feeding conditions are the bacteria capable of rapidly multiplying and produc-

ing the toxin that causes the disorder. The production of the toxin and the absorption across the intestinal ways are rapid, such that lambs can die within hours of consuming too much feed. Treatment using antitoxins has been effective when supervised by a veterinarian. Vaccines (toxoids) are available and are used as a preventive measure. Feedlot lambs need to be vaccinated twice at least 10 days apart with *C. perfringens* type D toxoid. For younger lambs, adequate protection can be achieved by immunizing breeding ewes 4 to 6 weeks before lambing.

***Escherichia coli* Complex.** *Escherichia coli* complex is believed to be caused by a combination and proliferation of *E. coli, Clostridium perfringens,* and rotavirus. The disease affects lambs from birth to a few days of age and can be prevented by vaccinating all pregnant ewes twice in later pregnancy with type C and D *C. perfringens* toxoid.

Enzootic Abortion of Ewes. Enzootic abortion of ewes (EAE) is caused by *Chlamydia psittaci* (a different strain from that which causes conjunctivitis). EAE is becoming increasingly more problematic in the western United States. Its symptoms include abortion, stillbirths, and the birth of weak lambs. Control consists of vaccination. Ewes should be immunized with a killed ovine chlamydial vaccine before breeding if they are in endemic areas. Oxytetracycline and oral tetracycline have both shown good results in the treatment of EAE.

Foot Rot. Foot rot is one of the most economically devastating diseases that affect sheep. Associated losses are from production, labor, and material costs for treatment. It is much easier to prevent than to control or to eradicate foot rot. Two main types of bacteria are blamed for this disorder in sheep: *Dichelobacter nodosus* and *Fusobacterium necrophorum*. Symptoms include a moist, reddened area between the toes, separation of the horny tissues, a characteristic foul odor from the foot, and lameness. The disease spreads easily from infected to noninfected sheep in moist, cool soil (soil temperatures of 40 to 70°F).

Suggested control measures include the following:

1. Never buy sheep infected with foot rot or from infected flocks.
2. Avoid the use of common trails and corrals where infected sheep have recently traveled for at least 2 weeks.
3. Disinfect commercial vehicles prior to transporting sheep.
4. Quarantine all new additions to the flock for a minimum of 2 weeks. Trim all feet immediately on arrival and treat feet after trimming.

Vaccination of flocks for *D. nodosus* is most beneficial if used in conjunction with other foot rot control measures (e.g., sanitation, quarantine, hoof trimming, and footbaths). Footbaths containing 10 percent copper sulfate or 5 percent formalin once a week greatly reduce the spread of foot rot to unaffected animals. Penicillin or tetracycline injected intramuscularly provides the highest cure rates. New Zealand farmers have found some success in feeding Zn-methionine to reduce the incidence of foot rot.

Grass Tetany (Grass Staggers). Grass tetany occurs in the spring when lactating ewes are turned out onto rapidly growing pastures. A low magnesium content of the feed or heavy fertilization of the pasture (pasture growing very fast) causes the symptoms of hypomagnesemia. Sheep show muscular tremors, nervous excitement, ataxia, tetanic spasms, and convulsions. Death can occur within hours. Animals at risk need to receive daily doses of magnesium, because magnesium is not stored in any appreciable quantity in the body. Treatment includes immediate intravascular injections of both calcium and magnesium followed by subcutaneous injection of magnesium sulfate.

Johne's Disease (Paratuberculosis).   Johne's disease is a chronic infectious disease of ruminants. In 1998, it was estimated to cost U.S. cattle producers $1.5 billion in losses. The causative organism is *Mycobacterium paratuberculosis*. This organism can survive for long periods in the environment and is even resistant to many commonly used disinfectants. Some evidence suggests that the human disorder Crohn's disease may be associated with Johne's disease.

In sheep, common symptoms include weight loss, intermittent fever, submandibular edema (bottle jaw), and lethargy. Microscopic examination of tissues of sheep dying from the disease can establish a positive diagnosis. Fecal cultures used for diagnosis in cattle have not proven as effective in diagnosis of Johne's disease in sheep.

A closed flock is recommended to control Johne's disease. The use of embryo transfer and artificial insemination can bring in new genetics to the flock and minimize the risk of Johne's disease and other contagious disorders. For infected flocks the following recommendations have been made to reduce the spread of Johne's disease:

1. Provide clean lambing area.
2. Wash between lambings.
3. Shear or crutch ewes before lambing to reduce contamination for nursing lambs.
4. House replacement ewe lambs separate from mature animals and their droppings.
5. Cull offspring from ewes that are known to be infected (in utero transmission is common).

There is no approved Johne's disease vaccine for sheep in the United States. The vaccine used for cattle can only be given by a licensed veterinarian; it causes severe tissue reactions in people accidentally injected.

Lamb Dysentery (Scours).   Lamb dysentery is one of the enterotoxemia disorders caused by *Clostridium perfringens* type B. It is most common during periods of wet weather and very early in life. The principal signs are diarrhea and fever the first few days after birth. Death losses are very high. The problem is seldom seen on the open range or pasture and is most common when sheep are kept in close confinement. Antibiotics may reduce death losses, but no treatment has proven effective. Lamb dysentery can be prevented by vaccinating with clostridial type BCD vaccine and through good sanitation in a well-sheltered, dry area.

Ovine Progressive Pneumonia (Maedi, Zwoegersiekte, La Bouhite, Graaff-Reinet Disease, Marsh's Progressive Pneumonia).   Ovine progressive pneumonia (OPP) is a slowly progressing viral disease of sheep. Various names have been given to this disorder since it was first reported in South Africa in 1915. It is caused by a lentivirus (the retrovirus family that causes slow, progressive diseases such as cancer in people). It affects sheep of all ages but most commonly those over 4 years of age. Once contracted, the disorder may last for several months to a year, with steady progression until death. Symptoms include weight loss (thin ewe syndrome), pneumonia, dyspnea, congested udder, arthritis, lameness, encephalitis, and death.

With the exception of Australia and New Zealand, OPP is found in all major sheep-producing countries of the world. Producers in Iceland have lost nearly 30 percent of their sheep trying to eradicate the problem. There is no effective treatment. Control is primarily by separating all seropositive animals (those that test positive for antibodies to the OPP virus in the blood). Transmission of the virus is through fluids and tissues of the sheep (e.g., milk, saliva, and placental fluids). Most sheep with OPP die of secondary bacterial pneumonia. Currently, there are no approved vaccines for OPP, but researchers are working to develop one.

Pinkeye.   Pinkeye also affects cattle (see chapter 5).

Pregnancy Disease (Ketosis, Acetonemia, Pregnancy Toxemia, Twin-Lamb Disease).   Called pregnancy disease in sheep, this metabolic disorder is the same as ketosis in cattle (see chapter 5). It is the most common metabolic disease of sheep. The condition usually strikes ewes during the last 2 weeks of gestation. It is seen most often in ewes carrying twins and triplets and in obese ewes. Trembling when exercised, weakness, and collapse are characteristic signs. Hypoglycemia, or low blood sugar levels, decreased liver glycogen levels, and fatty liver are the causes. Unlike with cattle, it is highly fatal in sheep if not treated. Because glucose is essential for normal central nervous system function, untreated animals will eventually fall into a coma and die. As with cattle, the feeding of a high-energy feed such as molasses at the time of stress is beneficial to both prevent and treat the disorder.

Inadequate nutrition is the most common factor. Feeding some high-energy grains (e.g., corn, barley, or milo) to undernourished ewes is one method of prevention. During the last 3 to 4 weeks of pregnancy, the amount of grains and molasses should be increased. A drench of 200 to 300 milliliters of propylene glycol can be used as an initial treatment when the first signs appear.

Scrapie.   The name *scrapie* is derived from the chief sign of scraping wool off by rubbing against fences and other objects because of intense itching (see figure 8–24). Scrapie is an infectious disease of both sheep and goats that causes degeneration of the central nervous system. It is one of the many transmissible spongiform encephalopathies. It is not transmitted to humans. Related disorders include the human diseases known as kuru and Creutzfeldt-Jakob disease and the bovine disorder known as mad cow disease (see chapter 5). Current evidence points to the feeding of inadequately treated sheep brain and spinal cord to cows as the probable source of mad cow disease.

There is usually no fever, but an unsteady, uncoordinated gait precedes paralysis and death. The cause is a transmissible agent, an abnormal form of the prion protein called PrPSc. This prion is highly resistant to heat and common disinfectants.

Scrapie is not a genetically transmitted disease. The highest known prevalence of scrapie transmission is from ewe to lamb. Scrapie was first described more than 200 years ago in England. The first case diagnosed in the United States was in 1947 in sheep originating from England. A total of 1,281 sheep

FIGURE 8–24
Scrapie is an infectious disease that attacks the nervous system of sheep. The name *scrapie* describes a main symptom of the disease—an infected animal scrapes off patches of wool as it rubs against objects to relieve intense itching. Courtesy of the U.S. Department of Agriculture.

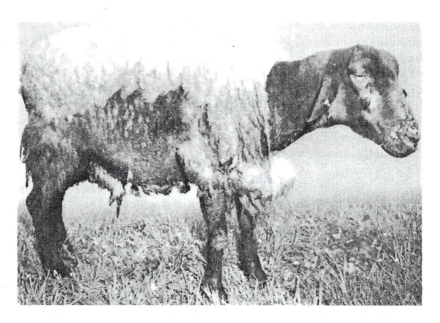

in 850 flocks (in the United States) have been diagnosed with scrapie in the past 50 years.[9]

Positive diagnosis of scrapie is made by microscopic examination of the brain. The disease is prevented by keeping a closed flock. The federal scrapie control program is called the Voluntary Scrapie Flock Certification Program (VSFCP).[10] It is run by the Animal Plant Health Inspection Service Veterinary Section (USDA-APHIS). There is no treatment for scrapie and no practical method of controlling the spread in endemic areas. Australia and New Zealand eradicated scrapie by forced slaughter of imported sheep and flock mates shortly after release from quarantine.

Shipping Fever (Pneumonia, Hemorrhagic Septicemia, Pasteurellosis).
Shipping fever is a highly infectious and contagious disease of lambs usually following transportation or other stressful events. The causative organisms are *Pasteurella hemolytica, P. multocida,* and *Corynebacterium pyogenes.* Symptoms include respiratory disease, septicemia, arthritis, meningitis, and mastitis. Exposure to various stressors facilitates the transition from infection to disease. Antibiotics and sulfonamides are used to treat this disorder.

Sore Mouth (Contagious Ecthyma, Contagious Pustular Dermatitis).
Lambs are most commonly affected by sore mouth. Small blisters appear on the mouth, nose, lips, and tongue. A scab soon develops, and pus drains from infected areas. The condition is sometimes misdiagnosed as bluetongue. Death losses are low, but economic losses because of failure to eat and gain are tremendous. The cause is a virus that also affects humans with a condition called *Orf.* Symptoms in humans include very painful lesions on the hands and face.

---

[9]American Sheep Industry (ASI) Association, 1999.
[10]USDA/APHIS/VS, National Animal Health Program, 4700 River Road, Unit 43, Riverdale, MD 20737-1231.

In sheep the virus causes lesions of the lips; vesicles form, followed by pustules and scabs. Sometimes the nostrils, eyelids, mouth, and vulva are affected. Eating is very painful. Occasionally lambs die from starvation or secondary pneumonia. A vaccine is available, but it should be used cautiously to avoid contaminating uninfected premises. Lambs should be vaccinated when they are 1 month old and then receive a booster 2 to 3 months later. All vaccinated animals should be segregated from unvaccinated sheep.

Stiff Lamb Disease (White Muscle Disease, Nutritional Muscular Dystrophy). Stiff lamb disease causes degeneration of the skeletal and cardiac muscles of lambs. It is caused by a deficiency of selenium and/or vitamin E in the diet. Stiff rear legs and a humped back are chief signs. Usually, only young lambs less than a month old are affected. Lambs may die or remain stunted. Treatment with injections of selenium or vitamin E has proved useful. The disease may be prevented by feeding ingredients such as linseed meal, which contains adequate vitamin E, or by injection of Se to pregnant ewes during the last trimester of pregnancy and to lambs at birth. The feeding of selenium and tocophoral to pregnant ewes 1 to 4 weeks prior to lambing has also proven to reduce the incidence in lambs.

Tetanus (Lockjaw). Tetanus is caused by *Clostridium tetani,* a bacterium commonly found in the soil and rich in horse manure. There is no effective treatment. If infection is likely, 200 to 300 units of antitoxin should be administered to provide passive immunity (protection) for about 2 weeks. Sheep are quite commonly affected following shearing, docking, castration, and even vaccination. The use of elastrators (rubber bands) for castration or docking is known to predispose lambs to the infectious disorder. In addition to the antitoxin being administered as described, tetanus toxoid vaccinations are recommended for flocks that have a history of tetanus.

Toxoplasmosis. The abortion disorder toxoplasmosis is caused by infection with *Toxoplasma gondii* early in gestation. It causes resorption and mummification of the fetus. If contact is made later in gestation, abortion generally occurs. Once ewes recover, they are immune to reinfection. The disease may be prevented by ridding the area of cats. Toxoplasmosis is a zoonotic disorder.

Urinary Calculi (Urolithiasis, Water Belly, Kidney Stones). Rams and wethers in feedlots or on high-concentrate rations are most often affected by urinary calculi. Stonelike pebbles composed chiefly of calcium develop in the urinary tract of rams or wethers, blocking passage. Males have difficulty urinating or complete blockage occurs, resulting in death through *uremic poisoning.* Rations high in phosphorus content, sheep grazing grain stubble or succulent pastures, and imbalances in calcium and phosphorus intake can increase the incidence.

When the condition arises, an increase in the salt content of the diet by 1.0 to 1.5 percent (normally 0.5 percent), a calcium–phosphorus ratio of 2:1, or use of 0.5 percent ammonium chloride (acid-forming salt) has been beneficial. The use of 20 percent alfalfa in rations may also be valuable as a preventive measure. In range sheep, the disease is associated with the consumption of high silica content of range forages. Increasing the water consumption in range sheep by an increase in dietary salt assists in reducing the incidence.

Vibriosis and Other Disorders Causing Abortion and Infertility.   Vibriosis in sheep is caused by the bacterium *Vibrio fetus intestinalis (Campylobacter fetus)*. Affected ewes abort in late pregnancy or produce a stillborn offspring. The fetus is autolyzed. Tetracycline may help prevent abortion. Vaccinations are available and recommended, along with strict sanitation, to prevent outbreaks. There are many other causes of abortion and infertility in sheep, however. Positive diagnosis should be made to rule out leptospirosis, ram epididymitis, and infection with *Salmonella, Listeria, Chlamydia,* or *Toxoplasma species,* all of which could contribute to the problem and are common in sheep.

Although leptospirosis is much less common in sheep than in cattle, it does occur. The most common serovars in sheep are *Leptospira pomona* and *L. hardjo.* Brucellosis in sheep primarily affects the ram, causing a swelling of the epididymis, tunica, and testis (*ram epididymitis*). Fertility is thus impaired through the ram and occasionally causes abortion in the ewe. Although producers in New Zealand vaccinate their rams for *Brucella ovis,* there is no recommended brucellosis vaccination for sheep in the United States.

Salmonellosis, caused by *Salmonella abortus ovis, S. dublin,* and *S. typhimurium,* is another common cause of abortion in sheep, particularly in Europe. In the United States, abortion due to salmonella is usually stress related. The fetus is autolyzed. Positive diagnosis is made by placental, fetal, or uterine culture.

Corynebacterium renale, a gram-positive bacterium, and urea cause *sheath rot (posthitis, pizzle rot, urine scald)* from the urine. The bacteria hydrolyze the urea and produce ammonia, which irritates the prepuce. Control includes ration modification because high-protein diets increase the concentration of urea in the urine. Thus, reducing protein levels to the lowest level needed should help reduce the incidence. Cleansing of the sheath and application of antibiotic ointments help reduce severity and incidence as well.

In mild cases, only a swelling of the prepuce is observed. In more severe cases, the condition causes severe pain and affected animals may die. In the ewe, signs include vulvar inflammation, redness, and secondary infections. *C. renale* is usually sensitive to penicillin and cephalosporin.

Parasites of Sheep

Because of their thick wool covering, sheep are both protected and victimized by external parasites. While providing warmth and protection from the elements, the wool also provides a favorable environment for maggots, ticks, mites, lice, and other parasites.

Ked wingless fly that lives off the blood (from the skin) of sheep

The sheep **ked** (*Melophagus ovinus*) is a wingless fly that resembles a tick. Keds spend their entire life cycle on sheep, transferring among animals by contact. Keds increase their numbers during the winter and early spring. To feed, they pierce the sheep's skin and suck blood, causing a condition known as *cockle.* Hide buyers downgrade sheep skins with cockle because it weakens the hide. The parasite is controlled by shearing and applying insecticides (either through a dip or pour-on or by dusting after shearing).

Lice small, flat, wingless parasitic insects

Sheep **lice** also feed on the skin. The lice cause the sheep to rub and scratch. Anemia is a common result of high populations of sheep lice. Low-pressure insecticide sprays and dusts are common treatments.

Wool **maggots** are the larvae of some kinds of *blowflies.* Blowflies lay their eggs in dirty wool, usually in the crotch area or on wounds. They are especially a nuisance when unsanitary conditions of the wool prevail because of hot, wet, muddy weather. Contamination by urine and manure in close confinement is also a chief cause. Sheep react to blowflies, houseflies, stable flies, and face flies by decreased performance. Care and medication of wounds, early shearing or clipping, and commercial chemical sprays and dips are very effective in controlling maggots.

Sheep nose botfly is a problem in which the female bot deposits larvae (maggots) into the nostrils of sheep. The larvae migrate to the head sinuses and migrate back out the nasal passages to the ground, where they complete their development into the adult fly. In 1997, 90 percent of sheep slaughtered from Wyoming, Colorado, Nebraska, and Idaho were infected with nose bots. Older and severely infested sheep can die as a result of these bots. Ivermectin, administered as an oral drench, has proven effective in treating this disorder.

Ticks are a problem in some areas. The *spinose ear tick* infects both cattle and sheep. It is found in arid range areas of the southwestern United States. The ticks suck blood or lymph from the inner folds of the outer ear. The irritation to the ear causes the animal to become unthrifty and decreases animal performance.

Mites cause mange or *sheep scab.* Researchers have identified at least five types that create problems in specific body areas. *Psoroptic scab (scabies)* is a highly contagious skin disease of sheep caused by microscopic mites. It causes large, scaly, crusted lesions on the wooly parts of the sheep. Untreated animals become emaciated and anemic. This disorder is included in the Federal Quarantine Act and is thus a reportable disease to both federal and state agencies. Although *Sarcoptic mange* is also a reportable disease in sheep, it is much rarer and usually affects the head and face of the animal. For all types of mange, sheep dip and/or treatment with ivermectin are practical control methods.

Although sheep may be less susceptible to health problems than are other farm animals, internal parasites are a constant threat to economical production. Because sheep have a cleft upper lip, allowing them to graze very close to the ground, they have close contact with the eggs and larvae of internal parasites that also can live in the soil.

Internal parasites cause the greatest damage to lambs and extremely old sheep but affect animals in all age groups. The most common internal parasites affecting sheep include the following:

1. Coccidiosis (Protozoa)
2. Stomach worms (*Haemonchus sp.*)
3. Nodular worms (*Oesophogostomum sp.*)
4. Liver flukes (Trematoda sp.)
5. Lungworms (nematodes; *Dictyocaulus filaria*)
6. Roundworms (*Ostertagia sp.*)
7. Tapeworms (cestodes)

Signs of internal parasite infection include paleness of eyelids, anemia, poor growth, potbellies, and bottle jaw or poverty jaw (swellings under the jaw). Treatment with a drench prescribed by a veterinarian is quite effective in restoring optimum health. New drugs are appearing on the market almost annually, so consultation with a veterinarian is necessary.

▓ **Maggots** larvae of a fly

Summary Vaccination Recommendations for Sheep and Goats

According to Dr. N. E. East, all sheep and goats should be vaccinated against the following[11]:

1. *Clostridium perfringens,* types C and D
2. Tetanus

In addition, sheep should be vaccinated for

1. Foot rot (*Bacteroides nodosus* bacterin)
2. *Chlamydial psittaci* (EAE)
3. *Campylobacter* spp. (vibriosis)
4. Bluetongue virus
5. Sore mouth or contagious ecthyma (in endemic areas only)
6. Additional disorders per local sheep veterinarian

PREDATORS

Predators have posed a serious threat to sheep for as long as they have been domesticated by humans. When the Environmental Protection Agency (EPA) banned the use of some chemicals that may be harmful to the environment, predators increased at an alarming rate and losses forced many sheep and goat producers out of business. Most shepherds have accepted some loss to predators as the cost of doing business. However, when losses exceed 40 percent, as is the case on some operations, it becomes a matter of survival to stop the predators. In 1994, more than 1,426 sheep per day were killed by predators in the United States, a total of 520,600 for that year alone. This loss cost the sheep industry in excess of $35 million in potential income.[12]

Some producers have been determined to find ways to control losses to predators. Guard dogs, donkeys, and llamas have been enlisted as means of protecting flocks.

Electric fences, electric guards, pens, and noise-making devices have also helped reduce losses to predators. Coyotes and other predators quickly adjust to noise-making devices, however, and have been known to attack in packs and kill guard animals.

Animal Damage Control Program

To assist U.S. farmers, the USDA established the Animal Damage Control (ADC) program. Its $27 million budget helps producers who are economically threatened by predators. In addition, the ADC program provides protection for endangered and threatened species (e.g., kit fox and desert tortoises), controls diseases (e.g., rabies, plague, and Lyme disease), controls populations of predators (e.g., coyotes), restores native habitats for endangered animals (e.g., black-footed ferrets), and protects citizens from potentially dangerous wildlife (e.g., bears, mountain lions, and coyotes). Without the ADC program, it is estimated that the agriculture industry would lose an additional $1 to 2 billion annually.

---

[11]N. E. East, D.V.M., University of California–Davis, personal communication, 1998.

[12]American Sheep Industry (ASI) Association, 1999.

SHEEP SELECTION, WOOL CARE, AND MARKETING

A sheep should never be caught by its wool because the skin is pulled away from the flesh, causing a bruise. Instead, the rear flank is grasped with one hand and under the chin with the other. The first hand is then switched to the dock area. Learning how to separate an animal from the flock without causing panic is extremely important for veterinary and animal science students. It is equally important to learn how to judge or assess differences between sheep for purposes of health, genetic selection, and basic identification.

Breeding flock replacements have different criteria than feeder lamb prospects, and wool producers have requirements that are quite different from both of the aforementioned. Selection is made using a variety of techniques that can be broken down into selection based on judging individuals, selection based on pedigree, selection based on animal performance, and selection based on production testing.

After becoming familiar with terminology, observations may be made for the purpose of selecting animals with the characteristics desired, most often put into practice by judging one animal against others (see figure 8–25).

Judging: Mutton Type

Sheep produce two major products, meat and wool. However, in mutton-type breeds, the emphasis is on carcass traits; wool is generally a minor consideration. The procedure used in livestock shows or judging contests is a systematic, orderly evaluation and selection of stock. Although the practical application of selection may vary considerably, the showring approach is adhered to for the purposes of illustration. Several divisions are used to equalize comparisons by age, sex, or use; however, condensing these divisions to slaughter classes and breeding classes will simplify this discussion.

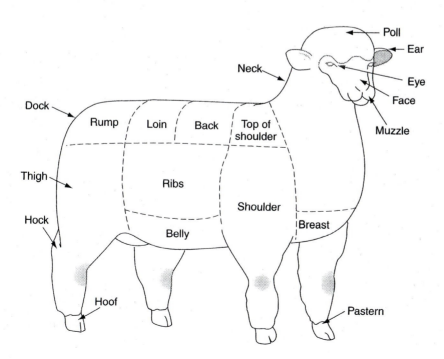

FIGURE 8–25
Parts of a sheep.

Wether castrated male sheep or goat

Slaughter Classes.   In slaughter classes, **wether** lambs (males castrated before reaching sexual maturity) are most often compared, although ewes or other classes may be substituted. The principle to keep in mind is that the live animal should be mentally stripped of its pelt, feet, and head. Once a decision is made on this projected carcass, the finer points (wool, structural defects, color, and so on) can be evaluated in the event of a close decision.

As indicated in figure 8–26, a class of slaughter sheep should be observed from a distance of 15 feet or more from the side (to evaluate depth and length of body, straightness of topline, depth of flank, size, and scale), front (to evaluate depth and thickness of forequarter and strength of bone), and rear (to evaluate thickness and depth of quarter, width of loin, and uniformity of width from dock to rack). Judges may order lambs somewhat based on visual appraisal, but more than any other class of mammals, sheep must be judged by the hands. Thickness of wool and degree of fat can camouflage many defects. The most important step is moving in for a systematic mental calibration of differences between pairs. Any system may be used in any combination. One common system is to first measure the loin (see figure 8–27) by placing all four fingers flat against the loin edge at a 90-degree angle to the topline. It is helpful to cross the thumbs for purposes of imagining this width. Second, the width of the rack is determined by using the edge of the arc made between the thumb and the index finger, not the palms. Next, the size of the leg is determined by placing the arc between the thumb and index finger of one hand in the flank of the lamb. The other hand is placed at the same level to complete a circle (see figure 8–28). A final check should be made for finish (fat covering) by feeling the backbone and ribs with fingertips as if feeling for the padding under a carpet.

FIGURE 8–26
Slaughter sheep should first be judged from a distance, viewed from the side, front, and rear.
Courtesy of the American Hampshire Sheep Association.

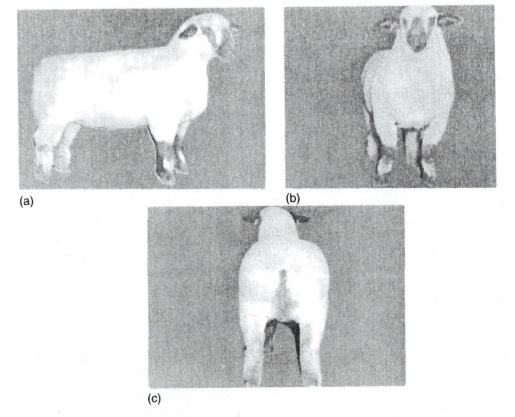

(a)

(b)

(c)

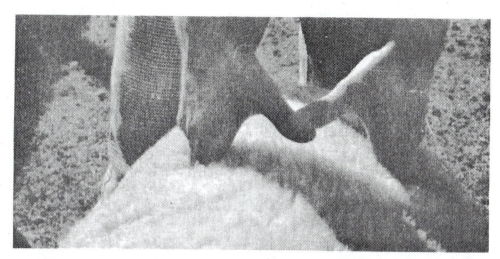

FIGURE 8–27
Measuring the loin.
Reprinted with
permission from
J. Blakely and D.H. Bade,
*The Science of Animal
Husbandry,* 6th ed.
(Englewood Cliffs, NJ:
Prentice-Hall, 1994).

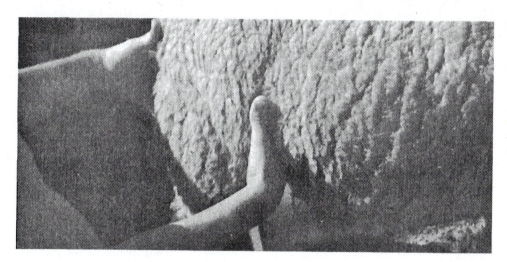

FIGURE 8–28   Size
of leg as determined
in judging.
Reprinted with
permission from
J. Blakely and D.H. Bade,
*The Science of Animal
Husbandry,* 6th ed.
(Englewood Cliffs, NJ:
Prentice-Hall, 1994).

It is essential that this procedure be conducted in the same order and as quickly as possible so that measurements of one animal are not forgotten before feeling the next subject. An experienced judge may spend 10 seconds or less feeling one lamb. However, the judge may return to repeat the process several times on one or all lambs before completing comparisons.

The visual observations combined with the physical handling determine the superior animals in the mind of the judge, and lambs are ranked in order. Note that breed type, color, feet and legs, and wool were not mentioned. These are minor points in a carcass class and usually disregarded unless the lambs are very close in muscling and other characteristics.

Breeding Classes.   Ewes and rams that are to be kept for production purposes are judged the same as slaughter classes to determine muscling, except that natural fleshing is looked for instead of fat. Sex classes are usually not mixed; rams are compared with rams, and ewes are compared with ewes. In addition to muscling, other characteristics become important, such as breed type, color, feet and legs, size, and wool.

Registered breeding stock must have the necessary characteristics of size, type, or color as required by the breed association, so these matters become all-important, whereas slaughter classes may disregard them almost entirely. Perhaps the most important and most often overlooked characteristics of breeding animals are the feet and legs. Because sheep are range animals, they must have sound structures to carry them long distances without difficulty. In addition, rams or ewes with leg defects are reluctant or unable to breed.

Wool, even in mutton-type breeds, now becomes an important consideration. The fleece thickness can be judged by feeling with both hands on both sides at the britch. The fleece is also parted at three points to visually estimate the average length and quality of fiber. A heavy-fleeced animal with exceptional muscling is the goal of sheep breeders.

Judging: Wool Type

When income from the sale of wool is the major objective, judges deemphasize body conformation. However, selection principles remain the same as those previously discussed, the only difference being a higher value placed on the wool. The fleece must be judged for weight and quality. Ideally, the fibers should be long, crimps and yolk (yellow color) should be noted, and black fibers should not be present.

As a general rule, wool-type sheep are larger, more angular, and less muscular in conformation but heavier fleeced. In selecting stock, the wool is considered first, then the meat. The opposite is true in selecting mutton-type animals.

The price of clean wool (that portion of grease wool that is free of foreign material, such as mineral matter, vegetable matter, dirt, polypropylene twine, tags, paint, and grease) is based on factors that determine its end use. To maintain clean wool, the shearing area must be kept clean with frequent sweeping.

Grease wool is wool removed from sheep prior to scouring. **Scouring** the wool (treating it with soap and soda solution) removes the soluble impurities, which include **yolk** (yellow coloring), **suint** (salts caused by sweating that produce the characteristic odor), and gland secretions that serve to protect the fibers while they are on the sheep. **Lanolin,** a common product used in cosmetics, is refined from this grease. The weight loss of these impurities is called *shrinkage.* Most wools shrink 50 to 65 percent, which influences their grade. Insoluble impurities, such as dirt, straw, burrs, insects, paint, and tar, must be removed by hand, by a chemical process called *carbonizing,* or by **carding** on huge machines built for this specific purpose.

Grade is the most important of the qualitative factors. An expert grader examines the scoured wool for length, diameter, fineness, crimp, pliability, color, luster, and other qualities before grading each bundle (see tables 8–4, 8–5, 8–6, 8–7, 8–8, and 8–9 for comparison of breeds). A minimum number of fibers is required in a cross section of yarn for it to be strong enough for weaving or knitting purposes. Typically, fine wools are those with a fiber diameter less than or equal to 22 microns. The English method of grading (spinning count system) estimates the number of hanks of yarn that can be spun from 1 pound of scoured wool. A hank is 560 yards. Grades range from 36 to 80. A grade of 40, for example, would mean 40 hanks could be spun from 1 pound of scoured wool.

The most important qualitative factors determining wool pricing are length, uniformity, strength, crimp, handle, color, character, purity, waste, and con-

Grease wool   wool as it comes from the sheep and prior to cleaning; it contains the natural oils and impurities present in the fleece

Scouring   treating wool with a soap and soda solution removes the soluble impurities, which include the yolk (yellow coloring), suint (salts caused from sweating that produce the characteristic odor), and gland secretions that serve to protect the fibers while they are on the sheep

Yolk   natural grease of wool that keeps the wool in good condition; lanolin is a common product used in cosmetics that is refined from this grease

Suint   salts of perspiration in the raw wool fleece usually removed when it is scoured

Lanolin   refined wool grease

Carding   process in which loose, clean, scoured wool is converted into untwisted strands ready for spinning

taminants. The ancient art of weaving wool fibers for apparel, carpets, and novelty fabrics continues despite the development of synthetics. Because of a worldwide shortage of oil and consequently its synthetic derivatives, wool will continue to be an important and necessary fiber to meet world market demands.

Domestic production and use of wool has declined over the years; synthetic fibers make up 66 percent of mill use and 56 percent of domestic fiber consumption. Wool makes up only 1.7 percent of domestic fiber consumption; cotton makes up 38 percent. The United States produces about 10 percent of the world's wool; the leading wool-producing states are Texas, Wyoming, California, Montana, Colorado, South Dakota, Utah, and New Mexico. Private or cooperative wool warehouses are the most important marketing methods used in these states. Typically, wool is pooled in a local market area and then sold through one of several wool warehouses.

Wool is classified into one of two categories—apparel or carpet wool. The apparel wool industry consists of **worsted** and *woolen* manufacturing processes. In the worsted process the longer wool fibers (2 inches or longer) are carded, combed, and drawn into a thin strand of parallel fibers and then spun together to form a stronger and thinner yarn. It is also known as combing wool. For the woolen system (clothing wool), shorter wool fibers are woven to give a fuzzy look. The French manufacturers have developed a process that can utilize fibers longer than those for clothing wool but shorter than those for combing wool to make worsted fabrics (after Worsted, now Worstead, England), thereby expanding wool utilization even further. Carpet wool varies in length but must be coarse and elastic to withstand heavy use. Most carpet is produced outside the United States.

> **Worsted** manufacturing process of wool using the longer wool fibers (2 inches or longer) and in which the wool is carded, combed, and drawn into a thin strand of parallel fibers spun together to form a strong but thin yarn

Preparing Clean Wool.   Sheep are usually shorn in the spring, before the hot weather months. The sheep is set on its dock and cradled between the shearer's legs, which are used to maneuver the sheep into the positions needed to ease the shearing. The fleece must be dry when shorn to prevent gumming of the clippers. Properly removed wool remains in one large piece (called the **fleece**) and is tied with paper twine. Second cuts (going over the same surface a second time) are objectionable because they create short fibers that are difficult to weave. The fleece is put into a huge sack. Wool preparation varies by breed, age, sex, and volume.

> **Fleece** wool from one sheep

Belly wool is usually sheared and packaged separately. Areas high in contamination with locks or excess tags are also sheared, and the wool is packaged separate from the fleece.

Each fleece is skirted to remove inferior wool, stains, tags, skin pieces, top knots, belly wool, crutchings, second cuts, shanks, and any heavy vegetable matter. This extra step in preparing quality wool may bring additional revenue per pound of wool sold. A wool marketing representative should be consulted prior to taking on this extra expense.

Pedigree

The importance of bloodlines must not be overlooked or underestimated in selecting sheep. The pedigree is a written record of past appraisals and an estimate of potential performance. For instance, a ram that has produced lambs with high weaning weights, has sired sons and daughters with heavy, quality fleeces, or has produced champions of carcass shows might well be expected to pass these qualities on to its offspring. To the trained breeder, reading the pedigree and making selections based on it is like having a blueprint of the foundation on which future designs may be built.

Animal Performance

No matter how good an animal looks in the showring or how illustrious its ancestors, it is without valid credentials until its own merit is proven. That measurement of merit is animal performance.

Simple Measuring Techniques.
Laptop computer or pencil and pad   The birth weight of lambs, regularity of lambing and incidence of twinning in ewes, weaning weight, fleece weight, and so on are easily quantified. These data can be transferred to permanent records for lifetime histories of ewes or rams for use in aiding selection of flock replacements.

Scales   Perhaps the most useful evaluation tool is scales, which provide an accurate appraisal of wool, lamb, or mutton production. Weight gain and feed efficiency data over a specified period require their use. Yet many sheep breeders do not know the value of their product until the buyer's scales announce their findings.

Complex Measuring Techniques.   Sheep readily lend themselves to sophisticated methods of evaluation such as ultrasonic measurement of loin-eye size and carcass cutout data, although these techniques are not as widely used in sheep as in other species. These findings are applied much the same way as those previously discussed for beef cattle (chapter 7).

Production Testing

Adequate records demand the initiation of meaningful individual evaluation tests through performance testing or progeny testing. To make breed improvements, certain characteristics must be measured, recorded, and used so that intelligent selection principles may evolve. Table 8–12 gives the characteristics that are of greatest economic importance in production testing and the corresponding heritability range, indicating rate of improvement.

TABLE 8–12 Heritability of Some Characteristics in Sheep	
Characteristic	**Heritability**
Prolificacy (rate of production, twinning)	Low
Birth weight	Medium
Weaning weight	Medium
Weaning conformation score	Low
Wrinkles or skin folds	Medium
Face covering	High
Fleece weight	Medium
Staple length	High

These characteristics are tested for the obvious reason of their strong economic importance. Prolificacy is important because sheep have the inclination and ability to twin regularly. Ewes that are more productive, although not a highly heritable trait, stand a better chance of passing on this ability to their daughters.

Birth weights, weaning weights, and conformation scores are easy to obtain, and improvement might be desired and expected at a somewhat faster pace because of slightly higher heritability. Wrinkles and skin folds are important because shearing is difficult in their presence and fibers lack uniformity. Smoother-bodied sheep are generally preferred in the United States for ease of shearing. Face covering may appear of minor importance, but sheep people know it to be a major concern. Wool-blind ewes graze less, have high labor requirements if clipped around the eyes, and wean lighter lambs than open-faced ewes.

Fleece weight and staple length are an estimate of both quantity and quality of fiber produced. About two-thirds of the emphasis in sheep selection is on lamb or mutton production, with the remaining one-third on wool production. A selection system therefore is somewhat different from that discussed for cattle and swine. The independent culling level method can be used if both wool and meat production minimums are established. Any individuals that fall below these standards are removed from the flock.

The selection index, in which each economically important trait is assigned a weighted value, is particularly adaptable to sheep selection. Extremely strong features tend to average out very weak faults to give a total score representative of all measurements. A minimum acceptable score is selected, and only animals that meet or exceed that value are kept for breeding purposes. A selection system for sheep, as with other species, varies with the changing times and market demands, but a combination of judging, pedigree, animal performance, production testing, and common sense makes improvement possible.

SLAUGHTERING PROCESS OF SHEEP

The slaughtering process of sheep is different from that of other species in several respects. The end product, the carcass, is unique because of an age distinction that categorizes young carcasses as lamb and mature carcasses as mutton.

After stunning with either a captive bolt gun or an electric stun gun, sheep are shackled by the hind leg. A double-edged knife is then used to swiftly pierce the neck, severing both jugular veins. This step allows rapid bleeding while the heart still beats to provide the necessary pressure to ensure proper drainage. The throat is not cut but pierced. The pointed, sharp knife is effective and humane; most sheep never show signs of emotion or pain.

The foreleg is severed at the pastern above the foot, and the joint is broken with the hands. This factor determines the lamb or mutton classification. Growing animals deposit calcium at the ends of the long bones at the epiphyseal cartilage, which does not ossify (harden) until about 1 year of age. If this break occurs in the cartilage, it exposes a *lamb joint* (break joint), indicating a young sheep. The carcass is then classified as lamb. If ossification

FIGURE 8–29 A lamb joint (left) indicates a young sheep, and the carcass is classified as lamb. A spool joint (right) indicates a more mature sheep, and the carcass is classified as mutton.
Reprinted with permission from J. Blakely and D. H. Bade, *The Science of Animal Husbandry*, 6th ed. (Englewood Cliffs, NJ: Prentice-Hall, 1994)

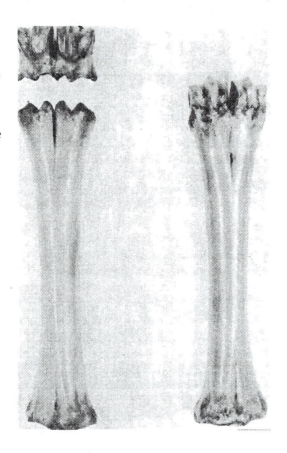

has occurred, the joint has fused, growth is ended, and the *spool joint* is exposed, indicating a mature mutton carcass (see figure 8–29).

The pelt is taken off carefully. Forcing the hands between the pelt and **fell** (a thin membrane that covers the carcass) is called **fisking.** Care is taken not to tear the fell because it protects the meat and reduces drying out.

The head and feet are removed, and the **offal** (intestines) are taken out for inspection. The breastbone is split, and the forelegs are folded back with a wooden or stainless steel pin, securing them in place to shape the carcass in an attractive style for marketing.

Washing takes place next, followed by a 24- to 72-hour cooling period. About 1 percent shrinkage usually occurs from hot weight to cool weight because of moisture loss. High-grade carcasses are often shrouded (wrapped in cheesecloth or light canvas) to further protect and hold them in shape.

The dressing percentage of sheep is calculated from live and cooled carcass weights and generally ranges from 46 to 53 percent. Fat influences the dressing percentage, and older sheep generally yield at the low end of the scale.

Fell   thin membrane that covers the carcass

Fisking   method of forcing the hands between the pelt and fell to remove the pelt from the carcass

Offal   organs, trimmings, and tissues that are removed at slaughter

The Carcass

If the carcass is passed as suitable for human consumption after a thorough inspection, a government grader determines the grade according to conformation and finish. The wholesale cuts are produced by the packing plants and enter the local butcher market for further breakdown into retail cuts (see figure 8–30).

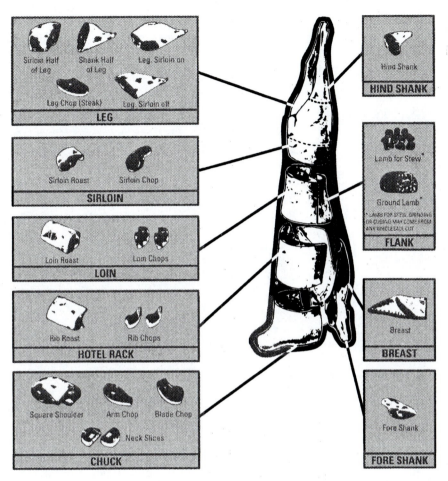

FIGURE 8–30
Mutton wholesale and retail cuts.
Courtesy of the U.S. Department of Agriculture.

By-Products

Recalling that the maximum dressing percentage of sheep is usually 53 percent, there remains 47 percent to be disposed of or utilized. Very little, if any, is wasted; this portion of the slaughtering process is utilized as by-products in a variety of forms. See table 8–2 for a detailed list of sheep by-products in the United States.

SUMMARY

Sheep were one of the first domesticated animals and to this day have no natural defenses against predators. Today most U.S. sheep are found in large flocks in the western states. The consuming public has replaced lamb with other cheaper and more easily prepared meats and replaced wool with synthetics. Sheep are still used to harvest dry rangeland feeds that are of little value to other livestock species. Many sheep farmers consider themselves range or grass farmers who market cheap forage through sheep.

For the sheep industry to survive in the next century, increased consumer awareness and education, as well as improved marketing, increased protection against predators, and continued use of cheap natural resources (rangelands), will need to be a focal point of the industry.

EVALUATION QUESTIONS

1. _____ is an example of a U.S. synthetic breed of sheep.
2. _____ is the season when most sheep come into estrus.
3. Ewes reach puberty at age _____.
4. _____ is the gestation period for sheep.
5. Sheep first lamb at age _____.
6. Clipping wool from the dock, udder, and vulvar regions of the ewe prior to breeding and lambing is called _____.
7. _____ is a relatively large group of range sheep.
8. _____ is a group of sheep.
9. The wool shorn at one time from all parts of the sheep is called _____.
10. _____ is a measurement of the fineness of wool.
11. Wool as it comes from the sheep and prior to cleaning is called _____.
12. A metabolic disease in late pregnancy affecting primarily ewes carrying twins or triplets is _____.
13. _____ is a recessive genetic abnormality common to black-faced sheep.
14. Inflammation of the prepuce in male sheep is called _____.
15. _____ refers to the length of wool fibers.
16. _____ is a disorder in sheep in which they cannot see because of the wool covering their eyes.
17. The natural grease (lanolin) of wool is called _____.
18. Undesirable large, chalky white hairs found in mohair are known as _____.
19. _____ is the most important characteristic of a meat breed.
20. Virgin wool is known as _____.
21. _____ is the management practice of increasing the nutritional status of the ewe just prior to breeding to increase the likelihood of multiple births.
22. _____ is an example of a feed high in phytoestrogen.
23. The genetic defect in which the upper jaw is longer than the lower jaw is _____.
24. _____ is the genetic defect in which the eyelids are turned inward.
25. Enterotoxemia is also known as _____.
26. _____ is a metabolic disorder caused by selenium deficiency in sheep.
27. _____ is also known as stiff lamb disease.
28. A metabolic disorder caused by a deficiency of magnesium is _____.
29. The state with the largest number of sheep is _____.

30. Ewes of the English breed _____ will breed out of season.
31. The small dual-purpose sheep breed _____ is noted for high lambing rates.
32. _____ is a large, black-faced English breed of sheep with no wool on its head or legs.
33. _____ is a large Asian breed of sheep used for making Persian rugs and fur.
34. _____ is the disease related to mad cow disease.
35. _____ is the bacterial cause of enzootic abortion of ewes.
36. _____ is also known as contagious ecthyma.
37. The largest fine-wool breed in the United States is _____.
38. _____ is the breed with the best lamb carcasses.
39. _____ is a castrated ram.
40. Forcing the hands between the pelt and fell is called _____.
41. _____ is the number of sheep in California; _____ is the number of sheep in Texas.
42. Paleness of the inside of the eyelid of sheep indicates _____.
43. A 100-pound wether will produce a _____-pound carcass.
44. Nearly all wild sheep have short _____.
45. Spain is the origin of the _____ wool breed of sheep called _____.
46. Sheep belong to the species _____.
47. The largest breed of sheep is the _____, which is a _____ wool breed.
48. A mature ram may be bred to _____ ewes.
49. Navels of lambs born in confined quarters are treated with _____ to guard against _____.
50. Lambs are weaned at about _____ months of age.
51. _____ is the difference between mutton and lamb.
52. A mutton carcass exposes a _____ joint.
53. The _____ is a membrane that covers the carcass.
54. Removal of yolk, suint, and other impurities is called _____.
55. _____ are salts caused by sweating that produce the characteristic odor of sheep.
56. Most wools shrink _____percent during scouring, which influences their grade.
57. The largest death losses in feedlot lambs are from _____.
58. A common itching-type disease is called _____.
59. Sore mouth is often confused with _____.
60. Urinary calculi result from an imbalance of _____ and _____.
61. A practical control method for external parasites is use of a recommended _____.
62. Sheep are susceptible to internal parasites because of a cleft _____.
63. _____ is also called lockjaw.
64. _____ is an old, wild, short-tailed sheep breed.
65. The approximate number of sheep in the United States is _____.
66. _____ and _____ are two examples of plants that are poisonous to cattle but that sheep voluntarily consume.

67. _____ is another name for bummers.
68. Cutting the wool above and below the eyes of sheep is called

    _____.
69. _____ hours of dark for 12 weeks will improve libido and fertility during the normal anestrous season.
70. _____ days after birth is the minimum age recommended for castration and tail docking.
71. The normal rectal temperature of a healthy ewe is _____ °F.
72. _____ means hidden testicle.
73. _____ causes a cheeselike accumulation in the lymph nodes and internal organs in sheep.
74. _____ is also known as circling disease.
75. _____ is caused by *Chlamydia psittaci.*
76. Copper sulfate (5 percent) administered to the _____ will help in the treatment of _____.
77. _____ is another name for grass tetany.
78. _____ is another name for paratuberculosis.
79. _____ is a viral disorder that causes thin ewe syndrome.
80. In people the disorder _____ causes orf.
81. The use of elastrators for castration or docking increases the likelihood of the clostridial disorder _____.
82. _____ is an abortion disease spread by cat fecal droppings.
83. _____ is also known as pizzle rot.
84. _____ are the larvae of blowflies.
85. _____ causes a condition known as cockle.
86. _____ and _____ are two reportable external parasites of sheep.
87. Ostertagia is also known as _____.
88. ADC is the abbreviation for _____.
89. _____ is the treating of wool with a soap and soda solution.
90. _____ is the weight loss of impurities of wool.
91. The length of wool measuring 560 yards is _____.
92. _____ is the process of manufacturing wool that uses the long wool fibers.

DISCUSSION QUESTIONS

1. What are the three major purposes of raising sheep in the United States?
2. What is a synthetic breed?
3. What is the purpose of a ewe breed? Ram breed? Provide examples of each.
4. What is the most important characteristic of a meat breed?
5. Define and describe cryptorchidism, dwarfism, entropion, parrot mouth, and spider syndrome.
6. What are some common methods of castrating and docking sheep? At what age are these procedures done?
7. List the most common sheep diseases and their symptoms and preventive measures.
8. What is the importance of the spool joint? Break joint?

9. Describe the symptoms of flukes, scrapie, and stiff lamb disease in sheep.
10. What are some distinguishing characteristics of the Dorset breed? Suffolk? Southdown?
11. What effect does environmental temperature have on the fertility of rams?
12. Give two examples of dual-purpose breeds.

RECOMMENDED READING

1. Botkin, M. P., R. A. Field, and C. L. Johnson. *Sheep and Wool: Science, Production, and Management.* (Englewood Cliffs, NJ: Prentice-Hall, 1988).
2. Ross, C. V. *Sheep Production and Management.* (Englewood Cliffs, NJ: Prentice-Hall, 1989).
3. *The Sheepman's Production Handbook.* (Englewood, Colo.: American Sheep Industry Association, 1997).

C H A P T E R

9

The Swine Industry

OBJECTIVES

After completing the study of this chapter you should know:

- The historical, current, and future trends of the U.S. swine industry.
- Distinguishing characteristics of common breeds of U.S. swine.
- Methods of selecting, breeding, and judging swine.
- Market classes and marketing of pork products, including carcass quality traits, dressing percentages, and methods of measuring quality of carcass.
- The new Maternal Sow Line National Genetic Evaluation Program and its potential effect on evaluating genetic lines in the swine industry.
- Common management procedures and goals suggested and used in the various swine production systems.
- The importance of the production registry and the Unified Meat Type Hog Certification Program to the future of the modern-day pig.
- Differences and similarities in breeding pigs compared with cattle, sheep, and goats, including the use of artificial insemination, the collection and processing of semen, and methods of heat detection.
- Protein quality and its importance in swine nutrition compared with that of ruminants.
- Common methods of meeting the nutritional needs of swine, preventing common deficiency and toxicity disorders, and feeding for maximum profit.
- Common swine diseases and their symptoms (in brief), prevention, treatment, and diagnosis.
- How the National Pork Producers Council's creation of its own animal welfare committee has improved the economics, health, and welfare of the U.S. hog industry.

KEY TERMS

Backfat
Barrows

Boars
Certified litter (CL)

Certified mating
(CM)

Certified meat sire (CMS)	Parakeratosis	Stags
Finished market hogs	Pegging down	Swine dermatosis
First limiting amino acid	Polytocous	Transport myopathy
Gilts	Production registry (PR)	Unified Meat Type Hog Certification Program
Malignant hyperthermia	Protein quality	
	Sows	

INTRODUCTION

Swine production in the United States is concentrated heavily in the midsection of the country known as the Corn Belt. This area produces most of the nation's corn, which is the principal feed used for swine. The hog carcass has one of the highest dressing percentages, and today's advertising stresses pork as a lean meat. In 1998, there were only 114,380 hog operations raising 62,156,000 hogs and pigs in the United States.[1] In 1954 there were more than 2.25 million hog farms in the United States. Most of the remaining operations are large, are mechanized, and use the most modern science for the selection, breeding, nutrition, and health care of their animals. Throughout the United States, the trend in pork production is toward the larger and more specialized. Improved production techniques, better business management, and improved pork quality keep the industry increasing. The number of hogs on farms is increasing, and the number of farms producing hogs is decreasing.

BRIEF HISTORY OF SWINE

Swine were domesticated much later than other livestock. Nomadic people could not travel with swine as easily as they could with cattle, sheep, or horses because of the confinement required and the pigsty odor. Hogs were regarded with contempt. Swine were domesticated in many regions because they could not travel. However, certain religions (Islam, Judaism, and Seventh Day Adventism) forbid their followers to eat them.

Two wild stocks contributed to the American breeds of swine. Hogs that evolved from the wild hog of Europe are of the species *Sus scrofa*. Hogs derived from the wild hog of eastern India are of the species *Sus vittatus*. The European wild hog had coarser hair (much like a mane along the neck) than present swine. It had large legs and feet, a long head and tusks, a narrow body, and great ability to run and fight. Wild descendants of the European wild hog found today have many of the same characteristics. The East Indian pig was smaller and more refined than the European wild hog. Instead of solid black, the East Indian pig had a white streak along its side. This breed was a scavenger and in domestication was used to eliminate the food wastes of people. Today, less than 1 percent of U.S. swine are fed kitchen wastes.

---

[1]National Agricultural Statistics Services (NASS), Livestock and Economics Branch, U.S. Department of Agriculture (USDA), Agricultural Statistics, 1999.

Both species gave rise to present-day swine, but the European wild hog was most influential. Swine revert very quickly to the wild state, as evidenced by the razorbacks and other wild hogs found in the United States.

TAXONOMY

As with the other animals discussed so far in this text, swine are part of the kingdom Animalia, phylum Chordata (backbone; vertebrates), class Mammalia (warm-blooded, hairy animals that produce their young alive and suckle them for a variable period from the mammary glands), order Artiodactyla (even-toed, hoofed mammals), family Suidae (nonruminant artiodactyl ungulates), genus *Sus* (restricted to European wild boars and domestic breeds derived from them), and species *Sus scrofa* (wild hog from continental Europe, from which most domestic swine are derived) and *Sus vittatus* (the race of the East Indian swine that also contributed to today's domestic swine).

BREEDS OF SWINE

Throughout swine production history in the United States, breeds have been developed and maintained to fulfill a demand for a certain type of retail product. The demand has moved from a lard-type hog (carcass high in fat) to the present lean-meat-type hog (carcass high in meat compared with lard). The lower demand and profit for lard as a final product of pork have brought about this switch in hog types. Today, only the lean-meat-type breeds as described here and those being raised for pets or research are maintained.

American Landrace

The Landrace breed was developed in Denmark and is responsible for the Danes' reputation as pork producers. The Danish Landrace breed originated in 1895. The Danish government protected this breed and did not allow exportation of it until an agreement signed in 1934 allowed shipment of surplus stock. The foundation stock of American Landrace carried a small infusion (1/16 to 1/64) of Poland China blood.

The American Landrace Association, Inc., was organized in 1950. The breed is white in color. Black-spotted pigs are not eligible for registration. They have medium lop ears, a very long side (making them known for bacon), square hams, trim jowls, and short legs (see figure 9–1). They are noted for good feed conversion and prolific reproduction. The American Landrace is longer than other breeds because it has an extra vertebra.

Berkshire

The Berkshire breed came from the county of Berkshire in England. It is black with white feet, face, tail, and switch (six white points). The Berkshire's face is dished, and its ears are erect (see figure 9–2). It is an excellent meat-type hog, with a long body and a high-quality carcass. Its reputation throughout the years has been for a meaty, well-balanced carcass with a high cutout value. Berkshires are relatively docile.

FIGURE 9–1 The Landrace Hog College boar at the 1999 National Barrow Show, exhibited by KDM Landrace, Chatsworth, Illinois. Courtesy National Swine Registry.

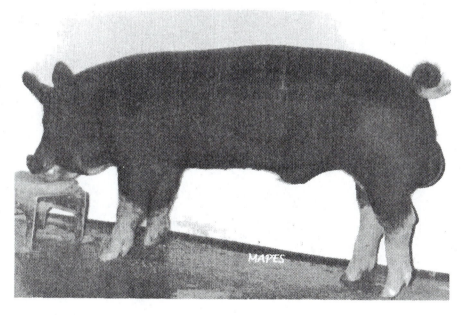

FIGURE 9–2 The Berkshire is black with white feet and is white in the face, tail, and switch. Courtesy of American Berkshire Association.

Chester Whites

The Chester White breed originated in Chester and Delaware Counties of Pennsylvania in 1848. It is a prolific breed that has large litters and is also noted for its good mothering ability. The carcass is high quality and lean and has large hams. The sow averages 500 to 600 pounds and is noted for length and good size, good breeding qualities, gentle disposition, coarse, drooped ears, large litters, and large dressing percentages (see figure 9–3).

FIGURE 9–3 The Chester White breed is noted for large litters, mothering ability, and a carcass with large hams. Courtesy Chester White Swine Record Association.

FIGURE 9–4 The red Duroc Jersey has an excellent rate of gain. Courtesy National Swine Registry.

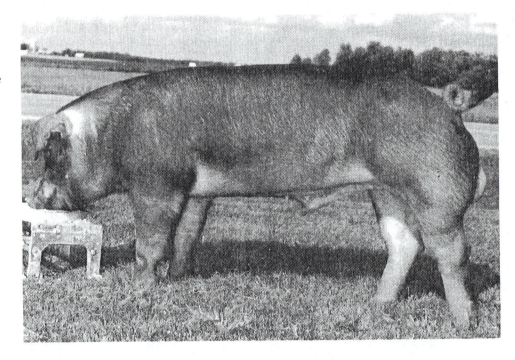

Duroc Jersey

The Duroc Jersey ranks second in annual registrations (behind Yorkshires) and is thus one of the most popular breeds of swine in the United States. It is a blend of Jersey Reds (New Jersey) and Durocs (New York) and originated sometime around 1860. Duroc Jerseys are light to dark in color with cherry red preferred. Duroc Jersey ears are medium sized and tip forward (see figure 9–4). Duroc Jerseys have excellent rate of gain and feed efficiency and mature early. The sows have large litters and are good mothers. The carcass is considered a good meat type.

FIGURE 9–5 The white-belted black Hampshire produces a long carcass that is high in muscling and low in backfat. Courtesy National Swine Registry.

Hampshire

Hampshire is a U.S. breed, developed in Boone County, Kentucky. The Hampshire has a white belt around the shoulders and front legs. The black color with the white belt is its distinguishing characteristic (see figure 9–5). It has a refined head and body and is known for producing a good, long carcass, high in muscle and low in backfat. It is a good gaining breed of swine and is known for good mothering ability.

Hereford

The color markings of the Hereford breed of swine are similar to those of Hereford cattle. The swine have a white face, red body, and at least two white feet, white underline, and white switch (see figure 9–6). The foundation stock included Chester Whites, Durocs, and Ohio Improved Chesters. The Hereford breed was founded in 1830 in La Plata, Missouri. It is smaller than other breeds of swine. A well-finished carcass and heavy shoulders with compact type are desirable characteristics for the Hereford.

Ohio Improved Chester

L. B. Silver developed the Ohio Improved Chester (OIC) in the 1860s in Salem, Ohio. In 1909, the Western O.I.C. Recording Co., Inc., was formed in Nebraska. In 1930, the Chester White Swine Record Association was reincorporated; it moved to Peoria, Illinois, in 1986. Its purpose was to gather and combine the various associations of Chester swine. Controversy persists over their existence separate from the parent breed, but the name is still recognized; the OIC is known for being more compact, larger, and fattening more quickly.

FIGURE 9–6 Hereford swine, like Hereford cattle, are red with white face and white on at least two feet, the underline, and the tail switch. (a), Reserve grand champion (PH Mr. Lean) at the National Hereford Hog Show and Sale, Greencastle, Indiana, September 12, 1999. (b), Champion-bred gilt (ICF Miss Super Shelly 9) at the same show.
Courtesy National Hereford Hog Record Association.

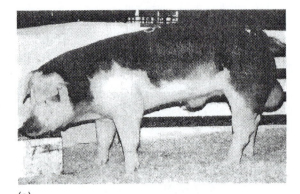

(a)

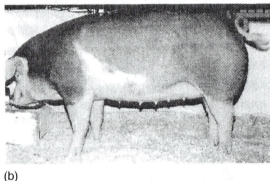

(b)

FIGURE 9–7 The Poland China is known for producing a heavy ham carcass and for its gainability.
Courtesy National Pork Producers Council.

Poland China

The Poland China originated in Warren and Butler Counties of the Miami Valley of Ohio in 1816. The name was officially established in 1872. The Poland China is black with white feet, face, and tip of tail (six white points). The breed was developed and is known for a heavy ham carcass and gainability, reaching the maximum weight at a given age (see figure 9–7).

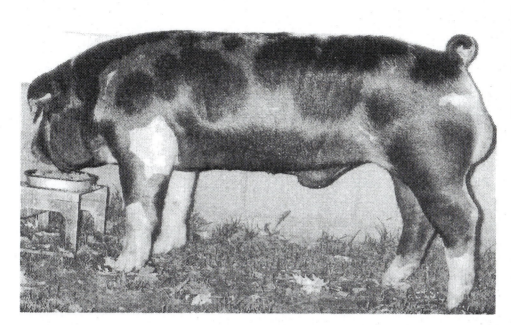

FIGURE 9–8 The Spotted swine breed developed from the Poland China breed, which is often spotted at birth.
Courtesy National Spotted Swine Record.

Spotted

Spotted swine, previously known as Spotted Poland China, are spotted at birth in a black and white pattern with about a 50 percent split (see figure 9–8). They were developed in Indiana and are similar in size and carcass cutout to the Poland China.

Tamworth

The Tamworth is one of the oldest breeds of hogs. Tamworths originated in Ireland, where they were known as Irish Grazers. They were imported into England in 1812 and into the United States in 1822. Tamworths are noted for their bacon type with a very deep side and a strong, uniform arch of back, a very muscular top, and a long rump. The ham is muscular and firm, although it lacks the bulk and size found in many of the other modern breeds. The color may vary from light to dark red. Tamworths are excellent mothers and have long legs, making them adaptable to rough country. They are the most prepotent of the swine breeds in fixing their type of offspring.

Vietnamese Potbellied

Vietnamese Potbellied pigs are a dwarf swine breed. The breed was developed in the 1960s from the I breed of Vietnam. The first potbellied pigs were brought into the United States through Canada in 1986. Originally, they were intended to be used as attractions in zoos, but a private buyer became interested in using them as porcine pets. A full-grown potbellied pig weighs an average of 70 to 150 pounds (with some reaching 200 pounds or more). They are only 3 feet long and 15 inches tall, compared with a normal domestic pig, which weighs 600 to 1,500 pounds when mature. Colors range from solid black to solid white, with a variety of spots in between.

Yorkshire

The Yorkshire originated in England, where it was known for more than 50 years as the Large White. Yorkshires were brought into the United States to Ohio in 1830. The Yorkshire carries the title of "The Mother Breed" because the Yorkshire sow is known for large litters and good mothering ability (see figure 9–9). The Yorkshire is a good feed converter and yields a carcass of high dressing percentage. After World War I, the breed led the change in breeding from a lard hog to a so-called bacon hog. This "progress" was short-lived, because at that time lard was selling for more money than bacon (muscle). Today, Yorkshires are leaner and more durable than ever; they lead the pork industry in numbers of registered swine (see table 9–1).

TABLE 9–1	**Annual Registration Numbers for Major U.S. Breeds of Swine**	
Breed	**Annual Registration**	**Numbers of Litters[b]**
Yorkshire	211,600[a]	19,007
Duroc	168,900	13,709
Hampshire	163,800	16,244
Chester White	55,700	3,077
Spotted	48,500	3,251
Landrace	33,800[a]	4,405
Berkshire	20,100	1,934
Poland China	17,800	1,600

Source: R. Taylor, *Scientific Farm Animal Production* (Upper Saddle River, NJ: Prentice-Hall, 1995).
[a]R. Taylor, *Scientific Farm Animal Production* (Upper Saddle River, NJ: Prentice-Hall, 1992).
[b]R. Taylor, *Scientific Farm Animal Production* (Upper Saddle River, NJ: Prentice-Hall, 1998).

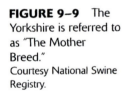

FIGURE 9–9 The Yorkshire is referred to as "The Mother Breed." Courtesy National Swine Registry.

SWINE SELECTION

Selection criteria for swine follow similar lines, with some minor variations because of the different species, to those already discussed for beef cattle (chapter 7). The broad selection bases are judging, pedigree, performance, and production testing.

Eye Appraisal: Judging

A thrilling sight is the action of the grand champion barrow auction at the local county fair; a barrow is a castrated male hog lacking sex characteristics. The exhibitor proudly guides the barrow with a cane to show the animal's strong points that won it the grand champion banner. How did the barrow become grand champion? A judge had to exercise judgment on the field of candidates, no matter how many were entered in the contest. However, judging is never more than comparing two animals at a time. Thus, through comparing two animals with an ideal hog in mind and selecting one over the other, animals are eliminated until a grand champion emerges. The desirable ideal hog is described in this chapter.

Parts of a Hog.   As explained in chapter 7, a judge must speak in terms of the class of livestock being judged. Veterinary and animal science students must also try to describe the animals they are evaluating. Figure 9–10 illustrates the locations and names of the different parts of a market hog.

Description of the Ideal Meat-Type Hog.   At one time, hog types were referred to as meat type, bacon type, or lard type. Because the lard type (overly fat) is no longer popular or necessary, all breeds have the goal of producing a lean, muscular carcass with little waste. Most describe this ideal as the meat-type hog.

Knowledge of the parts of a hog is helpful only when animal scientists can use the terminology to describe faults or strong points of one hog as compared with the ideal meat-type hog. In viewing the swine from the side, back, and rear, the desirable points described in the following sections should be noted.

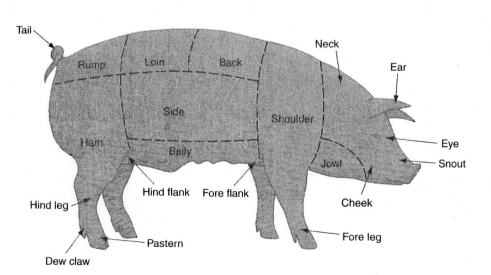

FIGURE 9–10 The parts of a market hog.

Side view   When judging a hog from the side, the judge looks at the size for age, balance, length of side, depth of body, topline, trimness of underline and jaw, legs, size of bone, finish, breed type, and teats in breeding classes. Because rapid gains are usually economical gains, a large, early-maturing hog is desired (see figure 9–11). A moderately long body that is fairly deep gives a balanced appearance to the hog. An evenly arched back with high tail setting is desired. The belly and jowl should be trim and the flanks deep. The ideal hog is smooth over the shoulders, free from wrinkles, and moderately finished. Legs are moderately long, straight, true, and squarely set with strong pasterns. Ample bone size is desired. Breed type and teats, well-developed and numbering 12 or more, are desired for breeding swine.

Rear view   The rear observation is an important view that shows muscling and finish. Width from front to rear is noted, with good, uniform width desired (see figure 9–12). Smoothness of the back of the shoulders with the shoulders "well laid in" (blending in the uniform thickness) should be present.

FIGURE 9–11   The side view should show a lengthy, moderately arched back, straight legs, and smoothness throughout, showing moderate finish. Courtesy of T. D. Tanskley, Texas A&M University.

FIGURE 9–12   The width along the topline, muscling, and finish are noted from the rear view. Courtesy of T. D. Tanskley, Texas A&M University.

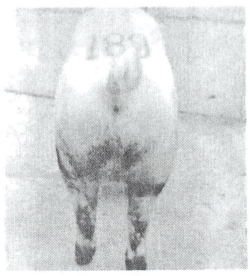

The ideal hog is wide and full throughout the loin area of the back. The rear of the hog should show a plump, full, trim, and firm ham with muscling well down to the hocks (see figure 9–13). The hind legs should be set well apart and be sound and straight. This side stance is referred to as *pegging down;* the wider the stance, the heavier the muscle along the backbone.

Front view   A shapely, trim head showing sex characteristics and having wide-set eyes is best seen from the front view of the ideal hog (see figure 9–14). In breeding classes, the head should show breed characteristics as well as sex. The front legs should be medium in length, straight, and correctly set.

Feel for proper finish   Under the current trend of producing meat-type hogs rather than lard-type hogs, proper finish with a minimum of fat is desired. The ideal barrow possesses proper conformation, finish, and quality for maximum

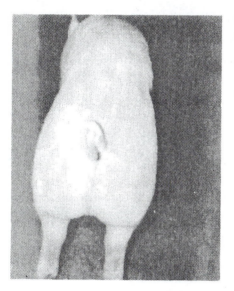

FIGURE 9–13   The rear of the hog should show a plump, full, trim, and firm ham with muscling down to the hocks.
Courtesy of T. D. Tanskley, Texas A&M University.

FIGURE 9–14   The trim head showing sex and breed characteristics is desirable from the front view.
Courtesy of T. D. Tanskley, Texas A&M University.

FIGURE 9–15
Determining backfat on a live animal using a probe is then checked on the carcass in the areas of the loin eye.
Courtesy of Michael De La Zerda, Texas A&M University.

development of the high-priced cuts and a minimum of lardiness. Although some fat is essential for the quality and cooking properties of the meat, large layers indicate waste, overfinish, and poor conformation. Well-finished hogs have a moderately thick, firm covering of flesh all over the body that is indicated by width of body, plumpness of hams, and firmness of jowls, belly, and flanks. A firm, uniform layer of backfat that is 0.9 inches or less is desired. This backfat can be measured on the chilled carcass or determined on the live animal by backfat probing (see figure 9–15).

▓ **Backfat** amount of fat over a pig's back and an indicator of the overall fat content of the animal. Used in the selection of breeding stock and in evaluation of carcass grade

Selection of Show Animals. Future show animals in either the market or breeding class are selected throughout the year. Selection traits are kept in mind at the time of breeding, and sows and boars are carefully selected to produce the next grand champion. Farrowing dates are often arranged so that market hogs are at the optimum stage of development for show. Show animals must be selected far enough in advance to allow time for fitting and training. Often, many animals are selected and fitted, but only the best are used at show time.

Selection by ShowRing and Pedigree

Animals of the breeding herd are selected by pedigree, showring record, or individual merit. The pedigree of an animal (record of the animal's ancestry) can be used to select animals used for breeding. Pedigrees that give the production records of closely related animals (parents and grandparents of the animal) can serve as an indication of the breeding value of a young prospect.

Swine breeders have always recognized showring winners as superior animals. Show auctions for the breeding classes are always filled with spirited bidding for the top hogs, which often bring premium prices. Amateur buyers may prefer this method of selection because an acknowledged expert (the judge) selects the best animals. Experienced swine buyers can judge breeding stock on the farm and can often obtain breeding stock of equal quality at more reasonable prices.

Pedigrees are especially useful in swine breeding because the generation interval is very short compared with that of other species and because more than one or two offspring are produced at one parturition. A large number of observations of a particular bloodline or results of crossing bloodlines are thus possible to achieve desired results. Similar results might be expected by selecting animals related to those with a desirable production record. The pedigree offers an excellent starting place. As in most other species, the recent ancestors such as parents and grandparents are most valuable because each generation may reduce potential improvement by one-half.

Fads and promotional advertising affect selection by pedigree. The pedigrees of swine are generally believed to be less susceptible to serious manipulation because swine have rarely demanded the extremes in prices that cattle and horses can attract because of speculative breeding. Therefore, demands for change are dictated at the market, and breeders scan their pedigree books for genetic material to produce quality and quantity, knowing that the latter is the key to profitable returns.

The best use of pedigree information is as a basis for selecting young animals before their performance or that of their offspring is known. Because the generation interval is so short, traits such as longevity are not critical, but many others, such as soundness, milk production, rate of gain, feed efficiency, and so on, are important. If these fast-developing traits can be predicted with some degree of probability based on other bloodline observations, then the value of the pedigree as a selection tool becomes obvious. Also, one line that is weak in a particular trait, such as weak pasterns, may be bred to a line strong in the same area. In this way, each may overcome a fault of the other.

Pedigrees serve as a useful tool, but their limitations must be realized. Once an individual's performance is measured or recognized, the pedigree may be relegated to obscurity or, at best, receive less weight in an individual's evaluation.

Animal Performance

Measuring an individual's ability to perform may be as simple as memory of past experiences or as complex as use of such atomic-age tools as radioactive isotopes. Practical measurements are at the disposal of nearly all swine breeders. Gain, feed conversion, backfat thickness, carcass cutout data, and loin-eye area are examples of valuable data that supplement other selection methods.

Measuring Techniques.

Laptop or pencil and pad   A laptop computer or a simple pocket notepad can be used to record such data as litter size at birth, litter size weaned, conformation score, and general remarks too valuable to be left to memory. The information may be transferred to permanent records for lifetime evaluation of bloodlines.

Scales   Litter weight at birth and at weaning is useful information obtained from a simple scale. Weighing of feed consumed over a period can also be used, along with weight gains, to calculate feed efficiency. Rapid, efficient gains would be difficult to select without a scale. Some breeders believe the scale is the most important tool at their disposal for developing the framework of a selection program.

Swine-testing stations   Since 1954 the Agricultural Extension Service and other organizations have devised centers in many states to evaluate individual swine performance. The number of central stations has declined since 1980. Some stations only test boars, whereas others test boars, gilts, and market hogs. Most stations screen proposed applicants for inherited defects such as swirls and hernia. Only good, genetically sound swine are desired for test and later breed improvement. Insofar as possible, nutrition and environment are kept identical to ensure accurate appraisals. In many instances, a boar and a barrow that are full brothers are tested together so that the barrow may be slaughtered and data collected to estimate the boar's potential breeding capabilities for passing on a similar trait.

Rate and economy of gain are important measurements. The goal is attainment of a weight of 200 pounds to be reached during the 150 days from weaning to slaughter, utilizing 320 pounds of feed per 100 pounds of gain. The net carcass value, also a valuable measurement, is based on backfat thickness, length of side, area of the loin eye, and percentage of four lean cuts of the chilled carcass. Females may also be tested. Prolificacy, the number of pigs produced over a long life, and weaning weight of the litter are the criteria of chief interest. As pointed out in chapters 7 and 8, the effectiveness of selection can be meaningful provided that traits of reasonably high heritability are tested for and used in evaluation of potential breeding stock. Breeding animals can only transmit unfailingly those genes that they themselves have been found to possess in relatively homozygous (pure) form. Higher heritability estimates indicate the most rapid improvement to be expected in a trait. Some traits commonly used in production testing and their heritability ratings are given in table 9–2. Faster progress can be expected when dealing with traits of medium to very high heritability.

TABLE 9–2	**Heritability of Some Economically Important Characteristics in Swine**
Characteristics	**Percentage Heritable**
Number of pigs born alive	10
Birth weight	5
21-day litter weight	15
Litter size at weaning	12
Age at puberty	35
Days to 230 pounds	30
Average daily gain	30
Loin lean area	50
Carcass length	60
Percentage ham	50
Percentage lean cuts	50
Nipple number	18
Conformation score	29
Feed efficiency	30

Source: Modified from M. E. Ensminger and R. O. Parker, *Swine Science,* 6th ed. (Danville, IL: Interstate Publishers, 1997).

Ultrasonic devices may be used to determine loin-eye area in swine using a method similar to that described for cattle in chapter 7. Other complex information such as cutout data may be obtained if facilities are available to the testing station.

Production Testing

In developing a production-testing program, permanent markings are essential. Ear notches at birth or shortly thereafter are the traditional and effective system used on swine. Observations are made from birth to slaughter or determination of useful life, and this information is recorded on lifetime records and used as an aid in selecting breeding stock.

Production Registry.   Some record associations in the United States have a system of production testing known as production registry. The associations are generally open only to registered purebreds with no outstanding defects. The swine must be ear notched and spring from or produce a minimum of eight or nine (depending on the association) pigs in a litter. Minimum litter weights are also specified and must be witnessed by a person acceptable to the registry association. To qualify, a sow must produce at least one litter that meets the requirements. Some associations require two litters before certification. A boar becomes certified when it produces at least 5 such litters (some associations require as many as 15) or at least 2 certified daughters (some require as many as 10).

Certified Meat Hogs.   The National Association of Swine Records adopted a program that adds carcass evaluation to production registry to designate certified meat-type hogs. A certified litter must meet certain minimum standards to qualify. The first step is qualifying for production registry. Then two gilts or barrows from the litter are slaughtered, provided they are between a 200- and 220-pound equivalent weight at 175 days of age; equivalent weights may be calculated by adding 2 pounds for each day under 175 days or by subtracting 2 pounds for each day over 175 days. The carcasses of both hogs must have at least 4 square inches of loin-eye areas and 29 inches of length (measured from the front of the first rib to the front of the aitchbone [hipbone]). No more than 1.6-inch average backfat thickness is allowed. If these standards are met, the entire litter is certified. See the later section "The Pork Carcass" for more modern methods of selection based on sophisticated mechanical instruments.

A certified boar is one that has produced five certified litters. However, each litter must be from a different sow, not more than two of which can be full sisters or direct descendants. A certified mating means the rebreeding of a sow and boar that have produced a certified litter.

A System of Selection

Finally, using a combination of all of the previously described techniques, swine breeders need to follow a system of selection that will result in maximum progress (see figure 9–16). Three common systems follow: tandem selection, independent culling level, and selection index.

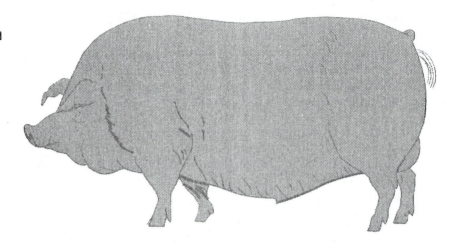

FIGURE 9–16 Selection has changed the ideal hog of the nineteenth century depicted here to our modern meat-type swine.

Tandem Selection.   Using tandem selection, breeders select one trait at a time until maximum progress has been achieved; then they select another trait for improvement. A good example of this method was the change from the lard-type hog to the long, lean variety. Because the heritability of length of side and backfat thickness is higher than that of most other traits, rapid progress was made. Then it was discovered that muscle, loin eye, and other traits were often substandard, so they became the target for improvement. The limitations of this system are obviously that many traits are interrelated, and selection for one may not always result in improvement of another.

Independent Culling Level.   Production testing fits well with the systematic procedure of the independent culling level. As already discussed, minimum standards are set for several traits. A failure to meet these standards by any individual results in slaughter or removal from the herd. Even with high standards, this system works well and is the one most used with swine. Large litters give many numbers to work with; even if only a few animals remain, the reestablishment of a herd is relatively rapid. For this reason, the standards are usually high, and the progress in swine breeding has been the envy of cattle breeders and others working with less flexible species.

Selection Index.   By using a selection index all important traits can be easily combined into one figure or score. Higher scores mean more valuable animals for breeding purposes. The weight assigned to each trait depends on its economic importance, heritability, and genetic linkage to other traits.

THE PORK CARCASS

The production of pork products that meet consumer demands keeps the swine market strong. No other livestock carcass is marketed to the consumer in such a variety of products, varying all the way from sowbelly to pigs' feet. The marketing, slaughtering, inspecting, grading, and retailing of pork carcasses fulfill the producers' goal of producing a high-quality product for a ready consumer market.

Market Classes and Marketing Pork

Most hogs are marketed either by direct sale (directly to a packer) or through a local auction. Market hogs are classified and sold in four market classes based on sex, the use to which the animal is best suited, and weight.

Barrows and Gilts. **Barrows** (castrated male hogs showing no sex characteristics) and **gilts** (young female hogs) form the class of most *finished market hogs.* Weights of barrows and gilts usually vary from 120 to 300 pounds; the most desirable weight is 200 to 250 pounds. Most pork for human consumption is derived from this class, depending on the U.S. carcass grade of the animal in the market class.

Sows. **Sows** (either pregnant or having had at least one litter) represent the second market class of swine. Usually they represent culled sows or bred gilts that range in weight upward of 220 pounds. Pregnancy reduces dressing percentage and results in a carcass that carries more fat than do barrows and gilts. Pork from sows is mainly marketed for human consumption as cured pork, and the carcasses are graded according to U.S. carcass grades.

Stags. *Stags* (castrated males showing sexual development) as a class are usually not marketed in sufficient numbers to warrant U.S. grading and are usually used for nonhuman consumption. Culled boars are generally castrated after culling and fed for a short time and thus fit into this class.

Boars. **Boars** (uncastrated males) are low in market value because a large portion of the carcass is not fit for human consumption because of the odor. Manufactured by-products such as inedible grease, fertilizers, hides, and others are the main use of this class.

Barrows castrated male swine

Gilts young female swine that have not farrowed

Sows mature female swine or female swine that have farrowed

Boars noncastrated male swine

Slaughtering and Dressing Hogs

On arrival at the packing plant hogs are penned in small pens, showered, and kept until slaughter (see figure 9–17). The slaughtering process is a set pattern of stunning, hoisting, sticking, scalding, dehairing, and dressing to yield a cooled pork carcass ready for inspection, grading, and further processing.

FIGURE 9–17
Veterinarian examining swine before slaughter. Courtesy of the U.S. Department of Agriculture.

Measurements of Carcass Quality

Carcass quality is measured in various ways, depending on the use of the carcass. Carcasses used primarily for bacon are measured by length and percentage of trimmed belly, whereas those used for other purposes are measured by percentage of lean cuts, minimum backfat thickness, and loin-eye area. In all evaluations, the emphasis is on the meat-type hog.

Carcass Yield or Dressing Percentage

One of the first calculations made in slaughtering hogs is the dressing percentage or carcass yield (the percentage yield of chilled carcass in relation to live weight). For example, if a hog slaughtered at 200 pounds live weight yields a 140-pound chilled carcass, the dressing percentage would be calculated as

$$\frac{140 \text{ pounds}}{200 \text{ pounds}} \times 100 = 70 \text{ percent}$$

Thus, the chilled carcass weighs 70 percent of the live weight. Normal dressing percentages vary from 60 to 70 percent. The other 30 to 40 percent of the live weight is by-products of pork slaughter. The dressing percentage of swine is higher than that of cattle or sheep because hogs do not have the large chest cavity and four-compartment stomachs of those animals.

Fat- or lard-type hogs have a higher dressing percentage than meat-type hogs. Yet because lard is low in retail price, the carcass value of lard-type hogs is less. Thus most packers use the cutout value of the carcass, especially on the maximum yield of the more important primal cuts.

Once weighed, the chilled carcass is halved and other measurements are taken. The first of these is carcass length, which is measured from the front of the aitchbone (the point at which the pelvis joins together) to the front of the first rib (see figure 9–18). Desired length is 29 inches or more.

FIGURE 9–18
Carcass length is from the front of the aitchbone to the front of the first rib. Location of backfat measurements is also shown here.
Courtesy of Michael De La Zerda, Texas A&M University.

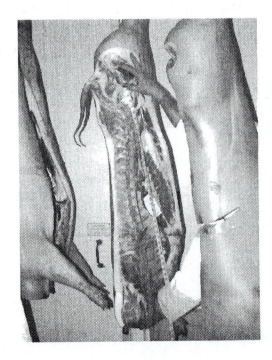

Backfat of the chilled carcass is measured at three locations: the first rib, the last rib, and the last lumbar (see figure 9–19). The average of these three measurements is reported as carcass backfat. Standards for backfat differ among sex and market classes, with the general rule of 0.9 inches or less desired.

Loin-Eye Area.   Because of the importance of pork chops in retailing pork, the lean meat area of the loin eye is often measured. For this measurement, the carcass is split between the 10th and 11th ribs, and the loin-eye muscle is traced and measured by use of grid paper (see figure 9–20). The loin-eye area should be 4.5 square inches or more in meat-type hogs. With several modern day hogs, averages have exceeded 6.5 sq. in. (Symbol II cross in 1996 for example).

FIGURE 9–19
Backfat on a carcass is the average of backfat readings measured at the first rib, last rib, and last lumbar.
Courtesy of Michael De La Zerda, Texas A&M University.

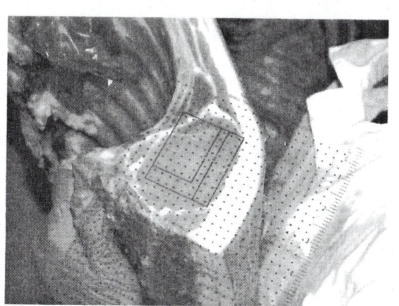

FIGURE 9–20
Loin-eye area is measured between the 10th and 11th ribs. Only the one large muscle is included in the measurement. The muscle is traced on a clear piece of acetate, with the area later determined using a planimeter.
Courtesy of Michael De La Zerda, Texas A&M University.

Percentage of Lean Cuts.   A measure of the retail value of the carcass is the percentage of the four major lean cuts (ham, loin, picnics, and Boston butts). These cuts, although only 45 to 50 percent of the hog's live weight, represent 75 percent of the carcass retail value. These four cuts are cut out, trimmed, and weighed. The percentage of lean cuts is determined in much the same way as dressing percentage.

Ham-Loin Index.   The ham-loin index is often used to evaluate carcasses in barrow shows. The score, based on the percentage of trimmed ham and loin-eye area, is determined as follows:

Percentage of ham above 10 percent $\times$ 10 = _____

Square inch loin-eye area $\times$ 10 = _____

Total score

For example, a hog yielding a 19.6 percent trimmed ham that had a 4.6-square-inch loin-eye area would have the following ham-loin score:

$$9.6 \times 10 = 96$$
$$4.6 \times 10 = \underline{46}$$

142 ham-loin index

Carcass Grades and Grading

Chilled carcasses are inspected for human consumption. Carcass grades are an indication of the excellence and quality of an animal in its specific market class. Carcass selling is commonly called grade and yield selling. U.S. carcass grades are based on the quality of lean, the amount of fat, and the expected yield of trimmed major wholesale cuts (ham, loins, picnics, and Boston butts). The U.S. carcass grades are U.S. No. 1, U.S. No. 2, U.S. No. 3, U.S. No. 4, and U.S. utility. Although live grades are not covered in this chapter, the same five grades exist, and live grades are based on the correlation of the live animal to yield the corresponding carcass grades. Skilled swine buyers recognize this correlation and through visual evaluation and backfat probing can accurately determine resulting carcass grade. Carcass grades are determined by inspecting both the fat and lean for quality (firmness, color, actual backfat thickness, fatness around the belly, and marbling of the loin) and the percentage of the four major wholesale cuts (ham, picnics, loins, and Boston butts). Grade is equal to (4.0 $\times$ backfat thickness over the last rib in inches) $-$ (1.0 $\times$ muscling score). Muscling scores used in the equation are thin, 1; average, 2; and thick, 3. See table 9–3 for comparison of dressing percentages by grades.

U.S. No. 1.   U.S. No. 1 grade carcasses have a high quality of lean meat, a high yield of lean cuts, and a low percentage of backfat (see figure 9–21). The yield of the four principal lean cuts is 60.4 percent or more; the maximum backfat is 1.0 percent with slightly firm loin eye, a slight amount of marbling, and pink to red coloring. The carcass has an intermediate degree of finish, is well muscled, and has good length.

U.S. No. 2.   Figure 9–21 illustrates a U.S. No. 2 grade carcass. Such a carcass has an acceptable quality of lean, a slightly high number of lean cuts (57.4 to

TABLE 9–3	Average Dressing Percentage for Barrows and Gilts by Grade	
Grade	**Dressing Percentage**	
U.S. No. 1	70	
U.S. No. 2	71	
U.S. No. 3	72	
U.S. No. 4	73	
Utility	69	

Source: Livestock Division, Agriculture Marketing Service, U.S. Department of Agriculture (USDA), 1996.

60.3 percent), and more thickness of backfat than U.S. No. 1 carcasses (1.0 to 1.24 inches backfat).

U.S. No. 3.   Carcasses that yield 54.4 to 57.3 percent of the four lean cuts and have 1.25 to 1.49 inches backfat and an acceptable quality of lean are graded U.S. No. 3. This carcass has a slightly low percentage of lean cuts and a high percentage of fat cuts.

U.S. No. 4.   The U.S. No. 4 grade is assigned to carcasses low in percentage of lean cuts (less than 54.5 percent of four lean cuts) and high in backfat (more than 1.5 inches) with a high percentage of fat cuts and yet an acceptable quality of lean.

U.S. Utility.   The U.S. utility grade includes carcasses with some lean quality but not enough to be graded U.S. No. 1 through U.S. No. 4. Carcasses with unacceptable belly thickness and all carcasses that are soft or oily or are pale, soft, and exudative (PSE) are also graded U.S. utility, as are unthrifty or unhealthy feeder pigs.

Wholesale and Retail Pork Cuts

Once carcasses are graded, they are usually cut up and sold by the packing plant in the form of wholesale cuts (see figure 9–22). The most common wholesale cuts are ham, bacon, loin, picnic shoulder, Boston butt, jowl, spareribs, and feet. Only about 30 percent (loins, shoulders, and spareribs) of the wholesale cuts are sold as fresh meat; the remaining 70 percent is sold as cured pork. The average lard production for hogs is 15 percent of the live body weight. Wholesale cuts are further processed into the retail cuts found at the common meat market or supermarket.

Pork By-Products

All products from slaughtering other than the carcass meat and lard are designated as by-products; by-products usually make up about 30 to 40 percent of the animal's live weight. The successful use of all by-products is a goal for which all packers aim. By-products vary from blood to glycerin for explosives, meat scraps to medicine, glue to leather, and violin strings to fertilizers.

Feeder Pigs
Official U.S. Grades

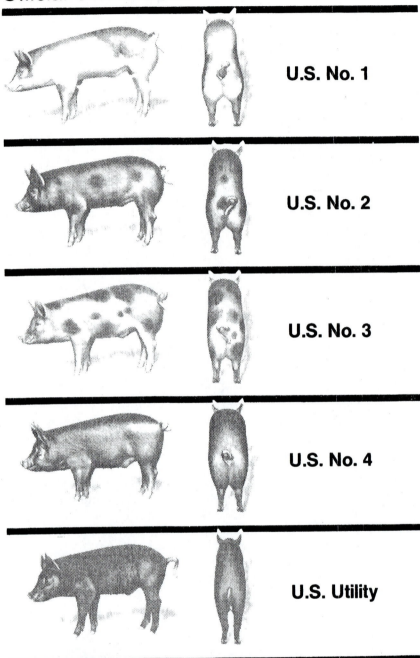

U.S. No. 1

U.S. No. 2

U.S. No. 3

U.S. No. 4

U.S. Utility

United States Department of Agriculture
Agricultural Marketing Service
Livestock and Seed Division
Washington, DC 20250

Copies of the Official United States Standards for Grades are available on request.

August 1988

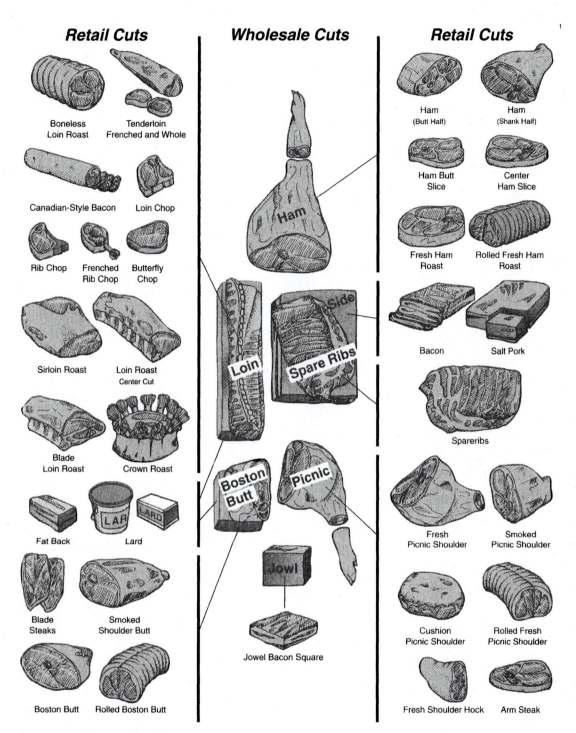

Retail Cuts

Boneless
Loin Roast

Tenderloin
Frenched and Whole

Canadian-Style Bacon

Loin Chop

Rib Chop

Frenched
Rib Chop

Butterfly
Chop

Sirloin Roast

Loin Roast
Center Cut

Blade
Loin Roast

Crown Roast

Fat Back

Lard

Blade
Steaks

Smoked
Shoulder Butt

Boston Butt

Rolled Boston Butt

Wholesale Cuts

Ham

Loin

Side

Spare Ribs

Boston
Butt

Picnic

Jowl

Jowel Bacon Square

Retail Cuts

Ham
(Butt Half)

Ham
(Shank Half)

Ham Butt
Slice

Center
Ham Slice

Fresh Ham
Roast

Rolled Fresh Ham
Roast

Bacon

Salt Pork

Spareribs

Fresh
Picnic Shoulder

Smoked
Picnic Shoulder

Cushion
Picnic Shoulder

Rolled Fresh
Picnic Shoulder

Fresh Shoulder Hock

Arm Steak

FIGURE 9–22   Wholesale and retail cuts of swine.
Courtesy of the U.S. Department of Agriculture.

Measuring Carcass Traits in Live Hogs

Hog producers of the future will require performance information on their animals prior to marketing and slaughter. There are several methods currently available to assist producers in obtaining accurate estimates of backfat thickness in the live hog. Backfat thickness, as previously mentioned, is the most useful single live indicator of carcass lean meat. Selection for decreased backfat thickness should result in improvement in leanness. The two main methods of measuring backfat thickness in live animals are the manual backfat probe and the use of ultrasonic machines.[2]

When probing live hogs for backfat thickness, the hog should be held with a nose snare or squeeze chute. The measurements are made 1 1/2 inches off the midline and above the elbow, last rib, and flank, corresponding to the first rib, last rib, and last lumbar vertebra, respectively, in the carcass. The probe is inserted at a slight angle through the fat down to the loin muscle after using a scalpel to penetrate the skin. The probe is very small, and no injury or infection should result if the procedure is done properly.

A much easier and more accurate method is to measure the backfat thickness using ultrasound. New machines can also measure loin-eye depth and thus estimate the loin-eye area.

New Genetic Evaluation Systems for Carcass and Live Animal Traits

In 1997, the National Pork Producers Council implemented the Maternal Sow Line National Genetic Evaluation Program.[3] Its purpose is to evaluate genetic lines for use in producing crossbred market hogs. Reproductive traits such as sow longevity, breeding, litter size, and milking ability, along with the progeny traits of growth, carcass, meat quality, and eating quality, are evaluated.

All females are weighed and backfat thickness measured when moved to a farrowing crate and again at weaning. The following measures of *sow performance* are recorded and used for genetic evaluation:

1. Pigs per day of life
2. Pounds of pigs per day of life
3. Number of pigs born per litter
4. Number of pigs born alive per litter
5. Birth weight of litters (born alive)
6. Litter weight at weaning
7. Number of pigs weaned at 21 days
8. Sow weight change (from farrowing to weaning)
9. Sow feed intake during lactation
10. Time to estrus after weaning
11. Total feed intake
12. Salvage value-weight at culling
13. Nonproductive sow days

---

[2]W. G. Luce, *Measuring Carcass Traits in Live Hogs*, OSU Fact Sheet no. 3662.

[3]R. Goodwin and D. Boyd, National Pork Producers Council, personal communication, June 1999.

Measurements of *sow longevity* include the following:

1. Sow age at culling in days
2. Body condition score based on backfat depth

Following are measurements of *growth:*

1. Days to 250 pounds (test females and progeny)
2. Daily gain on test, progeny
3. Feed efficiency, progeny
4. Daily feed consumption, progeny
5. Leg soundness score, progeny

Measurements of *carcass merit* (progeny) include the following:

1. Dressing percentage
2. Off-midline backfat thickness at the 10th rib
3. Midline backfat thickness at the last rib
4. Midline backfat thickness at the last lumbar vertebra
5. Loin muscle area at the 10th rib
6. Carcass length

Measurements of *loin muscle meat quality* (progeny) include the following:

1. Minolta chromameter reflectance and Hunter scale color
2. Ultimate pH of loin
3. Total loin lipid content
4. Loin muscle color score
5. Loin muscle marbling score
6. Loin muscle firmness score
7. Water-holding capacity and drip loss of loin muscle

The first public report of this genetic testing was released in April 2000. Producers will now be able to use this information to help choose replacement stock and make culling decisions.

So far we have talked about evaluating the hog based on phenotype (judging its appearance and measurable performance), its genotype (based on backfat probes, carcass evaluation of many progeny or ancestors, and so forth), and systems used to reach various selection goals. In 1992, the *Pork Industry Handbook* outlined selection goals for the producer based on economic and practical considerations of most hog operations (see table 9–4) and standards for selection of replacement boars (see table 9–5) .

Since that time, advances in swine genetics and selection have included the incorporation of selection based on similar production testing that is used in cattle evaluation. Estimated breeding value (EBV) and estimated progeny difference (EPD) since 1994 have been calculated to determine the genetic worth of swine. The EPD values currently used and available include

- EPD days/230
- EPD backfat
- EPD terminal sire index (TSI)
- EPD number born alive (NBA)
- EPD litter weight (LW)
- EPD sow productivity index (SPI)
- EPD maternal sire index (MSI)

TABLE 9–4 Guidelines for Selection of Gilts by Age and Weight

Birth	Identify gilts born in large litters. Hernias, cryptorchids, and other abnormalities should disqualify all gilts in a litter from which replacement gilts are to be selected. Record birth dates, litter size, and identification. Equalize litter size by moving boar pigs from large litters to sows with small litters. Pigs should nurse before moving. Keep notes on sow behavior at time of farrowing and check on disposition, length of farrowing, any drugs (oxytocin) administered, condition of udder, and extended fever.
3 to 5 weeks	Take 21-day weight of litter. Wean litters. Feed balanced, well-fortified diets for excellent growth and development. Screen gilts identified at birth by examining underlines and reject those with fewer than 12 well-spaced teats. If possible, at this time select and identify about two to three times the number of gilts needed for replacement.
180 to 200 pounds	Weigh and measure backfat to determine potential replacement gilts. Evaluate for soundness. Select for replacements the fastest-growing, leanest gilts that are sound and from large litters. Save 25 to 30 percent more than needed for breeding. Remove selected gilts from market hogs. Place on restricted feed. Allow gilts to have exposure to a boar along the fence that separates them. Observe gilts for sexual maturity. If records are kept, give advantage to gilts that have had several heat cycles prior to final selection.
Breeding time	Make final cull when the breeding season begins and keep sufficient extra gilts to offset the percentage of nonconception in the herd. Ensure that all sows and gilts are earmarked, ear tagged, or otherwise identified.

Source: Modified from *Pork Industry Handbook, PIH-27, Guidelines for Choosing Replacement Females,* 1992. Distributed by Agriculture Communication Service Media Distribution Control, Lafayette, IN 47901-1232.

TABLE 9–5 Suggested Selection Standards for Replacement Boars

Trait	Standard
Litter size	10 or more pigs farrowed; 8 or more pigs weaned
Underline	12 or more fully developed, well-spaced teats
Feet and legs	Wide stance both front and rear, free in movement, good cushion to both front and rear feet, and equal-sized toes.
Age at 230 pounds	155 days or less
Feed/gain boar basis (60 to 230 pounds)	275 pounds of feed per 100 pounds of gain
Daily gain (60 to 230 pounds)	2.00 pounds per day or more
Backfat probe (adjusted to 230 pounds)	1.00 inches or less

Source: From *Pork Industry Handbook, PIH-9, Boar Selection for Commercial Production,* 1988. Distributed by Agriculture Communication Service Media Distribution Control, Lafayette, IN 47901-1232.

See chapter 7 for a review of EPD and accuracy (ACC) of genetic merit. These values are increasingly important because more producers are using artificial insemination (purchasing the semen of boars from across the country) and need a means of evaluating the genetic merit of dozens of potential sires in a quick, systematic fashion. EPD values can also be used when choosing replacement boars for natural service. The more reliable genetic information one has the greater the accuracy available for real genetic improvement.

To compare animals raised in different environments, ratios similar to those used for cattle are used. The National Swine Improvement Federation recommends the use of ratios in production testing to compare the level of performance for various traits. Ratios are calculated as follows:

$$\frac{\text{Animal's performance} \times 100}{\text{Average performance of all animals in a group}}$$

An average ratio for a group would thus be 100 percent. A ratio of 115 for average daily gain (ADG) would indicate that the boar gained 15 percent above the test group in ADG. Ratios allow contemporaries to be compared, thus eliminating differences caused by housing, feed, weather, and management.

SWINE MANAGEMENT

Swine require different methods of handling than cattle, sheep, or horses; methods found successful in those animals may not work with swine. Swine management techniques vary somewhat among swine breeders. However, all swine operations fall into one of the production systems discussed in this chapter. The type of production system a producer follows depends on management skill, available capital, feed supply, available labor, and personal preference.

Feeder Pig Production

Feeder pig production systems produce weaned pigs that are sold to be grown and finished out on other farms (see figure 9–23). This system has a comparatively low feed requirement but requires a high level of management for success. Detailed management in the selection and care of the breeding herd at

FIGURE 9–23
Feeder pig production systems produce weaned pigs.
Courtesy of National Pork Producers Council.

breeding time, at farrowing, and in the care of the young pigs is essential. Goals for this system include 2.2 litters per sow per year with 10 or more pigs farrowing per litter and 8 or more of them being weaned and sold at 45 pounds or more. Not all of the 8 weaned pigs will be sold. The manager must select replacement gilts and boars from this group.

Management of the Breeding Herd

The quality of the breeding herd determines the quality of the marketed feeder pigs. Selection of replacement gilts and boars, therefore, must be based on visual appraisal, records of parents, EPD values, ratios, and other predetermined criteria. The desired gilts and boars should be from large litters and selected on number of nipples, growth weight and rate (days to reach 200 pounds or more for gilts or 220 pounds or more for boars), backfat probe (1.0 inch or less adjusted to a 230-pound basis), and litter mate carcass cutout data (60 percent or more lean cuts, 40 percent or more ham and loin, 1.2 inches or less backfat, and 29 inches or more length). It should be noted that selection for all these factors is almost impossible, but the goal should be to approach these standards in as many areas as is practical. Because litter size has a low heritability, it cannot be improved on very much by selection. Therefore, management becomes as or more important than selection.

Success depends on litter size, so a critical management time is during breeding and gestation. Maximum litter size results when sows are bred on their second or third heat period, when gilts are flushed, and when sows and gilts are bred with more than one service and are bred to a tested, well-cared-for boar. More eggs are ovulated in the second and third estrous cycles, and maximum fertilization is achieved when breeding occurs 12 to 24 hours after the onset of estrus. Flushing of gilts increases the number of eggs ovulated. Feed is increased 7 to 10 days prior to and 4 to 8 days after breeding to flush a gilt by conventional methods. Flushing of older sows does not seem to have any significant benefit, probably due to the short period between weaning and rebreeding. The benefit of larger litter size with gilts, however, is much more pronounced. Proper nutrition, disease prevention, and control of extreme environmental conditions during gestation ensure large litters as well.

Care of the Sow at Farrowing. Sows should be dewormed 2 to 3 weeks before farrowing. They should be sprayed or dusted for external parasites at this time as well. The farrowing house should be disinfected before farrowing each group of sows. Sows are generally moved into the farrowing crates (stalls) 4 to 5 days prior to the farrowing date to allow for proper adjustment to new conditions before farrowing (see figure 9–24). At this time, each sow should be washed with warm water and soap and then rinsed with a mild disinfectant. Pens should be dry and free of drafts, and fresh, clean water should be provided at all times. A laxative ration may be helpful to prevent constipation at the first postfarrow feeding.

Efficient swine managers are present at farrowing, no matter what time of day or night it may be. Managers make sure that newly born pigs are freed from their embryonic membranes so they will not suffocate and help sows that do not farrow in 6 to 8 hours by an injection of oxytocin (a hormone that causes uterine contractions and milk letdown). After farrowing, the sows must be properly cleaned out and be giving milk for the young pigs. Sows that do not give milk may require an oxytocin injection for proper milk letdown.

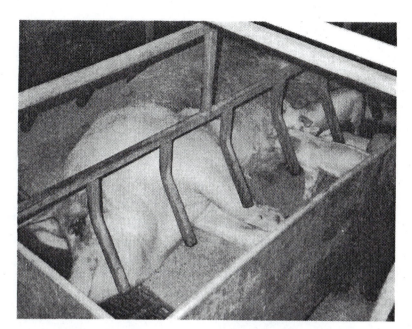

FIGURE 9–24
Farrowing crates are used at farrowing to allow for proper care of both sow and young pigs.
Courtesy of L. A. Pierce College Teaching Farm Laboratory.

Management of Young Pigs after Birth

The membrane should be removed from around the piglet's nose if necessary to prevent suffocation. The navel cord is disinfected with tincture of iodine solution to help reduce the incidence of navel illnesses. Supplemental heat prevents the pigs from becoming chilled. The ideal room temperature for farrowing houses is 70 to 75°F. However, the creep area should be closer to 85 to 95°F during the first week of life. The creep area should be zone heated by the use of heat lamps, gas brooders, and/or heat pads to prevent overheating of the sow and still provide necessary warmth for its young.

Most feeder pig producers follow a set schedule for management of young pigs. Although the order varies, the following management duties are generally performed during the first 4 days after parturition:

1. Clip the tips of needle teeth to prevent pigs from injuring each other or irritating the sow's udder.
2. Ear notch pigs. Ear notching is the most common method of identifying both the pig and its litter. Litter numbers are usually placed in the pig's right ear. The left ear identifies the individual pig within the litter.
3. Adjust the litter size if possible. A higher percentage of pigs will survive when litter size is equalized among the sows. Equalizing should be done during the first 24 to 36 hours.
4. Inject each piglet with 100 to 150 milligrams of iron dextran in the neck or ham muscle. A second injection 14 days later may be needed. Iron prevents anemia.
5. Dock tails at 3 days of age to prevent tail biting of pigs fed later in confinement.
6. Castrate at 3 to 14 days of age.
7. Provide starter rations at 7 to 10 days of age.

By 3 weeks of age, if they are not weaned at that time, feeder pigs should be creep fed. Diets should be gradually changed to prepare them for the feedlot.

FIGURE 9–25 A rotational cross-breeding program for swine.

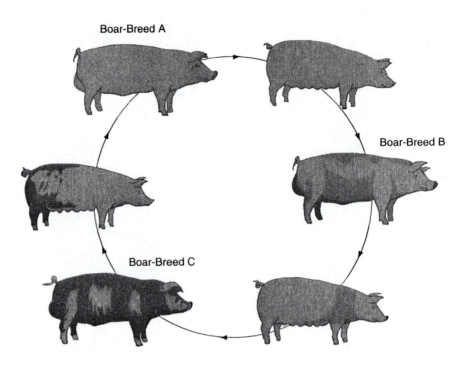

Feeder pigs should be weaned when they are at least 25 pounds and are usually marketed when 40 pounds or more. In many areas, feeder pigs are the major income producer in swine herds. Feeder pig producers require more skilled labor but less feed than those with finishing operations.

Use of Crossbreeding in Commercial Swine Operations

The advantages of crossbreeding are well-known, and approximately 90 percent of the sows are crossbred in commercial herds. Crossbred pigs are stronger at birth and grow faster; thus, more and heavier pigs are marketed. Crossbred sows produce more pigs. Most crossbreeding is not a single cross of two purebred parents but some type of crossing of crossbred animals. This crossbreeding is accomplished by using two breeds in a crisscross or backcross system or using more than two breeds in a rotational crossbreeding program (see figure 9–25). By rotating breeds without backcrossing to one of the original breeds, high levels of hybrid vigor are maintained.

Farrowing Sows on Pasture

With increasing concern of animal rights activists, some producers have considered farrowing sows on pasture or dirt lots. Researchers at Texas Tech University, Lubbock, Texas, compared indoor and outdoor systems of pork production. The litter size weaned is generally much smaller in pasture operations than in confinement operations. Labor increases and disease control are added concerns (see table 9–6). The main advantage is lower initial investment per sow unit.

TABLE 9–6	Effects of Farrowing Environment of Sow and Litter Performance	
	Indoor	**Outdoor**
Piglets born per litter	10.99	11.01
Piglets born live per litter	9.91	10.59
Birth weight per piglet (lb.)	4.14	4.25
Pigs weaned per litter	8.6	7.5
Preweaning mortality (%)	12.3	27.5
Weaning weight per pig (lb.)	14.3	15.3
Total weaning weight (lb.)	120.5	114.6

Source: *National Hog Farmer,* January 15, 1995.

For sows farrowing on pasture, stocking rates of 10 sows per acre are recommended. Suggested pastures include any of the small grains in the winter and legumes in areas where they can be grown. An individual farrowing house should be provided for each sow in the pasture. In the northern states farrowing in pasture or dirt lots is not recommended during the extreme winter months. In the summer, adequate ventilation should be provided to all farrowing houses. Pigs farrowed on pasture and dirt lots are generally weaned at 6 to 8 weeks of age. Postfarrowing and preweaning procedures similar to those already mentioned should be followed for farrowing pigs on pasture.

Multiple Farrowing

In recent years, swine producers have gone from seasonal farrowing twice a year to multiple farrowing (farrowing pigs throughout the year). This approach allows for maximum use of buildings and equipment and provides a steady supply of feeder pigs to a ready market. Multiple farrowing also prevents sharp price increases, which consumers despise. The main disadvantages are the possible buildup of pathogens in buildings and facilities with no rest period and the more frequent observation periods for producers.

Multiple farrowing requires a planned management schedule for rotating sow groups to fit the farrowing house schedule. Sows that will farrow within 7 to 10 days are rotated together throughout the year. The group is moved into the farrowing house about 3 to 5 days before the farrowing date. With a 48-day rotating schedule, sows are given 35 days from farrowing to weaning, 3 days for adjustment of pigs, 5 days for cleanup, and 5 days for bringing in new sows prior to their farrowing.

Finishing Program

The finishing program covers the growth and finishing of feeder pigs for slaughter (see figure 9–26). No breeding stock are kept, so less management and time are required for a finishing operation. A large quantity of feed at a reasonable cost must be present for the finishing program. The feeder is purchased from a feeder pig producer at approximately 45 pounds and must reach 200 to 240 pounds in 10 to 13 weeks with only 600 to 800 pounds of feed. Feed conversion averages 3.2 pounds per pound of gain. Thus three to four crops of pigs can be

FIGURE 9–26   The finishing swine program covers the growth and finishing of feeder pigs for slaughter.
Reprinted with permission from J. Blakely and D. H. Bade, *The Science of Animal Husbandry,* 6th ed. (Englewood Cliffs, NJ: Prentice-Hall, 1994).

finished out in a year provided feed and feeder pigs are available. Because the growth rate of a pig declines after 240 pounds, the efficiency of producing additional weight is reduced and hogs are generally marketed prior to reaching that weight.

Confinement versus Pasture Finishing.   Finishing programs on good cereal grain and legume pasture reduce both feed requirements and building and equipment needs. However, with more specialization in finishing swine, confinement is used more and more in growing, finishing, farrowing, and at the nursery stage as well. Confinement systems allow for mechanized handling of feeds and waste as well as better sanitation and disease control.

Farrow-to-Finish Operations.   Farrow-to-finish operations combine the feeder pig production and finishing operation into one. A producer finishes the feeder pigs produced, so a large quantity of reasonably priced feed and a high level of management are required. Large investment in buildings and equipment is also required. Maximum profits are obtained because profits are not shared with other producers.

Purebred Operations.   Purebred breeders supply the seed stock for the commercial swine herd. These producers sell genetics and not pork. They have similar requirements as the farrow-to-finish operations except that their final product is replacement stock and not finished feeders. Generally, these types of operations are smaller and require a high investment in buildings and breeding stock and a highly skilled manager to produce, record, and market purebred hogs. Master purebred breeders make maximum use of all of the tools available to test and improve the herds.

Breed associations have set up methods of evaluating purebred sows. The *production registry (PR)* and the *Unified Meat Type Hog Certification Program* are two such methods. Only registered purebreds may be included in the PR. A

PR litter must be nominated and must consist of eight or more pigs farrowed live in the same litter. All pigs must meet the breed association standards for rate of gain (typically 175 days to 220 pounds).

A *certified litter (CL)* is a PR litter that has carcass evaluations of two pigs slaughtered at 242 pounds or less. The standards for a certified litter adjusted to 220 pounds are an age of 175 days or less, carcass length of 29.5 inches or more, backfat of 1.45 inches or less, and a loin-eye area of 4.75 square inches or more.

These methods and standards have led to the terminology of a five-star PR sow (sow farrowing five PR litters). A pig from a certified litter has CL on its records. A *certified meat sire (CMS)* is one that has sired five CL litters, and *certified mating (CM)* is a mating that in the past has produced a certified litter. The figures presented vary somewhat depending on the association office and its minimum standards.

SWINE REPRODUCTION

General swine reproduction was covered in chapter 3. However, some specific additional considerations need to be addressed at this point.

Swine are one of the most prolific animals raised for the farm market. Litter-bearing (*polytocous*) capacity and short generation intervals make large numbers of marketed animals possible. An outstanding sow can easily produce two litters of 10 market pigs, or 4,000 pounds of pork, per year. The swine industry is thus an "in-and-out" business: it requires about as short a time to get into the swine business as to get out. Table 9–7 reviews some reproductive parameters of swine.

For maximum fertilization of the ova, breeding is usually done 12 to 24 hours after the onset of heat. Because ova are ovulated on the second day of heat, sperm must be present to fertilize the ova before they die (within 12 hours of ovulation). A practice of allowing two services during mating, at 12 and 24 hours after heat starts, results in higher conception rates, more ova fertilized, and thus larger litters.

Hand Mating, Lot Mating, and Artificial Insemination

The most common method of mating is by lot mating. Lot mating involves penning sows or gilts and mating one or more boars to them during expected heat periods. This method reduces labor requirements. Hand mating (putting a gilt

TABLE 9–7	Reproductive Parameters of Swine		
		Range	**Average**
Age at puberty (months)		4–7	6
Weight at estrus (lb.)		150–250	200
Duration of estrus (days)		1–5	2–3
Length of estrous cycle (days)		18–24	21
Time of ovulation (hours after onset of estrus)		12–48	24–36
Best time to breed		Second day of estrus	
Gestation period (days)		111–115	114
Weaning to first estrus (days)		3–7	5

or sow and boar in a pen, allowing service, and removing) has been used to regulate breeding, to increase litter size, and to keep accurate breeding records for registration purposes. Breeders of purebred stock usually practice hand mating. Small commercial swine producers usually practice pen mating. Large commercial swine operations use both hand mating and, increasingly, artificial insemination.

Both raw and frozen semen is used in commercial operations. The disadvantage of using raw semen is that the boar of choice must be present or in close proximity to the sow being bred. There must also be several sows in heat at the same time collection is made. Commercially available frozen semen comes in both pellet and straw form. Frozen semen is thawed by adding the semen to extenders.

Artificial insemination has long been practiced in European and Asian swine operations. Only in the past 20 years has interest in the use of this procedure picked up in the United States. In the early 1980s, fewer than 15,000 sows (<1 percent) were bred using artificial insemination in the United States. In 1995, it was estimated that 15 percent of the bred sows and gilts in the United States were artificially inseminated.[4] In Europe more than 400,000 sows (in some countries 20 to 30 percent) were bred using artificial insemination.

The main advantages of using artificial insemination for swine are the same as those for cattle. They include the following:

- Using genetically superior boars
- Bringing in new genetics to the herd with minimal disease risk (closed herd can be maintained)
- Reducing the danger of keeping boars on the premises
- Increasing diversity, leading to increased heterosis

When three or more females are bred in a day, the labor is significantly less using artificial insemination than natural service.[5]

Collection and Processing of Semen

Semen is collected primarily through the use of an artificial vagina inserted into a dummy sow. Some operations use a gloved hand instead of the vagina. A boar ejaculates three different fractions of semen:

- *Prespermal fraction:* Originating in the Cowper's glands, this gel fraction should be discarded because it contains few spermatozoa.
- *Sperm-rich fraction:* This fraction is opaque and milky and contains the highest concentration of spermatozoa.
- *Sperm-poor fraction:* This fraction should also be discarded because it contains primarily a gel from the accessory glands of the boar.

Semen has been stored primarily by freezing in pellets (0.15 to 0.2 milliliters). Boar semen does not freeze as well as bull semen. Using frozen semen reduces the litter size by as much as one pig. Lately, boar semen has also been

---

[4]M. E. Ensminger and R. O. Parker, *Swine Science,* 6th ed. (Danville, IL: Interstate Publishers, 1997).

[5]D. G. Levis, *1995 Proceedings Nebraska Whole Hog Days,* Artificial Insemination. Compiled by Dept. of Animal Science University of Nebraska, Lincoln, pages 155–175.

frozen in macrotubes. The semen is thawed by mixing 10 milliliters of pellets in 25 milliliters of a thawing solution. Several commercial organizations are now marketing frozen boar semen (see list at end of chapter).

Heat Detection

When using artificial insemination, breeders must monitor heat detection carefully. Breeders detect when a sow is in heat by applying pressure to the loin region. If the sow assumes the mating stance, it is in standing heat. Heat can also be detected by use of a teaser boar, observing sows walking fences near the boars, observing the restlessness in sows, and knowing that sows come into heat 3 to 5 days after weaning. Sows are inseminated with 30 to 50 milliliters of extended semen (fresh or frozen). This large volume needed for artificial insemination in swine compared with cattle (1/2 milliliter) limits the number of sows that can be bred from one collection of semen. Sows are usually inseminated two times, at 12 hours after the onset of estrus and 12 hours later. An average of 6 to 12 sows can be bred from one collection of semen through artificial insemination.

SWINE NUTRITION

Because 55 to 80 percent of the total cost of producing pork is feed, pork producers need to be aware of all aspects of swine nutrition. Modern hog operations are able to obtain feed efficiencies of 3.3 pounds of feed to 1 pound of gain from 40 pounds to market. For the producer to achieve these goals, a well-balanced ration must be designed and followed.

As discussed in chapter 4, the pig has a digestive system similar to that of humans; it is a simple-stomached animal. Swine require all of the basic nutrients that cattle and sheep require and in addition require 10 essential amino acids, B complex vitamins, and additional fatty acids that are not required of ruminants (or at least not in as great a quantity). Because swine are omnivores, without the benefit of large populations of microorganisms in a rumen, their diet must be low in fiber (usually 5 percent or less). Roughages usually serve to bulk up a ration, provide vitamins, and control feed intake.

Protein quality refers to the amino acid content; the higher the level and the more the variety, the higher the quality. Pigs require 10 essential amino acids for normal growth: arginine, histidine, isoleucine, leucine, lysine, methionine, phenylalanine, threonine, tryptophan, and valine. Cereal grains are too poor in protein quality to meet the needs of growing pigs. Therefore, protein supplements must be added to meet the precise requirements of pigs. Lysine, in particular, is usually the *first limiting amino acid* in most practical swine rations. Occasionally, tryptophan and threonine may also be deficient. Common protein supplements used in commercial swine operations include soybean meal, peanut meal, fish meal, and meat and bone scraps or tankage.

Although some of the energy in a swine diet is provided for by fat, most comes from the cereal grains. Corn is the most common cereal grain fed to swine in the United States. In addition to corn, milo, barley, wheat, and by-products from these grains are used to feed U.S. swine.

Minerals required by swine include calcium, phosphorus, sodium, chlorine, copper, iron, iodine, zinc, manganese, and selenium. Next to common salt (sodium chloride), calcium is the most needed mineral in a pig's diet. Calcium deficiencies in swine appear gradually. Lameness is among the first observable symptoms, with a gradual deterioration ending in paralysis of the hindquarters. After 3 or 4 months of a deficient ration, the paralysis will set in. Steamed bone-meal, ground limestone, and ground oyster shell flour are common sources of calcium. An excess of calcium or phosphorus or an imbalance between the two can interfere with the absorption and utilization of all the other minerals. It is therefore critical to balance the ration.

An iodine deficiency in bred sows may cause the farrowing of weak or hairless pigs (see figure 9–27). Iodine is needed for growth, gestation, and lactation. It is used by the thyroid gland in the manufacture of the hormone thyroxin, which regulates body temperature, among other things.

It is especially important to provide zinc when using soybean meal as the chief plant protein supplement. A mangelike skin disease called *swine dermatosis* or *parakeratosis* can be prevented by supplementation with zinc (see figure 9–28). High levels of calcium in the diet interfere with the absorption of zinc and thus can also cause this disorder, even when adequate amounts of zinc are fed.

FIGURE 9–27
Iodine deficiency. The hairless pigs shown in (a) were stillborn. The hairless pigs shown in (b) were born alive. Reprinted with permission from J. Blakely and D. H. Bade, *The Science of Animal Husbandry,* 6th ed. (Englewood Cliffs, NJ: Prentice-Hall, 1994).

(a)

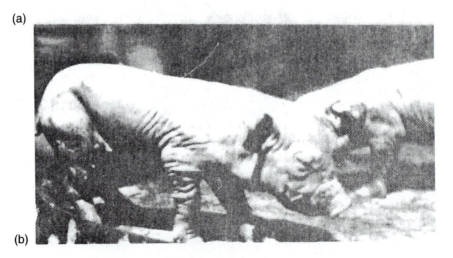

(b)

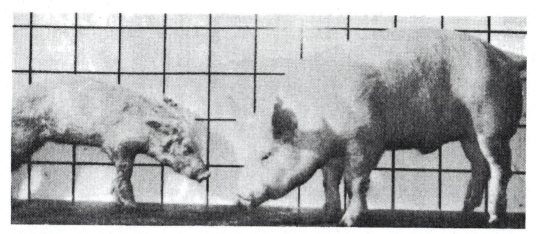

FIGURE 9–28   Zinc deficiency. The pig on the left received 17 parts per million of zinc and gained 1.4 kilograms in 74 days. Note the severe dermatosis and parakeratosis. The pig on the right received the same diet as the pig on the left except that the diet contained 67 parts per million of zinc. Reprinted with permission from J. Blakely and D. H. Bade, *The Science of Animal Husbandry,* 6th ed. (Englewood Cliffs, NJ: Prentice-Hall, 1994).

FIGURE 9–29   Iron deficiency. Note the listlessness and wrinkled skin of this anemic pig. Reprinted with permission from J. Blakely and D. H. Bade, *The Science of Animal Husbandry,* 6th ed. (Englewood Cliffs, NJ: Prentice-Hall, 1994).

Iron and copper are needed for the formation of hemoglobin and for the prevention of nutritional anemia. Anemia occurs in pigs during the suckling period when they are kept on drylot. The inability of hemoglobin to carry sufficient oxygen through the bloodstream creates an accelerated heartbeat that compensates for the reduced oxygen level by increasing the blood flow. The accelerated heartbeat results in a readily detectable pounding of the heart known as baby pig thumps or simply thumps. Anemic pigs may lose their appetite and become weak and inactive (see figure 9–29). In severe cases they die. The most effective means of preventing anemia, in addition to iron and copper in the ration, is administering two injections of iron dextran to baby pigs at 3 and 21 days of age. Injections of 100 milligrams per pig for the first injection and 50 milligrams for the second are recommended. A level of 125 to 250 parts per

FIGURE 9–30   Magnesium deficiency. The pig at the left was fed 413 parts per million of magnesium for 3 weeks; the pig at the right was fed 70 parts per million for the same period. Note the extreme leg weakness, arched back, and general unthriftiness of the pig on the right.
Reprinted with permission from J. Blakely and D. H. Bade, *The Science of Animal Husbandry,* 6th ed. (Englewood Cliffs, NJ: Prentice-Hall, 1994).

million of copper in the diet of growing-finishing hogs has shown a chemotherapeutic effect of improving the rate of gain and of feed efficiency.[6]

The symptoms of magnesium deficiency (see figure 9–30) in pigs include weakness of the pasterns, uncontrollable twitching, a staggering gait, and tetany; magnesium deficiency leads to death 50 percent of the time. Magnesium is necessary for normal muscle control, contraction, and equilibrium. Practical rations usually have sufficient magnesium to make deficiency unlikely.

Manganese deficiency symptoms in swine have been produced experimentally by feeding unnaturally low levels (see figure 9–31). A diet low in manganese in growing pigs can cause excess fat deposits and lameness. In sows it may result in the birth of weak pigs with a poor sense of balance. The optimum level for growth is 18 milligrams per pound of feed.

Phosphorus is also needed for proper skeletal growth and development. Typical deficient symptoms are weak legs and crooked leg bones (see figure 9–32). Steamed bonemeal and dicalcium phosphate are good sources and also supply calcium.

Diets deficient in selenium may cause necrosis of the liver and death in young pigs. A selenium deficiency is more severe when the diet is also deficient in vitamin E.

Vitamin A is necessary for growth, to prevent lameness or stiffness of the joints, and for normal reproduction (see figure 9–33). A deficiency causes incoordination of movement, lameness, swollen joints, and night blindness because of constriction of the optic nerve. Pregnant deficient sows often bear pigs that are weak, deformed, blind, or dead. Yellow corn is the only cereal grain that contains an appreciable amount of carotene (which pigs can convert to vitamin A provided the corn has not been stored for any length of time). Alfalfa leaf meal and synthetic vitamin A are both good sources of supplemental vitamin for pigs.

---

[6]W. G. Luce and C. V. Maxwell, Swine Nutrition Fact Sheet no. 3500, Oklahoma State University, 1999.

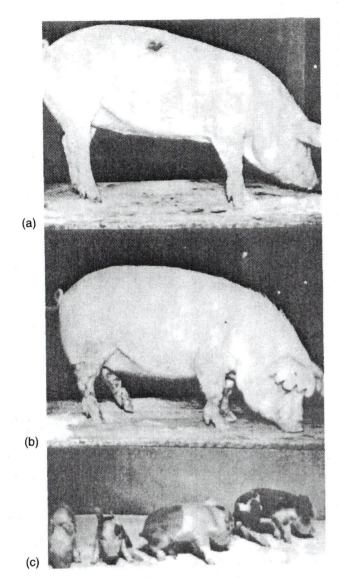

(a)

(b)

(c)

FIGURE 9–31
Manganese deficiency.
(a), Gilt, 132 days old,
that has been fed 40
parts per million of
manganese since it
weighted 3.9
kilograms.
(b), Littermate of the
gilt shown in (a). This
gilt was started at the
same weight and has
been fed 0.5 parts per
million of manganese.
Note the increased fat
deposit due to a low-
manganese diet.
(c), Litter from a sow
that was fed 0.5 parts
per million of
manganese. The pigs
showed weakness and
poor sense of balance
at birth.
Reprinted with
permission from
J. Blakely and D. H. Bade,
*The Science of Animal
Husbandry,* 6th ed.
(Englewood Cliffs, NJ:
Prentice-Hall, 1994).

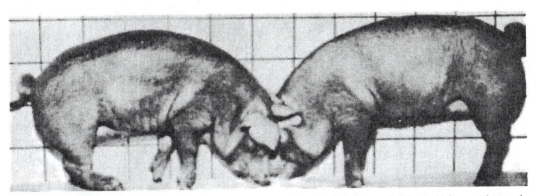

FIGURE 9–32 Phosphorus deficiency. On the left is a typical phosphorus-deficient pig in an advanced state of deficiency. Leg bones are weak and crooked. The pig on the right received the same ration as the one on the left, except that the ration was adequate in available phosphorus.
Reprinted with permission from J. Blakely and D. H. Bade, *The Science of Animal Husbandry,* 6th ed.
(Englewood Cliffs, NJ: Prentice-Hall, 1994).

FIGURE 9–33

Vitamin A deficiency. The pigs shown in (a) are 8 weeks old. The one on the right is deficient in vitamin A and shows a slow rate of growth. The pig on the right in (b) is deficient in vitamin A and has xerophthalmia. The pig in (c) is paralyzed in its hind legs.

Reprinted with permission from J. Blakely and D. H. Bade, *The Science of Animal Husbandry,* 6th ed. (Englewood Cliffs, NJ: Prentice-Hall, 1994).

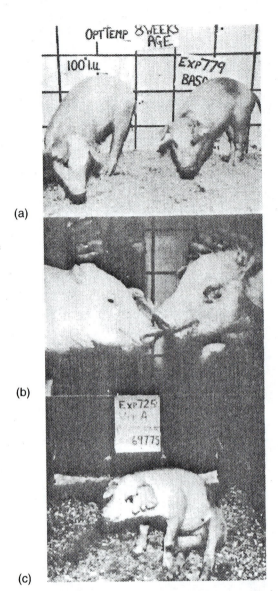

(a)

(b)

(c)

Vitamin D (the sunshine vitamin) is necessary for normal growth and bone development. When exposed to the sun, swine convert 7-dehydrocholesterol in the skin to vitamin D and thus seldom develop any deficiency problems. However, confinement housing permits little sunshine to fall directly on the skin of the hog, creating a condition in which dietary supplementation is required. Deficiency symptoms include rickets, enlarged joints, and weak bones. Vitamin D is associated with the utilization of the minerals calcium and phosphorus, which have similar deficiency symptoms (see figure 9–34). Any sun-cured forage, such as alfalfa, can be used in small amounts as a supplement. Synthetic sources are also available.

An increase in artificial drying of grains and the decline in use of pastures by hogs has resulted in an increased occurrence of vitamin E deficiencies. Vitamin E deficiency is more severe if selenium is also deficient. Symptoms in-

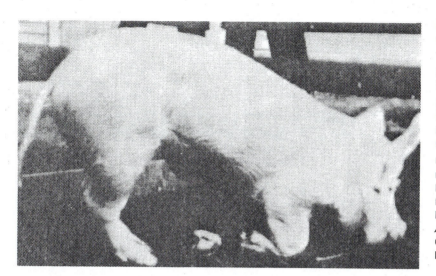

FIGURE 9–34

Vitamin D deficiency. This pig has advanced rickets resulting from lack of vitamin D. The pig was fed indoors. Because of leg abnormalities, it was unable to walk. It later responded to supplementary vitamin D. Reprinted with permission from J. Blakely and D. H. Bade, *The Science of Animal Husbandry,* 6th ed. (Englewood Cliffs, NJ: Prentice-Hall, 1994).

clude increased embryonic mortality and muscular incoordination in suckling pigs and sudden death, jaundice, edema, white muscles, and liver necrosis in growing-finishing pigs.

When moldy feeds are fed, the amount of vitamin K required increases. Supplementation with vitamin-K, under those conditions, reduces hemorrhagic symptoms. Under normal conditions, pigs produce all the vitamin K needed by microorganisms found in their intestinal tract.

The B complex vitamins that should be added to swine rations are riboflavin (B$_2$), niacin (B$_3$), pantothenic acid, choline, and vitamin B$_{12}$. Biotin supplementation may improve reproductive performance, including the number of pigs farrowed and weaned and the litter weaning weight.

Recent research also indicates improved litter size born and weaned, reduced early embryonic death, and increased conception rate with the supplementation of folic acid. Niacin is necessary for normal hair and skin health. A deficiency may result in slow growth, occasional vomiting, pig pellagra (skin lesions, gastrointestinal disturbance, and nervous disorders), and impaired reproduction (see figure 9–35). Natural sources that have been used for niacin supplement are distillers' dried solubles, condensed fish solubles, and alfalfa leaf meal.

Pantothenic acid is necessary for proper muscle and nerve function. The most obvious deficiency symptom is the classic goose-stepping gait that is manifest in severe ration shortages (see figure 9–36). The hog walks with an uncoordinated, wobbly motion and kicks upward with the hind legs because of impairment of nervous function. Dried brewer's yeast, dried whey, and alfalfa leaf meal are all good insurance as natural supplements.

Riboflavin is an important vitamin needed in many functions of growth and reproduction. Dietary deficiency results in the most obvious symptom—crooked legs. Other symptoms include slow growth, lameness, diarrhea, anemia, and impaired reproduction. (see figure 9–37). Any milk product, such as dried skim milk, is usually considered a good natural source. Thiamin is associated with growth, normal functions of the heart, and temperature regulation. A deficiency results in slow pulse rate and low body temperature, and autopsies have revealed flabby hearts.

FIGURE 9–35
Niacin deficiency. The pig shown in (a) has received adequate niacin; the pig in (b) has not. The difference in growth and condition is due to the addition of niacin in the diet containing 80 percent ground yellow corn.
Reprinted with permission from J. Blakely and D. H. Bade, *The Science of Animal Husbandry*, 6th ed. (Englewood Cliffs, NJ: Prentice-Hall, 1994).

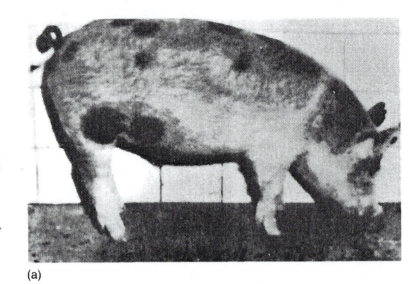

(a)

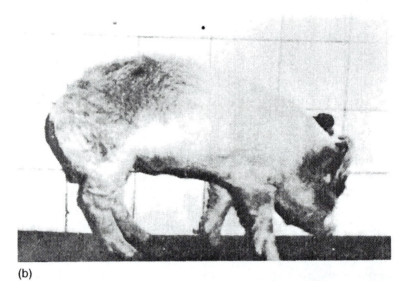

(b)

FIGURE 9–36
Pantothenic acid deficiency. Locomotor incoordination (goose stepping) was produced by feeding a ration (corn-soybean meal) low in pantothenic acid.
Reprinted with permission from J. Blakely and D. H. Bade, *The Science of Animal Husbandry*, 6th ed. (Englewood Cliffs, NJ: Prentice-Hall, 1994).

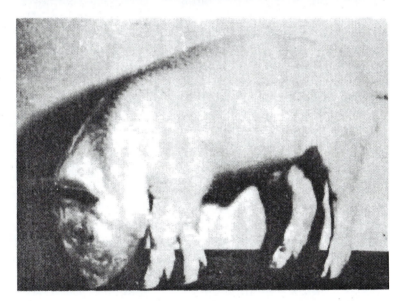

FIGURE 9–37
Riboflavin deficiency.
The pig shown at the
top received no
riboflavin. Note the
rough hair coat, poor
growth, and
dermatitis. The pig at
the bottom received
adequate riboflavin.
Reprinted with
permission from
J. Blakely and D. H. Bade,
*The Science of Animal
Husbandry,* 6th ed.
(Englewood Cliffs, NJ:
Prentice-Hall, 1994).

Vitamin B_{12} is important for growth and normal appetite. Major vitamin B_{12} deficiency symptoms are hyperirritability and lack of coordination of the posterior. The mineral cobalt is a part of the B_{12} molecule. The value of tankage, meat scraps, and dairy by-products for swine rations is related in part to their vitamin B_{12} content.

Feeding Swine Classes

Swine are classified according to the purpose and age for which they are fed. Each class has unique nutrient requirements and is fed in a different way. In some cases, rations are simplified to four categories—a prestarter ration, a starter ration, a grower-finisher ration that is also fed to lactating sows, and a gestation ration that is also fed to boars.

Feeding Young Pigs

Young pigs are given small amounts of creep feed in the farrowing stall at 7 to 10 days of age. Very small amounts are given two or three times daily to ensure that fresh feed is available. When pigs begin eating (at about 2 weeks of age), enough starter is left to last 2 or 3 days at a time. Pigs less than 25 pounds should be fed a 20 to 22 percent protein ration containing at least 1.30 percent lysine. As the pigs grow, the percentage of lysine and protein may be reduced to meet National Research Council (NRC) guidelines. A fresh water source other than that used by the sow should be provided. Only very palatable, high-quality feed sources are used in formulating the starter ration. Iron in the form of iron sulfate is included to prevent anemia. Antibiotics are added to starter rations to prevent disease and promote growth.

Growing and Finishing Pigs

Swine are usually grown and finished on a full feed of high-quality growing-finishing ration. A different ration is usually used for growing from 40 to 120 pounds than that used in the finishing stage of 120 pounds to market weight. The grower ration is higher in protein (15.5 to 17 percent crude protein with 0.75 percent lysine) for formation and growth of body tissue and bone, and the finishing ration (13.5 to 15 percent crude protein and 0.62 percent lysine) is higher in energy for fattening and finishing for market. Full feed requires some method to limit wastage, such as pelleting, using feed troughs with lids on them, and maintaining a low level of feed in the feeders. Many producers hand-feed (give a certain amount each day) rather than self-feed finishing swine to maximize utilization of feed and control wastage.

Feeding Replacement Gilts

Replacement gilts for the breeding herd should be separated from the market hogs when they weigh about 200 pounds (usually at 5 to 6 months of age) and fed only 5 to 6 pounds per day. This separation will keep them in good condition until breeding time at 7 to 8 months of age. Litter sizes have increased due to flush feeding (as previously discussed) 7 to 10 days prior to breeding. Feed is increased to 8 pounds per day and maintained 4 to 8 days postbreeding and then reduced to 5 to 6 pounds per day until farrowing.

Feeding Gestating Sows

Limited feeding of gestating sows and gilts ensures large litters. Sows that receive too much energy and put on too much weight during gestation have higher embryonic mortality rates (the unborn pigs die). The amount of gain desired during gestation is 60 to 80 pounds for a sow and 75 to 100 pounds for a gilt. The amount of feed required under drylot conditions is 1.5 to 2 percent of the body weight in gilts (about 5 pounds) and 1 to 1.5 percent of the body weight in sows (about 4 pounds). Many producers save feed costs by using legume and cereal grain pastures for gestating sows. Good pastures supplemented with 0.5 pound of protein supplement and 2 to 3 pounds of grain can cut the feed costs of gestating sows by 30 percent. Total crude protein requirements of gestating sows is 14 to 15 percent (0.62 percent lysine). The stocking rate on pasture is 8 to 10 sows per acre or 10 to 12 gilts per acre. To minimize the labor involved in hand-feeding gestating sows, a skip-day feeding method is commonly used. In this type of feeding, gestating sow groups are given access to self-feeders for 6 to 12 hours every third day. Thus they receive their pasture supplement with less investment and less labor, but more management is required to keep sows and gilts in desired condition.

Feeding Breeding Boars

Boars in the breeding herd are fed the same ration as gestating sows. In drylot, feeding 3 to 4 pounds when the boar is not used for breeding and 6 to 7 pounds per day when used will keep the boar in excellent condition. The boar should be kept on legume or cereal grain pastures, requiring 1 percent of its body weight of the concentrate ration.

Feeding Lactating Sows

To ensure adequate milk production by lactating sows, a high-quality ration is full fed throughout the nursing period. The only exception to this rule is when small litters occur and sows gain excessive weight during lactation. The amount of feed then rests on the judgment of the breeder.

Ration Suggestions

Although experienced herdsmen may have the background necessary to completely formulate rations using raw ingredients, most of them prefer to use a premix. Commercial premix supplements contain antibiotics, protein, vitamins, and minerals in the proportions required by swine. All that remains, then, is to add the small premix to the other ingredients, as illustrated in table 9–8.

TABLE 9–8   Ration Suggestions (Rations Using Premixed Commercial Supplements)

	18% Starter		16% Grower		14% Finisher	
Grain (corn or milo)	50	60	42.5	38.5	66.5	62.5
Cereal grain (oats, barley, wheat)	16.5	12	35	35	15	15
35% supplement	33.5	28		21.5		17.5
40% supplement			17.5		13.5	
Alfalfa leaf meal			5.0	5.0	5	5
	100 lbs.	100 lbs.	100 lbs.	100 lbs.	100 lbs.	100 lbs.

	14% Gestation				16% Lactation	
	Drylot		On Pasture			
Grain (corn or milo)	78.5	76.5	78.5	82.5	42.5	38.5
Cereal grain (oats, barley, wheat)					35	38
35% supplement		18.5	21.5			21.5
40% supplement	16.5			17.5	17.5	
Alfalfa leaf meal	5	5			5	5
	100 lbs.	100 lbs.	100 lbs.	100 lbs.	100 lbs.	100 lbs.

TABLE 9–9   Daily Nutrient Requirements (or percent of ration) for Swine

Live Weight (lb) Expected Daily Gain (lb) Expected FE (feed/gain)	Growing-Finishing (fed ad libitum)			Breeding Swine	
				Bred Gilts and Sows (4 lb daily air- dry feed intake)	Lactating Gilts and Sows (10.5 lb daily air-dry feed intake)
	11–22 0.55 1.84	48–110 1.54 2.71	110–242 1.81 3.79		
ME (kcal, daily)	1,490	6,200	10,185	5,836	15,320
Crude protein, lb(%)[a]	0.20(20)	0.63(15)	0.89(13)	(12)	(13)
Indispensable amino acids					
Lysine g (%)	5.3(1.15)	14.3(0.75)	18.7(0.60)	(0.43)	(0.60)
Arginine g (%)	2.3(0.50)	4.3(0.05)	3.1(0.10)	(0)	(0.40)
Histidine g (%)	1.4(0.31)	4.2(0.22)	5.6(0.18)	(0.15)	(0.25)
Isoleucine g (%)	3.0(0.15)	8.7(0.46)	11.8(0.38)	(0.30)	(0.39)
Leucine g (%)	3.9(0.85)	11.4(0.60)	15.6(0.50)	(0.30)	(0.48)
Methionine plus cystine g (%)	2.7(0.58)	7.8(0.41)	10.6(0.34)	(0.23)	(0.36)
Phenylalanine plus					
tyrosine g (%)	4.3(0.94)	12.5(0.66)	17.1(0.55)	(0.45)	(0.70)
Threonine g (%)	3.1(0.68)	9.1(0.48)	12.4(0.40)	(0.30)	(0.43)
Trptophan g (%)	0.8(0.17)	2.3(0.12)	3.1(0.10)	(0.09)	(0.12)
Valine g (%)	3.8(0.68)	9.1(0.48)	12.4(0.40)	(0.32)	(0.60)
Minerals (selected)					
Calcium g (%)	3.7(0.80)	11.4(0.60)	15.6(0.50)	(0.75)	(0.75)
Phosphorus g (%)	3.0(0.65)	9.5(0.50)	12.4(0.40)	(0.60)	(0.60)
Sodium g (%)	0.5(0.10)	1.9(0.10)	3.1(0.10)	(0.15)	(0.20)
Chlorine g (%)	0.4(0.08)	1.5(0.08)	2.5(0.08)	(0.12)	(0.16)
Magnesium g (%)	0.2(0.04)	0.8(0.04)	1.2(0.04)	(0.04)	(0.04)
Iron mg	46	114	124	144	380
Zinc mg	46	114	155	90	238
Manganese mg	2	4	6	18	48
Vitamins (selected)					
A IU	1,012	2,470	4,043	7,200	9,500
D IU	101	285	466	360	950
E IU	7	21	34	18	47.5
Riboflavin mg	1.61	4.57	6.22	5.4	14.2
Niacin mg	6.90	19	21.77	18	47.5
Pantothenic acid mg	4.60	15.20	21.77	21.6	57
B$_{12}$ mg	8.05	19	15.55	27	71.2

[a]Pounds per day or percentage of the ration.
Source: National Research Council, *Nutrient Requirements of Swine,* 1988.

More detailed minimum daily nutrient requirements for swine can be found in table 9–9. Nutritionists, veterinarians, and herdsmen use this and other tables to formulate maximizing-profit rations that meet the minimum daily requirements, as specified by the NRC. Linear formulas and computers assist in the formulation, but the eye of the herdsman is necessary in assessing individual animal and herd needs.

SWINE HEALTH MANAGEMENT

Most people think that a hog's dream of the perfect environment is a mud wallow. They consider the hog a filthy animal; however, given a choice, the hog is quite sanitary, preferring clean, cool, well-kept quarters. The reason hogs are often found in mud wallows is because they have no efficient way of dissipating body heat except by conduction, so they lie against something cooler than their body temperature.

A swine health management program should prevent disease and parasites from becoming established in the herd rather than wait to control diseases after they develop. Disease prevention involves isolating the herd from disease pathogens and parasites and providing proper sanitation of buildings and equipment and a planned immunization program. Disease-free herds can remain so only if no diseases are introduced by humans or replacement stock. Visitors should be required to wear clean footwear and should not be allowed in the farrowing house because small pigs are very susceptible to diseases. A working routine followed on most farms involves caring for the farrowing house and small pigs first and the nursery and the older pigs (finishing and breeding herd) last. Thus diseases are not transmitted to the young pigs.

Sanitation is very important in swine management because larger numbers of hogs are kept in close confinement than other species. The usual procedure is to steam clean farrowing houses before each group enters. The sows are washed and scrubbed thoroughly and proper disinfectants are used. It is also recommended that farmers move sows and litters to clean pastures by hauling them rather than letting them walk through areas that might be contaminated. Whenever possible, swine should be vaccinated against diseases prevalent in the area of production.

All animal caretakers should be able to take the vital signs and recognize major symptoms (abnormalities) to assess the seriousness of any health disorder. The normal vital signs of swine are as follows:

1. Rectal temperature of 101.6 to 103.6°F (average of 102.5°F [39.2°C])
2. Heart rate of 60 beats per minute (55 to 85 average)
3. Respiration rate of 16 breaths per minute (8 to 18 average)

In addition to disease causing a change in these vital signs, environmental temperatures, humidity, and stress can all significantly affect these parameters. Each livestock handler needs to determine breathing rate and heart rate without exciting the animal. A veterinarian who specializes in swine should be consulted for all diagnosis and treatment. Only the common problems and those of serious economic concern are covered here.

Anemia

See discussion in the section "Swine Nutrition."

Atrophic Rhinitis

Atrophic rhinitis (AR) is not a disease that causes many deaths, but it creates a tremendous economic loss because of irritation to the nose, inability to eat, and a generally deteriorated physical condition. About 80 percent of the nation's swine herds are affected. AR lowers the body's resistance to secondary lung infections brought about by the reduction of the normal filtering activity of the nasal bones. The snout is affected by decreased growth of the small, thin bones that make up the nasal cavity (turbinate bones). The turbinate bones are reduced in size and are misshapen by this disease. Failure of normal turbinate bone growth causes a misshapen snout and severe pain.

AR develops first in young pigs about 3 weeks of age or less and becomes chronic as the pig grows older. Initially, there is sneezing, coughing, blowing, and snorting. When severe degeneration occurs on both sides, the nose is shorter than the lower jaw. When one side is affected, the nose turns to the affected side in a crooked-nose condition. The tear ducts may also be plugged up, creating tear discharges or dirty rings under the eyes. The nose will occasionally appear bloody on the surface.

The cause is primarily the bacterium *Bordetella bronchiseptica.* A secondary bacterium, *Pasteurella multocida,* has also been listed as a possible cause. Dust and other irritants contribute to the severity of the attack. The infecting bacterium is found in most of the pig herds in the United States. It is very difficult to control because other mammals, such as dogs and especially farm cats, carry it.

Prevention is the recommended method of dealing with AR. It is recommended that farmers purchase only specific pathogen free pigs (see later section "Specific Pathogen Free Pigs") and provide adequate ventilation in swine houses and good care of nursing pigs because any other disease (e.g., scours, flu, and pneumonia) can make an AR outbreak worse. A vaccine containing a bacterin-toxoid mixture of *B. bronchiseptica* and *P. multocida* shows the best results. It is given during the first week the pig is born and again at 28 days after birth to provide protection against the disease. Sows are injected with the same preparation 2 to 4 weeks prior to farrowing, and boars are injected twice annually to provide similar protection. The practices of segregated early weaning (see the later section "Segregated Early Weaning") and improved sanitation have both been associated with significant reduction in AR levels.

Recommended treatment is to use sulfathiazole in the water and sulfamethazine in the feed continuously for 5 weeks. Tylosin, tetracycline, and ceftiofur have also been used successfully to treat this disorder. Sows should be treated before farrowing to reduce the organisms transferred from the sow to its newborn litter.

Aujeszky's Disease (Pseudorabies, False Rabies, Mad Itch)

The signs of Aujeszky's disease, commonly called pseudorabies, are seen in young pigs. Baby pigs that have no maternal antibodies develop high temperatures, convulsions, and paralysis and usually die. In pigs less than 2 weeks old, death losses approach 100 percent. After 3 weeks, young pigs develop some resistance, and losses are considerably reduced. Pseudorabies is a reportable disease.

In adult swine, signs may be mild to nonexistent, or they may include abortion, stillbirth, infertility, mummified fetuses, mild respiratory signs, and lack of appetite. Death is rare for adults, but recovered hogs serve as a silent reservoir to carry the disease.

Pregnant and nursing swine are most susceptible. Depending on the stage of gestation, pregnant sows may have stillborns and weak pigs (late gestation), mummification (midgestation), and embryonic death and infertility (early gestation). Abortion may occur at any stage but is most prevalent during the last one-third of pregnancy. Piglets may develop signs as early as 12 hours after farrowing, but generally the first signs are seen at 5 to 7 days of life.

Pseudorabies is caused by a DNA herpes virus. Because the disease can occur in adult swine with few clinical signs and can persist in their systems for long periods, these swine serve as a natural reservoir for the disease. The virus causes a fatal encephalomyelitis (inflammation of the nervous system) with signs of severe itching and self-mutilation—hence the name mad itch. Other than eradication, no treatment is known to be effective. Secondary infections in adult animals can be suppressed by the use of injectable antibiotics, and the subcutaneous administration of hyperimmune serum to piglets markedly reduces their mortality rate.

The disease is spread via nasal discharges; from saliva contaminating water, bedding, and feed; and from fomites. Although pigs are the only natural host, cattle, sheep, goats, cats, and dogs can also be affected.

Intranasal vaccination of sows and neonatal piglets followed by intramuscular vaccination of all other swine on the farm helps reduce viral shedding and improve survival rates. Breeding herds should be vaccinated quarterly and finishing hogs after levels of maternal antibodies decrease. Vaccinated swine shed the virus for a while, so dogs, cats, and cattle should be kept away from swine. Segregated early weaning, removing the piglets (18 to 21 days old) from their dams, and raising them at a completely neutral site, along with individual testing using the serum neutralization test, have shown some success at stopping the spread of the disease.

Brucellosis

Abortion is the key sign of brucellosis. It can occur at any stage of gestation but normally happens at 2 to 3 months or later. Retained placenta, temporary sterility due to inflammation of the uterus, lowered conception rate, pigs born weak, inflammation of the mammary gland, and arthritis are additional signs. Males with brucellosis have signs of orchitis (swelling of the testes) and lameness. Although the disease is primarily spread by ingestion of infected tissues or wastes, it can also be transmitted in natural breeding by the boar. People working in packinghouses have been infected when handling pigs.

A bacterium causes the disease. Swine are most often infected with *Brucella suis* but are also occasionally infected with *B. melitensis* and the bovine variety, *B. abortus.* This bacterium is very resistant and hard to kill with any disinfectant unless its protective organic material is removed. It remains in the soil, water, or aborted fetus for up to 60 days under natural conditions. Cresylic acid compounds and sodium orthophenylphenate will kill the organism provided the area is well scrubbed with soap and water prior to disinfection.

There is no vaccine and no practical recommendation for treatment of brucellosis in swine. Infected animals should be slaughtered. To prevent the disease, herds should be tested annually and reactors eliminated to control the disease. Replacement animals should be from brucellosis-validated herds or animals that have had a negative brucellosis test.

Cholera (Swine Fever)

Although not an explosive disease, hog cholera (see figure 9–38) is a devastating problem with swine. It is a highly contagious viral disease with high morbidity and mortality rates. It is found in endemic proportions in South

FIGURE 9–38
Listless victims of hog cholera.
Courtesy of the U.S. Department of Agriculture.

FIGURE 9–39
Button ulcers in the large intestine are a common sign of hog cholera.
Courtesy of the U.S. Department of Agriculture.

America, Africa, and Asia. In February 1997, it swept into Holland.[7] Swine fever has been absent from Canada, Australia, New Zealand, South Africa, and Great Britain and was eradicated from the United States in 1976.

Several stages may be present in a herd due to the often slow, chronic spread; the disease requires 6 to 20 days to fully develop. The first warning is several hogs dying with no clinical signs observed. After the initial unexplained deaths, the major signs are lack of appetite, elevated temperatures (104 to 109°F), involvement of the central nervous system, depression, diarrhea, dehydration, weight loss, and a high death rate (see figure 9–39). Conjunctivitis (severe inflammation of the eyelids), in which the eyelids may be glued together due to exudates, is common. Recovered pigs often do poorly for the rest of their lives.

The cause is a ribonucleic acid (RNA) virus that is specific for swine. Besides direct transmission between pigs, blood-sucking insects can also spread infection. It is very often passed in uncooked garbage, and for this reason it is illegal to feed raw, uncooked garbage in the United States.

---

[7] *National Hog Farmer,* February 15, 1998.

There is no treatment other than administration of hyperimmune serum during the very early stages of the illness. The disease is under U.S. government control by means of federal quarantine and slaughter. Vaccines are no longer available for cholera in the United States. A good preventive practice is to cook all garbage fed to hogs. Any animals suspected of having cholera should be quarantined and a 30-day isolation period maintained by a veterinarian. Although hog fever originated in the United States in the early nineteenth century, it is now found worldwide.

Colibacillosis *Escherichia coli* (Baby Pig Scours)

Baby pig scours is a common disease of nursing and weanling pigs. It is caused by strains of *Escherichia coli* with pili that adhere to the intestinal tract of the baby pig. Symptoms include profuse, watery diarrhea, which leads to death due to dehydration; highest death losses are in piglets under 4 days old. Treatment with antibacterial drugs and restoration of electrolytes and fluids is usually recommended. Preventive measures include improved sanitation and vaccinating gestating sows with pilus-specific vaccines (which increases antibody protection in the colostrum).

Dysentery (Bloody Scours)

Bloody scours is one of the most important intestinal diseases of weaned pigs. It is described as an acute, infectious, and contagious disease associated with animals that pass through public auctions. Swine dysentery is caused by a large spirochete, *Serpulina (Treponema) hyodysenteriae,* that produces a hemolysin. Carrier pigs are the most likely source of infection. These pigs show no signs of the disease, having developed immunity, but spread it to others that have not been previously exposed. Birds, dogs, people, and equipment have also been identified as mechanical carriers.

Signs of dysentery usually occur at 10 to 16 weeks of age. Both unweaned animals and adults are susceptible. Reduced appetite is often the first noticeable sign and is accompanied by soft, off-colored bowel movements that have been described as appearing like wet cement or milky coffee. The stool becomes streaked with blood as the disease progresses and usually contains undigested food as well as mucus from the intestinal wall. The rear of affected pigs becomes wet and stained, and pigs take on a characteristic gaunt appearance. Death losses are not high, but some sudden losses can occur early in the outbreaks.

Treatment consists of using organic arsenic preparations. The preferred method of drug delivery is via the drinking water rather than in the feed. After the initial outbreak is controlled, medicated feed can be used to hold the disease in check until the pigs are marketed.

Tylosin has proved beneficial in injectable form but does not work in feed or water at currently approved levels. However, individual pigs that show extreme weakness may be injected with tylosin. Drugs to be mixed with feed are available, such as bacitracin, carbadox, nitroimidazoles, virginiamycin, and Lincomix.

Preventive methods pay big dividends. Clean, dry floors and uncontaminated feed and water reduce losses. Steam cleaning and reduction of stress factors such as overcrowding, chilling, and poor ventilation have reduced outbreaks of dysentery.

Erysipelas

Erysipelas is caused by the bacterium *Erysipelothrix rhusiopathiae (insidiosa).* The organism is very difficult to control because it lives on a wide range of hosts including rats, birds, insects, and the soil and may be carried by other swine that recover from the disease. Erysipelas also affects sheep, rabbits, turkeys, and humans. The organism is excreted by infected pigs and spread by ingestion. Death rates range from 50 to 70 percent. In humans erysipelas causes erysipeloid, which produces disabling, woundlike infections. Clinical signs vary considerably because of the three forms: acute, subacute, and chronic. In the acute form, some pigs may die very suddenly. Others may develop a high temperature (104 to 108°F) and lack of appetite, appear physically sick, develop sore muscles or tender feet, have an arched back, and walk in a shuffling gait, often squealing because of pain in the feet and joints. Affected pigs may remain lying down or may protest loudly when they are forced up. About one-third of those with the disorder develop red, diamond-shaped skin lesions (square patches) on the skin. This sign is considered a diagnostic clinical sign of erysipelas. Septicemia, purplish-colored ears and snout, and endocarditis are additional symptoms.

Subacute forms produce milder signs with fewer deaths. Pigs may show only reluctance to move and reddish skin discoloration of the ears, tail, jowl, and legs. In the chronic form (see figure 9–40), there is a persistent arthritis and swelling of joints, and the tips of the tail or ears may blacken and fall off. Although the acute form would appear to be the most dangerous type, probably more economic loss is caused by the subacute and chronic forms because of developing lameness and arthritis, a condition that is very difficult to reverse once started.

Preventive measures include controlling rats, predators, and flies and cremating dead pigs. Vaccines are available and give immunity for about 6 months after administration. Revaccinating every 6 months is therefore required. Vaccination raises the level of immunity in the pig but does not provide complete protection. Elimination of carriers, improved sanitation, and reduction in stress are needed as well. Antiserum and penicillin are used to treat acutely affected pigs. Treatment of chronic erysipelas is not effective.

FIGURE 9–40
Chronic swine erysipelas, showing sloughing of skin. Courtesy of the U.S. Department of Agriculture.

Foot-and-Mouth Disease

Foot-and-mouth disease (FMD) is a highly infectious disease of pigs, cattle, sheep, goats, buffalo, and many wildlife species. In March 1997, FMD killed more than 4 million pigs on 6,100 farms in Taiwan. FMD is found in endemic proportions in Africa, the Middle East, parts of South America, and parts of Asia. FMD is caused by a virus and has high morbidity and mortality rates.

Symptoms include vesicular lesions on the feet and snout; fever as high as 106°F; anorexia; and vesicles on the tongue, dental pad, gums, teats, and udder. Piglets may die before showing any symptoms. In both pigs and cattle, the symptoms of FMD are indistinguishable from those of vesicular stomatitis. Both diseases are therefore reportable to state and federal authorities.

Prevention includes vaccination every 4 to 6 months in endemic areas. In nations free of FMD, preventive measures include quarantine, testing, slaughtering, and burning carcasses.

Greasy Pig Syndrome (Exudative Epidermitis)

Greasy pig syndrome is caused by the bacterium *Staphylococcus hyicus*. It has a morbidity rate of 10 to 90 percent and a mortality rate of 5 to 90 percent (highest rates are among pigs 1 to 6 weeks old). Pigs develop a resistance to infection with age. Predisposing factors include mites and skin abrasions. Symptoms include listlessness, reddening of the skin, depression, anorexia, vesicles and ulcers on the nasal disk and tongue, and erosions at the coronary band. Treatment includes administration of antibiotics for 7 to 10 days in addition to antiseptics applied to the body surface areas. Clipping of the needle teeth and improved sanitation help to minimize infections.

Hypoglycemia (Baby Pig Disease)

Hypoglycemia means low blood sugar level. This condition is due to inadequate food intake and may develop 24 to 36 hours after birth. Stress also plays a significant part, particularly cold stress during the first 10 days of life. Baby pigs cannot produce glucose as adults do and therefore must depend on daily intake from milk. If milk intake is inadequate or if demands are too high because of a cold barn, hypoglycemia develops. Death losses may be very high during the first few days of life. The first signs are shivering, erection of hair, squealing, weakness, rolling of the eyeballs, and coma. Without treatment, death usually occurs within 36 hours.

The most common causes are lack of milk production in the sow and diarrhea in baby pigs. Exposure to cold and an inherited weakness may be contributing causes. Treatment involves providing a warm environment through heat lamps and supplemental feeding and if signs are severe, an injection of 5 percent glucose solution every 4 to 6 hours. Preventive measures include selecting breeding stock for high-milking qualities, providing heat lamps during cold times of the year, and controlling litter size by redistribution of piglets.

Leptospirosis

The main signs of leptospirosis are abortion; small, weak pigs at farrowing; and a high incidence of stillbirths. In adult hogs, the only other signs are a mild rise in temperature, occasional blood-stained urine, and a brief loss of appetite. In some herds that have been exposed to the disease for the first time there is an abortion 'storm.' In other herds, the disease spreads slowly, causing abortions over a span of a month or more. The frequency of abortions then diminishes as immunity develops. Most abortions occur 2 to 4 weeks before parturition. Leptospirosis was observed in humans (1915) long before it was recognized in pigs (1952). Human infection is primarily through skin abrasions—when handling or slaughtering infected animals or by swimming in contaminated water—or through consumption of uncooked contaminated foods.

There are more than 200 serovars of leptospirosis. Five serovars are common in pigs: *Leptospira bratislava, L. pomona, L. grippotyphosa, L. canicola,* and *L. icterohaemorrhagiae.* Cattle, rats, raccoons, and other wildlife act as carriers of the organism and spread the disease. Active organisms are carried in the urine of infected animals and excreted for up to 6 months after recovery.

Treatment consists of two injections of streptomycin given 24 hours apart. Other antibiotics such as oxytetracycline and chlortetracycline may be used in the feed at 400 to 500 grams per ton for 4 weeks to stop the shedding of the leptospiral organisms from infected swine.

Preventive measures can be incorporated into the management system. Owners should vaccinate for all common serovars 2 to 3 weeks prior to breeding. Any new animal brought to the premises should be isolated and tested for leptospirosis prior to admission to the herd. Routine antibiotic treatment and vaccination of new additions have also been recommended.

Mycoplasma Pneumonia (Enzootic Pneumonia, Virus Pneumonia)

Viral pig pneumonia is actually misnamed because the condition is not caused by a virus but rather a bacterium, *Mycoplasma hyopneumoniae.* The most common sign is a dry, rasping, persistent, nonproductive cough that is most noticeable when pigs are quiet. During the first 2 to 3 days of infection, young pigs 3 to 10 weeks of age may have mild diarrhea accompanied by this cough, but the diarrhea is very mild and short-lived and generally goes unnoticed by the producer. Slow growth and unthriftiness of litters are the major results of the disease. The original infection is mild, but secondary infections with other organisms, particularly *Pasteurella* spp., may cause severe disease losses. Mycoplasma pneumonia is spread by direct contact from infected to uninfected animals.

Prevention consists of isolating infected animals from the rest of the herd, purchasing all new animals from specific pathogen free herds, and farrowing old sows in isolation and ascertaining that the pigs are free of disease before adding them to the herd. Treatment with lincomycin (200 grams per ton of feed for 21 days), tetracycline, tylosin, or tiamulin reduces the severity and improves pig performance.

Parasites

Other than sheep, hogs are affected by parasites more than any other class of livestock. Infected hogs become unthrifty, unproductive, and more susceptible to other disorders. More than 50 different species of worms and protozoan parasites are found in swine. The most common ones found in U.S. swine operations are included in the following section. In addition, several external parasites commonly infect U.S. swine and require brief mention regarding their prevention and treatment.

Roundworms (Ascarids). Ascarids are the most common and destructive parasite of swine. Roundworm infections reduce growth rate and feed efficiency and in severe cases cause death of the animal. Damage to the animal's liver results in liver condemnations, and damage to the lung causes increased development of bacterial and viral pneumonia. Prevention consists of improved sanitation and the worming of sows and gilts 10 to 30 days before farrowing.

Lungworms *(Metastrongylus elongates, M. pudendotectus, M. salmi)*. Three common species of lungworms infect swine. Lungworms infect the bronchi of the lungs, causing lowered feed efficiency and growth rate; the animals experience spasmodic coughing. Prevention consists of keeping hogs away from earthworms, which are the intermediate host. Various anthelmintic products are effective at controlling lungworms, including fenbendazole, ivermectin, and levamisole hydrochloride.

Trichinosis *(Trichinella spiralis)*. Trichina, or trichinosis, is a parasitic disease of human beings. The main source is infected pork eaten raw or improperly cooked. Only 0.1 percent of grain-fed hogs and 1 percent of garbage-fed hogs are infected with trichina. There are no specific symptoms in infected pigs. In humans infection causes fever, digestive upset, and myositis. Prevention of trichinosis in swine consists of destruction of all rats on the farm (which also carry the worm), proper carcass disposal of hogs that die on the farm, and cooking of all garbage and offal from slaughtering houses at a temperature of 212°F for at least 30 minutes.

Lice. Small, flattened, wingless insects include both sucking lice and biting lice. Of these two groups, the sucking lice cause the most damage to the hog. Lice cause loss of condition, irritation, severe itching, and hair loss and their bites can lead to secondary infections. Treatment consists of dusting all hogs or spraying and dipping (weather permitting) with an approved insecticide. State extension specialists and a swine veterinarian should be contacted for the most current control methods.

Mites (Mange). Mites produce a host-specific contagious disease known as mange (scabies). Mange costs the average hog producer $84 to $115 per sow per year.[8] Two types of mites infect swine—sarcoptic mange and demodectic mange. Of these two, the sarcoptic is the most common. Symptoms include the formation of scabs, skin irritations, reduced growth rate, and general unthrifty appearance. Prevention includes isolating new and infected animals, improving sanitation, and spraying for mange control with approved insecticides.

Myiasis (Fly Strike). Myiasis is an infestation of wounds with maggots. Tail docking of baby pigs seems to be the best method of preventing tail biting and thus helps to reduce the incidence of myiasis. Any means of reducing injuries to pigs, reducing fly populations in general, and treating all wounds with fly repellent will further reduce myiasis.

---

[8]J. J. Arends, C. M. Stanislaw, and D. Gerdon, "Effects of Sarcoptic Mange on Lactating Swine and Growing Pigs," *Journal of Animal Science,* 68:1495–1499, 1990.

Ringworm (Barn Itch).   Ringworm is caused by a microscopic mold or fungus (*Trichophyton, Achorion,* or *Microsporon* spp.). Ringworm infects all farm and domestic animals and humans. Symptoms include round, scaly, crusty areas of the skin devoid of hair. Treatment includes removing the scabs and hair around them and treating with tincture of iodine or salicylic acid and alcohol every 3 days. Oral treatment with griseofulvin may also be beneficial.

Porcine Reproductive and Respiratory Syndrome

Porcine reproductive and respiratory syndrome (PRRS) was first reported in 1997 in Minnesota, Iowa, and North Carolina. It has since been reported in Europe, Canada, and across the United States. In 1995, it was declared the most difficult and costly health problem affecting swine,[9] costing up to $350 per inventoried sow.[10] The virus that causes PRRS was isolated in 1991. Diagnosis involves enzyme-linked immunosorbent assay (ELISA) or an antibody test that measures immunoglobulin G (IgG) antibodies to the PRRS virus. There is also a new polymerase chain reaction (PCR) 48-hour test commercially available.[11]

PRRS symptoms include late-term abortions (107 to 112 days), stillbirths, mummification, and weak born and postweaning respiratory diseases (deficient oxygenation of the blood). Transmission is by contact with infected animals. There is no effective treatment. Prevention involves segregation, quarantine, and vaccination of pigs less than 16 weeks of age. The vaccine is not approved for older animals.

Porcine Stress Syndrome

Porcine stress syndrome (PSS) is an acute, shocklike syndrome, quite common in the 1960s and 1970s. Stress was induced through routine management procedures such as handling or moving. *Transport myopathy* and *malignant hyperthermia* were two other names originally given to the disorder. Clinical signs often include open-mouthed breathing, elevated body temperature, muscle and tail tremors, and sudden death. In addition, PSS is related to low-quality pork products, called pale, soft, exudative (PSE) pork, when the stress-susceptible pigs survive until slaughter.

The stress-induced condition occurs most often in heavily muscled pigs, especially shorter legged, more compact pigs that resulted from intensive genetic selection. The condition is most common in the Poland China and some lines of the Landrace breed. Partial or total confinement of pigs appears to increase the incidence of PSS. Pigs suspected of having PSS should be provided immediate rest; it may be advisable to tranquilize them if they are known to be susceptible to PSS.

Prevention of stress inducement through careful handling, guarding against overheating or overcrowding, and sudden changes of management is the gen-

---

[9]Dr. Steve Dritz. "Porcine reproductive and respiratory syndrome (PRRS)-Strategies for Control," *Swine Update,* 16(2):1–3, 1995.

[10]J. Carlton, "20 Questions and Answers about PRRS Virus," *Swine Practitioner,* March–April 1997, pp. 10–14.

[11]J. Collins, "New PRRS Diagnostic Test Available," *Large Animal Practice,* Vol 19(1):Page 12, January–February 1998.

eral recommendation. Because PSS is a known inherited characteristic, proper breeding practices and the use of deoxyribonucleic acid (DNA) probe tests have been effective in reducing the incidence of PSS.

Segregated Early Weaning

Segregated early weaning (SEW) is an infectious disease-control procedure with the primary objective of preventing the transfer of disease through segregation of early weaned piglets from their dams. Piglets are weaned at 10 to 17 days rather than the normal 21 to 28 days. Weaning age is critical because the maternal immunity (passive immunity via colostrum consumption) decreases at different rates for different diseases. Segregating piglets from their mothers prevents vertical transmission of infectious diseases from sows to their piglets. In addition, SEW provides increased performance of growing and finishing hogs due to the decreased burden of fighting infectious diseases.

Soft Pork

When finishing hogs are fed high levels of unsaturated fat (fat that is liquid at room temperature), soft pork results. Pigs allowed to "glean the feeds," eating large amounts of peanuts or soybeans, are more prone to this condition than other pigs. Soft pork is very undesirable from the standpoint of both the consumer and the processor. The cuts do not stand up and are unattractive in the meat showcase.

To prevent soft pork, the amount of soybeans and peanuts fed should be kept under 10 percent of the diet for all pigs 85 pounds or greater. Changing the type of fat to saturated fat will also help alleviate the problem.

Specific Pathogen Free Pigs

Certain diseases are so costly and so widespread that a method was sought to raise swine that are free of the disease in an area. Disease-free pigs were raised by strict control measures to clean up a farm—letting it lay out of swine production until the disease organisms died off or were killed. After eradication, the farm is quarantined to keep people off unless special clothing and treated footwear are worn. The next step is obtaining swine that are free of the diseases that are to be controlled. It is not practical to expect to produce swine that are free of all disease, but it is possible and practical to rid an area of specific pathogens, thus the name specific pathogen free (SPF) pigs. SPF pigs are produced by qualified veterinarians who perform cesarean operations on thoroughly disinfected sows and take the pigs just before natural birth, rearing them on sterilized milk by artificial means in sterile confinement. The developing swine then serve as a foundation herd and are allowed to breed and farrow naturally on the clean farm. The specific pathogens that are usually controlled by this method are atrophic rhinitis, mycoplasma pneumonia, and swine dysentery. In addition, SPF herds must be validated to be free of brucellosis, leptospirosis, lice, and mange. However, the pigs are not free from the effects of all diseases. The supervision of SPF herds is the responsibility of the National Swine Repopulation Association, Lincoln, Nebraska.

Transmissible Gastroenteritis

Transmissible gastroenteritis (TGE) is an infectious, transmittable disease that causes a high death rate in pigs less than 10 to 14 days of age. It affects older swine also, but it is seldom fatal in adult animals. Surviving animals may shed TGE virus for up to 40 days in manure and up to 120 days from the respiratory tract following recovery. Signs of the disease are poor appetite, vomiting, scours, and weight loss. Lactating sows stop giving milk, and there are high death losses in pigs under 2 weeks of age.

In young pigs, vomiting and diarrhea are constant. Whitish, yellowish, or greenish fecal material is passed through the digestive system. Ingested milk often appears in the manure unchanged. There is rapid dehydration and weight loss, with a high mortality rate in young pigs. The disease can occur at any time of the year, and the major losses are due to poor performance and production of some runt pigs on recovery from the initial outbreak.

The cause is a corona virus usually passed through manure and the respiratory tract of infected swine. There is no effective treatment. Outside exposure during farrowing time should be avoided. One form of prevention—exposing pregnant sows to the disease to create antibodies in the milk and to pass on the immunity to nursing pigs—has been effective. This method of prevention is not without risk; caution and veterinary assistance are advised. This procedure should only be used if you are sure sows will be exposed at farrowing time and there is no danger of spreading the disease to neighboring farms. A TGE vaccine is now available and is considered to be the most effective method of prevention. Prevention must rely on strict sanitation and disinfection and possibly vaccination of pigs while they are suckling the sow. Planned infection of the sow herd and an all-in/all-out management of farrowing, nursery, and grower rooms can eliminate TGE from the herd. With the all-in/all-out management system, pigs are moved in groups through each of the various stages of development (e.g., farrowing, nursery, and growing-finishing stages). After each group is moved out of a barn, the facility is completely sanitized before a new group is allowed to be moved in.

Vesicular Exanthema (San Miguel Sea Lion Virus Disease)

Vesicular exanthema of swine (VES) is an acute, highly infectious viral disease that affects pigs, sea lions, fish, snakes, cattle, mink, horses, primates, and humans. Symptoms of this disorder include fever (105 to 106°F) and blisters on the snout, oral mucosa, soles of the feet, coronary band, and between the toes (see figures 9–41 and 9–42). Secondary infections of the blister drains may cause swelling of the mouth and feet areas, creating temporary lameness and reduced feed intake. The mortality rate is low, but occasionally sows may abort. In pigs, these clinical signs are indistinguishable from foot-and-mouth disease, vesicular stomatitis, and swine vesicular disease.

Originally the disease was confined to California. The disease became widespread in the 1950s. By 1959, the disease was eradicated from the United States through quarantine and slaughter. In 1972, the virus was identified in a sea lion in California and several other marine mammals. It has also been cultured from fish and other marine mammals as far north as Alaska, and recently outbreaks have been reported in other countries.

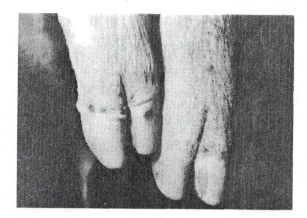

FIGURE 9–41 Feet blisters from vesicular exanthema. Courtesy of the U.S. Department of Agriculture.

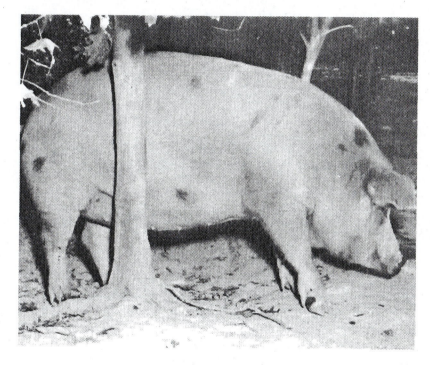

FIGURE 9–42 Hog affected with vesicular exanthema disease. Animals affected with VES may show blisters on the soft tissue just above the hoof and on the pads of the feet, causing lameness. Courtesy of the U.S. Department of Agriculture.

VES is a reportable disease. Prevention consists of cooking of garbage and fish feed (the principal source of the virus) and the use of a 2 percent sodium hydroxide disinfectant solution to clean the premises before restocking the herd. These preventive measures are essential because there is no known treatment.

ANIMAL WELFARE

Earlier in this text, we pointed out the need to explain methods of production to the nonagricultural community (the consumers of the products produced). The National Pork Producers Council (NPPC) has created its own animal welfare committee composed of leading veterinarians and animal scientists. Its purpose was to create a swine care handbook that would outline recommended methods of raising swine with considerations of their well-being. Much of the research used by this committee came from industry standards already in existence *(Pork Industry Handbooks).*

Items of concern included the following:

1. *Building ventilation:* Purpose is to remove foul air and replace it with fresh air, to remove moisture, and to remove odors and excess heat without causing drafts.
2. *Swine nutrition.*
3. *Housing design:* No wallowing in a muddy pen, biting insects, overheating, cold stress, or predators.
4. *Welfare and health intimately associated with peak efficiency:* Scientists have demonstrated that the abuse of animals in confinement systems with intensively raised livestock leads to lower overall production and profits. Still, many animal welfare activists challenge modern practices as being unnatural. Animal rights activists go a step further, demanding that animals be accorded the same moral protection as humans. Because most Americans are far removed from animal agriculture, the understanding and demands of animal husbandry, animal behavior, and animal needs are often based more on emotions than on facts.

The United Kingdom Farm Animal Welfare Council lists five freedoms as essential guarantees of animal welfare in any livestock operation:

- Freedom from hunger and thirst
- Freedom from discomfort
- Freedom from pain, injury, and disease
- Freedom from fear and distress
- Freedom to display normal behavior

Unfortunately, these rights are not yet being demanded for all humans.

SUMMARY

The American swine industry is concentrated heavily in the midsection of the country, the nation's Corn Belt. Although many animal activists have claimed that the grain fed to animals could be used to feed people, the swine industry is built on the premise that we have excess grains and grain by-products that can be marketed through swine. The availability of swine helps to alleviate what would otherwise be great fluctuations in grain prices when weather, soil, or other conditions suddenly change projected levels of grain production.

The pork industry has met consumer demands in producing a leaner and more wholesome meat and one with industry standards for animal welfare. Continued marketing and research in reproduction, nutrition, and food safety will enable the swine industry to maintain its strong foothold into the next century.

EVALUATION QUESTIONS

1. _____ is the breed of swine used as a household pet.
2. The most common metabolic disorder affecting baby pigs is _____ .
3. The most common preventive measure for question 2 is _____ .

4. _____ is caused by zinc deficiency.

5. Vitamin _____ is essential for confinement hogs.

6. _____ is caused by deficiency of Ca, P, or vitamin D in piglets.

7. _____ is also known as fly strike.

8. _____ is a disease transmitted to humans by an uncooked carcass.

9. The average length of estrus of a sow is _____ .

10. The average gestation of a sow is _____ .

11. A sow does not normally return to heat until after _____ .

12. Frozen boar semen is stored in _____ .

13. The breed of swine with an extra vertebra is _____ .

14. A _____ is a castrated boar.

15. The number of swine in the United States is _____ .

16. The number of swine operations in the United States is _____ .

17. _____ , _____ , and _____ are the top three swine states.

18. A group of hogs is referred to as a _____ .

19. Bacon comes from the _____ (part of the body).

20. The backfat on a hog should ideally not exceed _____ inches.

21. The loin-eye area, in square inches, of a desirable pork carcass should be _____ .

22. _____ results in colored patches on the skin, high temperatures, and low death rates.

23. _____ results in a deformed snout, irritation, and sneezing.

24. A 200-pound hog should dress out at _____ pounds.

25. _____ is a genetic defect in swine inherited as a simple or autosomal recessive trait. It is associated with heavily muscled animals that may suddenly die when exposed to stressful conditions. Their muscle is usually pale, soft, and exudative.

26. _____ is a term meaning litter-bearing animal.

27. _____ produces weaned pigs that are sold to be grown and finished on other farms.

28. Male pigs are castrated at age _____ .

29. SEW is the abbreviation for _____ .

30. Above average feed conversion of swine is _____ pounds feed or less per pound of gain.

31. _____ is commonly called thumps.

32. Baby pig disease, or _____ , often affects entire litters; the pigs shake uncontrollably, do not nurse, and are so cold their hair stands on end.

33. Vesicular exanthema in swine has the same symptoms as _____ disease in cattle.

34. _____ refers to pigs and herds that are free from specific diseases and pathogens.

35. The thigh of a hog prepared for food, or the hind leg of a swine from the hock upward on a live animal, is called _____ .

36. The cross section of the pork chop muscle, usually measured between the 10th and 11th ribs, is called _____ .

37. _____ is fat rendered from fresh pork tissue.

38. _____ is the meat from the cheeks of hogs.

39. The eight small and sharp teeth on the upper and lower corners of a piglet's mouth are called _____ .

40. _____ is a piglet of small size in relation to its litter mates.
41. _____ is a method using sound waves to measure depths of backfat and muscle.
42. NPPC is the abbreviation for _____ .
43. The procedure of early weaning (at 10 to 17 days) is called _____ .
44. _____ is another name for exudative epidermitis.
45. _____ is the most costly and difficult health problem of swine.
46. _____ is caused by mites.
47. _____ are also known as ascarids.
48. _____ is also known as swine fever.
49. _____ is the most popular breed of hog.
50. A very popular American breed of hog that is red in color (second in total U.S. numbers) is _____ .
51. _____ is an organization that created its own animal welfare committee composed of leading swine veterinarians and animal scientists.
52. _____ is one of the oldest breeds of swine; also known as Irish Grazers.
53. The term used to describe hog stance in which the hind legs are set well apart is _____ .
54. _____ , _____ , and _____ are three economic traits of pigs with a heritability greater than 50 percent.
55. The traditional method of permanent identification in hogs is _____ .
56. A castrated male hog showing no sex characteristics is called a _____ .
57. A castrated male hog showing sexual development is called a _____ .
58. Because of the importance of pork chops in retailing pork, the _____ area is often measured for genetic evaluation.
59. The five U.S. carcass grades are _____ , _____ , _____ , _____ , and _____ .
60. The U.S. carcass grade that indicates a high quality of lean, high yield of lean, and lowest percentage of backfat is _____ .
61. Five of the eight most common wholesale cuts of meat are _____ , _____ , _____ , _____ , and _____ .
62. EBV is the abbreviation for _____ .
63. _____ increases the number of eggs ovulated in gilts but not significantly in older females.
64. For maximum fertilization of the ova, breeding of sows takes place _____ hours after the onset of heat.
65. Pigs require _____ essential amino acids in their diet.
66. The most limiting of the amino acids in the diet of a pig is _____ .
67. Next to sodium chloride, _____ is the most needed mineral in a pig's diet.
68. The normal rectal temperature of a healthy hog is _____ °F.
69. _____ is caused by *Bordetella bronchiseptica*, the same organism responsible for kennel cough in dogs.
70. _____ is also known as pseudorabies.

71. _____ is the most common and destructive internal parasite in swine.

72. _____ causes scabies.

73. The feeding of large amounts of unsaturated fats causes _____ to develop in swine.

74. _____ is a stress-induced (genetic) disorder affecting heavily muscled pigs.

75. _____ is also known as San Miguel sea lion virus disease.

76. The five freedoms, as essential guarantees of animal welfare in any livestock operation, are _____ , _____ , _____ , _____ , and _____ .

DISCUSSION QUESTIONS

1. Briefly identify and list characteristics of the major breeds of U.S. swine.
2. What is the difference today between meat-type and lard-type hogs?
3. Describe the parts of the carcass of swine and the characteristics looked for in judging.
4. What is myiasis and how can you prevent it?
5. Identify the body parts of the pig.
6. Describe the following briefly: thumps, atrophic rhinitis, hog cholera, enterotoxemia, leptospirosis, and erysipelas.
7. Which major cuts of meat are found in swine?
8. How is measurement of growth evaluated using the new NPPC Maternal Sow Line National Genetic Evaluation Program?
9. What are the main advantages in using artificial insemination in swine?
10. What are the major differences in processing boar semen from that of bull and buck semen?

SWINE BREED ASSOCIATIONS

American Berkshire Association
PO Box 2346
West Lafayette, IN 47906
765-497-3618

American Landrace Association
PO Box 2340
West Lafayette, IN 47906-2340
765-497-3718

American Yorkshire Club
PO Box 2417
West Lafayette, IN 47906-2417
765-463-3593

Chester White Swine Records
PO Box 9758
Peoria, IL 61612-9758
309–691-0150

Hampshire Swine Registry
PO Box 2807
West Lafayette, IN 47906-2807
765-497-4123

National Hereford Hog Record
    Association
Route 1, Box 37
Flandreau, SD 57028
605-997-2116

National Spotted Swine Records
    Association
PO Box 9758
Peoria, IL 61612-9758
309–691-0150

Poland China Record Association
PO Box 9758
Peoria, IL 61612-9758
309–691-6301

United Duroc Swine Registry
PO Box 2397
West Lafayette, IN 47906
765-497-4084

SOURCES OF BOAR SEMEN

International Boar Semen
PO Box 538
Eldora, IA 50627
1-800-247-7877

Swine Genetics International
Rt. 1, Box 3
Cambridge, IA 50046
1-800-247-3958

Birchwood Swine Farm
Rt. 1, Box 130
West Manchester, OH 45382
513-678-9313

Stoney Creek Farms
Rt. 2, Box 262
Farmland, IN 47340
317-468-6099

CHAPTER

10

The Poultry Industry

OBJECTIVES

After completing the study of this chapter you should know:

- The historical, current, and future trends of the U.S. poultry industry.
- Where the poultry industry is located in the United States and its numbers and relative importance to U.S. agriculture.
- The importance of *The American Standard of Perfection,* classes, breeds, varieties, and strains of poultry (chickens, ducks, geese, and turkeys).
- The distinguishing characteristics of the most common U.S. poultry breeds.
- How the skeletal anatomy of the chicken differs from that of other livestock, including the connection of the skeletal system with that of the cardiorespiratory systems.
- Why roosters crow.
- How the vital statistics (temperature, pulse, and respiration) of the chicken compare with those of other livestock.
- The functions of the uropygial gland, preening, and molting.
- The unique features of the reproductive anatomy of the hen and functions of egg development and the stigma.
- The unique characteristic of determining the sex of offspring in poultry as compared with mammals.
- The quantitative and qualitative traits of the different types of poultry.
- Special considerations in feeding chickens, ducks, turkeys, and geese.
- The most common diseases of poultry and their symptoms (in brief), diagnosis, treatment, and prevention.
- Management techniques used to lessen cannibalism and hysteria in poultry operations.
- Management considerations in both the broiler and layer industries, including housing, incubation, feeding, the hatchery, and the use of lighting.
- Methods of grading broilers and other classes of poultry.
- Methods of grading eggs.

KEY TERMS

Antibiotics	Electrolytes	Qualitative traits
Antioxidants	Fryer	Quantitative traits
Arsenicals	Gander	Roaster
Blood spot	Goose	Rooster
Breed	Grit	Stag
Broiler	Hen	Stewing chicken
Candling	Molting	Strain
Capon	Necropsy	Syrinx
Class	Pellet binders	Uropygial gland
Coccidiostats	Poultry	Variety
Cock	Poults	Voice box
Cornish game hen	Preening	Xanthophylls

INTRODUCTION

The term *poultry* applies to chickens, turkeys, geese, ducks, pigeons, peafowl, and guineas. Chickens and turkeys dominate the world poultry industry. However, in parts of Asia, ducks are commercially more important than broilers (young chickens), and in some areas of Europe, there are more geese because they are more economically important than other poultry.

The earliest record of poultry dates back about 5,200 years in India. The Bible indicates that fowls of the air were domesticated very early in human history. In Egypt, chickens, have been bred in captivity, their eggs artificially incubated using rather crude ovens, and the chicks grown for sale of meat and eggs since 1400 B.C.E.

In 1607, the English brought chickens to America. During the past 200 years, more than 300 pure breeds and varieties of chickens have been developed. The commercial poultry industry did not develop until the early 1940s. From 1940 to 1960, a broiler was marketed at 16 weeks of age. Since that time, broilers have been attaining a market weight of 3 1/2 pounds in just 8 weeks' time, saving the producer close to 100 percent in production costs. Feed efficiency has also dramatically improved, contributing to the economical production of chicken.

In 1998, there were approximately 7,682,000 pullets (not of laying age), 320,694,000 laying hens (producing 79,717,000,000 eggs), 7,760,260,000 broilers, and 301,251,000 turkeys being raised in the United States. Total chicken population (excluding commercial broilers) at any one time in the United States was estimated at 424,094,000 head, valued at $1,139,984,000.[1] The financial contribution to U.S. agriculture with respective numbers of eggs and birds is outlined in table 10–1.

Whereas the overall consumption per capita of beef, lamb, and pork has decreased, consumption of poultry has increased markedly. Comparisons in world poultry consumption, by numbers, pounds consumed, and per capita consumption, are summarized in table 10–2.

The first small flocks of chickens were brought to Jamestown Colony in 1607. Although turkeys are native to the United States, the original wild turkeys

---

[1]National Agricultural Statistics Service (NASS), Livestock and Economics Branch, U.S. Department of Agriculture (USDA), Agricultural Statistics, 1999.

TABLE 10–1 U.S. Poultry Income and Numbers by Leading States (1997)

State	Broilers	Eggs	Turkeys
Arkansas	1,164,600,000	3,215,000,000	30,000,000
California	237,300,000	6,663,000,000	21,000,000
Georgia	1,182,800,000	4,867,000,000	175,000
Mississippi	720,000,000	1,327,000,000	—
North Carolina	665,000,000	2,794,000,000	53,500,000
Pennsylvania	135,200,000	5,900,000,000	11,600,000
U.S. total	7,760,260,000	77,532,000,000	301,251,000
U.S. income	$14.1 billion	$4.5 billion	$2.9 billion
Total U.S. income from poultry	$21.5 billion		

Source: National Agricultural Statistics Service (NASS), Livestock and Economics Branch, U.S. Department of Agriculture (USDA), Agricultural Statistics, 1999.

TABLE 10–2 World Poultry Numbers, Production, and Consumption

Nation	Number of Poultry (Head)	Production (lb.)	Per Capita Consumption (lb.)
China	3,136,000,000	14,200,000,000	—
United States	1,624,000,000	29,400,000,000	96
Russian Republic	623,000,000	3,000,000,000	—
Indonesia	667,000,000	—	—
Brazil	696,000,000	7,700,000,000	—
Hong Kong	—	—	89
Israel	—	—	83
World total	12,927,000,000	108,100,000,000	19

Source: Modified from U.S. Department of Agriculture (USDA), (www.fas.usda.gov) Foreign Agricultural Service (FAS), (World Poultry Situation), 1994 FAO Yearbook; and R.E. Taylor and T. G. Field, *Scientific Farm Animal Production,* 6th ed. (Upper Saddle River, NJ: Prentice-Hall, 1998).

do not resemble the commercially raised turkey meat birds of today. The first South Carolina White Leghorns were imported into the United States in 1828. The first record of American use of an incubator was in 1842, and the first commercial hatchery opened in 1870.

The American Poultry Association, Inc., was formed in 1873. It was the first American livestock organization. Its major goal has been to improve domestic poultry by maintaining purebred breeding stock. It publishes *The American Standard of Perfection,* which describes the recognized classes, breeds, and varieties of poultry and recognizes newly developed breeds and varieties.

Following are additional poultry historical highlights:

1889   Artificial light first used to stimulate egg production.
1909   Electric candler developed.
1935   Artificial insemination of poultry introduced.
1941   Newcastle disease first reported in the United States.
1942   Debeaker developed.
1956   Colonel Harland Sanders began franchise operations of Kentucky Fried Chicken.

1957   Poultry Products Inspection Act (PPIA) signed by President Eisenhower. The PPIA mandated compulsory inspection by the U.S. Department of Agriculture (USDA) of poultry and poultry products moving in interstate or foreign commerce. In the United States more than 90 percent of commercially processed poultry meat is inspected for wholesomeness by the USDA.

   The PPIA's functions include checking all poultry products to make sure they are (1) diagnosed for avian disease, (2) processed under sanitary conditions, and (3) inspected either directly or indirectly under supervision of a veterinarian. In addition, all carcasses, parts of carcasses, and meat products found to be adulterated or unwholesome are condemned and destroyed under supervision of an inspector, unless they are made wholesome and unadulterated by reprocessing.

1971   Newcastle disease discovered in southern California, requiring the destruction of 3,000,000 birds.
   Marek's vaccine approved.

1998   Preferential treatment was cited for poultry versus beef inspection; charges included[2]

   a. *Zero tolerance:* at slaughter, any visible contamination from feces or ingesta must be removed from beef, pork, or lamb carcasses. Poultry carcasses are not allowed to have any visible fecal contamination before they are immersed in water for chilling. The birds are inspected before they are eviscerated.

   b. *Cross contamination:* Poultry is chilled collectively in an ice water bath, and carcass-to-carcass contamination is possible. With beef, pork, and lamb, all fecal and ingesta contamination must be removed prior to processing to reduce the chance of cross contamination.

   c. *Wash versus trim:* Fecal contamination may be washed from poultry carcasses.

   d. *Added water:* Poultry can retain up to 8 percent water from washing and chilling.

   e. *Water temperature:* 180°F water must be used by meat processors to clean floors, walls, and equipment. No similar sanitation requirement exists for poultry.

   f. *Content of processed products:* The Meat and Poultry Acts require different minimum amounts of meat and poultry in processed foods. For example, baby food containing meat must have at least 61 percent meat or 43 percent poultry; meat pies must have 25 percent meat or 18 percent poultry.

      Besides competition at the grocery or meat stand, poultry and red meat producers are now complaining of unfair and unequal governmental regulations.

TAXONOMY AND BREEDS

In chapter 1 we briefly discussed taxonomy and its purpose. To summarize, classification is the arrangement of objects, ideas, or information into groups. The members of each group have one or more characteristics in common. Latin

---

[2]*California Farmer,* 281(11):WB9, 1998.

and Greek words are used in scientific classification. Every known animal and plant has a Latin or Greek name that has two parts (binomial system of nomenclature). An international commission of scientists establishes the rules for adopting scientific names. Seven chief groups make up the system of scientific classification: kingdom, phylum, class, order, family, genus, and species.

Kingdom: Animalia
Phylum: Chordata
Subphylum: Vertebrata
Class: Aves (Avian)
Subclass: Neornithes (true bird)
Order: Galliformes (short beak, feet for scratching)
Suborder: Carinatae (keel-like breastbone)
Family: Phasinanidae (for chickens)
Family: Meleagrididae (for turkeys)
Genus: *Gallus* (for chickens)
Genus: *Meleagris* (for turkeys)
Species: *Gallus domesticus* (most numerous domestic fowl in the world, descendant of *Gallus bankiva,* Wild Red Jungle fowl)
Species: *Meleagris galiopavo* (turkeys)

Poultry has some additional subdivisions that are not found in other livestock species. *Class* is an additional classification that describes a group of breeds originating in the same geographic area. Breeds of chickens are arranged first according to their class (Asiatic, Mediterranean, American, or English) and then alphabetically by breed within each class.

Breed is a term used to describe animals of common origin and having characteristics, such as body shape, skin color, and feathered or nonfeathered shanks, that distinguish them from other groups within the same species. Individuals within a breed, when mated, transmit these same features to their offspring.

Variety is a subdivision of a breed. Differentiating characteristics, including plumage color, comb type, and presence of a beard and muffs, distinguish one variety from another within a breed.

The **strain** is a family or breeding population within a variety that possesses common traits. Many commercial strains have been developed for specific purposes. Today, the commercial poultry industry is based primarily on strain crosses. However, foundation breeders are constantly looking for additional material for new gene pools.

In the United States, it is evident that there is less concern over maintaining purebred strains in chickens than there is in producing large animals. *The American Standard of Perfection* recognizes more than 300 pure breeds and varieties of chickens. Only a handful of these breeds seem to be of much commercial importance in the United States. In this chapter we summarize basic information on the most common and interesting breeds of chickens.

The National Poultry Improvement Plan (NPIP) is responsible for supervision of breeding practices and techniques in the United States to develop the high-quality, high-producing form of poultry in existence today. The NPIP was started in 1935 and administered jointly by the USDA and an official of each state agency. Coordination of the plan is the responsibility of the USDA. The objectives are to improve production and market qualities of chickens and to reduce losses from diseases. Associated with the disease control section is a

Breed animals of common origin and having characteristics that distinguish them from other groups within the same species

Variety subdivision of a breed used to describe poultry

Strain family or breeding population within a variety that possesses common traits

TABLE 10–3	Important Characteristics of Breeds of Chickens						
	Standard Weight (lb.)						
Breed	**Cock**	**Hen**	**Cockerel**	**Pullet**	**Skin Color**	**Egg Color**	**Varieties**
Leghorns	6	4.5	5	4	Yellow	White	12
Plymouth Rocks	9.5	7.5	8	6	Yellow	Brown	7
Rhode Island Reds	8.5	6.5	7.5	5.5	Yellow	Brown	2
New Hampshire	8.5	6.5	7.5	5.5	Yellow	Brown	0
Cornish	10.5	8	8.5	6.5	Yellow	Brown	4
Ross Breeder	10.75	8.5	9.25	7.5	Pink	White or brown	2

Source: L.S. Shapiro, *Principles of Animal Science*, 3rd ed. (Simi Valley, CA: Ari Farms, 1997).

program of blood testing to see if any breeders are carriers of any of four diseases associated with hatchery and breeder flocks (pullorum disease, fowl typhoid, *Mycoplasma gallisepticum* infection, and *M. synoviae* infection). Divisions of the plan are set forth in the USDA publication *The National Poultry Improvement Plans*, available from the Animal Plant Health Inspection Service—Veterinary Services (USDA/APHIS-VS), National Poultry Improvement Plan, Room 828 FB, Hyattesville, MD 20782. With the intensification of the poultry industry, traditional breeds have lost their commercial importance. See table 10–3 for a summary of breed characteristics. The vast majority of chickens in the United States are hybrids of one kind or another. Strain crosses, breed crosses, or crosses between inbred lines make up the bulk of modern poultry. Interestingly, Single-Comb White Leghorn is only one of several varieties of Leghorns but the only one used for commercial egg production.

Leghorns

Leghorns come from the city of Leghorn in Italy. There are 12 common varieties. Their primary use is for egg laying. They are the most common egg layer in the United States, averaging 240 to 250 eggs per year per bird. They are also the most numerous breed in America (see figure 10–1).

Plymouth Rocks

There are seven common varieties of Plymouth Rocks. This breed is used for both meat and eggs. White Plymouth Rock females are used as the female side of most of the commercial broilers produced today. The breed originated in America during the latter part of the nineteenth century (see figures 10–2 and 10–3).

Rhode Island Reds

Two main varieties exist of Rhode Island Reds: Single Comb and Rose Comb varieties. Rhode Island Reds are used as dual-purpose, medium-heavy fowl and are used more for egg production than for meat production because of their dark-colored pin feathers and good rate of lay. Today a large percentage of

FIGURE 10–1
White Leghorns.
Reprinted with permission from J. Blakely and D. H. Bade, *The Science of Animal Husbandry,* 6th ed. (Englewood Cliffs, NJ: Prentice-Hall, 1994).

FIGURE 10–2
White Plymouth Rocks.
Reprinted with permission from J. Blakely and D. H. Bade, *The Science of Animal Husbandry,* 6th ed. (Englewood Cliffs, NJ: Prentice-Hall, 1994).

FIGURE 10–3
Barred Plymouth Rocks.
Reprinted with permission from J. Blakely and D. H. Bade, *The Science of Animal Husbandry,* 6th ed. (Englewood Cliffs, NJ: Prentice-Hall, 1994).

FIGURE 10–4
Rhode Island Reds.
Reprinted with permission
from J. Blakely and D. H.
Bade, *The Science of
Animal Husbandry*, 6th
ed. (Englewood Cliffs, NJ:
Prentice-Hall, 1994).

FIGURE 10–5
New Hampshires.
Reprinted with permission
from J. Blakely and D. H.
Bade, *The Science of
Animal Husbandry*, 6th
ed. (Englewood Cliffs, NJ:
Prentice-Hall, 1994).

the commercial brown-egg layers are the result of crossing the Rhode Island Reds with the Barred Plymouth Rocks. Rhode Island Reds were developed in Massachusetts and Rhode Island (see figure 10–4).

New Hampshire

A dual-purpose chicken, the New Hampshire was selected more for meat production than for egg production. New Hampshire chickens dress a nice, plump carcass as either a broiler or a roaster. They were developed in New Hampshire and Massachusetts by intensive selection for rapid growth, fast feathering, early maturity, and vigor out of the Rhode Island Red breed (see figure 10–5).

FIGURE 10–6
Dark Cornishes.
Reprinted with permission
from J. Blakely and D. H.
Bade, *The Science of
Animal Husbandry,* 6th
ed. (Englewood Cliffs, NJ:
Prentice-Hall, 1994).

FIGURE 10–7 The
Ross male weighs 0.3
pound at 1 week of
age and 10.75 pounds
at 65 weeks.
Courtesy of Ross
Breeders, Huntsville, AL.

Cornish

The Cornish breed was developed as the ultimate meat bird in Cornwall,
England. Cornish hens have a very wide and muscular breast and short legs.
Their meat quality is considered very desirable, but their few small eggs do not
make them a sought-after egg bird (see figure 10–6).

Ross

The Ross male (see figure 10–7) is the industry standard for broiler growth rate,
feed conversion, and meat yield. The Ross 308 female (figure 10–8) crossed
with the Ross male produces a broiler with rapid growth rate, excellent feed

FIGURE 10–8
Ross female.
Courtesy of Ross
Breeders, Huntsville, AL.

FIGURE 10–9
Ross broiler.
Courtesy of Ross
Breeders, Huntsville, AL.

efficiency, and good meat yield; it is typically used for rotisserie, fast food, deboning, or a mix of products. The Ross 508 female crossed with the Ross male produces a broiler (figure 10–9) with maximum white meat (breast meat and wings) yield; it is typically used by broiler integrators that are primarily deboning white meat for additional revenue.

ANATOMY AND PHYSIOLOGY OF THE CHICKEN

Skeleton

The skeleton of the chicken is shown in figure 10–10. The bones of almost all fowl are pneumatic (hollow). The hollow bones are both strong and lightweight for purposes of flying.

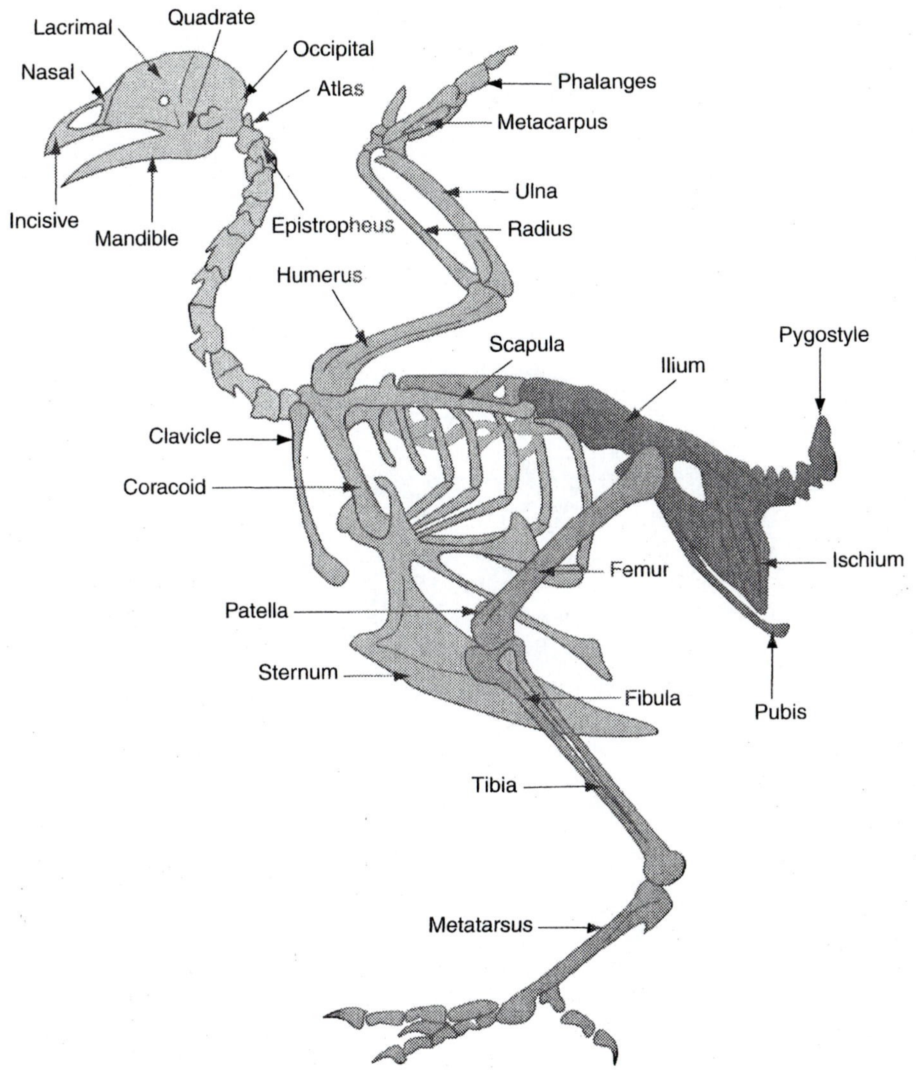

FIGURE 10-10
The skeleton of a fowl.

The hollow spaces connect with the respiratory system, making it possible for a bird with a broken wing to actually breathe through the wing, a fact about wounded birds long noted by hunters. Twelve percent of the bone structure in chickens is of the unique medullary type of bone. This type of bone has a network of tiny spines that lace the hollow structure, along with marrow, which serves the wild bird as a substance for egg formation when dietary calcium levels are low. In general, poultry have fewer bones than found in mammals, and many of them are fused together. In addition, the bones of chickens have high mineral content, and the pneumatic bones are filled with air instead of marrow.

Muscles

Although there are smooth muscles in the intestine and cardiac muscles, skeletal muscle is the most important for poultry producers. All muscles work in pairs or are antagonistic to each other. The breast is the largest skeletal muscle because of the need for flight in the wild bird. Through genetic selection the breast muscle is larger than normal, which hinders flight of the bird. The large breast muscle in turkeys has made them incapable of breeding except through the use of artificial insemination.

Birds have red muscle and white muscle, corresponding to dark meat and light meat, respectively. This differentiation in color is due to the myoglobin contained in red muscle. Myoglobin is the oxygen-carrying red pigment in the muscle of the chicken. Leg and thigh muscles make up the dark meat. Dark meat comes from muscles that do more work. In nonflying birds the leg and thigh muscles do more work and hence are darker (contain more blood capillaries, blood, pigment, and fat). In flying birds, movement of the wing requires a great amount of energy; the breast muscles of these animals are therefore darker (contain more blood capillaries, fat, pigment, and so forth). Birds that paddle and fly (ducks) have all dark meat.

Digestive System

The monogastric (simple stomach) digestive system of the fowl is divided into the various parts shown in figure 10–11. Chapter 4 briefly describes the function of each part of the digestive tract. Information in that chapter will provide the basic knowledge necessary to understand the discussion of introductory poultry nutrition in this chapter. The order or flow pattern of ingested feed through the avian digestive tract is summarized in figure 10–12.

FIGURE 10–11
The digestive system of a fowl.

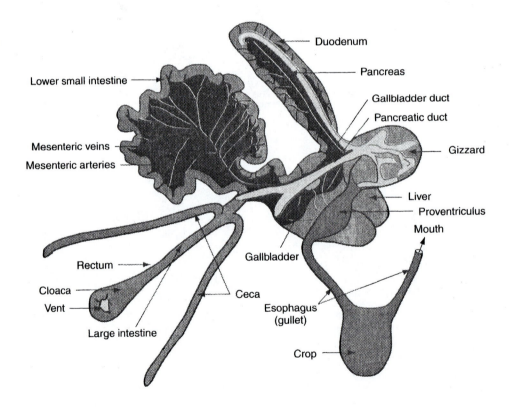

FIGURE 10–12
Flow pattern of feed through the avian digestive tract.

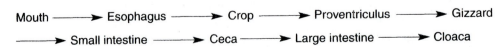

Cardiorespiratory Systems

The respiratory system consists of the lungs, which are relatively immobile, four pairs of air sacs, and one median air sac. Nine air sacs are located in the front of the body cavity near the clavicle, and extending into the hollow core of each humerus bone is a single air sac. The air sacs are given names based on the pneumatic bones or area of the body they enter (two cervical, two anterior thoracic, one interclavicular, two abdominal, and two posterior thoracic). The lungs of birds are relatively small and inexpansible. Birds have no diaphragm. Breathing is accomplished by muscular action of the entire thoracic and abdominal regions. It occurs primarily due to pressure changes in the air sacs rather than action of the lungs themselves. These four pairs of air sacs connect to a single median sac in the thorax. The entire system connects the lungs, passages, sacs, and hollows of the bones to create an interrelated respiratory system. The respiratory apparatus is important not only for breathing but also for losing heat through evaporation. Because they lack sweat glands, when the temperature exceeds 80°F chickens have to rely on evaporative cooling, which takes place through the air sacs and lungs. In addition, breathing creates a buoyancy to body air movement into and out of the lungs, allowing birds to float (even if only momentarily).

The *syrinx (voice box)* is located at the posterior end of the trachea. Voice is produced by air pressure on a sound valve and is modified by muscle tension. Hens do not crow because of female hormonal effects and the absence of the male hormone testosterone.

The Skin and Feathers

The skin of poultry is without glands except for the **uropygial gland** (oil gland, preen gland), which secretes oil used by the chicken to "dress" the feathers with a protective coating through the act of **preening** (combing the feathers with the beak). This gland is located at the base of the tail. The bird squeezes oil from the gland and spreads it over the body feathers during preening. Not all birds have these oil glands. Pigeons and parrots, for example, lack them completely. Poultry have many types of feathers, all originally used to aid in flight, to protect, and to provide warmth. Noting the types of feathers is not relevant to our discussion except in the case of the annual renewal (**molting**), which is an expense to the bird's net physical production. A general molt in hens involves tail and flight feathers. Molting is related to reproduction in that the ovary remains active until the time of molting. In the female, this feather replacement signals the end of egg production for the season. Molting in the male does not take place all at one time, as it does in the female. The male continually molts a few feathers; unlike the female, its sexual activity is not diminished by molting.

▓ **Uropygial gland** found in many birds and aids in the waterproofing of their feathers; also known as preen gland

▓ **Preening** combing the feathers with the beak

▓ **Molting** process of losing feathers from the wings and body

Chicken Vital Signs

Chickens have a relatively high body temperature of 104 to 109°F (40 to 43°C). Their hearts are also proportionally larger in size and beat faster, pumping more blood per unit of time than the typical mammalian heart. The basal heart rate of chickens is 300 beats per minute compared with a human's average rate of 60 to 70 beats per minute. Under conditions of stress, the bird's heart rate can climb to 1,000 beats per minute.

REPRODUCTIVE SYSTEM AND THE EGG

Please review the anatomy and physiology of the hen and rooster reproductive systems in chapter 3. The information will assist you in understanding the special reproductive management required for poultry as compared with that required for other farm animal species.

The egg contains all the nutrient material necessary to create and maintain early life. The hen creates an egg every 24 to 26 hours. When the ovum is released, hormones trigger the follicle to rupture along a line devoid of blood vessels, the stigma. Because this stigma is free of blood vessels, normally the yolk is free of hemorrhage. On occasion a small blood vessel may be ruptured, causing blood to appear on the yolk as a *blood spot.*

Classifying Baby Chicks

In hatcheries that are producing replacement pullets, disposal of surplus cockerels can be a major problem. Determining the sex of baby chicks as early as possible is a requirement to sound economic management decisions. Baby chicks can be classified by looking for the rudimentary copulatory organs in the vents of the baby chicks. Other methods of sexing baby chicks include the following:

- *Machine sexing:* Specialized sexing machines (proctoscopes) are used to see the sexual organs (testes) of the chicks.
- *Color sexing:* Color pattern is sex linked. The genes for silver and gold are also sex linked. Day-old female chicks possessing these genes are gold or buff colored compared with males, which are generally light yellow or white.
- *Feather sexing:* Rate of feathering is sex linked. At hatching time, the female chick's feathers are longer than those of the male chick.

Puberty

We discussed puberty in mammals in chapter 3. Sexual maturity in chickens can be reached by 21 weeks of age. Maturity will be delayed if the birds are subjected to reduced light (8 to 9 hours per day instead of 16 hours). Once chickens have reached puberty, sorting of birds by sex is important. The normal stocking rate for males to females, under natural mating systems, is 1 to 12 for chickens. Mating should be allowed for 10 to 14 days to be relatively certain of maximum fertilization in mass-mated flocks. This system permits males to mate with all females at least two to three times. Most males are not good breeders past 3 years of age.

In the beginning of this text we discussed sex determination in mammals, in which the male determines the sex of the offspring, and chromosome numbers for each species. In poultry, the normal chromosome number is 39 pairs, and sex is determined by the female. We symbolize the sex chromosomes in birds as follows:

Female = ZW

Male = ZZ

Quantitative versus Qualitative Traits in Poultry

From our discussion of qualitative traits in chapter 2, we learned that these traits are relatively simple and generally have higher heritability than do the quantitative traits. **Qualitative traits** include such characteristics as color, comb type, abnormalities, and sex-linked characteristics. The **quantitative traits** are expressed as the result of several genes working together; hence, there is no possibility of making a simple genetic cross to improve the likelihood of any one desired trait. Constant selection over several generations using culling methods and breeding to superior birds will increase the possibility of phenotypic expression of quantitative traits. Examples of quantitative traits and their respective inheritance percentages are fertility (5 percent), hatchability (10 percent), egg production (15 percent), and egg characteristics (egg shape is 60 percent heritable). The difference between the heritability percentages and 100 percent is environment (of which nutrition and management are the main factors).

> ▩ **Qualitative traits** traits controlled by one or a few pairs of genes and whose phenotypes (e.g., coat, color, and size) are clearly seen

> ▩ **Quantitative traits** traits usually controlled by many pairs of genes and whose phenotypic expressions (e.g., milk production and fertility) are more varied and influenced more profoundly by environmental and other factors than are qualitative traits

NUTRITION

Please review digestive anatomy and physiology of the bird and basic nutrition from chapter 4. There is from time to time an outcry against feeding livestock grains that could be used directly by people. There is a starving population somewhere in the world at all times. The problem, however, is rarely that we are feeding grain to livestock that could be fed to humans. The problem is a matter of payment. Who is going to buy this grain? Generally speaking, it is only when we have a surplus of grain production and its by-products that we feed these excesses to livestock, including poultry.

Feed Ingredients

Corn and soybean meal are generally produced in excess throughout the world and form the backbone of modern poultry rations. In general the major sources of other feeds are meat-packing by-products, fish meal, wheat, barley, millet, sorghum, rice, and corn-milling by-products. Others that are often used include yeast, distillery by-products, milk by-products, alfalfa leaf meals, and synthetic ingredients.

Corn is the most important grain used in poultry rations; sorghum grains (milo) rank second, and wheat ranks third. Both animal and vegetable protein supplements are used for protein in poultry rations. Common calcium supplements include ground limestone, crushed oyster shells, and bonemeal. The demand for calcium in egg-laying hens is extremely high. The average hen produces 250 (2-ounce) eggs per year and thus requires approximately 1.25 pounds of calcium for the shell only. This requirement is about 25 times the amount of calcium in the hen's skeleton. Most laying-hen rations contain from 3 to 4 percent calcium to meet this requirement.[3] Phosphorus is also supplied in the form of bonemeal and dicalcium phosphate.

---

[3]M. O. North and D. D. Bell, *Commercial Chicken Production Manual*, 4th ed. (New York, NY: Chapman & Hall, 1990).

Salt (sodium chloride) is a mineral supplement that should not exceed 0.5 percent of a ration. Too much salt causes an increase in water consumption in poultry and creates the undesirable management problem of very wet feces.

▓ **Antibiotics** class of drugs produced by living microorganisms that inhibit the growth of other microorganisms

Some nonnutritive additives in poultry rations include **antibiotics** (promote growth), *arsenicals* (promote growth and improve yellow color of the skin and shanks), *xanthophylls* (impart yellow color to the fat deposits, skin, and egg yolks), *antioxidants* (protect fat-soluble vitamins and prevent fat rancidity in the feed), *coccidiostats* (suppress coccidiosis), *electrolytes* (help regulate osmolarity of some body fluids), *grit* (helps the gizzard grind feed materials), and *pellet binders* (increase hardness of pellets to prevent crumbling). Many of these additives are regulated by the U.S. Food and Drug Administration (FDA).

Essential Nutrients

Chickens are extremely susceptible to deficiencies of every sort under the modern system, which keeps them caged or housed away from natural pasture or barnyard where many essential elements are consumed by accident. In general, the essential nutrients for poultry are the same as those for other classes of livestock previously discussed (water, carbohydrates, fats, proteins, minerals, and vitamins). Poultry require fresh, clean water at all times because of the high water content of their eggs and meat. Carbohydrates in poultry rations are generally provided by feeding high levels of grains (corn or sorghum). Fats are often added at levels of 3 to 8 percent to increase the dietary concentration of energy and thus promote fast growth or high energy levels for broiler production. Proteins are used in both the animal and vegetable form to maintain a balanced diet of essential amino acids. Minerals are required at adequate levels. Vitamins are needed for adequate health, growth, and reproduction. Table 10–4 gives the minimum nutrient requirements of Leghorn-type chickens for growth, laying, and breeding. Table 10–5 lists the minimum nutrient requirements of broilers.

Recommended vitamin and mineral levels for chickens are also included in these tables. Because poultry rations are highly concentrated, the carbohydrate content is normally sufficient for adequate growth; if it is not, it can be easily modified by adding fat to increase the energy level. Examples of typical broiler, replacement pullet, and layer and breeder hen diets are given in tables 10–6, 10–7, and 10–8, respectively.

About two-thirds of the total cost of producing eggs and meat from chickens is the cost of feed. Although many people in the world still feed relatively small numbers of birds on the farm by hand, for the most part poultry feeding has become very mechanized, highly specialized, and quite complex. Balancing rations of poultry is thus beyond the scope of this introductory text on animal science.

Special Considerations in Feeding Birds Other than Chickens

Turkeys.   Turkey poults have very poor eyesight and, for the first week or so, may not find food and water troughs and thus die of starvation. Bright strip lights are generally fitted over the troughs, and glass marbles are often used in

TABLE 10–4	Nutrient Requirements of Leghorn-Type Chickens (as Percentage or Amount per Kilogram of Diet)

Energy Base in kcal ME/kg Diet[a]	Growing			Laying		Breeding
	0–6 wks 2,900	6–14 wks 2,900	14–20 wks 2,900	2,900	Daily per Hen, mg[b]	2,900
Protein, %	18	15	12	14.5	16,000	14.5
Arginine, %	1.00	0.83	0.67	0.68	750	0.68
Glycine and Serine, %	0.70	0.58	0.47	0.50	550	0.50
Histidine, %	0.26	0.22	0.17	0.16	180	0.16
Isoleucine, %	0.60	0.50	0.40	0.50	550	0.50
Leucine, %	1.00	0.83	0.67	0.73	800	0.73
Lysine, %	0.85	0.60	0.45	0.64	700	0.64
Methionine + cystine, %	0.60	0.50	0.40	0.55	600	0.55
Methionine, %	0.30	0.25	0.20	0.32	350	0.32
Phenylalanine + tyrosine, %	1.00	0.83	0.67	0.80	880	0.80
Phenylalanine, %	0.54	0.45	0.36	0.40	440	0.40
Threonine, %	0.68	0.57	0.37	0.45	550	0.45
Tryptophan, %	0.17	0.14	0.11	0.14	150	0.14
Valine, %	0.62	0.52	0.41	0.55	600	0.55
Linoleic acid, %	1.00	1.00	1.00	1.00	1,100	1.00
Calcium, %	0.80	0.70	0.60	3.40	3,750	3.40
Phosphorus, available, %	0.40	0.35	0.30	0.32	350	0.32
Potassium, %	0.40	0.30	0.25	0.15	165	0.15
Sodium, %	0.15	0.15	0.15	0.15	165	0.15
Chlorine, %	0.15	0.12	0.12	0.15	165	0.15
Magnesium, mg	600	500	400	500	55	500
Manganese, mg	60	30	30	30	3.30	60
Zinc, mg	40	35	35	50	5.50	65
Iron, mg	80	60	60	50	5.50	60
Copper, mg	8	6	6	6	0.88	8
Iodine, mg	0.35	0.35	0.35	0.30	0.03	0.30
Selenium, mg	0.15	0.10	0.10	0.10	0.01	0.10
Vitamin A, IU	1,500	1,500	1,500	4,000	440	4,000
Vitamin D, ICU	200	200	200	500	55	500
Vitamin E, IU	10	5	5	5	0.55	10
Vitamin K, mg	0.50	0.50	0.50	0.50	0.055	0.50
Riboflavin, mg	3.60	1.80	1.80	2.20	0.242	3.80
Pantothenic acid, mg	10.0	10.0	10.0	2.20	0.242	10.0
Niacin, mg	27.0	11.0	11.0	10.0	1.10	10.0
Vitamin B_{12}, mg	0.009	0.003	0.003	0.004	0.00044	0.004
Choline, mg	1,300	900	500	?	?	?
Biotin, mg	0.15	0.10	0.10	0.10	0.011	0.15
Folacin, mg	0.55	0.25	0.25	0.25	0.0275	0.35
Thiamin, mg	1.8	1.3	1.3	0.80	0.088	0.80
Pyridoxine, mg	3.0	3.0	3.0	3.0	0.33	4.50

Source: Reprinted with permission of P. R. Cheeke, *Applied Animal Nutrition,* 2nd ed. (Upper Saddle River, NJ: Prentice-Hall, 1999).

[a]These are typical dietary energy concentrations.

[b]Assumes an average daily intake of 110 g of feed/hen daily.

TABLE 10–5	Nutrient Requirements of Broilers (as Percentage or Amount per Kilogram of Diet)		
Energy Base is 3,200 kcal ME/kg Diet[a]	Starter, 0–3 wks	Grower, 3–6 wks	Finisher, 6–8 wks
Protein, %	23.0	20.0	18.0
Arginine, %	1.44	1.20	1.00
Glycine + Serine, %	1.50	1.00	0.70
Histidine, %	0.35	0.30	0.26
Isoleucine, %	0.80	0.70	0.60
Leucine, %	1.35	1.18	1.00
Lysine, %	1.20	1.00	0.85
Methionine + cystine, %	0.93	0.72	0.60
Methionine, %	0.50	0.38	0.32
Phenylalanine + tyrosine, %	1.34	1.17	1.00
Phenylalanine, %	0.72	0.63	0.54
Threonine, %	0.80	0.74	0.68
Tryptophan, %	0.23	0.18	0.17
Valine, %	0.82	0.72	0.62
Linoleic acid, %	1.00	1.00	1.00
Calcium, %	1.00	0.90	0.80
Phosphorus, available, %	0.45	0.40	0.35
Potassium, %	0.40	0.35	0.30
Sodium, %	0.15	0.15	0.15
Chlorine, %	0.15	0.15	0.15
Magnesium, mg	600	600	600
Manganese, mg	60.00	60.0	60.0
Zinc, mg	40.0	40.0	40.0
Iron, mg	80.0	80.0	80.0
Copper, mg	8.0	8.0	8.0
Iodine, mg	0.35	0.35	0.35
Selenium, mg	0.15	0.15	0.15
Vitamin A, IU	1,500	1,500	1,500
Vitamin D, ICU	200	200	200
Vitamin E, IU	10	10	10
Vitamin K, mg	0.50	0.50	0.50
Riboflavin, mg	3.60	3.60	3.60
Pantothenic acid, mg	10.0	10.0	10.0
Niacin, mg	27.0	27.0	11.0
Vitamin B_{12}, mg	0.009	0.009	0.003
Choline, mg	1,300	850	500
Biotin, mg	0.15	0.15	0.10
Folacin, mg	0.55	0.55	0.25
Thiamin, mg	1.80	1.80	1.80
Pyridoxine, mg	3.0	3.0	2.5

Source: Reprinted with permission of P. R. Cheeke, *Applied Animal Nutrition,* 2nd ed. (Upper Saddle River, NJ: Prentice-Hall, 1999).
[a]These are typical dietary energy concentrations.

the feed to catch the light and attract the birds' attention. Pecking at the marbles encourages them to feed. Some growers sprinkle feed on corrugated cardboard floors to make poults notice and peck at it.

Because of peculiar behavior in poults, granite grit must be used with caution. It takes 3 to 6 weeks to gradually introduce poults to grit and not have them

TABLE 10–6	Percentage Composition of Typical Broiler Diets		
Ingredient	**Starter, 0–3 wks**	**Grower, 3–6 wks**	**Finisher, 6–9 wks**
Ground yellow corn	55.10	62.75	67.50
Soybean meal	29.00	25.00	20.25
Meat and bonemeal	5.00	4.00	4.00
Fish meal (menhaden)	2.00	—	—
Dehydrated alfalfa meal	2.00	2.00	2.00
Limestone	0.25	0.70	0.60
Dicalcium phosphate	0.30	0.70	0.80
Stabilized fat	5.00	3.50	3.50
Trace mineralized salt	0.25	0.25	0.25
Vitamin premix	1.00	1.00	1.00
Methionine	0.10	0.10	0.10

Source: Reprinted with permission of P. R. Cheeke, *Applied Animal Nutrition,* 2nd ed. (Upper Saddle River, NJ: Prentice-Hall, 1999).

TABLE 10–7	Percentage Composition of Typical Diets for Replacement Pullets		
Ingredient	**Starter, 0–6 wks**	**Grower, 6–14 wks**	**Developer, 14–20 wks**
Ground yellow corn	61.25	59.00	61.60
Wheat middlings	4.00	3.00	6.80
Oats	8.00	18.00	20.00
Dehydrated alfalfa meal	2.50	2.50	2.50
Fish meal, menhaden	2.50	—	—
Meat and bonemeal	2.50	—	—
Soybean meal	16.30	14.40	6.00
Dicalcium phosphate	1.00	1.00	1.00
Limestone	0.50	0.75	0.7
Salt	0.25	0.25	0.25
Trace mineral premix	0.10	0.10	0.10
Vitamin premix	1.00	1.00	1.00
Methionine	0.10	—	—

Source: Reprinted with permission of P. R. Cheeke, *Applied Animal Nutrition,* 2nd ed. (Upper Saddle River, NJ: Prentice-Hall, 1999).

gorge on it. After about 6 weeks, grit may be fed free choice. Table 10–9 lists minimal nutrient requirements of turkeys. Table 10–10 outlines a typical turkey grower diet.

A high-energy turkey starter diet (about 28 percent protein) is fed for about 4 weeks. The protein content can be reduced 2 percent during each 4-week period until the bird reaches market weight at about 20 weeks of age.

Ducks and Geese. Ducks and geese are fed diets that are similar to those of chickens. Water and feed should be kept some distance apart to prevent ducks from wasting feed by alternating between the feeder and the waterer. Tables 10–11 and 10–12 outline the nutrient requirements of ducks and geese, respectively.

TABLE 10–8 Percentage Composition of Complete Layer and Breeder Diets

Ingredient	Phase 1 Layer Diet, 22–40 weks		Phase 1 Breeder Diet	Phase 2 Layer Diet (after 40 wks)		Phase 1 Breeder Diet
Ground yellow corn	60.2	66.75	55.3	58.7	26.7	66.3
Wheat shorts	—	—	—	5.0	—	—
Ground sorghum	—	—	—	—	38.0	—
Hominy feed	2.5	—	10.0	—	—	—
Stabilized grease	2.0	—	1.5	2.5	2.5	—
Soybean meal, 49% CP	21.0	12.5	15.0	20.5	14.5	14.5
Fish meal, 60% CP	—	1.5	3.5	—	—	2.5
Dried fish solubles	0.5	—	1.0	—	—	—
Meat and bonemeal, 50% CP	—	5.0	—	—	5.0	2.5
Corn distillers dried grains	2.5	5.0	5.0	—	—	2.5
Alfalfa meal, 17% CP	1.5	1.5	2.0	2.5	4.0	2.5
Dicalcium phosphate	1.5	—	1.0	1.5	—	0.5
Limestone	5.0	5.0	4.5	4.5	4.5	4.0
Oyster shell	2.0	2.0	2.0	4.0	4.0	4.0
Trace mineralized salt	0.25	0.25	0.25	0.25	0.25	0.25
DL-methionine	0.05	0.05	0.025	0.05	50.08	0.025
Vitamin premix	0.5	0.5	0.5	0.5	0.5	0.5

Source: Reprinted with permission of P. R. Cheeke, *Applied Animal Nutrition,* 2nd ed. (Upper Saddle River, NJ: Prentice-Hall, 1999).

TABLE 10–9 Nutrient Requirements of Turkeys (as Percentage or Amount per Kilogram of Diet)

		Age in Weeks							
Energy Base in kcal ME/kg Diet[a]	Males	0–4	4–8	8–12	12–16	16–20	20–24	Holding	Breeding Hens
	Females	0–4	4–8	8–11	11–14	14–17	17–20		
		2,800	2,900	3,000	3,100	3,200	3,300	2,900	2,900
Protein, %		28	26	22	19	16.5	14	12	14
Arginine, %		1.6	1.5	1.25	1.2	0.95	0.8	0.6	0.6
Glycine + serine, %		1.0	0.9	0.8	0.7	0.6	0.5	0.4	0.5
Histidine, %		0.58	0.54	0.46	0.39	0.35	0.29	0.25	0.3
Isoleucine, %		1.1	1.0	0.85	0.75	0.65	0.55	0.45	0.5
Leucine, %		1.9	1.75	1.5	1.3	1.1	0.95	0.5	0.5
Lysine, %		1.6	1.5	1.3	1.0	0.8	0.65	0.5	0.6
Methionine + cystine, %		1.05	0.9	0.75	0.65	0.55	0.45	0.4	0.4
Methionine, %		0.53	0.45	0.38	0.33	0.28	0.23	0.2	0.2
Phenylalanine + tyrosine, %		1.8	1.65	1.4	1.2	1.05	0.9	0.8	1.0
Phenylalanine, %		1.0	0.9	0.8	0.7	0.6	0.5	0.4	0.55
Threonine, %		1.0	0.93	0.79	0.68	0.59	0.5	0.4	0.45
Tryptophan, %		0.26	0.24	0.2	0.18	0.15	0.13	0.1	0.13
Valine, %		1.2	1.1	0.94	0.8	0.7	0.6	0.5	0.58
Linoleic acid, %		1.0	1.0	0.8	0.8	0.8	0.8	0.8	1.0
Calcium, %		1.2	1.0	0.85	0.75	0.65	0.55	0.5	2.25
Phosphorus, available, %		0.6	0.5	0.42	0.38	0.32	0.28	0.25	0.35

(continued)

| Energy Base in kcal ME/kg Diet[a] | Males Female | Age in Weeks | | | | | | Holding | Breeding Hens |
		0–4 0–4 2,800	4–8 4–8 2,900	8–12 8–11 3,000	12–16 11–14 3,100	16–20 14–17 3,200	20–24 17–20 3,300	2,900	2,900
Potassium, %		0.7	0.6	0.5	0.5	0.4	0.4	0.4	0.6
Sodium, %		0.17	0.15	0.12	0.12	0.12	0.12	0.12	0.15
Chlorine, %		0.15	0.14	0.14	0.12	0.12	0.12	0.12	0.12
Magnesium, mg		600	600	600	600	600	600	600	600
Manganese, mg		60	60	60	60	60	60	60	60
Zinc, mg		75	65	50	40	40	40	40	65
Iron, mg		80	60	60	60	50	50	50	60
Copper, mg		8	8	6	6	6	6	6	8
Iodine, mg		0.4	0.4	0.4	0.4	0.4	0.4	0.4	0.4
Selenium, mg		0.2	0.2	0.2	0.2	0.2	0.2	0.2	0.2
Vitamin A, IU		4,000	4,000	4,000	4,000	4,000	4,000	4,000	4,000
Vitamin D[b], ICU		900	900	900	900	900	900	900	900
Vitamin E, IU		12	12	10	10	10	10	10	25
Vitamin K, mg		1.0	1.0	0.8	0.8	0.8	0.8	0.8	1.0
Riboflavin, mg		3.6	3.6	3.0	3.0	2.5	2.5	2.5	4.0
Pantothenic acid, mg		11.0	11.0	9.0	9.0	9.0	9.0	9.0	16.0
Niacin, mg		70.0	70.0	50.0	50.0	40.0	40.0	40.0	30.0
Vitamin B_{12}, mg		0.003	0.003	0.003	0.003	0.003	0.003	0.003	0.003
Biotin, mg		0.2	0.2	0.15	0.125	0.100	0.100	0.100	0.15
Folacin, mg		1.0	1.0	0.8	0.8	0.7	0.7	0.7	1.0
Thiamin, mg		2.0	2.0	2.0	2.00	2.0	2.0	2.0	2.0
Pyridoxine, mg		4.5	4.5	3.5	3.5	3.0	3.0	3.0	4.0

Source: Reprinted with permission of P. R. Cheeke, *Applied Animal Nutrition,* 2nd ed. (Upper Saddle River, NJ: Prentice-Hall, 1999).

[a]These are typical ME concentrations for corn-soy diets. Different ME values may be appropriate if other ingredients predominate.

[b]These concentrations of vitamin D are satisfactory when the dietary concentrations of calcium and available phosphorus conform with those in this table.

TABLE 10–10 Percentage Composition of Turkey Grower Diets

| Ingredient | Age of Birds | | | | | |
	0–4 wks	5–8 wks	9–12 wks	13–16 wks	17–20 wks	20+ wks
Ground yellow corn	48.15	51.00	62.30	70.50	75.95	80.65
Soybean meal	35.65	37.00	26.60	18.90	13.75	10.00
Meat and bonemeal	7.50	7.50	7.50	7.50	6.45	4.00
Fish meal, menhaden	4.00	—	—	—	—	—
Dehydrated alfalfa meal	2.00	—	—	—	—	—
Stabilized fat	1.00	2.00	1.50	1.50	2.00	2.50
Dicalcium phosphate	—	0.35	0.50	0.20	0.50	1.20
Limestone	0.25	0.70	0.15	—	—	0.30
Salt	0.25	0.25	0.25	0.25	0.25	0.25
Trace mineral mix	0.10	0.10	0.10	0.10	0.10	0.10
DL-methionine	0.10	0.10	0.10	0.05	—	—
Vitamin premix	1.00	1.00	1.00	1.00	1.00	1.00

Source: Reprinted with permission of P. R. Cheeke, *Applied Animal Nutrition,* 2nd ed. (Upper Saddle River, NJ: Prentice-Hall, 1999).

TABLE 10-11	Nutrient Requirements of Ducks[a] (as Percentage or Amount per Kilogram of Diet)		
Energy Base is 2,900 kcal ME/kg Diet[b]	Starting (0–2 wks)	Growing (2–7 wks)	Breeding
Protein, %	22.0	16.0	15.0
Arginine, %	1.1	1.0	—
Lysine, %	1.1	0.9	0.7
Methionine + cystine,%	0.8	0.6	0.55
Calcium, %	0.65	0.6	2.75
Phosphorus, available, %	0.40	0.35	0.35
Sodium, %	0.15	0.15	0.15
Chlorine, %	0.12	0.12	0.12
Magnesium, mg	500	500	500
Manganese, mg	40.0	40.0	25.0
Zinc, mg	60.0	60.0	60.0
Selenium, mg	0.14	0.14	0.14
Vitamin A, IU	4,000	4,000	4,000
Vitamin D, ICU	220	220	500
Vitamin K, mg	0.4	0.4	0.4
Riboflavin, mg	4.0	4.0	4.0
Pantothenic acid, mg	11.0	11.0	10.0
Niacin, mg	55.0	55.0	40.0
Pyridoxine, mg	2.6	2.6	3.0

Source: Reprinted with permission of P. R. Cheeke, *Applied Animal Nutrition,* 2nd ed. (Upper Saddle River, NJ: Prentice-Hall, 1999).
[a]For nutrients not listed, see requirements for chickens as a guide.
[b]These are typical dietary energy concentrations.

TABLE 10-12	Nutrient Requirements of Geese[a] (as Percentage or Amount per Kilogram of Diet)		
Energy Base is 2,900 kcal ME/kg Diet[b]	Starting (0–6 wks)	Growing (After 6 wks)	Breeding
Protein, %	22.0	15.0	15.0
Lysine, %	0.9	0.6	0.6
Methionine + cystine, %	0.75	—	—
Calcium, %	0.8	0.6	2.25
Phosphorus, available, %	0.4	0.3	0.3
Vitamin A, IU	1,500	1,500	4,000
Vitamin D, ICU	200	200	200
Riboflavin, mg	4.0	2.5	4.0
Pantothenic acid, mg	15.0	—	—
Niacin, mg	55.0	35.0	20.0

Source: Reprinted with permission of P. R. Cheeke, *Applied Animal Nutrition,* 2nd ed. (Upper Saddle River, NJ: Prentice-Hall, 1999).
[a]For nutrients not listed, see requirements for chickens as a guide.
[b]These are typical dietary energy concentrations.

DISEASES AND PARASITES OF POULTRY

A healthy flock is attained through a total disease control program. Chickens will not grow or lay eggs up to potential if they have parasites or diseases. Reducing the chance of a disease outbreak is the key to maintaining health.

Treatment of diseases, as opposed to preventive medicine, is costly and often unsuccessful.

In this text *disease* is defined as any departure from normal health. Harmful microorganisms, nutritional deficiencies, and unusual environmental stress can cause disease. The discussion of diseases in this text is limited to problems caused by microorganisms, environmental conditions, external and internal parasites, and behavioral problems. The strategy of disease control includes stimulating the body's natural defense mechanisms (immunization), sanitation, quarantine, and good environmental management.

Diseases are transmitted horizontally and vertically. *Horizontal transmission* refers to the spread of disease by contact with litter; inhaling dust particles containing organisms; and infestation by vectors such as insects, wild birds, or parasites. Horizontal transmission is the most common form of disease transfer. However, some diseases are also transmitted through *vertical* means—that is, the passage of disease-producing agents from the dam to the offspring through the egg. A number of diseases are spread by both horizontal and vertical transmission (e.g., pullorum disease, lymphoid leukosis, and mycoplasmosis).

Poultry farmers should be aware of the deviation from normal health that signals the onset of disease. Often the first sign of a problem is a decrease in feed and water consumption. Other specific signs may include diarrhea, paralysis, respiratory difficulty, poor skin condition, bloody or unusually wet droppings, or other signs that distinguish healthy from unhealthy fowl. Many states have diagnostic laboratories that confirm the suspicion of a disease outbreak by the posting (**necropsy**) of a dead bird or the sacrifice of a bird for analysis. Birds that die, apart from the normal expected losses, should be preserved through refrigeration or packing in ice until a qualified veterinarian or diagnostic laboratory scientist can conduct a postmortem examination.

▧ **Necropsy** postmortem examination of an animal's internal organs

Diagnostic laboratories in regions throughout the United States recognize more than 80 separate diseases or parasitic problems of poultry. Inspectors reject millions of pounds of poultry for human consumption annually due to diseased birds that do not succumb to an outbreak but are a loss to the industry nonetheless. Leukosis, airsacculitis, and septicemia are some of the most important causes of rejection by poultry inspectors during the slaughtering procedures. *Septicemia* is a nonspecific condition referring to rejection of the carcass because of such things as anemia, edema, dehydration, or other indications of disease. Septicemia, however, does not specify the disease. *Airsacculitis* is a term used to describe symptoms seen with respiratory infection in general by chicken- and turkey-processing plant inspectors. Again, it does not define a specific disease.

In a successful laying flock, a 10 to 12 percent death rate is considered normal in an average production year. With broiler flocks, the normal maximum annual death loss is about 4 percent. Poultry farmers should take any death losses in excess of these two figures as a serious condition that should receive prompt attention.

Aspergillosis

Aspergillosis is a respiratory disease characterized by gasping and rapid breathing. It is mainly a disease of young chicks. Often there is a loss of appetite and an increase in thirst. Birds lose weight and exhibit some central nervous system (CNS) signs. A fungus (*Aspergillus fumigatus*) causes the condition. Outbreaks usually occur only when moist conditions in litter will support the growth of mold. Spores from the fungi enter the air and are breathed in by the birds, resulting in aspergillosis. There is no treatment. The disease is prevented by keeping feed and litter low in moisture content to prevent mold.

Avian Influenza

Avian influenza is a viral disease that affects the respiratory, enteric, and nervous systems of poultry. Its symptoms include coughing, sneezing, sinusitis, low egg production, diarrhea, edema of the face, and up to 90 percent mortality. It is transmitted by waterfowl and other wild and domestic species. Improved flock security, including protecting birds from contaminated clothing, crates, filler flats, equipment, or any other fomites that may have made contact with water used by waterfowl, is one primary means of control. There is no known treatment. In 1983–1984, there was an outbreak on the East Coast of the United States; it killed 17 million birds and cost $63 million to eradicate.

Coccidiosis

Coccidiosis is one of the oldest known and most destructive protozoan diseases of poultry. It is normally a controllable disease, but it is one of the most expensive to the poultry industry. Affected birds are obviously ill. The coccidia parasites are host specific but found worldwide in droppings that contaminate feed, dust, water, litter, and soil. Symptoms include decreased growth rate, severe diarrhea, bloody droppings, and a high mortality rate. The coccidia cause ulceration of the cells lining the intestine. Bacteria can then invade the damaged areas to cause infection. Intestinal bleeding shows up in the droppings. The coccidia spend 4 to 7 days in the bird before passing out in the feces. The external stage then starts. If the litter is sufficiently warm and moist, a process called sporulation takes place within 24 to 48 hours, and the infection repeats itself.

Coccidiosis is controlled by maintaining poultry at all times on wire floors to separate birds from droppings (see figure 10–13). Wire cages only prevent in-

FIGURE 10–13
Wire cages used to house laying hens help to prevent coccidiosis and greatly increase the efficiency of operation in the egg business.
Photo by author.

fection of birds not already exposed to coccidia. Hence, birds transferred from a litter floor to a wire cage should be checked and treated for coccidia. Anticoccidial drugs (lasalocid, monensin sodium, and oxytetracycline) can also be used to control coccidiosis. Virtually all commercial rations for broilers now routinely use an anticoccidial compound.

Curled-Toe Paralysis

Chicks with a vitamin B_2 (riboflavin) deficiency become lame with an ailment known as curled-toe paralysis. Supplementing all breeding-hen diets with vitamin B_2 easily prevents this nutritional disorder.

Duck Virus Enteritis (Duck Plague)

Duck plague is an acute, highly contagious, viral enteric disease of ducks, geese, and swans. Its symptoms include sudden death, high mortality rates, extreme thirst, droopiness, ataxia, soiled vents, watery or bloody diarrhea, and hemorrhages and necrosis of internal organs. There is no effective treatment. Preventive measures include maintaining birds in a disease-free environment and preventing contact with wild, free-flying waterfowl. A new live virus vaccine modified by chicken embryo is now available for domestic ducks.

Fowl Pox

Fowl pox is seen in many areas of the world but has been brought under control in commercial operations through the use of vaccines. Black, raised scabs on the comb, wattles, face, earlobes, shanks, and feet make this disease easy to recognize. Birds become very ill, egg production is reduced, growth is retarded, and hatchability and fertility are reduced. Mortality rates are high in the generalized or internal form.

The cause is a DNA virus. Although the signs are similar to those of chicken pox in people, the virus is not the same one, despite the name, and will not infect people. The disease is spread horizontally by other chickens and by mosquitoes.

Treatment of noninfected birds with a live virus vaccine protects birds against infection. Chickens and turkeys should be vaccinated well before egg production commences. However, routine vaccination is not recommended except in areas where fowl pox is known to be a threat. A practical treatment is to remove and isolate infected birds at the first sign of fowl pox and vaccinate the remainder of the flock. A more recently developed turkey pox vaccine has been developed to control pox in turkey flocks in which use of the fowl pox vaccine has been ineffective.

Fowl Typhoid

Signs of fowl typhoid are quite similar to those of pullorum disease. The cause is another of the salmonella organisms, *Salmonella gallinarum*. Although the drug furazolidone has been effective in reducing losses to this organism, good hatchery sanitation and elimination of infected adults are keys to control. Furazolidone cannot be fed to pullets older than 14 weeks of age or to laying hens. A vaccine is now available and is most effective if given when fowl are 9 to 10 weeks of age and before natural exposure occurs. The disease is spread by equipment, feed, people, and rats or can be spread from hen to chick through the egg. Symptoms include a high mortality rate (50 percent or more) preceded by dehydration; swollen, friable, bile-stained liver; anemia; and enteritis. Death occurs at any age and is not confined to young chicks as it is in the case of pullorum disease.

Infectious Bursal Disease

Infectious bursal disease (IBD) is a highly infectious viral disease of poultry. It used to be called Gumboro disease because it was first discovered in Gumboro, Delaware, in 1962. Symptoms include whitish diarrhea, inflamed vents, listlessness, nervousness, sleepiness, and dehydration. These signs may persist 5 to 7 days, and death losses may reach 30 percent. Mortality rates increase with age up to about 10 weeks.

The disease is transmitted horizontally only (no egg transmission). There is no effective treatment for IBD. Prevention is by improved sanitation and, on farms on which IBD has been identified, vaccination. A combined Marek's disease and IBD vaccine is currently being marketed in some areas.

Infectious Coryza

Signs of infectious coryza, a respiratory disorder, include discharges from the nostrils and eyes, sneezing, lowered egg production, and loss of appetite. Coryza is spread rather slowly but can be spread more quickly by contact with birds using the same drinking water. This condition is similar to the common cold in humans. Infectious coryza is caused by the bacterium *Hemophilus gallinarum*.

Treatment with drugs such as erythromycin and oxytetracycline and the use of various sulfonamides and sulfonamide-trimethoprim have been beneficial. However, medication is not the answer to control. Separation of affected birds, disposal of old hens at the end of the year, and isolation rearing in a controlled, clean environment are considered keys to prevention of infectious coryza. A killed, polyvalent bacterin is now available for growing birds. It must be given in two vaccinations, one at 8 to 10 weeks of age and a booster 4 weeks later.

Lice

There are about 40 species of lice, a whitish to brownish yellow, wingless, flattened insect known to infest poultry. These external parasites lower reproductive potential in males, lower egg production, irritate the skin, lower weight gain, and provide sites for secondary bacterial infections. Adult lice lay their eggs around the vent at the base of the feathers. Young lice appear transparent and reach full size about 2 weeks after hatching. They spend their entire life cycle on the host and generally spread to other chickens by direct contact. The biting and chewing of the insects on feathers, scales, and skin damage the host.

Control of lice by the use of approved pesticides is quite effective. Several applications or spraying with pyrethroids, carbaryl, coumaphos, malathion, or stirofos may be required to kill the insects that may have hatched since the time of the first application. Treatment should include the body of the bird as well as roosts and house. Young chicks placed in a clean house with no older birds should therefore remain free of infestation. Lice are the most prevalent external parasite in poultry.

Lymphoid Leukosis (Avian Leukosis, Big Liver Disease, Visceral Lymphomatosis)

Lymphoid leukosis is a viral (retroviral) disease that produces tumors in chickens 16 weeks of age and older. Nodular lymphoid tumors (liver, spleen, and bursa), precede death The disease is found in all but specific pathogen free (SPF) flocks.

Avian leukosis virus is primarily shed by the hen into the albumen or yolk. Infection occurs after onset of incubation. Congenitally infected chickens remain viremic for life and fail to produce neutralizing antibodies.

This disease is controlled by eradication of infection, standard disease control, and sanitation. These measures can keep chicken flocks free of the disease. Vaccination against tumor production has not proven successful. Infected hens can be identified by a blood test.

Marek's Disease (Range Paralysis)

Marek's disease is a viral (herpes virus) disease that affects chickens from 6 to 16 weeks of age. Chickens are the primary hosts of this acute neural leukosis. Tumor production and mortality continue through the lifetime of the flock. Additional symptoms include lameness, ataxia, and enlarged feather follicles (skin leukosis). Marek's disease is found worldwide in all chicken flocks except SPF flocks. It is highly contagious and is easily spread in dust or dander from infected chickens.

Control consists of a vaccine (effective if given early, before birds leave the hatchery). Shedding of the infectious virus can be reduced but not prevented by prior vaccination. There is no known treatment for this disorder.

Mites

Although mites are not the most prevalent external parasite of chickens, they are the most economically important one. Mites are smaller than lice and are barely visible to the unaided eye. The two most common species affecting poultry are the red mite and the northern fowl mite. Red mites spend part of their life cycle living and breeding in the cracks and crevices of poultry houses, feeding on poultry only during the night. Northern fowl mites are found on the bird and in the house at all times; they are the most common vector of caged hens in the United States. The chicken is damaged by the sucking of the mite, which lives primarily off the blood and lymph. The blood sucking usually causes scabs to form on the skin. Loss of blood, anemia, and carcass blemishes or condemnation are causes of poultry losses. Control of mites requires the application of approved pesticides to both house and bird to eliminate the pest from habitat and host.

Mycoplasmosis (Chronic Respiratory Disease)

Mycoplasmosis is a disease of the air sacs in which the sacs become filled with exudates (mucous fluid). The lungs harden, and breathing is obviously impaired. The death rate may be quite high. A reduced growth rate is evident during and shortly after recovery. Broilers that recover may be condemned at carcass inspection. The principal cause of the disease is the primitive organism

Mycoplasma gallisepticum, in association with other viruses and bacteria. The disease is transmitted both vertically and horizontally. Treatment of hatching eggs with antibiotics has proven successful in breaking the vertical transmission. Isolation and proper sanitation thereafter to establish mycoplasma-free breeding flocks, similar to SPF hog operations, prevent horizontal transmission. A live vaccine has been developed in the United States but may be used only with permission of the state veterinarian. A newer, nonpathogenic live vaccine has also been approved and should offer advantages of improved safety as well as protection from *M. gallisepticum.*

Newcastle Disease (Avian Pneumoencephalitis)

Newcastle disease is a viral disease found worldwide in domestic poultry and other birds. The disease affects chickens, turkeys, pheasants, and many other birds. The virus is present in exhaled air, respiratory discharges, feces, eggs laid during the clinical disease, and all parts of the carcass. Chickens are readily infected by aerosols and by ingesting contaminated water or feed.

Symptoms include respiratory or nervous signs such as gasping, coughing, drooping wings, dragging legs, twisting of head and neck, circling, depression, paralysis, watery albumen in eggs, and death (see figure 10–14). Death rates can be high in broilers. In 1971, more than 3,000,000 birds died of Newcastle disease in southern California. In laying hens, it is not uncommon to observe a complete failure of the flock to produce eggs. Although egg production may recover to normal levels following an outbreak, the eggs are often of poor quality for some time thereafter.

Transmission of Newcastle disease is by horizontal means. The disease was first encountered near Newcastle, England. Newcastle disease is controlled primarily by inoculation. Vaccines may be used in the drinking water, as a spray

FIGURE 10–14
Newcastle disease is characterized by muscular incoordination and a particular paralysis of the neck muscles, as shown. Courtesy of the U.S. Department of Agriculture.

or dust to treat the entire flock, in wing web inoculation, or through other direct methods. Vaccination should start after parental immunity subsides (usually between days 14 and 21). Some researchers recommend vaccinating at 1 day of age with maternal antibodies and then revaccinating between days 14 and 21. Repeated vaccinations are required to protect chickens throughout their lives. There is no effective treatment for Newcastle disease.

Perosis (Slipped Tendon)

Perosis is another nutritional disorder caused by a manganese deficiency. The condition may also be caused by deficiencies in choline, biotin, nicotinic acid, or folacin. The disease affects young chicks with symptoms of a twisting of the leg and lameness.

Poult Enteritis Mortality Syndrome

Poult enteritis mortality syndrome (PEMS) is a transmissible, infectious, enteric disease of turkey poults. It affects birds 7 to 28 days of age and has a 9 percent mortality rate. It is believed to be caused by a combination of spiking mortality of turkeys (SMT) and excess mortality of turkeys (EMT).

Symptoms include diarrhea, dehydration, weight loss, anorexia, and growth depression of greater than 50 percent. Control of this disorder involves strict biosecurity measures to prevent transmission of infected litter and birds before 7 days of age.[4]

Pullorum Disease *(Salmonella pullorum)*

Pullorum disease is caused by another salmonella bacterium, *Salmonella pullorum*. Pullorum disease occurs in chickens, turkeys, pheasant, quail, pigeons, and many wild birds. Transmission of the organism is chiefly through the egg. In the United States, it was once the most serious disease of young chickens. It is still a very serious problem in many countries. Rodents, lizards, and many other cold-blooded animals can spread the organism. A few days after hatching affected chicks huddle together, lose appetite, may show labored breathing, and often develop a whitish diarrhea. Death rates can be very high. Posted birds reveal lesions in many organs including the heart, spleen, kidney, lungs, and digestive tract. Adult birds that have recovered from pullorum disease carry the disease but do not show signs of it, although they are responsible for the vertical transmission of the infection. Both vertical transmission through infected eggs and horizontal spread are possible while chicks are still in the incubator. Furazolidone added to the feed is one of the most effective treatments.

The disease is controlled through routine testing of breeding stock and disposal of all reactors. Where pullorum disease has been a problem or is suspected, the incubator and eggs are commonly fumigated with formaldehyde gas followed by strict sanitation measures between hatches. Because people are very susceptible when they inhale formaldehyde gas, the Occupational Safety

---

[4]R. Chin, Branch Chief of California Veterinary Diagnostic Laboratory System (CVDLS) Laboratory, Fresno, Calif., 1999.

and Health Administration (OSHA) has set up standards for concentration levels and methods of application. Consultation with a poultry extension agent and/or veterinarian should precede application.

The national poultry improvement plan in the United States set in motion the annual blood testing of breeding flocks to eliminate infected breeder birds. By eliminating the possibility of vertical transmission in this way, pullorum disease has almost been eradicated in the United States.

Trichomoniasis

Trichomoniasis is a protozoan disease that affects turkeys. Young and growing birds have the highest incidence and greatest losses. Symptoms include anorexia, depressed growth rate, and droopiness. Preventive measures include improved sanitation and the addition of copper sulfate to the drinking water (to prevent the spread among the flock).

Worms

Ascaridia galli, a large roundworm, is a common intestinal parasite of the chicken. Large roundworms are unsegmented, white, round, internal parasites that may measure 2 to 5 inches in length. Infestation begins when foraging chickens eat the worm eggs, which hatch in the digestive system. Larvae enter the intestinal wall, where they live to maturity. Eggs pass out in the droppings and start the cycle again. When chickens are infested with many roundworms, the birds become unthrifty, and growth retardation and reduced feed efficiency occur. Worm eggs shed in the feces are many times picked up on workers' shoes, equipment, truck tires, and so forth and then taken into other poultry houses.

Control consists of good sanitation and management practices, which will upset the life cycle of the worms, either by causing them to be expelled prematurely from the intestines or by preventing worm egg production. Two common anthelmintics used to treat birds for roundworms are piperazine and hygromycin B (Hygromix).

Other worms that infest chickens include the cecal worm (*Heterakis gallinarum*), a small, white roundworm that inhabits the cecum of the chicken and is known to carry disease-producing protozoans, and the tapeworm (*Railleitina cesticullus*). Several species of tapeworms may inhabit the intestinal tract of chickens. It is important to know that tapeworms require an intermediate host such as snails, slugs, beetles, crayfish, or flies; thus both external intermediate hosts and the tapeworms themselves must be eradicated. Signs of heavy internal parasite infestation in the chicken may include lowered production, unhealthy appearance, anemia, inactivity, ruffled feathers, and drooping wings. Tapeworm treatment includes a combined piperazine and dibutyltin dilaurate anthelmintic.

Behavioral Problems in Poultry

Cannibalism and hysteria are the two most common social problems of rearing chickens in high concentrations in confinement. The habit of birds pecking one individual or each other to death is called cannibalism. The exact cause of cannibalism is not well understood, but it occurs frequently under confinement conditions. Some speculative causes, such as nutrient deficiencies, overcrowding, insufficient feeding or water space, or too much light in the house, have been suggested. Whatever the cause, the only correction and standard prevention is debeaking, the removal of a part of the upper and lower mandible. Debeaking is usually done at 6 to 8 days of age to produce the minimum stress. A commercially available electric debeaking machine cuts the beak and cauterizes the wound (cauterizing blade has a temperature of 1,500°F) at the same time to aid clotting and speed healing. The high temperature of the blade destroys the tissue responsible for regenerating beak growth.

Some producers are forced to trim the beak a second time when chickens are between 6 and 12 weeks of age. This second trimming results in a more permanent cutback of the beak. Another method of beak trimming involves notching the upper beak 6 days after hatching with a hot blade trimmer. By 10 days the tip of the upper beak will separate, leaving a trimmed upper beak. This method leaves the tip of the beak intact as the chick learns to eat and drink, thus reducing early stress.

Recent research shows that available poultry stock differ in their beak-trimming requirement[5]; therefore, when possible, replacement birds that require either minimal or no beak trimming should be selected. Still, until feasible alternatives exist, beak trimming is recommended to reduce feather pecking and cannibalism.

Beak trimming is also standard practice for turkeys. Strains of turkeys have been identified as less likely to be cannibalistic[6] and should also be sought to lessen the need for beak trimming when feasible.

Under some conditions, such as excessive, unusual noise, rapid change in light, or quick movement, broilers in open houses fly into one corner of the house, causing many deaths due to suffocation. In caged layers, birds may try to fly, which results in broken wings or necks.

The exact causes of hysteria are unknown, but care should be taken not to frighten birds. In addition, the number of hens per cage should not exceed 8. When group size increases to 12 hens or more in high-density cages, there is greater incidence of hen hysteria.[7] Sudden loud noises have also been credited with causing hysteria in some strains of chickens.[8] Some poultry managers use radio broadcasts piped into houses to accustom birds to the sound of human voices and noise. Others knock several times before opening a door to the poultry house. This approach supposedly draws the birds' attention to the door, lessening the surprise of door movement.

---

[5]J. V. Craig and H. Y. Lee, "Beak Trimming and Genetic Stock Effects on Behavior and Mortality from Cannibalism in White Leghorn-Type Pullets," *Applied Animal Behavior Science* 25:107–123, 1990.

[6]D. O. Nobel, F. V. Muir, K. K. Krueger, and K. E. Nestor, "The Effect of Beak Trimming on Two Strains of Commercial Tom Turkeys," *Poultry Science* 73:1850–1857, 1994.

[7]J. V. Craig and G. A. Milliken, "Further Studies of Density and Group Size Effects in Caged Hens of Stocks Differing in Fearful Behavior: Productivity and Behavior," *Poultry Science* 68:9–16, 1989.

[8]A. Mills and J. M. Faure, "Panic and Hysteria in Domestic Fowl: A Review," *Current Topics in Veterinary Medicine and Animal Science* 53:248–272, 1997.

THE BROILER AND LAYER INDUSTRY

Although the management of chickens for either meat or egg production is highly complex and scientific, a few basic management suggestions may be helpful in orienting beginning poultry farmers. The profit margins in the broiler industry are so small that large volumes are necessary to stay in business.

The term *fryer* was originally used to mean a young frying chicken. Today, the term has been replaced by **broiler.** Chicks are normally bought from companies that specialize in producing special strains of chickens for either meat or egg production (see figure 10–15). Straight run broilers (unsexed) chicks are usually ordered for broiler production. As previously mentioned, the all-in/all-out system—in which only one age of broilers is on the farm at any one time—helps break the cycle of an infectious disease. Some large farms are keeping multiple ages on the farm but isolating them in separate housing areas. Broilers are marketed at a live weight between 4 and 4 1/2 pounds, usually between 6 and 8 weeks of age.

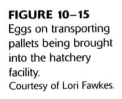

■**Broiler** a young chicken under 13 weeks of age (of either sex); it provides tender meat and has a soft, pliable, smooth-textured skin and flexible breast-bone cartilage; also known as a fryer

Size of Operation

With mechanized feeders and waterers, one full-time operator can care for 40,000 to 50,000 broilers at one time. As a contractor, the farmer works every day until the birds are sold and then takes off while the premises are being cleaned. Downtime ranges from 7 to 14 days between broods. Part-time producers may have only a fraction of that number of animals, and some large establishments house 100,000 or more at one time. About 18 minutes per 1,000 birds per day is required to feed using automatic feeding equipment; without automation the time required is 30 minutes.

FIGURE 10–15
Eggs on transporting pallets being brought into the hatchery facility.
Courtesy of Lori Fawkes.

Housing

Generally speaking, broiler houses are 30 to 40 feet wide and as long as necessary to accommodate the number of birds desired (see figure 10–16). The minimum space requirement is 0.6 square feet per bird in completely environmentally controlled houses and 0.75 square feet per bird in conventional houses. The house should be constructed so that birds are protected from heat, cold, and other inclement weather, thus providing a comfortable environment for maximum growth potential (see figure 10–17). Most houses today are well ventilated and insulated and have controlled temperature and automatic lights, feeders,

FIGURE 10–16
Broiler houses are generally 30 to 40 feet wide and as long as necessary to provide a minimum of 0.75 square feet per bird.
Courtesy of Lori Fawkes.

FIGURE 10–17
Modern poultry houses protect birds from heat, cold, and other inclement weather conditions.
Courtesy of Lori Fawkes.

FIGURE 10–18
Litter (e.g., sawdust, shavings, or ground corncobs) should be 4 inches deep. Note the automatic feeders and waterers.
Courtesy of Lori Fawkes.

and waterers. Litter (e.g., sawdust, shavings, and ground corncobs) should be 4 inches deep (see figure 10–18).

Production Expectations

Broilers should be fed to reach market weight and condition by 8 to 9 weeks of age. The goal is to have a feed conversion of 1.85 pounds of feed per 1 pound of gain for males and 1.95 pounds of feed per 1 pound of gain for females. Death losses should not be more than 1 percent to be considered normal.

Grades of Broilers (Fryers) and Other Classes

The grades established by the USDA are A or No. 1 quality, B or No. 2 quality, and C or No. 3 quality for live birds. Dressed birds are graded U.S. grade A, U.S. grade B, and U.S. grade C. Criteria for the grades are based on conformation, fleshing, fat covering, and defects. The USDA lists seven classes of live chickens:

Capon castrated male chicken less than 8 months of age

Stag male chicken under 10 months of age that shows developing sex characteristics intermediate between those of a roaster and those of a rooster; term is also used to describe livestock that have been castrated after reaching maturity

Rooster mature male chicken; also known as a cock

Hen adult female poultry

1. *Rock Cornish game hen or Cornish game hen:* A 5- to 7-week-old immature female weighing no more than 2 pounds ready to cook; selected from Cornish or Cornish cross matings. Cornish hens are very popular at convention banquets.
2. *Roaster:* A young chicken, usually 3 to 5 months of age, of either sex.
3. **Capon:** A castrated male, under 8 months of age.
4. **Stag:** A male chicken under 10 months of age but showing developing sex characteristics; not to be confused with the term *stag* used in describing livestock that have been castrated after having reached sexual maturity. A stag has a degree of maturity intermediate between that of a roaster and a rooster.
5. *Cock or* **Rooster:** A mature male.
6. **Hen** *or stewing chicken:* A mature hen, usually older than 10 months of age, often culled from laying operations.

7. *Broiler or fryer:* A young chicken under 13 weeks of age (of either sex). Broilers have more tender meat than other chickens, a soft, pliable, smooth-textured skin, and a flexible breastbone cartilage.

Normal and Abnormal Signs of Broilers

A good sign of health and proper social adjustments is even distribution of chicks or pullets in the house. Crowding against the walls is a sign that they are too hot; huddling closely and crowding around the heat source (brooder) or at corners of the house means that they are too cold. Gasping or panting could mean a disease or too much heat. If litter is packed, wet, or giving off a strong ammonia odor, there could be a ventilation problem that should be corrected before a more serious problem such as respiratory disease develops.

Producers should watch for telltale signs such as blood in the droppings (disease), rough feathers (parasites), pale combs (anemia), and any other abnormalities that may signal the beginnings of health problems. A few minutes of careful observation each day will soon train the eye to distinguish between normal and abnormal signs. For preveterinary students, this task should be performed for each species of livestock while preparing skills for admission into veterinary school. For poultry farmers, this skill is a requirement to stay in business; they must recognize a small problem before it results in total destruction of the flock.

Layers

Except for a few areas where premium markets exist for brown eggs, it is recommended that layers of white eggs (White Leghorns or Leghorn crosses) be used for commercial egg production. The key to success in the competitive egg business is to use high-producing layers regardless of breed, cross, or strain. The performance records of selected types of laying tests should be emphasized.

Housing

The construction of layer houses depends on the size of the operation, the environmental conditions, and individual preferences. Houses may even be multiple-story arrangements. The most common housing designs are the floor type, slat or wire floor type, and cage.

The floor- or litter-type house is an open house with the floor covered with 6 to 8 inches of litter (as opposed to 4 inches for broilers) to absorb the moisture from the droppings over an entire laying year. Nests and roosts are provided in addition to feeders and waterers. The slat or wire floor house uses either wooden slats turned edgewise, wire floors, or a combination of the two. This type of flooring allows for easier manure removal and requires no litter. Nests, roosts, and automatic feeders and waterers are used.

Some variation of one or more hens in wire cages is probably the most popular design in the United States. The most common type is one hen per cage (8 by 16 inches) or two hens per cage (also 8 by 16 inches) with hardly more than room enough to turn around. A few colony cages (about 25 hens per large cage) have gained acceptance. The advantages of the cage system are that droppings are well placed, more birds can be housed in a given area, eggs roll on the slanted cage bottom to any easy collection point, and internal parasites are eliminated. Disadvantages, such as difficult manure removal and outbreaks of lice, have been noted but are easy to resolve through automated manure removal and completely enclosed housing, respectively.

Incubation

Incubation of eggs involves keeping them at the right temperature and humidity for hatching (see figure 10–19). Incubators in commercial hatcheries have high-speed electric fans with automatic egg-turning devices, spray-nozzle humidifiers, and solid state-of-the-art computer controls. Many of the large hatcheries can handle 100,000 eggs at a time. Eggs should be set large end up and turned 6 to 12 times a day during the first 14 days but never after 18 days of incubation. Turning the eggs the last 2 days of incubation may injure the embryo. The ideal temperature is about 99.5°F (37.5°C). The relative humidity greatly affects the survivability of the egg. When humidity is too high, the metabolic water produced by the developing embryo will not be expelled through the shell and the embryo will drown in its own by-product (the water). When the relative humidity is too low, too much water will be removed from the egg and the embryo will either die or hatch premature, small, and weak. The average relative humidity should thus be set at 50 to 60 percent.

Incubation times vary with poultry species (see table 10–13). Even within species there can be significant variability. For example, Muscovy ducks have an incubation period of 33 to 35 days, whereas most other duck breeds have incubation requirements of only 28 days. In addition, ducks require turning of the eggs for the first 22 days to prevent the embryo from sticking to the shell.

FIGURE 10–19
Incubation of eggs involves keeping them at the right temperature and humidity.
Courtesy of Lori Fawkes.

TABLE 10–13	Incubation Times for Various Birds
Bird	**Incubation Period (days)**
Chicken	21
Turkey	28
Duck	28
Goose	28–34

Hatchery

Hatcheries are modern buildings that require isolation and proper sanitation measures to maintain strict disease control. They should be placed at least 1,000 feet from houses that contain chickens and should have their own entrance and exit that are unassociated with those of the rest of the poultry operation. Hatching eggs should be taken in at one end and the chicks removed from the other (never crossing paths). In addition , workers should never backtrack. Forced air should be used to ventilate the hatchery, and each room should be independently ventilated and have its own air filter. At hatching, chicks require an ambient temperature of 85 to 87°F. As the chicks grow and feather out, their ability to regulate their own body temperature increases and thus ambient temperatures can decrease.

Use of Lighting

During the growing period there should be a maximum of 11 to 12 hours per day of light. Light restriction should be started no later than 12 weeks of age. For Leghorn pullets, alternating 15 minutes of light and 45 minutes of darkness for a total of 6 hours of light per 24 hours reduces feed consumption by 5 percent without reducing body weight at sexual maturity. With laying hens, 14 hours of light maximizes egg production. Most programs call for 16 hours of light to provide a safety factor. The light causes the release of follicle-stimulating hormone (FSH) and luteinizing hormone (LH), which increase the growth and release of ova, respectively.

Production Expectations

Initiation of egg production between 19 and 20 weeks is now considered average. A good average production rate is 20 eggs per month for commercial high-producing layers. Flock mortality rates can greatly affect these numbers. Hens of the White Leghorn breed commonly produce 250 to 300 eggs per year. Standards in the United States now rate birds having the capacity to produce 289 eggs to 76 weeks of age with an average egg weight of 25.4 ounces per dozen. Death losses of 1 percent per month are to be expected as normal; unhealthy or diseased birds should be culled. Layers should not be kept in the flock over 19 months of age because declining production is inevitable by this age. Many birds are force molted at this time and brought back into production for another 6 to 8 months. A feed conversion goal of less than 4 1/2 pounds of feed per dozen eggs is considered attainable and in keeping with good management practices. Eggs should be collected five times daily, cleaned immediately, and stored under refrigeration at 13°C (55°F), 75 to 80 percent relative humidity.

Production indexes help determine the profitability of a flock. Two formulas commonly used in this regard are the following:

$$\text{Number eggs produced/Number live hens} \times 100 =$$
$$\text{Percentage of hen-day production for 1 day}$$

$$\frac{\text{Average daily number of eggs produced/}}{\text{number of hens housed}} \times 100 = \begin{array}{c}\text{Percentage of hen-housed}\\ \text{egg production}\end{array}$$

Grades of Eggs

Grading of eggs is the method used to determine their quality. The *Egg Grading Manual, Agricultural Handbook No. 75.* (Washington, D.C. 1977. USDA, Agricultural Marketing Service), serves as a guide to all egg grades. Basically, eggs are graded according to similar characteristics of weight and quality for three marketing outlets:

1. *Consumer grades:* Grade AA or fresh fancy, grade A, and grade B, and a size classification of jumbo, extra large, large, medium, small, and peewee
2. *Wholesale grades:* U.S. specials, U.S. extras, U.S. standards, U.S. trades, U.S. dirties, and U.S. checks (used in wholesale channels of trade and may be resorted to conform to consumer grades)
3. *U.S. procurement grades:* Special designations for institutions and the armed forces

Candling process of examining an intact egg by holding the egg in front of a strong beam of light while checking for shell soundness, interior quality, and stage of embryonic development

Eggs for the consumer trade are candled so that the inside contents may be seen to determine quality based on size and condition of the white, yolk, and air cell. Candling is the process of examining an intact egg so as to determine its interior quality, its shell soundness, and its stage of embryonic development (if it is a fertile egg). **Candling** is accomplished in a dark room by holding the egg in front of a strong beam of light and reciprocally turning the egg. Soundness of the shell is also estimated and weight (or size) is expressed in ounces per dozen. Weight, however, is distinctly different from quality and has no bearing on grade. Color does not influence grades but sometimes enters into descriptive marketing channels as whites or browns. An AA quality egg based on candled appearance is described as having an unbroken, clean shell with an air cell of 1/8 inch or less in depth and 3/4 inch in diameter. It should have unlimited movement and be free of bubbles, the white should be clear and firm, and the yolk outline should be slightly defined and practically free from defects. In the United States the size of the air cell is used to estimate the age of the egg. The diameter and depth of the air cell increase as the egg ages. The USDA standards are explicit for each of the grades and can be found in detail in the previously mentioned guide.

Normal and Abnormal Signs of Layers

Alert birds with red color to comb and wattles, well-bleached beaks and shanks (indicating good laying production), roughened feathers (old feathers, lack of molt), good appetite, and condition are signs of healthy, producing flocks. Abnormal conditions include early molt, shrunken head parts, yellow shanks, excessively soiled feathers, poor eggshell thickness, reduced feed consumption, and lowered egg production. These signs can be early indications of disease, parasites, environmental problems such as excessive heat or cold, improper ration, poor management of light, and so on.

SPECIAL CONSIDERATIONS FOR TURKEYS, DUCKS, AND GEESE

Turkeys

In 1997 some 301,251,000 turkeys producing 7,225,059,000 pounds of meat and yielding $2,884,377,000 were produced in the United States.[9] Ducks and geese brought in another $40 million in sales. Turkeys, which are native to America, have been making strides in recent years to increase their share of the fowl meat market at times other than Thanksgiving and Christmas.

By producing eggs year-round, specialists using several flocks and artificial lighting can produce large numbers of day-old **poults** (young birds under 8 weeks of age), which are fattened by other specialists. The integration of breeding, rearing, and marketing has led to turkey production groups and through proper advertising has created an increasing demand for oven-ready birds.

Poults young turkeys of either sex

The five principal breeds of turkeys commercially raised in the United States are

Broad Breasted White: White-feathered breed, 14 to 22 pounds for the hens and 25 to 40 pounds for the males. The hens are poor layers (50 to 60 eggs per season).

Broad Breasted Bronze: Similar to Broad Breasted White. Feathers have terminal edging of white, and the tail and wings have copper bronzing.

American Mammoth Bronze: Similar to the Broad Breasted Bronze but with less developed breast muscles.

White Beltsville: A small, white bird, with females averaging 10 pounds and males 14 pounds. Egg production is good (100 to 120 per season). This bird was developed by the USDA to meet the market demand for a 6- to 8-pound carcass and helped create year-round acceptance.

Hybrids: Inbred lines from various breeds are crossed to produce faster-growing, fleshier, more efficient poults than the parent stock.

Turkey stags, toms, or cocks (male adults over 26 weeks of age) are placed with 10 to 15 hens for breeding (usually starting as early as 34 weeks of age). The larger, broad-breasted varieties are not capable of natural breeding (because of their large breasts) and thus need to be bred using artificial insemination. The cost of keeping breeding turkey hens more than one season is usually prohibitive. About 50 to 120 eggs per breeding season may be expected in the first clutch. The second season usually yields only 40 to 75 eggs.

Ducks

Ducks have been produced for thousands of years in China and were mentioned as early as 2,000 years ago by the Romans. In the United States, which produces about 10 million ducks annually, commercial production is concentrated around several urban areas including Long Island, New York (see table 10–14). Ducks are excellent producers of meat and eggs and are much less susceptible to diseases than are chickens. Some breeds of duck will reach market weight

---

[9]National Agricultural Statistics Service (NASS), Livestock and Economics Branch, U.S. Department of Agriculture (USDA), Dairy and Poultry Statistics, 1999.

TABLE 10-14	Leading Areas of U.S. Commercial Duck Production	
State	**Ducks**	**Farms**
Indiana	1,179,062	325
Wisconsin	832,573	590
California	817,765	584
Pennsylvania	245,533	575
New York	217,568	482

Source: National Agricultural Statistics Service, U.S. Department of Agriculture (USDA), 1999.

(7 pounds) in 8 weeks and lay 150 or more eggs per year. Geese are great weed eaters and are vegetarians. Some farmers who keep geese around to weed their crops add a few ducks to control snails, slugs, and other pests. The maximum number of ducks per acre of water is 100. Excessive ducks will rapidly pollute the water and destroy the edges of the ponds as the ducks dabble around the sides.

For maximum production ducks must have limited feed from 3 weeks of age until they are laying eggs (no more than 0.33 pounds of feed per duck per day for the larger strains). Overweight ducks exhibit poor fertility and hatchability. Once ducks are laying well they can be fed free choice.

Ducks will produce more eggs with increased lighting. Artificial light maximizing at 17 hours a day will yield greater quantities of eggs. This increase should be started when ducks are 20 to 23 weeks of age to allow time for birds to reach sexual maturity. Table 10–15 summarizes incubation and hatching temperatures, humidity, and numbers of turns per day required when hatching duck eggs.

Turning of the duck eggs is critical, especially during the first week of incubation. Commercial incubators turn the eggs every hour. During incubation a typical duck egg will lose about 13 percent of its weight between the time it is laid and the 25th day of incubation. If significantly more or less weight is lost, hatchability of the egg is reduced.

Ducks may be used for a variety of purposes. Commercial producers use them for meat, down, and eggs. Table ducks are in demand in hotels and restaurants and are considered a delicacy in many parts of the United States with high Chinese populations, such as San Francisco. The down feathers are in demand for quilts, bedcovers, pillows, and winter clothing of various sorts. Standard performance of a typical Pekin meat duck from a commercial producer in California is summarized in table 10–16.

With the exception of the Muscovy, it is thought that most ducks originated from the wild Mallard. Although no one is sure of its ancestry, the Muscovy, which originated in South America, is thought to have descended from a different line entirely. The major distinguishing characteristic of the Muscovy is that it has very sharp claws in addition to its webbed feet, it roosts in trees, and it has a very mean disposition. There are many breeds of ducks, both ornamental and domestic, but we shall concentrate on the most common and utilitarian ducks.

The Khaki Campbell breed is one of the best for egg production. It was originally bred by Mrs. Campbell in England at the beginning of the twentieth century. It can lay up to 300 eggs per year and has great popularity worldwide.

TABLE 10–15	Incubation and Hatching of Duck Eggs	
	Incubation Hatching	
	(Day 1–25)	**(Day 26–28)**
Temperature	99.5	98.5
Humidity	86	94
Turns per day	3–7	0

Source: Metzer Farms, Gonzales, CA 93926, 1999.

TABLE 10–16	Standard Performance of a Pekin Meat Duck
Age at slaughter (days)	49–56
Live weight (lb.)	6.6–8.6
Eviscerated weight (lb.)	4.6–6.05
Feed conversion (lb.)	2.5–2.71
Mortality rate (%)	2.5
Breast fillet (%)	19.7–21.4
Skin and % fat in carcass	29.1–30.5
Egg production in 40 weeks	190–225
Fertility rate (%)	89–92

Source: Metzer Farms, Gonzales, CA 93926, 1999.

The Indian Runner breed originated in Malaya in about 1870. Varieties include the Buff, the Fawn, and the White. The breed has a most peculiar upright stance, from which the name *Runner* is derived. The ducks lay about 180 eggs per year, on the average.

The Welsh Harlequin is a very popular breed in Great Britain, derived from Khaki Campbell stock. It is similar in egg production to the Khaki Campbell but superior as a table bird.

The Aylesbury breed is a snow white, heavy bird that produces 100 eggs per year and weighs up to 10 pounds.

Pekin ducks were introduced into the United States from China in 1870. They are the most desirable table duck in Australia and in North America. Their average egg production is about 130 per duck annually.

The American Buff (Buff Orpington) breed originated in Great Britain. The ducks average 240 eggs per year and are considered good dual-purpose ducks (eggs and meat).

The Cayuga and Black East Indie are ornamental breeds that are in high demand as table ducks by gourmets because of their game bird flavor. They are poor layers but have an excellent record as good hatchers and mothers.

The Crested duck is named for its characteristic top knot. It is of British origin and medium weight; however, its breeding stock is not readily available in the United States.

Geese

The goose is said to be one of the oldest domesticated animals. Geese were among the few animals discovered in drawings on the walls of King Tut's tomb.

Looking back to customs in ancient China, in 4000 B.C.E., the favorite parental gift to the bride and groom was a pair of live geese, symbolizing a long, faithful marriage. Geese in the wild are monogamous and mate for life. Later the goose came to be known as the wedding feast bird, cherished not only for faithfulness and fidelity but also for good luck.

Not much else is known of the history of the goose. Even during colonial times, the goose was classified as "a silly bird, too much for one and too little for two." However, today's marketed goose is far more meaty and tender; most are packaged and frozen fresh for greater convenience and wider availability. Today's commercial goose is "too much for two" and actually enough for a whole family to enjoy.

Our colonial forefathers made great use of goose and goose products. Early New England settlers had to struggle to keep warm during the grim winters. They quickly learned to rely on the soft feathers and down of the goose to keep them comfortable in bed. Goose grease was used to clean and polish the black boots of early settlers, and, when the Declaration of Independence was signed, a goose quill declared a nation on July 4, 1776.

As strange as it may seem, some animal authorities equate geese with cattle in usefulness and compatibility. If the world should have to divert the usage of grain from animals, there would be only three surviving species adaptable to animal agriculture: the ruminant (those species that chew a cud), the rabbit, and the goose. All are forage consumers.

Unlike other forms of poultry, the goose is basically a roughage burner. This fact was well-known in the pioneer days because nearly every small farm had its flock of geese to provide down for pillows, eggs, and a delicate meal for special occasions. Geese were so practical and useful during the Revolutionary days that Benjamin Franklin made a strong appeal to the founding fathers to appoint the goose as the symbol of the United States. Little is known of the history of geese, but we do know that they were extremely important in the development of the United States. A record of the first Thanksgiving meal by the pilgrims listed some venison, 2 turkeys, and 70 geese.

Geese are capable of taking care of themselves on very rough land. Although geese like water, it is not necessary to have lakes, ponds, or running streams for them to be of benefit in making use of grasses and weeds. In the wild, one **gander** (mature male adult of geese; *geese* is the plural referring to males, females, adults, and young) mates with one female (**goose**) for life. However, under domestication, a gander may take care of four or five families.

A good example of how geese are managed on a large scale today can be found in Minnesota. Low-lying areas that cattle, sheep, or goats would find unattractive can serve the web-footed goose perfectly. One grower ran several thousand goslings (a gosling is a young goose or gander prior to adulthood) on nothing but Reed's canary grass (a rather undesirable roughage). The goslings were turned on to the grass at 4 weeks of age, left for 14 weeks, and marketed without any form of supplementary feed. In another case, a small acreage of cattails was cleared for other uses by 5,500 goslings (see figure 10–20).

The ability of the so-called weeder geese has long been recognized for its use in the cultivation of certain crops. All geese have some dislikes, including mint, strawberries, asparagus, potatoes, and cotton. Consequently, before the advent of herbicides, weeder geese kept the rows spotless in areas where these crops were grown. Because of herbicides and residues, weeder geese are not used much today except in mint and strawberry fields and in areas specializing

▓ Gander mature male goose

▓ Goose mature female goose

FIGURE 10–20
Weeder geese.
Reprinted with permission from J. Blakely and D. H. Bade, *The Science of Animal Husbandry,* 6th ed. (Englewood Cliffs, NJ: Prentice-Hall, 1994).

in organically raised produce. With the growing concern about the environment and health problems associated with the use of agricultural chemicals, perhaps we will see an increase of the biofriendly weed eaters.

In addition, geese do not compact the soil as they weed (compared with heavy machinery or people); they work in the rain 7 days a week, and at the end of the season the farmer can harvest the geese for meat and feathers. Geese readily consume young Bermuda grass, johnsongrass, sedge, nut grass, puncture vine, clover, chickweed, horsetail, and fallen fruit under orchards. Approximately six geese are needed to weed 1 acre of strawberries (temperature and humidity dependent). Weeder geese are generally kept in a state of slight hunger to encourage them to maximize weed control. Supplemental feed should be provided only at night. A constant supply of fresh water should be placed in the area of greatest weed concentration to encourage more weed consumption. Geese should be removed from crops such as berries and tomatoes before the fruit ripens.

Geese need to be protected from predators and thus should be enclosed at night in a well-constructed, covered shelter. The U.S. Forest Service has developed and uses a trailer into which geese are herded and then driven to various fields for weed control. In addition, shade should be provided to allow the geese protection from hot midday sun. A simple lean-to or shade tree is usually adequate.

Another interesting use of geese involves their unusual habit of honking when strangers encroach on their territory. A large distillery in Scotland was having difficulty with theft of its products stored in open yards, where valuable kegs of liquor were aged. Although the area was fenced and watch dogs were used, thieves were able to outwit the distillery by dropping female dogs in estrus over the fence. Losses continued to mount until the distillery got rid of the dogs and put in a flock of geese. The geese quickly became familiar with the people on duty and were silent except when a stranger neared the storage area. The geese then let out a clatter of protest, calling attention to the potential thieves. Thereafter, theft from breaking and entering was completely eliminated.

Commercial goose production supplies birds for special occasions in many parts of the world. In the United States, geese are served as a substitute for turkey during major holidays such as Thanksgiving and Christmas. In Europe, especially England, the most festive occasion is St. Michael's Day, September 29. Legend has it that if you eat goose on Michaelmas, you will never want for money all year. In some parts of the United States, especially Pennsylvania, St. Michael's Day is also celebrated in this way.

The most unusual method of producing a goose product has to be claimed by the French. For many years, the French have practiced the art of stuffing of geese, force-feeding by hand with a funnel more than the goose cares to consume. This practice causes an enlarged liver. Normally the liver will easily fit in one hand, but stuffing increases the size of the liver to 1.5 to 2.5 pounds—and a price approximately 10 times the price of an average goose carcass. The giant livers are then flavored with truffles, a fungus that grows primarily in southern France, to produce the gourmet delight known as goose-liver pâté. It is world famous and very expensive. An interesting side story is the way in which French farmers discover truffles, which grow about the depth of potatoes underground. Truffles have defied domestication and so must be hunted in the wild. Hogs have a keen taste and nose for them, so some French farmers train a few hogs to work on a leash like a dog and scour the countryside in search of truffles. When the hog makes a find, the farmer pulls it off the area and ties it to a tree while he or she digs the truffles (which normally sell wholesale for about $150 per pound) for later inclusion with the goose liver and numerous other foods.

One geese producer located in Sleepy Eye, Minnesota, markets his geese when they are only a few days old. He ships them off to farmers and then buys them back for processing in his slaughter plant after they have weeded a field or cleared an area of grass, weeds, and so on. Approximately 100,000 geese are hatched, sold, and bought back in this manner.

Legends of the goose abound but often are based on fact. Take the case of the long life spans attributed to geese. In the nineteenth century, it was traditional, especially in European countries, for a mother to give her daughter a prize goose so she would have a start on eggs, down for pillows, and meat for the table when offspring were numerous enough. Many daughters held a sentimental attachment to retaining the same goose in the family and would give their daughters the family goose. Reports indicate that some geese were handed down over several generations so that records of geese living to be 100 years or longer were not unusual.

Brooding Goslings.   Goslings must be kept warm and dry and protected from drafts. The temperature of the brooder should be 85 to 90°F the first week and then lowered approximately 5° each week. Fresh drinking water should be available in such a manner that goslings cannot get into it and get wet. Goslings should be started on a chick crumble and gradually switched over to pellets or cracked grains after 3 weeks of age. At 6 weeks old a gander can be used as an adoptive parent to teach young goslings how to forage in the field and to protect them from predators.

Geese and swans are the largest waterfowl and are broadly characterized by their long necks and noniridescent coloration. Among the geese, there are two major genera: *Anser* and *Branta.* Geese make up only 0.5 percent of the poultry raised in the United States. There are only 10 recognized breeds of commercial importance.

The Toulouse is the largest and most popular breed of geese in the United States, weighing from 20 to 25 pounds. Plumage is dark gray on the back, fading to light gray, turning to white on the breast and abdomen. The Toulouse averages 15 to 35 eggs per year.

The Emden is pure white and closely feathered; adults weigh 18 to 20 pounds. The Emden lays fewer eggs than the Toulouse but is a better sitter. In Australia, the Emden is now the most popular breed.

The Pilgrim, Roman, Buff, Sebastopol, African, Chinese (excellent foragers and lay more eggs than any other breed), Canada, and Egyptian breeds are of less economic importance but might be expected to perform in much the same way as the Toulouse and Emden.

SUMMARY

The poultry industry in the United States is divided into egg and meat production. Although considerable numbers of hobbyists and small farming operations exist in other avian enterprises (ducks, geese, quail, peafowl, and guineas) the majority in numbers and in dollar value are found in chicken egg and meat production. With Colonel Sanders's creation of Kentucky Fried Chicken, the production and availability of chicken to the consumer increased markedly.

Biosecurity of poultry houses and mechanization in feeding, hatching, and collection of eggs have enabled poultry farmers to increase the numbers of birds per worker and thus cheapen the cost of production while improving on the health and safety of the flock. The scientific raising of poultry is unsurpassed in the livestock industry. The most modern techniques in feeding, balancing of rations, disease prevention, and breeding are practiced on American poultry operations.

EVALUATION QUESTIONS

1. _____ is a young chicken (usually under 8 weeks of age) of either sex (usually under 6 pounds in ready-to-cook weight); term is used interchangeably with *fryer*.
2. _____ is a sexually immature female chicken.
3. _____ is a young chicken (usually under 15 weeks of age) of either sex (usually 5 pounds or more in ready-to-cook weight).
4. The period of growth in the life of a young bird in which it must be provided a source of heat in addition to that generated by its own body is called _____ .
5. _____ is the birth of a chicken.
6. A young goose of either sex is called a _____ .
7. The condition of a hen when prepared to sit on eggs for the purpose of incubation is called _____ .
8. A _____ is a castrated male chicken.
9. The process of examining an intact egg to determine interior quality, shell soundness, or stage of embryonic development is known as

_____ .
10. A _____ is a young turkey, of either sex, from day of hatch to 8 to 10 weeks of age.

11. _____ is the most common egg layer (breed) in the United States.
12. _____ , _____ , and _____ are the leading U.S. egg-producing states.
13. _____ is the most heritable trait in a chicken.
14. The white of the egg is produced in the _____ .
15. _____ is another name for the white of the egg.
16. An annual loss of feathers is referred to as _____ .
17. _____ is caused by a deficiency of manganese.
18. _____ is caused by a deficiency of vitamin B_2.
19. _____ is also called glandular stomach.
20. The egg spends most of its time while being formed in the _____ .
21. _____ is the hormone responsible for a rooster crowing.
22. The lung of a chicken has _____ air sacs (number of pairs).
23. _____ is the disease responsible for the destruction of more than 3 million birds in southern California in the 1970s.
24. _____ are eggs laid by a hen on consecutive days.
25. An immature male chicken is called a _____ .
26. _____ is also known as a syrinx.
27. Chickens have a total of _____ chromosomes.
28. _____ are the sex chromosomes of a rooster; _____ are the sex chromosomes of a hen.
29. _____ is the most destructive protozoan disease of poultry.
30. _____ and _____ are the two most common external parasites in chickens.
31. _____ helps prevent cannibalism in poultry.
32. At 8 weeks of age a broiler should weigh _____ pounds.
33. Hatching eggs should be stored at _____ °F.
34. When an ovum does not break right at the stigma, a _____ is often created.
35. The eggshell is formed in the _____ .
36. Turkey hens are commercially kept for _____ breedings.
37. The recommended stocking rate for breeding turkeys is _____ .
38. The drake-hen ratio for breeding should be _____ .
39. Broilers should gain 1 pound of weight per _____ pounds of feed.
40. Laying hens average _____ eggs per year.
41. _____ is used to prevent or eliminate coccidiosis in poultry.
42. A deficiency in _____ causes curled-toe paralysis.
43. An average turkey hen produces _____ eggs per season.
44. The copulatory organ of the male chicken is located in the _____ .
45. Mating between the male and female chicken is mainly a matter of joining _____ .
46. The closest chickens can come to having teeth is the action simulated by the _____ .
47. Laying hens should never be subjected to decreasing amounts of _____ .
48. The most harmful external poultry parasite is _____ .

49. If you see blood in the feces the chicken may have the ailment
    _____ .

50. _____ is the number of chickens in the United States.
51. The state that produces the most chicken eggs is _____ .
52. The nation with the largest number of chickens is _____ .
53. _____ is the oil gland of poultry.
54. _____ is the act of combing the feathers with the beak.
55. The basal heart rate of a chicken is _____ .
56. _____ is the most important grain used in poultry rations.
57. _____ is sometimes added to poultry diets to promote growth and improve the yellow color of the skin and shanks.
58. _____ is sometimes added to the diet of chickens to help the gizzard grind feed materials.
59. _____ refers to the spread of disease by contact with litter, inhaling dust particles containing organisms, and infestation by vectors such as insects, wild birds, or parasites.
60. _____ refers to the spread of disease from the dam to the offspring through the egg.
61. _____ is a poultry respiratory disease that affects young chicks, caused by the inhalation of a fungus.
62. One of the oldest known and most destructive protozoan diseases of poultry is _____ .
63. _____ is also known as duck plague.
64. _____ is caused by *Salmonella gallinarum*.
65. _____ used to be called Gumboro disease.
66. A respiratory disorder in chickens that is compared with the common cold in humans is _____ .
67. _____ is the most prevalent external parasite in poultry.
68. _____ is a retroviral disease in poultry that causes tumors.
69. _____ is a herpes viral disorder of chickens that causes tumors.
70. The most economically important external parasite of chickens is
    _____ .

71. _____ is also known as chronic respiratory disease.
72. A transmissible, infectious, enteric disease of turkey poults is
    _____ .

73. The salmonella disorder _____ occurs in poultry and is passed through the egg. It once was the most serious disease of young chickens and was associated with high mortality rates.
74. _____ is a common unsegmented, white, round internal parasite of chickens.
75. The habit of birds pecking one another to death is called _____ .
76. A management system in which only one age of broiler is on the farm at any one time to help break the cycle of an infectious disease is
    _____ .

77. _____ is a young chicken, usually 3 to 5 months of age, of either sex.
78. A male chicken under 10 months of age but showing developing sex characteristics is called a _____ .
79. _____ °F is the ideal temperature for incubating chicken eggs.
80. The incubation period of most breeds of ducks is _____ .

81. The optimum number of hours of light for egg production is
   _____ .
82. _____ is the highest grade of eggs in the United States.
83. The number of turkeys in the United States is _____ .
84. With the exception of the _____ , it is thought that most ducks
   originated from the wild Mallard.
85. _____ is the best breed of ducks for egg production.
86. _____ is the most desirable breed of ducks for meat (table duck).
87. _____ is the largest and most popular breed of geese in the
   United States.

DISCUSSION QUESTIONS

1. What are the most common breeds and their distinguishing
   characteristics?
2. The bones of the chicken have what unusual characteristic?
3. What causes the meat of a bird to be either light or dark?
4. Define the terms *clutch, candling, incubation period, class, variety,
   breed,* and *strain.*
5. What is the importance of the Poultry Products Inspection Act?
6. What genus does the chicken fall under?
7. Describe the importance of Marek's disease.
8. What is the difference between broodiness and brooding?
9. What is hysteria?
10. What is a stag?
11. What is the cuticle? Why is it important?
12. Cite differences in the anatomy of male and female chickens (compared
    with that of mammals).
13. In 1992 the average net return on a dozen of eggs was 1.9 cents. How
    many dozen of eggs would you need to support a family of four?
14. In 1992 the average net return on a pound of chicken (broiler) was 3.3
    cents. How many pounds of chicken and how many broilers would you
    need to support a family of four?

RECOMMENDED READING

Moreng, R. E., and J. S. Avens. *Poultry Science and Production.* (Prospect
Heights, Illinois: Waveland Press, 1991).
North, M. O., and D. D. Bell. *Commercial Chicken Production Manual,* 4th ed.
(New York, NY: Chapman & Hall, 1990).

CHAPTER 11

The Equine Industry

OBJECTIVES

After completing the study of this chapter you should know:

■ The historical, current, and future trends of the U.S. equine industry.
■ Distinguishing characteristics of the most common U.S. horse breeds.
■ Reproductive and general management procedures unique to the equine.
■ The benefits of artificial insemination in the equine industry, including the concerns of various breed organizations.
■ About horse selection, breeding, and judging.
■ What special considerations must be made in feeding horses compared with other livestock species.
■ The most common diseases of horses and their symptoms (in brief), diagnosis, treatment, and prevention.

KEY TERMS

Anaerobic
Blistering
Cribbing
Draft horses
Float
Foal heat
Galvayne's groove

Intense work
Light horse breeds
Light work
Meconium
Medium work
9-day heat

Osteochondritis dissecans (OCD)
Overo
Ponies
Tobiano
Waxing

INTRODUCTION

Domesticated more than 5,000 years ago, the horse was one of the last farm animals to be domesticated. Horses were first used as food, then for war and sports, and also for draft purposes. Horses became important in transportation, farming, mining, and forestry. There are an estimated 58 to 60 million horses in the world today.

Today's horse evolved from a small creature, no bigger than a fox, ranging from 18 to 19 inches tall at the withers; it had four toes and a splint on its forelegs and three toes and a splint on its hind legs. It had small, alert ears, a doglike, furry coat, and a swishing tail (see figures 11–1 and 11–2).

About 40 million years ago, the evolutionary process began to change the horse even more. At this time, during the Oligocene epoch, Mesohippus appeared (see figure 11–3). Mesohippus developed larger than its ancestors, standing 24 inches high at the shoulders, about the size of a Collie dog. Its feet were still clinging to pads, but tiny hooves had developed on three toes on the front and three toes on the back feet.

The one-toed horse evolved about 10 million years ago (see figure 11–4). About 2 million years ago, the horse as we know it today, *Equus caballus*, emerged as a rather large, magnificent creature (see figure 11–5). *Equus caballus* became extinct in North America about 8,000 years ago and in South America about 6,000 years ago; it was not to return until the Spanish brought horses to the New World in the fifteenth century. Today, there is clearly a

FIGURE 11–1 Artist's conception of the hypothetical prehorse, fossil remains of which have not been found. This earliest ancestor of the horse is believed to have had five toes with claws, to be covered by part fur and part hair, and to stand 6 to 7 inches high at the shoulder.
Reprinted with permission from J. Blakely and D.H. Bade, *The Science of Animal Husbandry,* 6th ed. (Englewood Cliffs, NJ: Prentice-Hall, 1994).

FIGURE 11–2
Left, Eohippus or the Dawn horse, originally called Hyracotherium when first discovered in Europe, is the first horse ancestor of which fossil remains have been found. It was no larger than a fox. Note the four toes on the front feet and three toes on the rear feet. Right, The same fossil remains were the basis for this reconstruction of Hyracotherium, which scientists opposed to evolutionary theories claim was not an ancestor of the horse. Reprinted with permission from J. Blakely and D.H. Bade, *The Science of Animal Husbandry*, 6th ed. (Englewood Cliffs, NJ: Prentice-Hall, 1994).

FIGURE 11–3
Mesohippus, stage three in the evolutionary development of the horse. The face and legs had begun to lengthen; feet still clung to pads, but tiny hooves had developed on each of the three toes on front and back. Size had increased to about 24 inches at the withers. Reprinted with permission from J. Blakely and D.H. Bade, *The Science of Animal Husbandry*, 6th ed. (Englewood Cliffs, NJ: Prentice-Hall, 1994).

429

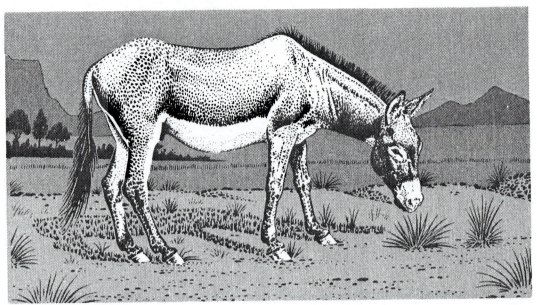

FIGURE 11–4   Pliohippus, the first truly one-toed horse, about 46 inches at the withers, was the nearest approach to modern-day *Equus* spp.
Reprinted with permission from J. Blakely and D.H. Bade, *The Science of Animal Husbandry*, 6th ed. (Englewood Cliffs, NJ: Prentice-Hall, 1994).

TABLE 11–1	World Horse, Donkey, and Mule Numbers (1994)		
Country	**Horses**	**Donkeys**	**Mules**
China	9,900,000	10,900,000	5,500,000
Mexico	6,200,000	5,200,000	3,200,000
Brazil	5,800,000	—	2,100,000
United States	3,900,000	—	—
Ethiopia	—	5,200,000	600,000
Argentina	3,400,000	—	—
World total	58,200,000	43,800,000	15,000,000

Source: 1994 Food and Agriculture Organization of the United Nations (http://www.fao.org/ag) Production Yearbook.

wide distribution of horses (see table 11–1); some nations use this magnificent creature for transportation and work and others simply for pleasure and companionship.

The early ancestors of the horse disappeared in pre-Columbian times, supposedly by crossing from Alaska to Siberia. These animals evolved in Asia and in Europe into the modern-day horse. The draft horses and Shetland ponies developed in Europe, and the lighter, more agile horses developed in Asia and the Middle East. In the early 1900s there were 25 million horses and mules in the United States. Their numbers rapidly declined after World War I.

Depending on the source, the total number of U.S. equine (including horses, ponies, mules, burros, and donkeys) ranges from 5.2 to 6 million (see table 11–2).

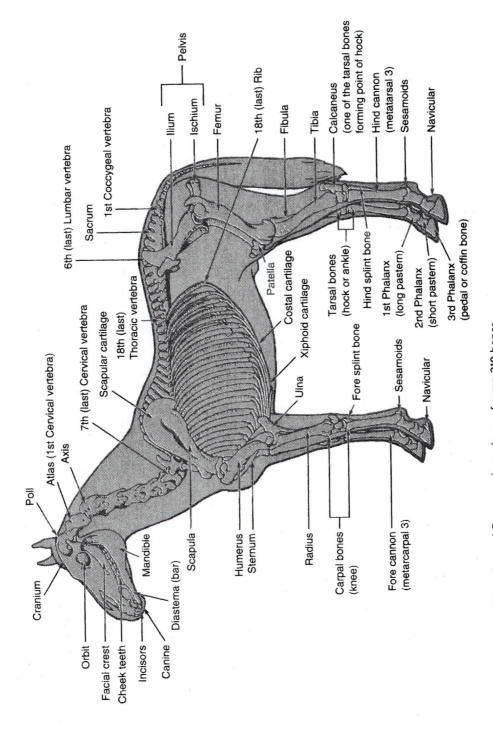

FIGURE 11–5 The skeleton of *Equus* spp. is made up of some 210 bones. Reprinted with permission from J. Blakely and D.H. Bade, *The Science of Animal Husbandry*, 6th ed. (Englewood Cliffs, NJ: Prentice-Hall, 1994).

TABLE 11–2	Leading U.S. Equine Populations		
		Number of Horses and Ponies	Total Number of Equine
Region	State	1997	1998
Central	Illinois	51,700	99,000
	Indiana	58,600	140,000
	Kansas	52,800	104,000
	Michigan	66,200	130,000
	Minnesota	55,900	155,000
	Missouri	85,700	140,000
	Wisconsin	52,400	115,000
Northeast	New Jersey	22,600	45,000
	New York	47,800	157,000
	Ohio	76,200	155,000
	Pennsylvania	65,100	165,000
South	Alabama	42,500	130,000
	Florida	54,900	170,000
	Georgia	35,300	69,000
	Kentucky	95,900	150,000
	Louisiana	30,100	65,000
	Maryland	22,500	45,000
	Oklahoma	93,700	165,000
	Tennessee	89,000	185,000
	Texas	242,000	595,000
	Virginia	50,300	145,000
West	California	113,100	235,000
	Colorado	81,700	140,000
	Montana	71,200	130,000
	New Mexico	38,800	64,000
	Oregon	68,300	120,000
	Washington	58,800	155,000
	Wyoming	50,600	61,000
Total (28 states)		1,873,700	4,029,000
Total United States (50 states)		2,427,300	5,250,400

Source: National Agricultural Statistics Service (NASS), March 2, 1999.

TAXONOMY

Like all of the species discussed in this text, the horse belongs to the kingdom Animalia and the phylum Chordata (please review chapter 1 for a description of each of these breakdowns). The equine falls under the class Mammalia, the order Perissodactyla (odd toed, nonruminating, and hoofed—includes tapir and rhinoceros), and the family Equidae (the horse family in the broadest sense, including evolutionary ancestors). The genus *Equus* includes all living members of the horse family and close ancestors (i.e., horses, zebras, and asses). The species *Equus caballus* includes the domestic horse and close wild relatives.

BREEDS OF HORSES

In no place in the world are there more varieties and breeds of horses than in the United States. These breeds came from the other parts of the world or were developed in the United States for a specific purpose. Some of them exist very much in the same form in which they originated. Others have changed considerably to meet modern demands. When evaluating breeds or individuals within a breed, it is important that we all speak the same language. Figures 11–5 and 11–6 describe the external anatomy of the horse in modern-day terms. Starting with figure 11–7, we include pictures of commonly found horse breeds in the United States. The size of the horse is expressed as height at withers in hands, not inches; 1 hand equals 4 inches.

Light horse breeds include those breeds commonly used for such activities as shows, racing, riding, and exhibition. They usually stand at 14.2 to 17 hands high and weigh anywhere from 900 to 1,400 pounds. The largest and most powerful draft horses are those generally used for heavy work. *Draft horses* stand

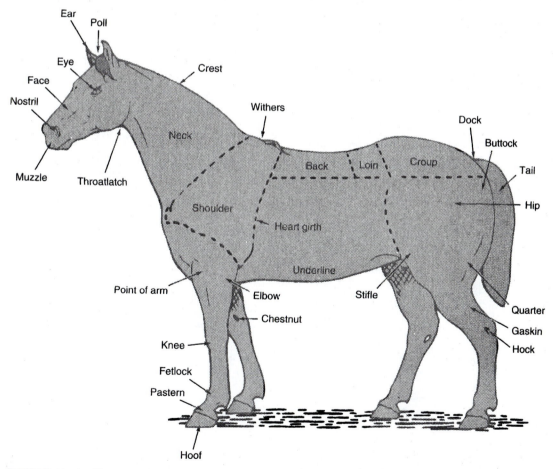

FIGURE 11–6   The parts of the horse.
Reprinted with permission from J. Blakely and D.H. Bade, *The Science of Animal Husbandry,* 6th ed. (Englewood Cliffs, NJ: Prentice-Hall, 1994).

14.2 to 17.2 hands high and weigh from 1,400 pounds to more than 2,000 pounds. Draft horses were originally developed for use in war and as beasts of burden. *Ponies* usually weigh between 500 and 900 pounds and stand under 14.2 hands high. These much smaller horses were developed to work in coal mines, to pull chariots, and to work cattle. Today they are primarily used for teaching young children how to ride and as companion animals.

Light Horse Breeds

American Quarter Horse.   The American quarter horse originated in the United States more than 300 years ago. At the beginning of America's settlement, towns were hacked out of the wilderness, and for entertainment people naturally migrated to these centers of supply distribution. Racing between neighbors became a familiar pastime. Most of the open areas led down the main street of town, and the distance was usually no longer than one-quarter of a mile. The farmers began to develop a breed of horse that would compete favorably in the local town races, and thus the quarter horse had its origin under these circumstances. There are now two types of quarter horses: The racing-type quarter horse, which developed through the infusion of Thoroughbred blood to improve speed over race track conditions, and the working-type quarter horse, which is used for roping and working cattle and pleasure riding. Therefore, there are two distinct types in the judging or selection of quarter horses. The breed registry association did not officially open until 1940. From 1940 until 1994, more than 3 million quarter horses were registered with the association.

The color of the American quarter horse is most often chestnut, sorrel, bay, and dun, but palomino, black, brown, roan, and copper are also seen. Quarter horses are well muscled and powerfully built (see figure 11–7). They average 14.2 to 15.2 hands high and weigh 1,000 to 1,250 pounds.

Artificial insemination is permitted provided insemination is completed within 72 hours. If cooled semen is to be transported, the stallion owner or lessee and the mare owner must sign a certificate. Any foal resulting from the

FIGURE 11–7
An excellent quarter horse. Close-coupled topline and long underline are characteristic of the breed. Reprinted with permission from J. Blakely and D.H. Bade, *The Science of Animal Husbandry,* 6th ed. (Englewood Cliffs, NJ: Prentice-Hall, 1994).

use of transported cooled semen must have its pedigree verified by deoxyribonucleic acid (DNA) testing, including sire, dam, and foal, and/or by other genetic testing as the American Quarter Horse Association deems necessary. Embryo transfer is permitted, but only one genetic offspring per mare per calendar year is allowed.[1]

American Saddlebred. The American saddlebred (see figure 11–8) was developed in the plantation states of the South. Almost all American saddlebred horses can be traced back to the great 4-mile race stallion Denmark, foaled in 1839. This breed's outstanding characteristic is that it is comfortable to ride for long distances and is able to work both in harness and under saddle. The coat color of the American saddlebred ranges from bay, brown, chestnut, gray, roan, and black to a golden color. Characteristics of this breed include an easy ride with great style and animation; long, graceful neck; proud action; and use as three- and five-gaited saddle horses, fine harness horses, pleasure horses, and stock horses. The American saddlebred stands 15 to 16 hands high and weighs 1,000 to 1,150 pounds.

Artificial insemination is accepted, and fresh, cooled, or frozen semen may be used. The use of frozen semen collected from a stallion that has died or has been castrated is allowed during the calendar year of death or castration. Embryo transfer is also accepted, but the blood type of the stallion and donor mare, as well as that of the foal resulting from embryo transfer, must be on file. Four foals per year per donor mare may be registered.

Arabian. The Arabian (see figure 11–9) breed originated from either Egyptian or Libyan domestication, with further development occurring in Saudi Arabia. The Arabian horse has been bred in the United States since the colonial period. George Washington rode a gray Arabian charger. The Arabian's outstanding

FIGURE 11–8 The American saddlebred horse is easy to ride for long distances and can work in a harness. Courtesy of the U.S. Department of Agriculture.

[1]M. E. Ensminger, *Horses and Horsemanship,* 7th ed. (Danville, IL: Interstate Publishers, 1999).

FIGURE 11–9 The
Arabian is noted for
speed, stamina, and
beauty.
Courtesy of The
International Arabian
Horse Association.

characteristics are speed, stamina, and beauty; it has a gentle disposition, seeming almost to prefer companionship with humans, a quality that endears it to its owner. Known as an easy keeper, the Arabian is able to maintain good condition on low levels of grain or pasture, which makes it an economical horse to own. The Arabian Horse Registry of America was established in 1908. More than 800,000 horses have been registered since its foundation. Currently, close to 20,000 Arabian horses are registered annually in the United States. The color of the Arabian horse is bay, gray, or chestnut, with an occasional white or black. Its skin is always dark. The Arabian is noted for its beautiful head, short coupling, docility, great endurance, and airy way for going; it is used for saddle horses, show purposes, pleasure, stock, and racing. The Arabian horse stands 14.2 to 15.2 hands high and weighs 850 to 1,000 pounds.

Artificial insemination is accepted, provided collection of semen from stallion and insemination of mare with this semen take place on the same premises and all semen is used within 72 hours of collection. Since January 1, 1995, semen collected in the United States or Mexico or semen imported into either country can be transported or stored provided stipulated requirements are met. Embryo transfer is accepted but limited to one foal registered per calendar year.[2]

Fjord Horse.   The Norwegian Fjord horse is a small, gentle but powerful all purpose horse. It is one of the oldest domestic breeds of horses dating back more than 2,000 years. Fjordings were first imported into the United States in 1900.

Fjordings are dun colored with a dorsal strip, dark bars on their legs and dark hooves, eyes and ear tips (see figure 11–10). They stand 13 to 14.2 hands at the withers. In Europe, the Fjord is used for riding, driving and draft purposes. In the United States, the Fjord is used for draft, carriage, pleasure riding, jumping, and dressage and for trail riding.

---

[2]M. E. Ensminger, *Horses and Horsemanship*, 7th ed. (Danville, IL: Interstate Publishers, 1999).

FIGURE 11–10 A Norwegian Fjord. Courtesy of Stephanie Abronson, Monte Nido, CA.

Missouri Fox Trotter.   The Missouri Fox Trotter breed developed in the Ozark hills of Missouri and Arkansas. It was primarily developed by crossing American saddlebred, Tennessee walker, and Standardbred horses. Sorrels predominate this breed, but any color is accepted. The main distinguishing characteristic is its unique fox-trot gait. It is used for pleasure, stock, and trail riding.

Artificial insemination is allowed. If semen is shipped, stipulations must be followed. However, the breed association does not accept embryo transfer foals for registration.

Morab.   The Morab breed was developed at the San Simeon Ranch, California. It evolved by crossing Arabian and Morgan horses. The Morab colors include bay, brown, black, chestnut, buckskin, dun, gray, grulla, palomino, and roan. Their characteristics are described as a combination of the ruggedness of the Morgan and the refinement of the Arabian (see figure 11–11). They are used for show purposes, pleasure riding, endurance rides, and ranch work. The Morab stands at 14.3 to 15 hands high and weighs 950 to 1,200 pounds.

Artificial insemination and embryo transfer require a special permit from the International Morab Breeders Association. DNA typing is required for foals born from transported semen or embryo transfer.

Morgan.   The Morgan horse is descended from one stallion named Justin Morgan. The stallion was named after its owner, an early-nineteenth-century New England farmer. Justin Morgan was a horse that never lost a race or contest. The Morgan was developed to outpull, outrun, outjump, or outdo any horse in any contest in which it was pitted. Today, the Morgan is used as a saddle horse, stock horse, and for driving. Morgans are considered to be quiet, reliable, all-purpose animals. They possess good endurance, are easy keepers, and are quite docile. Morgan horses are bay, brown, black, and chestnut in color (see figure 11–12). They stand 14.2 to 16 hands high and weigh an average of 800 to 1,200 pounds. Artificial insemination and embryo transfer are permitted with stipulations.

FIGURE 11–11
Morab stallion, Gentle
Ben, Box LT Ranch,
Ava, Mo.
Courtesy International
Morab Breeders
Association.

FIGURE 11–12 All
Morgan horses today
are the descendants of
the outstanding
stallion Justin Morgan.
Courtesy of The
American Morgan Horse
Association, Inc.

Peruvian Paso.   The Peruvian Paso breed originated in Peru some 450 years ago. Its origin has been traced to the Andalusian, Barb, Spanish Jennet, and Friesian breeds of horses. Any color is accepted, but solid colors are preferred. Its distinguishing characteristic is a broken gait; the legs on the same side move together, but the hind foot strikes the ground a fraction of a second before the front foot, producing a four-beat gait. It is used for pleasure, cutting, parades, endurance riding, and drill team work (see figure 11–13). The Peruvian Paso stands 14.3 to 15 hands high and weighs an average of 900 to 1,200 pounds. Both artificial insemination and embryo transfer are accepted, according to rules detailed by the registry association.

FIGURE 11–13

Peruvian Paso mare, (a) E.R.B. Ranch, Elk Grove, California. A second offshoot of the Paso horse is the Paso Fino horse (b), which orginated in the Caribbean Islands more than 400 years ago. Paso Fino horses range in height from 13.2 to 15.2 hands and are noted for their lateral gait. Shown here is a 4-year-old bay gelding; Coronado de la Corzaon, Simi Valley, CA.
Photo (a) courtesy of American Association of Owners and Breeders of Peruvian Paso Horses, photo (b) by author.

(a)

(b)

Standardbred.   The Standardbred breed (see figure 11–14) originated in the United States by the crossing of a Thoroughbred stallion named Messenger on native mares that were natural trotters or pacers. Standardbred blood contains the blood of Arabians, Spanish Barbs, Hackneys, and Morgans. The Hackney is also descended from a Thoroughbred stallion that was crossed, however, with native English mares. The Hackney was developed primarily as a harness or carriage horse. The Barb is actually more of a type versus a breed of horse that originated from Arabian breeding in northern Africa. There is a Spanish Barb Breeders Association, however, and thus breeders who continue to try to preserve this historic "breed" (see figure 11–15).

FIGURE 11–14
The Standardbred horse is used strictly as a harness race horse. Courtesy of The United States Trotting Association.

FIGURE 11–15   A Spanish Barb Mare, A. H. Amiga Magaju, owned by Marie and Paul Martineau, Anthony, Florida. Amiga Magaju is a combination of Spanish and Lakota Sioux, meaning Friend Rain. The Spanish Barb is a deep-bodied breed of rounded croup and hip, full neck, refined head, and strong, correct legs. It has great endurance and is a natural cow horse. This highly intelligent breed is smooth gaited and possesses a gentle, personable disposition. Courtesy of Marie Martineau.

The name *Standardbred* came from the demand that the horse make a standard mile time of 2 minutes and 30 seconds for a trot and 2 minutes and 35 seconds for a pace before it could be registered. Accepted colors for Standardbred are bay, brown, chestnut, and black and some grays, roans, and duns. Standardbreds are smaller, are less leggy, and have more substance and ruggedness than Thoroughbreds. They are used for harness racing—either trotting or pacing—and as harness horses in horse shows. Standardbreds are to the harness-racing industry what the Thoroughbred is to flat racing. Standardbreds stand 14.2 to 16.2 hands high and weigh an average of 850 to 1,200 pounds. A foal conceived by fresh semen is eligible for registration. No frozen or transported semen is allowed.

Tennessee Walking Horse.   The Tennessee walking horse (see figure 11–16a, b) was developed in the state whose name it bears. Some Standardbred, Thoroughbred, American saddlebred, and Morgan blood went into its development. The Tennessee walking horse is called the gentleman of the equines because of its gentle disposition and way of going. Perhaps its most

(a)

(b)

FIGURE 11–16
(a), Tennessee walking horse, Iron Works, owned by Gloria and Randall Dixon, Dacula, Georgia. (b), Although the Tennessee walking horse was established as a plantation horse, it is now used primarily as a pleasure horse. This 4-year-old Tennessee walking horse known as Wings of Gold of Simi Valley, CA provides its owner with easy gaits and a gentle ride.
Photo (a) by Larry Cowman.
Courtesy Tennessee Walking Horse Breeders and Exhibitors' Association. Photo (b) by author.

endearing quality is the gliding sensation felt by the rider. The Tennessee walking horse's gaits include a very easygoing flat foot walk, a running walk that covers a great deal of territory but is very comfortable for the rider (the Tennessee walking horse does not trot), and the canter. Tennessee walking horses are used as plantation walking horses, pleasure walking horses, and show horses. They stand 15 to 16 hands high and weigh an average of 1,000 to 1,200 pounds. Accepted colors include sorrel, chestnut, black, roan, white, bay, brown, gray, and golden. Artificial insemination is accepted, provided both parents are blood typed. Embryo transfer is accepted on one foal a year per donor.

Thoroughbred.   The Thoroughbred (see figure 11–17) originated in England during the seventeenth century. The royal families developed the Thoroughbred, creating a horse noted for speed and stamina. Thoroughbreds were imported into the United States around 1730. They are bay, brown, or chestnut and black, with some roan and gray. Noted for refined conformation with long, straight, and well-muscled legs, Thoroughbreds are used for racing, stock, saddle horses, polo mounts, and hunters. Thoroughbreds stand at 15.2 to 17 hands high and weigh an average of 1,000 to 1,300 pounds. Any foal that is the product of either artificial insemination or embryo transfer is not eligible for registration.[3]

FIGURE 11–17
Thoroughbred horses.
Courtesy of Kentucky
Horse Park.

---

[3]Jockey Club, July 3, 1999.

Breeds of Draft Horses

American Cream Draft Horse.   The American Cream draft horse (see figure 11–18) had its origin in Iowa (1911) and descended from one mare (Old Granny) of unknown ancestry that was cream colored and produced all cream-colored offspring. This breed is unusual because it stems not from a stallion but from a mare that had a great deal of prepotency (ability to stamp a characteristic on off-spring). Being the only draft horse developed in the United States is the out-standing characteristic of this medium-heavy draft breed. The American Cream Horse Association was formed in 1944; in 1993 the name changed to American Cream Draft Horse Association. The American Livestock Breeds Conservancy placed the breed on the endangered species list in the 1980s. The average weight is 1,600 to 1,800 pounds for mares and 1,800 to 2,000 pounds for stal-lions. They stand 15 to 16.3 hands high. Accepted colors are cream with white mains and tails, pink skin, and amber eyes. American Cream draft horses are used for harnessing, hitching, and driving. There are no current rules on the use of artificial insemination and embryo transfer.

Belgian.   The Belgian (see figure 11–19) is descended directly from a great wild horse of the Flanders area of Belgium and was first imported for draft use in the United States in 1866. Belgians were developed for farm work and

(a)

(b)

FIGURE 11–18
The only draft breed developed in the United States is the American Cream Draft Horse. (a), C. W. Sour Cream with her foal, Cream De L Creame. (b), C.W. Rich & Creamy with her foal, C.W. Cream Easter Surprise.
Courtesy of Colonial Williamsburg Foundation, VA.

FIGURE 11–19
The Belgian breed is
the most popular draft
horse in the United
States. (a), Belgian
team, and (b), Baby
Belgian.
Courtesy of Kentucky
Horse Park.

(a)

(b)

exhibition purposes. They are the lowest set and most massive of the draft horses, weighing 1,900 to 2,400 pounds and standing 15.2 to 17 hands high. Accepted colors include bay, chestnut, and roan.

Artificial insemination is accepted, provided the association is given advanced written notice and report of collection, shipment, and insemination is filed in writing within 120 days. When chilled semen is used, the registry requires blood typing of dam and foal. When frozen semen is used, all mares and foals must be blood typed. Embryo transfer is accepted provided both the sire and donor mare are blood typed prior to application for an embryo transfer permit.

Clydesdale.   The Clydesdale (see figure 11–20), which has become familiar as the beer wagon horse, came from the valley of the River Clyde in Scotland. This breed is now quite well-known in the United States because of its massive size, carrying style, distinctive coloring, feathers or hair on its legs, and great action. The breed was developed for farm work and exhibition purposes. The Clydesdale is considered a real show horse in the area of draft animals. Accepted colors include bay and brown with white markings. Clydesdales stand 15.2 to 17 hands high and weigh an average of 1,700 to 2,000 pounds.

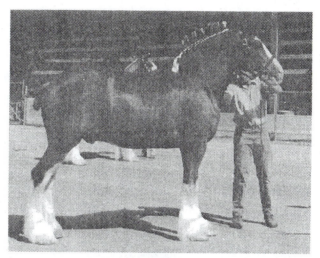

(a)

(b)

FIGURE 11–20
Among draft horses, the Clydesdale is noted for its style, beauty, and action. These two Clydesdale geldings are owned and exhibited by Carson Farms and Auction Services, Ltd., Listowel Ontario, Canada.
Courtesy of Albert Hewson, Secretary, Clydesdale Horse Association of Canada.

For transported or frozen semen the stallion from which it is collected must be blood typed, and a copy of the blood type document sent to the registry association; report of collection, shipment, and insemination resulting in conception must be forwarded to the association within 120 days of breeding. To register embryo transfer offspring, the association must be notified of the intended collection of the fertilized egg at least 15 days in advance of collection. In addition, an association-approved veterinarian must be present during the collection and transfer; the resulting foal must have its pedigree verified by blood test of the mare, stallion, and foal. The association must be notified in writing if the embryos are frozen.

Percheron.   The Percheron (see figure 11–21) horse originated in the northern French district of La Perche. Flemish and Arabian blood appear to have had some influence on the Percheron, and its quality was improved later through selection. Almost exclusively black or gray in color, the Percheron, a popular draft breed, generally ranks behind the Clydesdale and Belgian in use today.

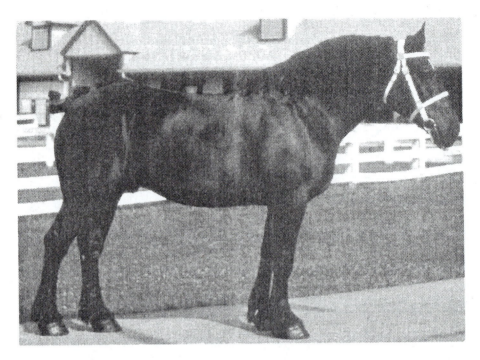

Like all the draft breeds, its numbers have declined since the mechanization of farms. Percheron horses stand 15.2 to 17 hands high and weigh an average of 1,600 to 2,200 pounds.

Artificial insemination is accepted if semen is obtained from a member of the Percheron Horse Association of America or a reputable artificial insemination establishment. DNA of the stallion must be on file with the association. Impregnation must be provided by a licensed veterinarian, and embryo transfer foals are accepted provided they meet the standard stipulations for registry. The association must further be notified within 90 days (in writing) of an attempted embryo transfer. The DNA of the sire, dam, and foal must be on file for all embryo transfer foal applicants.

Shire.   The Shire (see figure 11–22) came from Lincolnshire, England. Robert Bakewell, the famous English breeder, often called the father of animal science, became interested in the breed and contributed greatly to its development. The unusual characteristic of the Shire is its height: it is the tallest of the draft breeds. Shires average 16.2 to 17 hands high and weigh 1,800 to 2,200 pounds. They also have feathers or hair on the legs and are used for farm work and exhibition purposes. Accepted colors include bay, brown, and black with white markings. Both artificial insemination and embryo transfer are accepted under certain rules.

Suffolk.   The Suffolk (see figure 11–23) came from the county of Suffolk, England. The Suffolk is all chestnut in color and is noted both for being the smallest (15.2 to 16.2 hands high and 1,500 to 1,900 pounds) of the common draft horses and for its courage and strength. Perhaps this horse exemplifies the desire an animal can have to please its master. A characteristic contest for this breed developed in England. The horses were hitched to a tree, and the winner

FIGURE 11–22
The Shire is characterized by its great size and heavily feathered legs.
Courtesy of Dr. James Blakely.

FIGURE 11–23
The Suffolk or Suffolk Punch is a rotund, muscular, "punched-up" breed of English origin as demonstrated by these two magnificent teams.
Courtesy of Kentucky Horse Park.

(a)

(b)

of the contest was determined by the number of efforts it would make in trying to move the immovable object on voice command. Horses were often on the verge of exhaustion before they would give up in this attempt to respond. Although the association has no specific rules for artificial insemination, the association members are favorable toward its use.

Color Registries

American Albino. The American albino is a color breed that originated on the White Horse Ranch in Naper, Nebraska. The foundation sires were Morgans. American albinos are not true albinos because they do not have pink eyes. Their coat color varies from snow to milky white hair with pink skin and brown, blue, or hazel eyes. The association registers American white horses separately from the American crème horses. The American albino is used in troupes, in parades, and as a pleasure horse.

American Buckskin. The American buckskin also originated in the United States from horses of Spanish descent. The colors accepted include buckskin, red dun, and grulla (mouse dun). The buckskin is used for pleasure riding, for show, and as a stock horse. It stands at 14 to 16 hands high and weighs an average of 900 to 1,250 pounds. The International Buckskin Association accepts both artificial insemination and embryo transfer provided the foals meet all other registry requirements.

Appaloosa. The Appaloosa (see figure 11–24) is a beautiful horse with a distinct color pattern. It was developed by the Nez Perce Indians in what is now Washington. The color pattern dates back to prehistoric times. Drawings in caves located in France record spotted horses being hunted for food more than 20,000 years ago. Similar records have been found in ancient China, Iran, and Egypt. The name *Appaloosa* came from *palouse*, a French word meaning grassy plain. The Nez Perce developed this horse from Spanish spotted horses and instilled great quality in it. It has been said that the Nez Perce were never defeated on horseback by the U.S. Cavalry during the Indian wars. It was not until the Nez Perce signed a peace treaty with the United States that they were subdued.

FIGURE 11–24
The Appaloosa has a distinct color pattern. Although some are nearly solid in color, spots over the hip are preferred.
Courtesy of Don Shugart and the Appaloosa Horse Club.

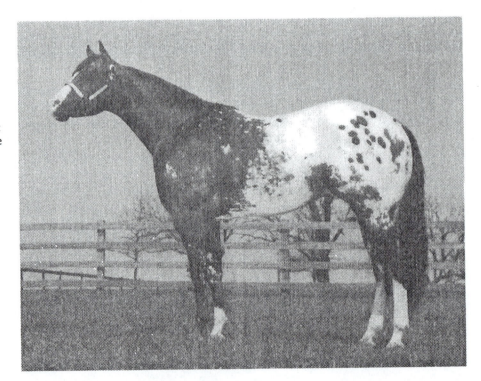

Legend has it that after the Nez Perce signed the peace treaty and moved to the reservation, the U.S. Cavalry herded all the Appaloosas they could find into a canyon and killed them, thus ending the threat of a fast-moving, proud nation. If this legend is true, it could account for a lack of Appaloosa blood today. The outstanding characteristic of this breed is its color pattern, preferably spots over the hip and loin, although some Appaloosas have no spots at all.

Although the color pattern is variable, it is usually seen as white over the loin and hips, with dark, round- or egg-shaped spots. The eye is encircled by white, the shin is mottled, and the hooves are striped black and white. Today the Appaloosa is used as a stock horse, pleasure horse, parade horse, and race-horse. Appaloosas stand 14 to 16 hands high and weigh 900 to 1,250 pounds. Artificial insemination is accepted, provided the stallion's blood type is filed with the Appaloosa Horse Club, Inc. and the mare is bred where the stallion is standing. Embryo transfer is limited to one foal per donor mare per year.

Paint and Pinto. The pinto color is characteristically brown and white, but the unique black on white color is referred to as piebald. Although not a breed in the conventional sense, the pinto and piebald are represented by the American Paint Horse Association (see figure 11–25). An interesting characteristic of the paint (pinto or piebald) is its "glass eyes," so called because they are white or milky in appearance. Glass eyes are not discriminated against in the association and do not indicate poor vision or lack of intelligence, as some elder horsemen used to believe. The Comanches favored the paint horse for its speed and durability. It is believed these "native horses" originated from those brought to Mexico in 1519 by Hernando Cortés. The American Paint Horse Association was formed in 1962. All registered paint horses must be sired by a registered paint, quarter horse, or Thoroughbred.

FIGURE 11–25
The American paint horse.
Courtesy of the American Paint Horse Association.

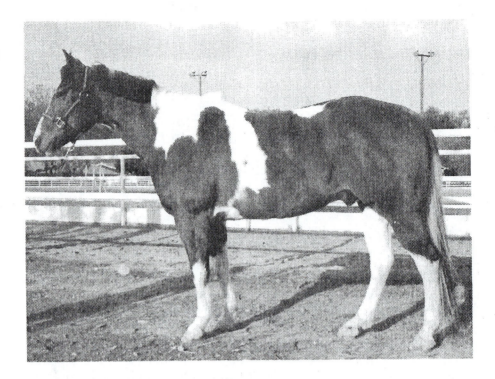

The Pinto Horse Association of America was formed in 1956. The pinto is primarily a color breed. Its registry accepts any pinto-colored horse of any background. Pinto horses are ideally half color or colors and half white with many well-placed spots (see figure 11–26). The two basic patterns are called *tobiano* and *overo*. Tobiano horses have color on their head, chest, flanks, and some in the tail. The legs are usually white, with white markings extending over their back. Overo horses have jagged white markings, mostly over the midsection of their body and neck area. Their legs are usually a color rather than white. The word *pinto* comes from the Spanish word *pintado*, meaning painted. Pinto horses also originated in the United States from horses of Spanish ancestry.

Both pintos and paints are used for stock, pleasure, show, and racing. Both horse types usually stand at 14 to 16 hands high and weigh an average of 900 to 1,250 pounds.

Artificial insemination is accepted for paints if the semen is used within 24 hours of collection and on the same premises as the collection. Special conditions must be met for transported semen. Embryo transfer is accepted with special stipulations. For pinto horses, artificial insemination is accepted if a licensed veterinarian certifies collection, insemination, and birth and if blood type evidence of parentage is furnished. The Pinto Horse Association of America has no rules on embryo transfer.

Palomino.   The palomino (see figure 11–27) came by way of California from Spanish stock. In Spain it was once known as the horse of queens, and commoners were forbidden to ride or own a palomino because of those royal connections. The palomino is not a true breed but rather a type whose color can always be produced by crossing a chestnut with an albino. A palomino bred to a palomino may produce a chestnut, an albino, or another palomino. Its most outstanding characteristic is its golden color, varying from light to dark with dark

(a)

FIGURE 11–27
(a), 1999 PHBA World Championship Horse Show, Tulsa, OK; Mr. Golden Zipper, Wise, Amber, Gause, TX, World champ Amateur Western Horsemanship, World Champ Amateur Showmanship, Reserve World Champ Amateur Hunt Seat Equitation.
(b), 1999 PHBA World Championship Horse Show, Tulsa, OK; Strawtown Rocky, Res World Champion Sr. Hunter Hack. Barbara J. Conroe, TX, Sharon Wellmann.
(c), 1999 PHBA World Championship Horse Show, Tulsa, OK; Two Socks for Te; Reserve World Champion, 1998 Amateur Yearling Stallions, Troy Wright, Oklahoma City, OK.
Courtesy of PHBA and Jeff Kirkbride.

(b)

(c)

skin and a light-colored mane and tail. The palomino's conformation, quality, and performance vary with the particular bloodlines involved. Palominos are used as stock, parade, pleasure, saddle, and fine harness horses. They stand 14 to 16 hands high and weigh an average of 900 to 1,250 pounds. The Palomino Horse Association, Inc. recognizes all rules of other registry associations in which palomino horses are registered.

Pony Breeds

Connemara.   The Connemara pony is the largest of the pony breeds. Connemaras stand approximately 14.2 hands high. The Connemara Pony Breeders Society was formed in 1923 in Ireland; the American Connemara Pony Society formed in 1956. Their colors include gray, black, bay, brown, dun, and cream. Although originally bred as utility animals for the harsh environment found in western Ireland, these animals are used today as jumpers, show ponies (saddle and harness), and mounts for children. Artificial insemination is accepted, and the semen may be transported fresh or frozen.

Pony of the Americas.   The Pony of the Americas (POA) breed was developed in 1954 in Iowa by crossing an Appaloosa mare and a Shetland stallion. Accepted colors are those similar to Appaloosas with white over the loin and hips and with dark, round- or egg-shaped spots (see figures 11–28 and 11–29). The POA stands 11.2 to 13.2 hands high; bloodlines now include quarter horse and Arabian with the Appaloosa color and characteristics. They are used mostly by children as a western-type pony. Artificial insemination and embryo transfer are accepted if the stallion is on the premises when mating occurs.

Shetland.   Shetland ponies originated in the Shetland Islands off the coast of Scotland. In the 1800s they were imported into the United States and England, where they were used to work in coal mines. Shetland ponies are of the draft type but with a small body, making them ideal for drawing up coal in small areas (see figure 11–30). The registry accepts all colors, either solid or broken. Today, Shetlands are used as children's mounts, harness horses, roadsters, and for racing. They usually stand 9.2 to 10 hands high and weigh 300 to 400 pounds.

Welsh Pony.   The Welsh pony (see figures 11–31 and 11–32) had its origin in Wales. Rough, mountainous terrain and sparse vegetation led to a natural selection system from which only the most fit survived. This rugged, agile breed

FIGURE 11–28
Pony of the Americas (POA) are a miniature version of the commingling of Arabian and quarter horse with Appaloosa color patterns.
Courtesy of Stephanie Abronson, Monte Nido, CA.

FIGURE 11–29
Pony of the Americas.
Note the spots and
striped hooves.
Courtesy of Stephanie
Abronson, Monte Nido,
CA.

FIGURE 11–30 The
American Shetland
pony has evolved
through selection as a
pleasure type.
Courtesy of Stephanie
Abronson, Monte Nido,
CA.

FIGURE 11–31
Leyeswick Pixie, Sec.
C Welsh Pony of Cob
type, Leyeswick Stud,
England.
Courtesy of Stephanie
Abronson, Monte
Nido, CA.

FIGURE 11–32
Section B Welsh mare
in Wales.
Courtesy of Stephanie
Abronson, Monte
Nido, CA.

is somewhat larger than the Shetland and is unexcelled as a mount for older, more experienced children. It is also popular as a show pony in the East, often crossed with the Thoroughbred. Frequently described as a miniature coach horse, it is also used, within size limitations, for roadsters, racing, trail riding, parade horses, stock work, and hunting. Any color except piebald or skewbald is accepted. Welsh ponies average 11 to 13 hands high and weigh between 350 and 850 pounds.

Artificial insemination is accepted with prior approval and if foal DNA is recorded. Embryo transfer is accepted with prior approval. The donor mare is limited to four genetic offspring per year. Either fresh or frozen embryos may be used.

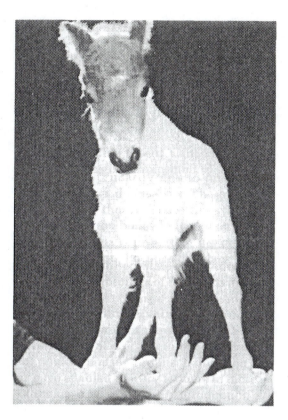

FIGURE 11–33
The miniature horse is bred as a pet; the smaller, the more prized.
Reprinted with permission from J. Blakely and D.H. Bade, *The Science of Animal Husbandry,* 6th ed. (Englewood Cliffs, NJ: Prentice-Hall, 1994).

Miniature Horses.   One of the most unusual breeds of horses is the true miniature (see figure 11–33), not to be confused with the ponies or dwarf breeds. The maximum height allowed by the International Miniature Horse Registry is 8.2 hands (34 inches). Most stand only 5.3 to 6.2 hands high. A revival of interest by hobbyists in the miniature horse has lead to a recent rediscovery of these small animals, which are found in increasing numbers throughout the United States, South Africa, England, Canada, and Australia.

Stables of miniature horses are kept by such notable personalities as Queen Elizabeth II, King Juan Carlos, the Kennedy family, and many Hollywood entertainers. One peculiar aspect related to miniature horses is that the smaller they are, the more valuable they become. The smallest miniature was recorded at 3.3 hands.

The miniature is kept strictly as a pet, for hobby, or for breeding purposes. It is generally thought that the miniature is too small to be ridden by even young children.

Sport Horse Breeds

Hanoverian.   The Hanoverian breed originated in the Hanover area of Germany more than 400 years ago. Hanoverians were developed and used as war horses. The American Hanoverian Society was started in 1978. Today Hanoverians are used for riding, driving (carriage horses), hunting, jumping, dressage, and utility purposes. They typically stand 16.2 hands high and weigh approximately 1,200 pounds. Artificial insemination is accepted, but a licensed veterinarian must issue and sign the insemination certificate.

FIGURE 11–34
Holsteiner mare.
Courtesy of Stephanie
Abronson, Monte
Nido, CA.

FIGURE 11–35
Trakehner breed was
developed in East
Prussia. Trakehner
Stallion.
Courtesy of Jean
Brinkman, Wellborn, FL.

Holsteiner.   The Holsteiner is one of the oldest breeds of warmbloods. The Holsteiner originated in Germany, and its breeding can be traced back to the 1300s. These large animals were also developed as war horses. Today they are used as jumping, dressage, and driving horses. The Holsteiner State Stud was formed in 1886. The Holsteiner stands 16 to 17 hands high. Bay is the dominant color, but grays and chestnuts are also accepted. (See figure 11–34.)

Trakehner.   The Trakehner was developed in Trakehner, East Prussia. The breed includes Thoroughbred and Arabian blood. Most of the breed was destroyed during five separate wars that crossed their primary breeding range (eastern Europe). In 1974, the North American Trakehner Association was formed. Although originally developed as a light cavalry horse, it is used today for dressage, for combined training, and for jumper and hunter classes. Trakehners stand 15.2 to 16.0 hands high (see figure 11–35).

Artificial insemination with fresh semen is accepted as long as the stallion remains on the premises. The collection and insemination must be under the direction of a licensed veterinarian. When fresh or frozen semen is transported, both the mare and foal must be blood typed. Embryo transfer is also accepted when both donor mare and foal are blood typed. Only one foal per donor per year is allowed.

DONKEYS: JACKS AND JENNETS

A jack is the male of the *Equus asinus* species, and a jennet is the female of the species. Compared with the horse, the donkey (see figure 11–36) is much smaller in stature, less subject to founder or injury, much more hardy, and less subject to hysteria under stress; it also has a longer gestation period (12 months).

MULES

Mules are a cross between a male donkey and a female horse. It has been said that the mule is without pride of ancestry or hope of posterity. Like most hybrids, it is seldom fertile (generally the mule is sterile; that is, it will not

FIGURE 11–36
Donkey.
Courtesy of Stephanie Abronson, Monte Nido, CA.

FIGURE 11–37   An excellent example of the standard mule. Courtesy of L.A. Pierce College Teaching Farm Laboratory.

reproduce). George Washington and Henry Clay first popularized the use of the mule. George Washington first imported a jack in 1787 as a gift of the king of Spain. This jack had a very docile temperament. By using another jack that had a mean disposition, Washington was able to combine the qualities of both of these superior jacks through his jennets, developing a strain of donkey that was eventually crossed with a stallion to develop the first good-quality mules in the United States. The state of Missouri is famous for its production of high-quality mules and is one of the few states today in which mules are actively shown. Compared with the horse, the mule (see figure 11–37) can withstand much higher temperatures, can eat irregular meals, or can be self-fed without digestive dangers. The mule is also much sounder of foot and will work in areas where most horses would panic. For instance, an object touching the head of a horse can cause it to rear up or to react in such a manner as to injure itself and perhaps its handler, whereas an object touching the head of a mule usually causes it to react in the opposite way—lowering its head very casually away from the sensation of touch.

HORSE REPRODUCTION AND MANAGEMENT

Although the reproduction of the equine was discussed in the general reproductive chapter (chapter 3), some additional specific horse reproductive parameters and management procedures need to be addressed. The horse is by nature a highly spirited, freedom-loving, nomadic animal. In the wild, the horse's reproduction approaches 90 percent efficiency or more. Under most domestic conditions, reproductive efficiency is closer to 50 percent at best. Lack of proper exercise, exposure to disease, poor nutrition, breeding at the wrong time of year, and many other factors contribute to this poorer reproductive performance. Therefore, knowledge of reproductive traits is especially important if horse breeders are to be successful. Please review basic reproductive anatomy and physiology from chapter 3.

Breeding Hints

Although fillies reach puberty at 12 to 15 months of age, they should not be bred before 2 years of age and preferably not before 3 years of age. Fillies bred at an earlier age traditionally have a poorer conception rate the following year. When bred at 3 years and properly cared for, mares can be expected to produce 10 to 12 foals during their lifetimes. Mares 20 years old or older can often produce foals.

Estrus

A mare's estrous cycle is normally 21 days (like that of a goat and cow), but it may range from 10 to 37 days and still be considered normal. The period of heat or estrus lasts 4 to 6 days, and diestrus lasts 17 to 19 days. Within 1 to 2 days preceding estrus the mare becomes more passive to the presence of the stallion. Signs of estrus in the mare are nervousness, a desire for company, frequent urination, nickering, and a swelling and movement of the vulva.

Teasing

Various schedules and techniques are used in the unique handling of mares to ensure maximum conception when breeding. To more effectively detect true estrus, mares suspected of "coming in" are allowed close proximity to a teaser stallion (see figure 11–38). The teaser may be on one side of a teasing board, restrained by halter in the field, or just led through the barn. Mares intended for breeding should be teased at least every other day. Some establishments use Shetland stallions as teasers because there is less chance, because of size differences, of an accidental mating. Cryptorchid stallions whose descended testicle has been removed make excellent teasers as well.

Actions of mares as observed by an experienced handler indicate the most logical time to breed. Mares empty their bladders frequently and exhibit restlessness. Most handlers think that talking to mares during the nervous time helps to calm them. This approach can be especially important in valuable brood mares to prevent injuries. By teasing mares to detect estrus, a good handler may get 75 percent of the mares in foal.

FIGURE 11–38
The mare suspected of "coming in" is allowed close proximity to a teaser stallion.
Photo by author.

Breeding

Because conception is low in domestic horses to begin with, the strict observance of sanitary measures and other precautions is important in successful breeding. The stallion and mare's genitals are washed with warm water and soap. A tail bandage made of flannel or gauze is applied to the mare, which may be restrained by use of a twitch and hobbles (see figure 11–39). These precautionary measures are needed only about 10 percent of the time but are recommended to prevent possible harm to the male from biting or kicking.

Sometimes a "phantom" is used to collect semen from a stallion (see figure 11–40). As the stallion mounts the phantom the penis is diverted into an artificial vagina, which is held at a fixed position while the stallion ejaculates. Use of the phantom greatly eliminates the potential injury to the stallion by the mare.

Time to Breed

Ovulation generally occurs on the third or fourth day of estrus. Ovulation is related to the end of estrus and not the onset. The ovum is viable for about 6 hours, and the stallion's sperm will live about 30 hours in the reproductive tract of the mare. It is thus recommended that mares be bred daily or every other day beginning with the third day of the estrous period.

Gestation

The average length of gestation varies from 337 days in the Arabian horse to 342 days in the Morgan.[4] Colts are usually carried 2 to 7 days longer than

---

[4]H. J. Bearden and J. W. Fuquay, *Applied Animal Reproduction,* 4th ed. (Englewood Cliffs, NJ: Prentice-Hall, 1997).

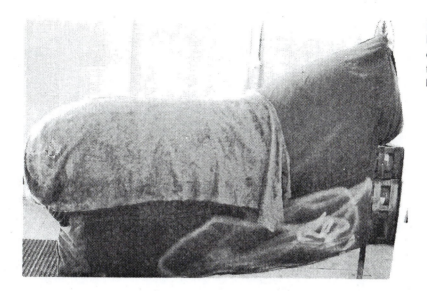

FIGURE 11–40
Phantom used to
collect semen from
the stallion.
Photo by author.

fillies. Nutrition also significantly affects gestation length. Mares on a high plane of nutrition foal 4 days sooner than mares that have poor-quality diets.

Foaling

A critical time for both the mare and foal occurs at parturition. Although some births require no assistance at all, other times life hangs by a thread and only human assistance can prevent a fatal situation. For unknown reasons, nearly all mares foal during dark hours; as inconvenient as it may be, experienced horse breeders are on hand to assist the mares.

Signs of Approaching Parturition.   One of the first signs that foaling is near is the mare's beginning to "make bag," the gradual enlargement of the mammary glands. Typically, *waxing* occurs (a waxy substance appears at the end of each teat) prior to foaling. Usually within 12 to 24 hours of birth, the wax softens and falls away; milk may then drip or even stream from the teats. It should be mentioned that these signs are not always on prompt schedule. Occasionally, a mare will wax two or three times and may have milk streaming from the teats 10 days before parturition. At about the same time, the mare will show a soft swelling and noticeable relaxation of the muscles of the vulva. The muscles and ligaments associated with the pelvis relax, making the mare appear loose about the hips.

The mare will usually leave other horses, if in the pasture, to seek solitude. It may appear in a bad temper, pinning its ears back and kicking when other horses approach. The mare carries its tail away from the body slightly, urinates frequently, may bite at its sides, and alternates lying and standing. Sweating profusely is normal at this time. The first water bag (formed by the chorioallantoic membrane) usually breaks at this point, discharging 2 to 5 gallons of fluid, which help to lubricate the birth canal. If the foal is in the right position (see figure 11–41), involuntary muscle contractions of the uterus and abdominal muscles (labor) may start and birth can occur with surprising ease.

FIGURE 11–41
Normal presentation of a foal.
Reprinted with permission from J. Blakely and D.H. Bade, *The Science of Animal Husbandry,* 6th ed. (Englewood Cliffs, NJ: Prentice-Hall, 1994).

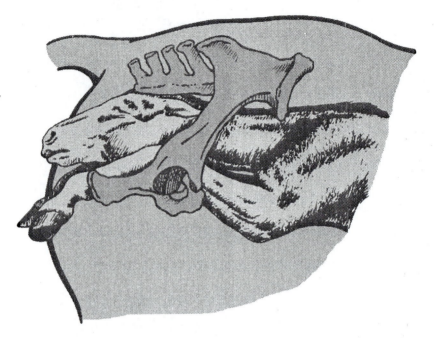

Care at Parturition.   Only a few items need be on hand at birth: a clean, well-bedded box stall; a source of light; clean, hot water and soap (Ivory is suggested); tail bandages; approved disinfectant; a suitable navel dressing (iodine); an enema bag; and a laxative (milk of magnesia).

When there is no doubt that the mare is about to foal, the attendant applies a tail bandage (flannel or gauze) and waits for nature to do the rest. No assistance should be given until it is obvious that help is needed. It is often necessary to tie a soft, thick rope around the foal's forefeet and pull to assist the mare. If both feet and the muzzle of the foal can be seen, then all is well. Birth may take 10 minutes to an hour.

If the muzzle or one or more legs cannot be seen after a reasonable amount of time, the attendant may need to wash his or her hands and the vulva with soap, water, and antiseptic solution and examine for complications. Some common difficulties are illustrated in figure 11–42 (head turned to one side), figure 11–43 (dorsotransverse position), and figure 11–44 (breech position). The foal must be rotated to the normal position (see figure 11–42) or suffocation may result. Nature supplies the foal with an umbilical cord of sufficient length to completely clear the birth canal only if birth occurs in the normal position. If the cord is crushed by a backward birth, oxygen is cut off and breathing may be stimulated in an **anaerobic** (without air) environment, resulting in suffocation. Suffocation may also occur in other positions if delivery of the head is delayed.

The new foal born with or without assistance should be checked for breathing restrictions. Any membranes or fluids covering the mouth or nostrils should be cleared. Breeders should allow 2 to 3 hours for the foal to gain strength and see that it nurses. It is absolutely necessary that the foal receive colostrum milk to supply its system with the necessary antibodies, vitamins, and energy required to start and maintain life. The antibodies that provide protection disappear from the milk within 24 to 36 hours. The foal—unlike

Anaerobic without oxygen

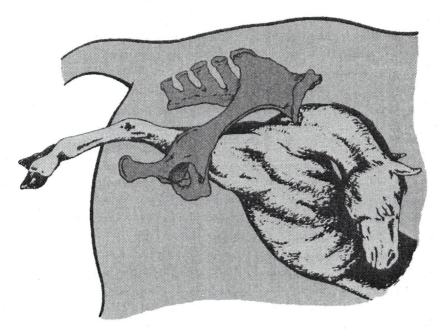

FIGURE 11–42 A foal with its head turned to one side. Reprinted with permission from J. Blakely and D.H. Bade, *The Science of Animal Husbandry*, 6th ed. (Englewood Cliffs, NJ: Prentice-Hall, 1994).

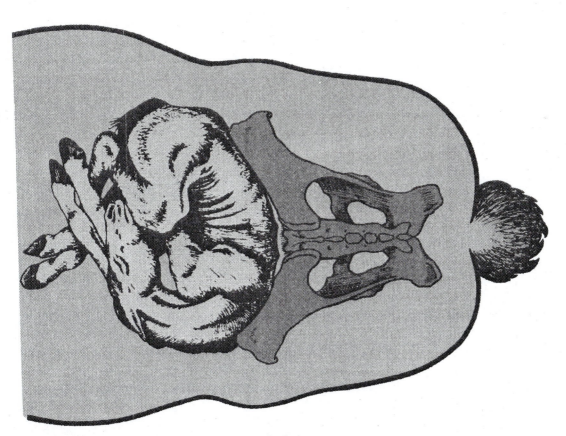

FIGURE 11–43 The dorsotransverse position of a foal.
Reprinted with permission from J. Blakely and D.H. Bade, *The Science of Animal Husbandry*, 6th ed. (Englewood Cliffs, NJ: Prentice-Hall, 1994).

FIGURE 11–44

The breech position of a foal.

Reprinted with permission from J. Blakely and D.H. Bade, *The Science of Animal Husbandry,* 6th ed. (Englewood Cliffs, NJ: Prentice-Hall, 1994).

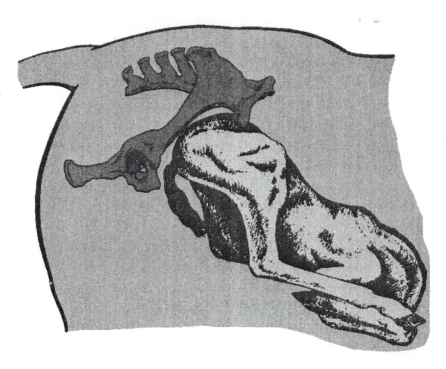

the calf, which receives some immunity through blood circulation—must rely almost entirely on colostrum. It is also a common practice to administer a tetanus antitoxin at this time.

An enema with warm, soapy water should be given the foal along with a dose of milk of magnesia. This is standard practice if the foal does not pass its meconium within 12 hours of birth. The **meconium** is a toxic substance developed in the large intestine just prior to birth and is sometimes absorbed if constipation occurs. After this passage, the problem does not develop again. Colic may also develop if early elimination is not induced.

Meconium first stools passed by a newborn

The navel cord should break by itself. It should not be tied because of the possibility of causing death due to navel ill. To prevent entry of harmful organisms to the bloodstream, as soon as the break occurs the navel stump and surrounding areas are treated with a 10 percent iodine solution daily until the area is dried.

Postfoaling Care.   The placenta, ideally shed within 3 hours, should be spread out and checked for missing pieces because retention of a piece as small as a hand can cause founder or infection of the mare. A normal placenta weighs 13 to 14 pounds. If all or part of the placenta is not shed within 6 to 8 hours, a veterinarian should be called for assistance.

Moderate exercise is important in stimulating the mare's uterus to tone up and in developing the foal's protective mechanism. A paddock should be provided for this purpose, and eventually the pair can be turned out to pasture during the day or night.

Rebreeding

Horses exhibit **foal heat,** more often called *9-day heat,* the first estrus at which rebreeding is possible. In actual practice, this foal heat may occur at 5 to 10 days after foaling or even later. Mares can be bred during this heat, but most modern equine breeders wait one additional estrous cycle to allow the mare to recover from the recent gestation and foaling. Conception rates at **foal heat** average only 25 percent. However, some mares will not cycle again for 50 to 60 days; thus, any chance at conception is thought to be worth trying for, no matter how slight.

Foal heat the first heat occurring after foaling in a mare, usually on the 9th day; also known as 9-day heat

Light-Induced Estrus.   Regardless of when a foal is born, its birth date is always January 1. The logic (or lack of logic, depending on your perspective) behind this practice is that horses can compete in racing and in showing based on their age. Everyone, because of this rule, breeds their mares so that they will foal as close to (but after) January 1 as possible. If a foal is born on December 31, it is automatically a year old the very next day (because of this rule). Many equine reproductive specialists have argued for return of the May 1 birthday, which is more in keeping with the mare's natural cycle.

Until such time and because mares in the United States are now typically bred 4 months sooner than they would in nature, many breeders have found it beneficial to use artificial lighting. Starting 90 days before breeding season, mares are gradually introduced to increasing amounts of light, so that they will be exposed to 16 hours of light and 8 hours of darkness by the time the breeding season starts.

Abortion

Mares are subject to a higher incidence of abortion than other species, further compounding the difficulties breeders face. It is important to have a veterinarian determine the cause of any aborted fetus (foal) by examining it because some forms are caused by microorganisms and thus may spread throughout the herd. Preventive vaccines or other sanitary measures should follow a positive diagnosis.

Two causes of abortion are not contagious and not uncommon. A twisted navel cord caused by fetal rotation that shuts off blood circulation is fatal. It is interesting to note that the other cause—the incidence of twinning—is of little concern in other species. However, in horses, only 9 percent of conceived twins are delivered live. Approximately 60 percent of mares with twin embryos abort or reabsorb one of the twins, and 31 percent lose both fetuses, usually by abortion.[5]

Pregnancy Examination

To get the highest efficiency possible, experienced breeders check mares suspected of conceiving by rectal palpation at approximately 35 days postbreeding. Mares should be properly restrained with breeding hobbles and nose twitch to prevent injury to the examiner. Chemical means of pregnancy diagnosis in mares is becoming more popular. Pregnant mare serum gonadotropin (PMSG) levels peak in mares pregnant 50 to 120 days but can be picked up in the blood as early as 40 days postconception. Progesterone levels in the blood or milk of

---

[5]H. J. Bearden and J. W. Fuquay, *Applied Animal Reproduction*, 4th ed. (Englewood Cliffs, NJ: Prentice-Hall, 1997).

mares can spot open (not-pregnant) mares but cannot guarantee pregnancy. Pregnancy can be determined using ultrasound as early as 18 to 20 days after breeding so that they can be rebred during the next heat period if they are open.

Artificial Insemination

Artificial insemination has some definite advantages in horses. Venereal disease, injuries to handlers and to stallions, and the costs of keeping a stallion or transporting the mare are all reduced or eliminated. However, the greatest advantage in the use of artificial insemination is the increase in genetic opportunities, from which the dairy, poultry, and more recently swine industries have benefited.

The artificial vagina is the normal method of semen collection from stallions. Semen is diluted, frozen, and stored as for cattle. Thawing and actual insemination are similar to the techniques used in cattle insemination (rectovaginal-cervical fixation method) (see figure 11–45). The equine industry has been reluctant to accept the use of artificial insemination, thus preventing the genetic progress seen in other species. As of July 1999, the Jockey Club did not approve the use of artificial insemination or the use of embryo transfer. Many equine breeders and reproductive specialists believe that with DNA typing and improved freezing techniques of stallion semen, there is no justification for not increasing their acceptance across the board. In other words, 95 percent of stallions in use should probably be gelded, with careful selection of the remaining 5 percent being used for artificial insemination and embryo splitting and transfer programs.

FIGURE 11–45
Rectovaginal-cervical fixation method of artificial insemination in a mare.
Photo by author.

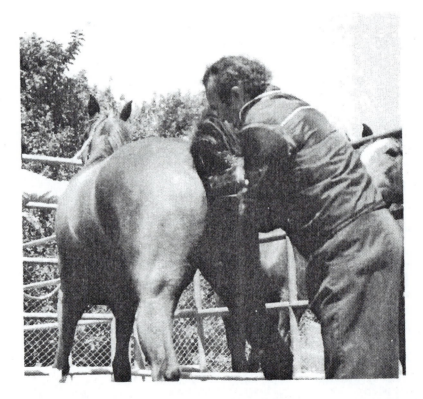

HORSE SELECTION

Judging a horse should be like looking at a beautiful painting; everything there should be pleasing to the eye. We can even go a little further in the selection of horses and say that everything we see has a direct relationship to the performance of that horse. It is with this thought in mind that the following examples are used to illustrate the direct relationship between conformation, type, and desired function.

There are basically three ways to select a horse—by pedigree, by performance, and by visual observation. The best method of selection would be to utilize all three types.

Horse Terminology

As discussed in judging cattle, sheep, and swine, there is unique terminology to describe the various parts of a horse's body. Before judging the equine, these terms must be learned (see figure 11–6).

Judging the horse varies depending on the breed and depending on the show (halter, model, or breeding class). The horse or class of horses of any breed should always be judged in a logical sequence. A common method of judging is to look at a horse first from the side, then from the front, and last from the rear. Then the accomplished judge will want to judge the moving horse. After taking a quick, overall view of the horse from all three positions, one can start looking for the finer points.

Because this is not a judging text we will concentrate here on only general equine judging terms and descriptions. The slope of the shoulder and the slope of the pastern are very important. The slope of the shoulder should be approximately 45 degrees, and it should match the slope of the pastern. The pastern acts as a shock absorber to the front end and gives a smoother ride. If this angle is much more than 45 degrees, the horse has a tendency to tire easily or go lame. If the slope is too straight it may develop into a bone-jarring type of ride, which no one desires for very long. The more highly developed the forearm muscle, generally the better movement the horse will have in the front end as well. A longer attachment is generally preferred to the heavier, more closely attached forearm muscle.

Figure 11–46 indicates what we should be looking for in a set of legs. Imagine dropping a plumb line down through the middle of the horse's leg, knee, and hoof and study the examples shown to determine the common deviations and the correct placement.

When judging the head and neck of a horse, the many breed variations or preferences should be kept in mind. In general, however, the ears are a good indication of temperament and intellect in a horse. Short, erect ears that often point forward indicate a horse that is in good condition, alert, and paying attention to what is going on. The nostrils are also important. It is desirable to have a very large nostril because intake of air is necessary for a hardworking horse. If the horse has sufficient lung capacity but is restricted at the nose, the lungs cannot be filled, thus reducing the horse's endurance.

Figure 11–47 shows the various parts of the head from the side view. The jaw is a good indication of strength of bone and constitution (hardiness). The desirable type is called a "dinner plate" jaw (for quarter horses), which means

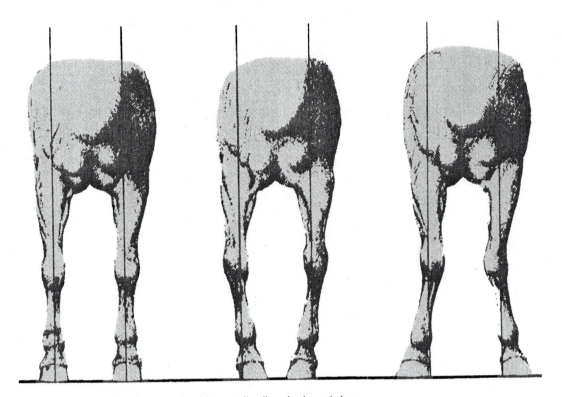

FIGURE 11–46 An observer should mentally align the horse's legs.
Reprinted with permission from J. Blakely and D.H. Bade, *The Science of Animal Husbandry,* 6th ed.
(Englewood Cliffs, NJ: Prentice-Hall, 1994).

FIGURE 11–47 A
"dinner plate" jaw,
shallow mouth, thin
neck, and trim
throatlatch are desired.
Reprinted with permission
from J. Blakely and D.H.
Bade, *The Science of
Animal Husbandry,* 6th
ed. (Englewood Cliffs,
NJ: Prentice-Hall, 1994).

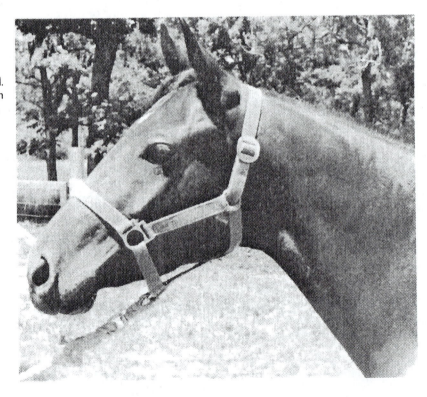

it is large and round. The mouth should be relatively shallow so that a bit placed in the mouth can control the horse with a minimum of pressure. A horse with a very deep mouth may be difficult to control. The throatlatch must be clean, with no excess muscling or finish because the throatlatch is the pivot point for the head and as the head goes, so goes the horse.

The side view (see figure 11–48) shows the withers, which should be prominent. The beginner may think this prominence unsightly, but its greatest function is to give the saddle a place to hang and prevent the saddle from turning. Thus a horse that has a full, flat withers is not desirable. Also from the side view the muscling commonly referred to as the quarter can be noted. It should be well developed with a bulge that lets down to the area of the stifle muscle. The leg settings from the side view, as illustrated in figure 11–49, are also observed using the same imaginary plumb line.

Perhaps the most important view of all is the rear view, as illustrated in figure 11–50. One should visualize the conformation lines of an apple turned with the stem down. Muscle that is quite similar to this imaginary example is desired. The horse should be very wide through the stifle muscle, and this area should be the widest part seen from the rear. The gaskin muscle is the most prominent feature on the way down. It should be full and bulging outward. Directly inside this muscle is another gaskin referred to as the inside gaskin muscle. Most breeders agree that the inside gaskin muscle is one of the most important muscles on a horse and one of the most difficult to develop through selection and breeding.

Figure 11–51 shows the rear view leg settings, both good and bad, using the techniques previously developed in figures 11–46 and 11–49. Every breed of horse should conform to the ideals illustrated in these figures.

After viewing the standing horse, the judge needs to view the horse walking away and jogging back. The judge is looking for its "way of going or moving," which means that its feet and legs move as smoothly and straightly as

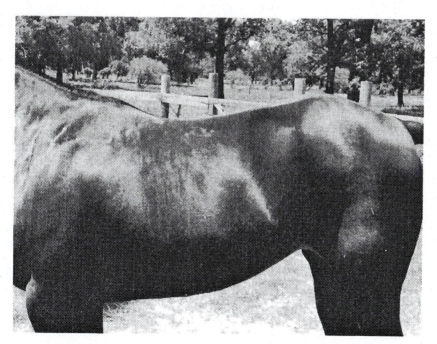

FIGURE 11–48
Prominent withers, short back, and long underline.
Reprinted with permission from J. Blakely and D.H. Bade, *The Science of Animal Husbandry,* 6th ed. (Englewood Cliffs, NJ: Prentice-Hall, 1994).

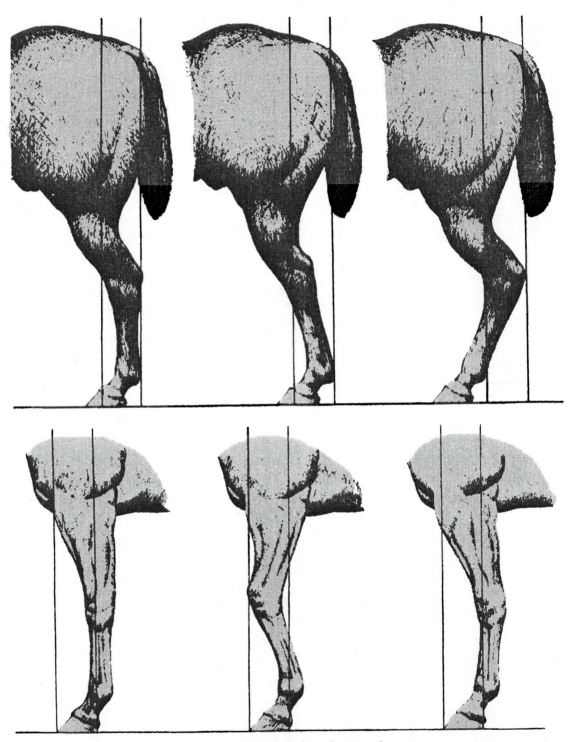

FIGURE 11–49 Leg settings should not vary drastically from the normal.
Reprinted with permission from J. Blakely and D.H. Bade, *The Science of Animal Husbandry*, 6th ed. (Englewood Cliffs, NJ: Prentice-Hall, 1994).

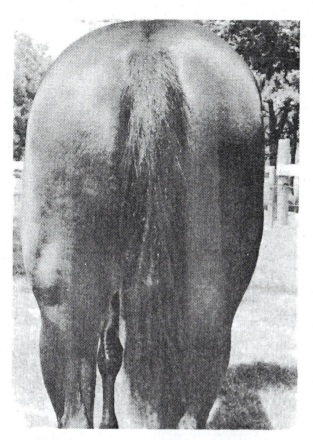

FIGURE 11–50
Lines of the rear
quarter resemble an
upside-down apple.
Reprinted with
permission from
J. Blakely and D.H. Bade,
*The Science of Animal
Husbandry,* 6th ed.
(Englewood Cliffs, NJ:
Prentice-Hall, 1994).

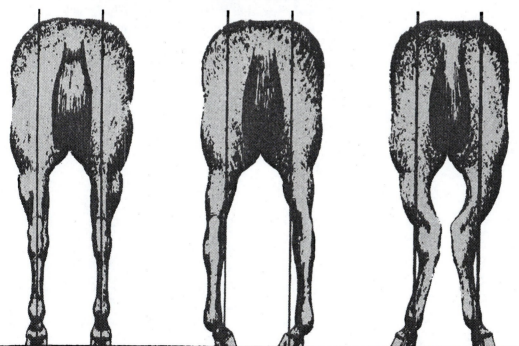

FIGURE 11–51 An example of leg settings from the rear view.
Reprinted with permission from J. Blakely and D.H. Bade, *The Science of Animal Husbandry,* 6th ed.
(Englewood Cliffs, NJ: Prentice-Hall, 1994).

expected from the standing plumb line observations. Imagine pushing someone on a swing. If the swing is well balanced, the movement will be like that
of a pendulum on a clock. Likewise, the legs of a horse (when viewed from the
front and rear) should move in a similar manner. Defects not detected at
slower speeds may show up at faster movement, which is why observations of
both walking and jogging are necessary. Some common defects include wing-
ing out or paddling (slinging the feet outward), scalping (front legs rubbing together), toeing in (slinging the feet inward), and forging (hind feet stepping on
front feet).

TEETHING

Nearly everyone has seen the horse trader on the late, late western cast a suspicious eye on a prospective horse being traded or bought. After examining the
horse's teeth, the trader calmly announces that the horse is much older than advertised. What did he see and what made him so sure? Horses' teeth come in
with such predictable accuracy that it is now quite common for race tracks to
employ a specialist to determine age, thus preventing unfair practices of running more mature horses against younger competitors.

Although age determination within a few months can be quite simple for
the expert, a more simplified approach is taken in this text. A young horse has
baby teeth that are replaced by permanent teeth. There are six upper front teeth
and six lower front teeth. The permanent teeth start erupting in pairs, starting
with the two middle incisors (the front teeth) when the horse is around 2 to
2 1/2 years of age. Both top and bottom middle incisors are complete by 3 years
of age (see figure 11–52). Note how much larger and longer these teeth are than
the baby teeth. At 4 years, the next pair is complete (see figure 11–53), leaving
only one pair of temporary incisors. Figure 11–54 illustrates the complete set of

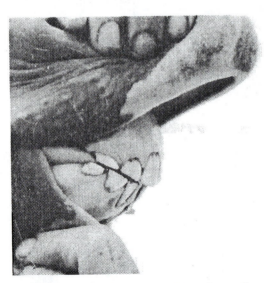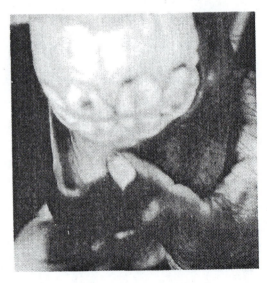

FIGURE 11–52 The 3-year-old horse's mouth.
Reprinted with permission from J. Blakely and D.H. Bade, *The Science of Animal Husbandry,* 6th ed.
(Englewood Cliffs, NJ: Prentice-Hall, 1994).

permanent teeth that exists for 5-year- old horses. An interesting point is the development of canine teeth at about this age (although it may be as early as age 3 1/2). These canine or wolf teeth are always seen in a stallion or gelding but only rarely in a mare. Figure 11–55 is a rare picture of an 8-year-old mare's mouth with a protruding canine tooth.

In summary, the average horse handler can tell the age of horses that are 3, 4, and 5 with a little practice. Beyond 5, more detailed observations are made

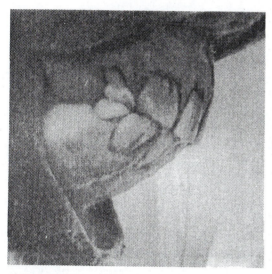

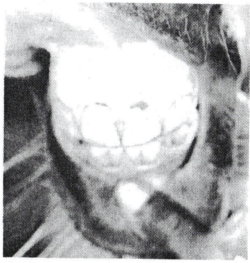

FIGURE 11–53   The 4-year-old horse's mouth.
Reprinted with permission from J. Blakely and D.H. Bade, *The Science of Animal Husbandry,* 6th ed. (Englewood Cliffs, NJ: Prentice-Hall, 1994).

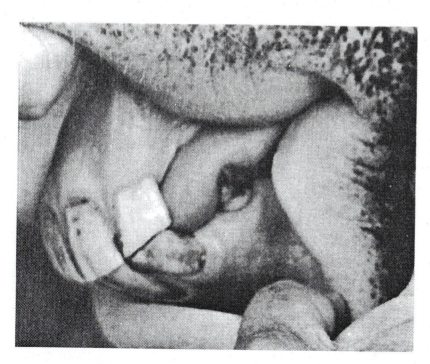

FIGURE 11–54   A complete mouth (age 5).
Reprinted with permission from J. Blakely and D.H. Bade, *The Science of Animal Husbandry,* 6th ed. (Englewood Cliffs, NJ: Prentice-Hall, 1994).

FIGURE 11–55 An 8-year-old mare's mouth with rare canine tooth. Reprinted with permission from J. Blakely and D.H. Bade, *The Science of Animal Husbandry,* 6th ed. (Englewood Cliffs, NJ: Prentice-Hall, 1994).

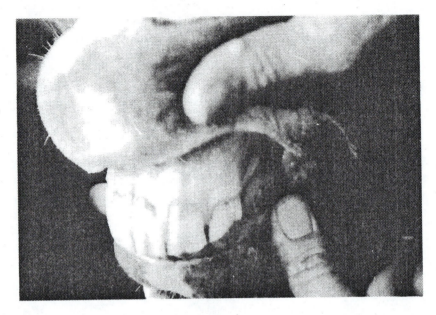

▓ **Galvayne's groove** longitudinal depression on the surface of the upper corner incisor teeth of horses 10 to 29 years of age; used to assist in age determination

with reference to the wearing of cusps, the angle of incisors, and **Galvayne's groove.** On the surface of the upper corner incisor teeth of horses 9 to 10 years of age a longitudinal depression begins to appear (Galvayne's groove). The groove grows with the tooth as the horse ages. At 15 years it appears halfway down the tooth. By 20 years, the groove extends the entire length of the tooth. By 25 years, it is absent at the top of the tooth, and by 30 years it disappears completely. An excellent reference with great detail is published by the American Association of Equine Practitioners, Golden, Colorado, for the serious student of this neglected art.

NUTRITION

The nutrient requirements of horses are not unlike those of cattle. In chapter 4 we discussed the basic digestive anatomy of the horse, the six basic nutrients required by all livestock, and some basic rules of thumb in feeding. Equines are used almost exclusively for sport, work, and recreation rather than for meat or milk. Owners should be cautioned against feeding prepared cattle feeds to horses because of the possibility of nonprotein nitrogen (NPN) content. Cattle and other ruminants can utilize some NPN, which cheapens the ration, but horses have a very low tolerance for it and may develop toxic reactions. Special prepared horse rations are commercially available to meet every safety and dietary need. Considerable skill and judgment are required to feed according to nutritive needs under fluctuating requirements.

The digestive efficiency of horses is not nearly as good as that of ruminants. Horses do not have large rumina to store and digest large amounts of forage; instead, horses must rely on the large cecum that follows the small intestine. The location of the functional cecum limits the absorption of many nutrients produced by the microorganisms there (because most digestion of nutrients other than energy takes place preceding the cecum).

Founder and colic are common disorders related to diet because horses require high energy but need some bulk to prevent compaction in the relatively small stomach, which leads to colic and/or founder. Consequently, horse breeders have long recognized oats as a safe, superior grain source for their animals because it is high in nutrition and moderately bulky. The horse owner's admiration for the grain is excelled only by the horse's desire to consume it.

Nutrient Needs of Horses

Specific feed requirements vary depending on the use of the horse. Idle horses require less energy than working horses, lactating mares require more protein than open brood mares, and nearly all requirements are higher for young horses than for mature horses. Nonworking mature horses are fed simply to maintain their body weight. Of course, this amount varies depending on whether the horse is in good condition to start with. A common mistake is to simply assume that a flack of hay morning and evening will satisfy all the nutrient needs of every single horse. Individual differences based on body condition scoring dictate individual requirements at the feed bunk.

Working horses also require varying amounts of feed based on intensity of work. *Light work* is defined as that performed by recreational or pleasure horses and show horses used for pleasure and in equitation classes. *Medium work* involves moderate exercise that usually results in various degrees of intensity (ranch work, roping, cutting, barrel racing, jumping, and hunting, for example). *Intense work* describes exposure to prolonged, intense exercise. These athletes would include polo ponies, endurance and competitive trail horses, racehorses, and three-day-event horses. Both the total amount of feed consumed and the ratio of grains to roughages must increase as work intensity increases (see table 11–3). The minimum forage requirement for an adult horse is 1.0 percent of its body weight per day. Feeding less than this amount may result in digestive disturbances that could lead to the death of the animal.

TABLE 11–3	Expected Feed Consumption in Horses (Percentage of Body Weight)*		
	Forage	**Concentrate**	**Total**
Mature horses			
Maintenance	1.5–2.0	0–0.5	1.5–2.0
Mares (late gestation)	1.0–1.5	0.5–1.0	1.5–2.0
Mares (early lactation)	1.0–2.0	1.0–2.0	2.0–3.0
Mares (late lactation)	1.0–2.0	0.5–1.5	2.0–2.5
Working horses			
Light work	1.0–2.0	0.5–1.0	1.5–2.5
Moderate work	1.0–2.0	0.75–1.5	1.75–2.5
Intense work	0.75–1.5	1.0–2.0	2.0–3.0
Young horses			
Nursing foal (3 months)	0	1.0–2.0	2.5–3.5
Weanling foal (6 months)	0.5–1.0	1.5–3.0	2.0–3.5
Yearling foal (12 months)	1.0–1.5	1.0–2.0	2.0–3.0
Long yearling (18 months)	1.0–1.5	1.0–1.5	2.0–2.5
Two-year-old (24 months)	1.0–1.5	1.0–1.5	1.75–2.5

Source: Edited by E. A. Ott, J. P. Baker, H. F. Hintz, et al. *Nutrient Requirements of Horses,* 5th ed. (Washington, D.C.: National Academy Press, 1989).

*Food consumption based on an air-dry feed (about 90 percent dry matter).

For gravid (pregnant) equines during the first 8 months of pregnancy, nutrient requirements are very similar to those of the maintenance horse. Pregnant mares require additional nutrients in the last 3 months, however, due to the rapid development of the fetus. Mares are expected to gain an average of 0.9 to 1.1 pounds per day during the final 90 days of gestation. Depending on the mare's body condition score, some breeders may start increasing nutrient density and quantity during the last 4 months of pregnancy.

The greatest nutrient demand of a mature horse occurs during the first 3 months of lactation. As milk production decreases, after the third month, nutrient requirements decrease as well.

Other Factors Affecting Feed Consumption and Dietary Requirements

Large horses have a greater mass to maintain and thus will have larger maintenance requirements than smaller horses. Cold temperatures mandate additional energy to provide for body heat; warm temperatures require less energy to maintain internal body temperatures. Forages and not concentrates should be provided to horses housed in cold environments. The metabolic breakdown of forages creates a greater internal heat than does the breakdown of concentrates or grains. The zone of thermal comfort is defined as the ambient temperature at which it is least difficult to warm and cool the body. For most horse breeds, the thermal comfort zone is approximately 62 to 79°F. Thus, horses housed at temperatures below 62°F should be provided extra forage; horses housed at temperatures above 79°F should have less forage and perhaps more grains in the diet.

Stress, internal and external parasites, age, degree of nervousness, and differences in individual horses' ability to process nutrients are additional modifiers of dietary needs. In other words, horses cannot be fed simply by some magic rule of thumb, computer program, or old wives' tale. Smart breeders look at each individual horse to modify basic diets set up for the group.

Although considerable energy is synthesized in the cecum and utilized by the horse, very little benefit results from protein created by the cecal microbes. Horses thus require high-quality or a variety of protein sources to meet their needs. In particular, the amino acid lysine is one that horse nutritionists tend to concentrate on when balancing equine diets. Soybean meal, cottonseed meal, and linseed meal are the most important sources of supplemental plant protein. Cottonseed meal should be used sparingly, if at all, due to its gossypol content. Horses are resistant to gossypol only at lower levels. Many animal protein supplements contain considerable amounts of fat, which tend to become rancid when stored for any length of time. Dried skim milk is a common protein source found in many foal feeds.

Although oats are the preferred grain among most horse owners, they are too costly for everyone to use. Corn and cob meal (ground ear corn), barley, and wheat are also used in horse rations. To meet the energy needs of growing, lactating, or working horses considerable quantities of grain are normally required. Tables 11–4 through 11–12 give requirements in terms of megacalories (Mcal) of digestible energy (DE) needed by the equine.

TABLE 11-4 Daily Nutrient Requirements of Ponies (200-Kilogram Mature Weight)

Animal	Weight (kg)	Daily Gain (kg)	DE (Mcal)	Crude Protein (g)	Lysine (g)	Calcium (g)	Phosphorus (g)	Magnesium (g)	Potassium (g)	Vitamin A (10³IU)
Mature Horses										
Maintenance	200		7.4	296	10	8	6	3.0	10.0	6
Stallions (breeding season)	200		9.3	370	13	11	8	4.3	14.1	9
Pregnant mares	200									
9 months			8.2	361	13	16	12	3.9	13.1	12
10 months			8.4	368	13	16	12	4.0	13.4	12
11 months			8.9	391	14	17	13	4.3	14.2	12
Lactating mares										
Foaling to 3 months	200		13.7	688	24	27	18	4.8	21.2	12
3 months to weaning	200		12.2	528	18	18	11	3.7	14.8	12
Working horses										
Light work[a]	200		9.3	370	13	11	8	4.3	14.1	9
Moderate work[b]	200		11.1	444	16	14	10	5.1	16.9	9
Intense work[c]	200		14.8	592	21	18	13	6.8	22.5	9
Growing Horses										
Weanling, 4 months	75	0.40	7.3	365	15	16	9	1.6	5.0	3
Weanling, 6 months										
Moderate growth	95	0.30	7.6	378	16	13	7	1.8	5.7	4
Rapid growth	95	0.40	8.7	433	18	17	9	1.9	6.0	4
Yearling, 12 months										
Moderate growth	140	0.20	8.7	392	17	12	7	2.4	7.6	6
Rapid growth	140	0.30	10.3	462	19	15	8	2.5	7.9	6
Long yearling, 18 months										
Not in training	170	0.10	8.3	375	16	10	6	2.7	8.8	8
In training	170	0.10	11.6	522	22	14	8	3.7	12.2	8
Two year old, 24 months										
Not in training	185	0.05	7.9	337	13	9	5	2.8	9.4	8
In training	185	0.05	11.4	485	19	13	7	4.1	13.5	8

Source: Reprinted with permission of T. W. Perry, A. E. Cullison, and R. S. Lowrey, *Feeds & Feeding,* 5th ed. (Upper Saddle River, NJ: Prentice-Hall, 1999).
Note: Mares should gain weight during late gestation to compensate for tissue deposition. However, nutrient requirements are based on maintenance body weight.
[a]Examples are horses used in Western and English pleasure, bridle path hack, equitation, etc.
[b]Examples are horses used in ranch work, roping, cutting, barrel racing, jumping, etc.
[c]Examples are horses used in race training, polo, etc.

TABLE 11-5 Daily Nutrient Requirements of Horses (400-Kilogram Mature Weight)

Animal	Weight (kg)	Daily Gain (kg)	DE (Mcal)	Crude Protein (g)	Lysine (g)	Calcium (g)	Phosphorus (g)	Magnesium (g)	Potassium (g)	Vitamin A (10³ IU)
Mature Horses										
Maintenance	400		13.4	536	19	16	11	6.0	20.0	12
Stallions (breeding season)	400		16.8	670	23	20	15	7.7	25.5	18
Pregnant mares	400									
9 months			14.9	654	23	28	21	7.1	23.8	24
10 months			15.1	666	23	29	22	7.3	24.2	24
11 months			16.1	708	25	31	23	7.7	25.7	24
Lactating mares										
Foaling to 3 months	400		22.9	1,141	40	45	29	8.7	36.8	24
3 months to weaning	400		19.7	839	29	29	18	6.9	26.4	24
Working horses										
Light work[a]	400		16.8	670	23	20	15	7.7	25.5	18
Moderate work[b]	400		20.1	804	28	25	17	9.2	30.6	18
Intense work[c]	400		26.8	1,072	38	33	23	12.3	40.7	18
Growing Horses										
Weanling, 4 months	145	0.85	13.5	675	28	33	18	3.2	9.8	7
Weanling, 6 months										
Moderate growth	180	0.55	12.9	643	27	25	14	3.4	10.7	8
Rapid growth	180	0.70	14.5	725	30	30	16	3.6	11.1	8
Yearling, 12 months										
Moderate growth	265	0.40	15.6	700	30	23	13	4.5	14.5	12
Rapid growth	265	0.50	17.1	770	33	27	15	4.6	14.8	12
Long yearling, 18 months										
Not in training	330	0.25	15.9	716	30	21	12	5.3	17.3	15
In training	330	0.25	21.6	970	41	29	16	7.1	23.4	15
Two year old, 24 months										
Not in training	365	0.15	15.3	650	26	19	11	5.7	18.7	16
In training	365	0.15	21.5	913	37	27	15	7.9	26.2	16

Source: Reprinted with permission of T. W. Perry, A. E. Cullison, and R. S. Lowrey, *Feeds & Feeding*, 5th ed. (Upper Saddle River, NJ): Prentice-Hall, 1999).

Note: Mares should gain weight during late gestation to compensate for tissue deposition. However, nutrient requirements are based on maintenance body weight.

[a]Examples are horses used in Western and English pleasure, bridle path hack, equitation, etc.

[b]Examples are horses used in ranch work, roping, cutting, barrel racing, jumping, etc.

[c]Examples are horses used in race training, polo, etc.

TABLE 11–6 Daily Nutrient Requirements of Horses (500-Kilogram Mature Weight)

Animal	Weight (kg)	Daily Gain (kg)	DE (Mcal)	Crude Protein (g)	Lysine (g)	Calcium (g)	Phosphorus (g)	Magnesium (g)	Potassium (g)	Vitamin A (10³IU)
Mature Horses										
Maintenance	500		16.4	656	23	20	14	7.5	25.0	15
Stallions (breeding season)	500		20.5	820	29	25	18	9.4	31.2	22
Pregnant mares	500									
9 months			18.2	801	28	35	26	8.7	29.1	30
10 months			18.5	815	29	35	27	8.9	29.7	30
11 months			19.7	866	30	37	28	9.4	31.5	30
Lactating mares										
Foaling to 3 months	500		28.3	1,427	50	56	36	10.9	46.0	30
3 months to weaning	500		24.3	1,048	37	36	22	8.6	33.0	30
Working horses										
Light work[a]	500		20.5	820	29	25	18	9.4	31.2	22
Moderate work[b]	500		24.6	984	34	30	21	11.3	37.4	22
Intense work[c]	500		32.8	1,312	46	40	29	15.1	49.9	22
Growing Horses										
Weanling, 4 months	175	0.85	14.4	720	30	34	19	3.7	11.3	8
Weanling, 6 months										
Moderate growth	215	0.65	15.0	750	32	29	16	4.0	12.7	10
Rapid growth	215	0.65	17.2	860	36	36	20	4.3	13.3	10
Yearling, 12 months										
Moderate growth	325	0.50	18.9	851	36	29	16	5.5	17.8	15
Rapid growth	325	0.65	21.3	956	40	34	19	5.7	18.2	15
Long yearling, 18 months										
Not in training	400	0.35	19.8	893	38	27	15	6.4	21.1	18
In training	400	0.35	26.5	1,195	50	36	20	8.6	28.2	18
Two year old, 24 months										
Not in training	450	0.20	18.8	800	32	24	13	7.0	23.1	20
In training	450	0.20	26.3	1,117	45	34	19	9.8	32.2	20

Source: Reprinted with permission of T. W. Perry, A. E. Cullison, and R. S. Lowrey, *Feeds & Feeding,* 5th ed. (Upper Saddle River, NJ: Prentice-Hall, 1999).
Note: Mares should gain weight during late gestation to compensate for tissue deposition. However, nutrient requirements are based on maintenance body weight.
[a]Examples are horses used in Western and English pleasure, bridle path hack, equitation, etc.
[b]Examples are horses used in ranch work, roping, cutting, barrel racing, jumping, etc.
[c]Examples are horses used in race training, polo, etc.

TABLE 11–7 Daily Nutrient Requirements of Horses (600-Kilogram Mature Weight)

Animal	Weight (kg)	Daily Gain (kg)	DE (Mcal)	Crude Protein (g)	Lysine (g)	Calcium (g)	Phosphorus (g)	Magnesium (g)	Potassium (g)	Vitamin A (10³IU)
Mature Horses										
Maintenance	600		19.4	776	27	24	17	9.0	30.0	18
Stallions (breeding season)	600		24.3	970	34	30	21	11.2	36.9	27
Pregnant mares	600									
9 months			21.5	947	33	41	31	10.3	34.5	36
10 months			21.9	965	34	42	32	10.5	35.1	36
11 months			23.3	1,024	36	44	34	11.2	37.2	36
Lactating mares										
Foaling to 3 months	600		33.7	1,711	60	67	43	13.1	55.2	36
3 months to weaning	600		28.9	1,258	44	43	27	10.4	39.6	36
Working horses										
Light work[a]	600		24.3	970	34	30	21	11.2	36.9	27
Moderate work[b]	600		29.1	1,164	41	36	25	13.4	44.2	27
Intense work[c]	600		38.8	1,552	54	47	34	17.8	59.0	27
Growing Horses										
Weanling, 4 months	200	1.00	16.5	825	35	40	22	4.3	13.0	9
Weanling, 6 months										
Moderate growth	245	0.75	17.0	850	36	34	19	4.6	14.5	11
Rapid growth	245	0.95	19.2	960	40	40	22	4.9	15.1	11
Yearling, 12 months										
Moderate growth	375	0.65	22.7	1,023	43	36	20	6.4	20.7	17
Rapid growth	375	0.80	25.1	1,127	48	41	22	6.6	21.2	17
Long yearling, 18 months										
Not in training	475	0.45	23.9	1,077	45	33	18	7.7	25.1	21
In training	475	0.45	32.0	1,429	60	44	24	10.2	33.3	21
Two year old, 24 months										
Not in training	540	0.30	23.5	998	40	31	17	8.5	27.9	24
In training	540	0.30	32.3	1,372	55	43	24	11.6	38.4	24

Source: Reprinted with permission of T. W. Perry, A. E. Cullison, and R. S. Lowrey, *Feeds & Feeding*, 5th ed. (Upper Saddle River, NJ: Prentice-Hall, 1999).
Note: Mares should gain weight during late gestation to compensate for tissue deposition. However, nutrient requirements are based on maintenance body weight.
[a]Examples are horses used in Western and English pleasure, bridle path hack, equitation, etc.
[b]Examples are horses used in ranch work, roping, cutting, barrel racing, jumping, etc.
[c]Examples are horses used in race training, polo, etc.

Animal	Weight (kg)	Daily Gain (kg)	DE (Mcal)	Crude Protein (g)	Lysine (g)	Calcium (g)	Phosphorus (g)	Magnesium (g)	Potassium (g)	Vitamin A (10^3IU)
Mature Horses										
Maintenance	700		21.3	851	30	28	20	10.5	35.0	21
Stallions (breeding season)	700		26.6	1,064	37	32	23	12.2	40.4	32
Pregnant mares										
9 months	700		23.6	1,039	36	45	34	11.3	37.8	42
10 months			24.0	1,058	37	46	35	11.5	38.5	42
11 months			25.5	1,124	39	49	37	12.3	40.9	42
Lactating mares										
Foaling to 3 months	700		37.9	1,997	70	78	51	15.2	64.4	42
3 months to weaning	700		32.4	1,468	51	50	31	12.1	46.2	42
Working horses										
Light work[a]	700		26.6	1,064	37	32	23	12.2	40.4	32
Moderate work[b]	700		31.9	1,277	45	39	28	14.7	48.5	32
Intense work[c]	700		42.6	1,702	60	52	37	19.6	64.7	32
Growing Horses										
Weanling, 4 months	225	1.10	19.7	986	41	44	25	4.8	14.6	10
Weanling, 6 months										
Moderate growth	275	0.80	20.0	1,001	42	37	20	5.1	16.2	12
Rapid growth	275	1.00	22.2	1,111	47	43	24	5.4	16.8	12
Yearling, 12 months										
Moderate growth	420	0.70	26.1	1,176	50	39	22	7.2	23.1	19
Rapid growth	420	0.85	28.5	1,281	54	44	24	7.4	23.6	19
Long yearling, 18 months										
Not in training	525	0.50	27.0	1,215	51	37	20	8.5	27.8	24
In training	525	0.50	36.0	1,615	68	49	27	11.3	36.9	24
Two year old, 24 months										
Not in training	600	0.35	26.3	1,117	45	35	19	9.4	31.1	27
In training	600	0.35	36.0	1,529	61	48	27	12.9	42.5	27

Source: Reprinted with permission of T. W. Perry, A.E. Cullison, and R. S. Lowrey, *Feeds & Feeding,* 5th ed. (Upper Saddle River, NJ: Prentice-Hall, 1999).
Note: Mares should gain weight during late gestation to compensate for tissue deposition. However, nutrient requirements are based on maintenance body weight.
[a]Examples are horses used in Western and English pleasure, bridle path hack, equitation, etc.
[b]Examples are horses used in ranch work, roping, cutting, barrel racing, jumping, etc.
[c]Examples are horses used in race training, polo, etc.

TABLE 11–9

Daily Nutrient Requirements of Horses (800-Kilogram Mature Weight)

Animal	Weight (kg)	Daily Gain (kg)	DE (Mcal)	Crude Protein (g)	Lysine (g)	Calcium (g)	Phosphorus (g)	Magnesium (g)	Potassium (g)	Vitamin A (10³IU)
Mature Horses										
Maintenance	800		22.9	914	32	32	22	12.0	40.0	24
Stallions (breeding season)	800		28.6	1,143	40	35	25	13.1	43.4	36
Pregnant mares										
9 months	800		25.4	1,116	39	48	37	12.2	40.6	48
10 months			25.8	1,137	40	49	37	12.4	41.3	48
11 months			27.4	1,207	42	52	40	13.2	43.9	48
Lactating mares										
Foaling to 3 months	800		41.9	2,282	81	90	58	17.4	73.6	48
3 months to weaning	800		35.5	1,678	60	58	36	13.8	52.8	48
Working horses										
Light work[a]	800		28.6	1,143	40	35	25	13.1	43.4	36
Moderate work[b]	800		34.3	1,372	48	42	30	15.8	52.1	36
Intense work[c]	800		45.7	1,829	64	56	40	21.0	69.5	36
Growing Horses										
Weanling, 4 months	250	1.20	21.4	1,070	45	48	27	5.3	16.1	11
Weanling, 6 months										
Moderate growth	305	0.90	22.0	1,100	46	41	23	5.7	18.0	14
Rapid growth	305	1.10	24.2	1,210	51	47	26	6.0	18.6	14
Yearling, 12 months										
Moderate growth	460	0.80	28.7	1,291	55	44	24	7.9	25.4	21
Rapid growth	460	0.95	31.0	1,396	59	49	27	8.1	25.9	21
Long yearling, 18 months										
Not in training	590	0.60	30.2	1,361	57	43	24	9.6	31.3	27
In training	590	0.60	39.8	1,793	76	56	31	12.6	41.2	27
Two year old, 24 months										
Not in training	675	0.40	28.7	1,220	49	40	22	10.6	35.0	30
In training	675	0.40	39.1	1,662	66	54	30	14.5	47.6	30

Source: Reprinted with permission of T. W. Perry, A. E. Cullison, and R. S. Lowrey, *Feeds & Feeding,* 5th ed. (Upper Saddle River, NJ: Prentice-Hall, 1999).
Note: Mares should gain weight during late gestation to compensate for tissue deposition. However, nutrient requirements are based on maintenance body weight.
[a]Examples are horses used in Western and English pleasure, bridle path hack, equitation, etc.
[b]Examples are horses used in ranch work, roping, cutting, barrel racing, jumping, etc.
[c]Examples are horses used in race training, polo, etc.

TABLE 11-10 **Daily Nutrient Requirements of Horses (900-Kilogram Mature Weight)**

Animal	Weight (kg)	Daily Gain (kg)	DE (Mcal)	Crude Protein (g)	Lysine (g)	Calcium (g)	Phosphorus (g)	Magnesium (g)	Potassium (g)	Vitamin A (10³ IU)
Mature Horses										
Maintenance	900		24.1	966	34	36	25	13.5	45.0	27
Stallions (breeding season)	900		30.2	1,207	42	37	26	13.9	45.9	40
Pregnant mares										
9 months	900		26.8	1,179	41	51	39	12.9	42.9	54
10 months			27.3	1,200	42	52	39	13.1	43.6	54
11 months			29.0	1,275	45	55	42	13.9	46.3	54
Lactating mares										
Foaling to 3 months	900		45.5	2,567	89	101	65	19.6	82.8	54
3 months to weaning	900		38.4	1,887	66	65	40	15.5	59.4	54
Working horses										
Light work[a]	900		30.2	1,207	42	37	26	13.9	45.9	40
Moderate work[b]	900		36.2	1,448	51	44	32	16.7	55.0	40
Intense work[c]	900		48.3	1,931	68	59	42	22.2	73.4	40
Growing Horses										
Weanling, 4 months	275	1.30	23.1	1,154	48	53	29	5.8	17.7	12
Weanling, 6 months										
Moderate growth	335	0.95	23.4	1,171	49	44	24	6.2	19.6	15
Rapid growth	335	1.15	25.6	1,281	54	50	28	6.5	20.2	15
Yearling, 12 months										
Moderate growth	500	0.90	31.2	1,404	59	49	27	8.6	27.7	22
Rapid growth	500	1.05	33.5	1,509	64	54	30	8.8	28.2	22
Long yearling, 18 months										
Not in training	665	0.70	33.6	1,510	64	49	27	10.9	35.4	30
In training	665	0.70	43.9	1,975	83	64	35	14.2	46.2	30
Two year old, 24 months										
Not in training	760	0.45	31.1	1,322	53	45	25	12.0	39.4	34
In training	760	0.45	42.2	1,795	72	61	34	16.2	53.4	34

Source: Reprinted with permission of T. W. Perry, A. E. Cullison, and R. S. Lowrey, *Feeds & Feeding*, 5th ed. (Upper Saddle River, NJ: Prentice-Hall, 1999).
Note: Mares should gain weight during late gestation to compensate for tissue deposition. However, nutrient requirements are based on maintenance body weight.
[a]Examples are horses used in Western and English pleasure, bridle path hack, equitation, etc.
[b]Examples are horses used in ranch work, roping, cutting, barrel racing, jumping, etc.
[c]Examples are horses used in race training, polo, etc.

TABLE 11-11   Nutrient Concentrations in Total Diets for Horses and Ponies (Dry Matter Basis)

	Digestible Energy[a]		Diet Proportions		Crude Protein (%)	Lysine (%)	Calcium (%)	Phosphorus (%)	Magnesium (%)	Potassium (%)	Vitamin A	
	(Mcal/kg)	(Mcal/lb)	Conc. (%)	Hay (%)							(IU/kg)	(IU/lb)
Mature Horses												
Maintenance	2.00	0.90	0	100	8.0	0.28	0.24	0.17	0.09	0.30	1830	830
Stallions	2.40	1.10	30	70	9.6	0.34	0.29	0.21	0.11	0.36	2640	1200
Pregnant mares												
9 months	2.25	1.00	20	80	10.0	0.35	0.43	0.32	0.10	0.35	3710	1680
10 months	2.25	1.00	20	80	10.0	0.35	0.43	0.32	0.10	0.36	3650	1660
11 months	2.40	1.10	30	70	10.6	0.37	0.45	0.34	0.11	0.38	3650	1660
Lactating mares												
Foaling to 3 months	2.60	1.20	50	50	13.2	0.46	0.52	0.34	0.10	0.42	2750	1250
3 months to weaning	2.45	1.15	35	65	11.0	0.37	0.36	0.22	0.09	0.33	3020	1370
Working horses												
Light work[b]	2.45	1.15	35	65	9.8	0.35	0.30	0.22	0.11	0.37	2690	1220
Moderate work[c]	2.65	1.20	50	50	10.4	0.37	0.31	0.23	0.12	0.39	2420	1100
Intense work[d]	2.85	1.30	65	35	11.4	0.40	0.35	0.25	0.13	0.43	1950	890
Growing Horses												
Weanling, 4 months	2.90	1.40	70	30	14.5	0.60	0.68	0.38	0.08	0.30	1580	720
Weanling, 6 months												
Moderate growth	2.90	1.40	70	30	14.5	0.61	0.56	0.31	0.08	0.30	1870	850
Rapid growth	2.90	1.40	70	30	14.5	0.61	0.61	0.34	0.08	0.30	1630	740
Yearling, 12 months												
Moderate growth	2.80	1.30	60	40	12.6	0.53	0.43	0.24	0.08	0.30	2160	980
Rapid growth	2.80	1.30	60	40	12.6	0.53	0.45	0.25	0.08	0.30	1920	870
Long yearling, 18 months												
Not in training	2.50	1.15	45	55	11.3	0.48	0.34	0.19	0.08	0.30	2270	1030
In training	2.65	1.20	50	50	12.0	0.50	0.36	0.20	0.09	0.30	1800	820
Two year old, 24 months												
Not in training	2.45	1.15	35	65	10.4	0.42	0.31	0.17	0.09	0.30	2640	1200
In training	2.65	1.20	50	50	11.3	0.45	0.34	0.20	0.10	0.32	2040	930

Source: Reprinted with permission of T. W. Perry, A. E. Cullison, and R. S. Lowrey, *Feeds & Feeding*, 5th ed. (Upper Saddle River, NJ: Prentice-Hall, 1999).

[a]Examples are horses used in Western and English pleasure, bridle path hack, equitation, etc.

[b]Examples are horses used in ranch work, roping, cutting, barrel racing, jumping, etc.

[c]Examples are horses used in race training, polo, etc.

[d]Examples are race training, polo, etc.

TABLE 11–12 Nutrient Concentrations in Total Diets for Horses and Ponies (90% Dry Matter Basis)

	Digestible Energy[a]		Diet Proportions		Crude Protein (%)	Lysine (%)	Calcium (%)	Phosphorus (%)	Magnesium (%)	Potassium (%)	Vitamin A	
	(Mcal/kg)	(Mcal/lb)	Conc. (%)	Hay (%)							(IU/kg)	(IU/lb)
Mature Horses												
Maintenance	1.80	0.80	0	100	7.2	0.25	0.21	0.15	0.08	0.27	1650	750
Stallions	2.15	1.00	30	70	8.6	0.30	0.26	0.19	0.10	0.33	2370	1080
Pregnant mares												
9 months	2.00	0.90	20	80	8.9	0.31	0.39	0.29	0.10	0.32	3330	1510
10 months	2.00	0.90	20	80	9.0	0.32	0.39	0.30	0.10	0.33	3280	1490
11 months	2.15	1.00	30	70	9.5	0.33	0.41	0.31	0.10	0.35	3280	1490
Lactating mares												
Foaling to 3 months	2.35	1.10	50	50	12.0	0.41	0.47	0.30	0.09	0.38	2480	1130
3 months to weaning	2.20	1.05	35	65	10.0	0.34	0.33	0.20	0.08	0.30	2720	1240
Working horses												
Light work[b]	2.20	1.05	35	65	8.8	0.32	0.27	0.19	0.10	0.34	2420	1100
Moderate work[c]	2.40	1.10	50	50	9.4	0.35	0.28	0.22	0.11	0.36	2140	970
Intense work[d]	2.55	1.20	65	35	10.3	0.36	0.31	0.23	0.12	0.39	1760	800
Growing Horses												
Weanling, 4 months	2.60	1.25	70	30	13.1	0.54	0.62	0.34	0.07	0.27	1420	650
Weanling, 6 months												
Moderate growth	2.60	1.25	70	30	13.0	0.55	0.50	0.28	0.07	0.27	1680	760
Rapid growth	2.60	1.25	70	30	13.1	0.55	0.55	0.30	0.07	0.27	1470	670
Yearling, 12 months												
Moderate growth	2.50	1.15	60	40	11.3	0.48	0.39	0.21	0.07	0.27	1950	890
Rapid growth	2.50	1.15	60	40	11.3	0.48	0.40	0.22	0.07	0.27	1730	790
Long yearling, 18 months												
Not in training	2.30	1.05	45	55	10.1	0.43	0.31	0.17	0.07	0.27	2050	930
In training	2.40	1.10	50	50	10.8	0.45	0.32	0.18	0.08	0.27	1620	740
Two year old, 24 months												
Not in training	2.20	1.00	35	65	9.4	0.38	0.28	0.15	0.08	0.27	2380	1080
In training	2.40	1.10	50	50	10.1	0.41	0.31	0.17	0.09	0.29	1840	840

Source: Reprinted with permission of T. W. Perry, A. E. Cullison, and R. S. Lowrey. *Feeds & Feeding*, 5th ed. (Upper Saddle River, NJ: Prentice-Hall, 1999).

[a]Examples are horses used in Western and English pleasure, bridle path hack, equitation, etc.
[b]Examples are horses used in ranch work, roping, cutting, barrel racing, jumping, etc.
[c]Examples are horses used in race training, polo, etc.
[d]Examples are race training, polo, etc.

Minerals and Vitamins

Calcium and phosphorus requirements are especially critical in horses because strong bones are so necessary in work. A lack of either may cause rickets in young equine or osteomalacia in mature horses. A calcium–phosphorus ratio of 1.1–2.5 to 1 is recommended to prevent the occurrence of this condition. Many equine nutritionists place an upper limit of 2.1 to 1 for all horses.[6] Ratios less than 1 to 1 (when phosphorus levels exceed those of calcium) may interfere with the absorption of calcium and may cause skeletal malformations. Horses require rations with 0.60 to 0.70 percent calcium and phosphorus. Bonemeal, calcium carbonate, and dicalcium phosphate are recommended sources and may be mixed 2 to 1 with salt and fed free choice. Horses can handle excess calcium so long as phosphorus intake is adequate.[7]

Vitamin requirements are affected by age, stage of production, and level of stress. The need for supplemental vitamins depends on the type and quality of feed and the amount of microbial vitamin synthesis and absorption in the intestinal tract. Horses grazing high-quality pastures or consuming green, leafy forages from well-fertilized lands are not likely to need vitamin supplementation.

Feeding Hints

Table 11–13 gives example horse rations for professional horse owners. Average horse owners doing average recreational riding may do just as well on a simple ration of 95 percent grain (oats, barley, and corn) and 5 percent protein supplement (linseed meal and soybean meal) for a concentrate source and mineral and vitamin supplement (if needed) along with a bright, clean, leafy hay for roughage.

Owners should be aware of sand colic, a very common disorder caused by feeding hay on the ground. This digestive disorder can easily be prevented by providing hay in bunks or in the characteristic hay nets seen at eye level in many horse stalls.

Each horse should be fed as an individual because some are gluttons and others are shy eaters. Feed boxes should always be kept clean of moldy feed. A horse should clean up its daily concentrate allowance in about 30 minutes and may be fed once, twice, or three times daily but always at the same time, even on weekends. Not eating is the first sign of illness. Fast eaters may be slowed down (to prevent digestive problems) by placing a few large round stones in the feed box so that the horses have to eat around them. Rations should never be changed abruptly; a week should be allowed to gradually switch to a new mixture to prevent digestive problems. Feeding is as much an art as it is a science.

Exercise is important in overall condition. Horses that are penned in a small place often resort to **cribbing** (eating and sucking on fences), which is not a sign of nutritional deficiency but of boredom. A spirited animal must be allowed adequate exercise. Care should be taken not to let horses take on a full fill of water after vigorous exercise because of the possibility of founder or colic. A horse should always walk to cool off before drinking.

▓ **Cribbing** is a vice in which a horse sets its teeth against an object (such as fencing), arches its neck, pulls backward, and swallows large quantities of air

---

[6]Dr. Robert Bray, California Polytechnic State University, Pomona, California.

[7]R. M. Jordan, V. S. Meyers, B. Yoho, and F. A. Spurrell, "Effect of Calcium and Phosphorus Levels on Growth, Reproduction, and Bone Development of Ponies," *Journal of Animal Science* 40:78, 1975.

| TABLE 11-13 | Some Example Grain Mixtures[a] for Horses |

	Nutrient Concentration (%) of Mixture[b]					
	1	**2**	**3**	**4**	**5**	**6**
Rolled oats	15.0	25.0	31.0	45.0	35.0	41.0
Cracked corn	15.5	20.5	29.5	25.5	25.0	35.0
Oat groats	15.0	—	—	—	—	—
Soybean meal	15.0	10.0	10.0	5.0	10.0	5.0
Linseed meal	10.0	10.0	5.0	5.0	10.0	—
Dried skimmed milk	5.0	5.0	5.0	—	—	—
Dehydrated alfalfa	5.0	10.0	10.0	10.0	—	—
Wheat bran	10.0	10.0	—	—	5.0	—
Molasses	7.0	7.0	7.0	7.0	5.0	10.0
Dicalcium phosphate	—	0.5	1.0	1.0	7.0	7.0
Ground limestone	1.5	1.0	0.5	0.5	1.0	—
Trace mineral salt	1.0	1.0	1.0	1.0	1.0	1.0
Vitamin premix	+	+	+	+	1.0	1.0
	100.0	100.0	100.0	100.0	+	+
Digestible energy meal/lb	1.4	1.4	1.4	1.4	100.0	100.0
Crude protein (%)	20.2	17.8	15.9	13.2	1.4	1.4
Digestible protein (%)	15.8	13.7	11.9	9.4	16.1	11.9
Calcium (%)	0.82	0.81	0.72	0.65	12.0	8.0
Phosphorus (%)	0.53	0.57	0.54	0.49	0.79	0.46
					0.58	0.37

Source: Reprinted with permission of J. Blakely and D. H. Bade, *The Science of Animal Husbandry*, 6th ed. (Englewood Cliffs, NJ: Prentice-Hall, 1994).
[a]To be fed with grass or grass-legume hay.
[b]Type of ration: (1) milk replacer; (2) creep feed; (3) postweaning; (4) grower; (5) broodmare; (6) maintenance and working.

As previously stated, each horse must be fed according to its needs as determined by observation of its condition, preferably by the same groom or stable manager over an extended period of time. The simple guidelines in table 11–3 serve as rules of thumb in feeding horses. A scoring system developed by Texas A & M University and published by the National Research Council is gaining popularity in describing body condition scores for assessing nutrient needs in horses. A summary of this nine-point system can be found in the 1989 edition of *Nutrient Requirement of Horses.*

DISEASES OF HORSES

Health programs include disease prevention, parasite control, and first-aid practice. It is important to make the correct diagnosis and have a working knowledge of medication. The best horse owners and breeders rely heavily on veterinarians. A brief description of the most common diseases, parasites, injuries, and health problems should aid in the successful management of horses. The first rule of management is cleanliness of both the stable and the horse itself. Feed boxes, bedding, and stable areas must be managed properly to prevent problems. The temperature of the stable should be adjusted to reduce the possibility of respiratory diseases.

FIGURE 11–56
Checking the vital signs of a horse starts with temperature, pulse, and respiration (TPR).
Courtesy Dr. Craig Barnett, Bayer Agriculture Division.

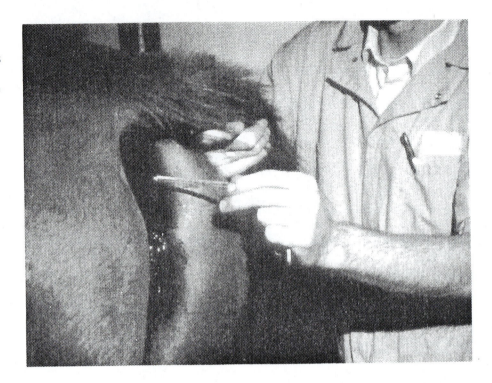

One of the first signs of any problem is a poor appetite or failure to eat at all, because a healthy horse is almost always hungry. A simple check of temperature by rectal thermometer (see figure 11–56) should verify any suspicions of illness. Any deviation above or below the average range of 99 to 100.8°F could indicate a serious problem. Proper diagnostic measures should be conducted by a qualified person (veterinarian or veterinary technician). Descriptions of some common diseases, parasites, and injuries follow.

Azoturia (Monday Morning Disease, Paralytic Myoglobinuria, Tying-Up Syndrome)

Horses that are normally worked hard every day, fed well, and rested on Sunday with the same level of feeding may develop lameness on Monday morning when work is resumed. Muscles of the quarter are extremely tight, legs are stiff, and extreme sweating occurs. Muscle tremors, pain, anxiety, and the production of reddish brown to black urine are further symptoms. The change in urine color is due to the release of myoglobin when the muscle tissue breaks down. To prevent the condition, idle horses should be fed at moderate levels. This condition is often seen in light horses. Once symptoms occur, all movement should be stopped immediately and the veterinarian called. Temporary shelter and slings may be necessary in severe cases. Medication varies with the intensity of the situation, and it is essential that a veterinarian prescribe sedatives, oily laxatives, alkalizing drugs, and so forth. Vitamin E and selenium therapy has also shown good results.

The cause is thought to be overproduction of lactic acid as a by-product of the metabolism of glycogen (animal starch). Excessive lactic acid production due to inactivity but continued glycogen metabolism is thought to cause deranged muscular activity. Symptoms of tying-up syndrome are less severe than but similar to those of azoturia. Tying-up syndrome does not occur until some time after muscular exertion, such as riding or racing. Victims of tying-up syndrome also sweat profusely, tremble, and have a rapid pulse. Most consider tying-up syndrome a mild form of azoturia.

Bowed Tendon (Tendonitis)

Tendonitis refers to enlarged tendons behind the cannon bone in both the front and hind legs. It is most common in the forelegs. The location of the injury is described as high (just under the knee), low (just above the fetlock), or middle (in between). Tendonitis may occur in the high or low areas alone but usually not in the middle area alone. When it does occur in the hind leg (rarely), it occurs in the low area only. As viewed from the side of the cannon, the swollen tendon is bulged or bowed out, giving rise to the name bowed tendon.

A severe strain is the cause, brought about by long, weak pasterns; toes that are too long; exertion from accident or forced training procedures; fatigue of muscles at the end of long races; muddy tracks; improper shoeing; and a horse that is too large for its foot structure. Signs of acute tendonitis appear swiftly. Shortly after injury, or even at the moment of injury, the horse pulls up lame, holding the heel elevated to relieve pressure and not allowing the fetlock to drop because of pain. Heat, swelling, and pain are evident on palpation.

In early acute stages, treatment by a veterinarian usually consists of an injection of corticosteroids and application of a light cast for about 2 weeks. If marked improvement is noted after this cast is removed, supportive bandages are applied for another 30 days. If there is little improvement, another cast may be tried. Finally, the horse is rested for at least 1 year, indicating the seriousness of the situation. The horse should be brought back to exercise slowly to prevent the formation of adhesions.

An alternative treatment some laypeople have used with moderate success on less valuable horses is **blistering**. Blisters are salvelike irritating substances that convert a chronic inflammation into an acute condition. This procedure brings more blood to the part, hastening nature's reparative process. The hair is closely clipped from the affected area and the blistering agent applied by hand and rubbed into the pores of the skin. The effectiveness of blistering agents is debated among veterinarians because the effect may be limited to the skin. In either case, if used, the horse should not be allowed to lick, rub, or bite the area. A rope halter is usually used to tie the head to prevent unwanted interference. Three days afterward, the affected area is bathed with warm, soapy water, dried, and treated with petroleum jelly or sweet oil to prevent cracking of the skin. A rest of at least 1 year is still required.

Blistering controversial treatment in which salvelike irritating substances are placed on the leg of a horse to bring more blood to the part, hastening nature's reparative process

Colic (Abdominal Pain)

Overeating, excessive drinking while hot, moldy feeds, feed impaction, sand impaction, enterolith, diaphragmatic hernias, and even infestation of roundworms can bring on this digestive problem. The intestine is blocked or impacted, creating pain where the horse is most sensitive. Signs are persistent movement, pain, kicking at the belly, sweating, laying down more than usual, rolling, and obvious discomfort. Rolling may lead to twisted gut, which is fatal. Horses should be haltered to prevent rolling. Other signs are curling up of the lip and refusal to eat. Treatment consists of walking the horse until the veterinarian arrives. Mineral oils are often given by stomach tube to relieve compaction. Intravenous administration of an analgesic; nonsteroidal, anti-inflammatory drugs; fluid therapy; and surgery may be needed.

Prevention of colic involves selecting quality hay, not overfeeding grain, regular deworming, not feeding on the ground, avoiding sudden dietary changes, and decreasing roughages fed to broodmares 2 to 3 weeks prior to foaling. Colic is usually preventable (created by poor management decisions).

Equine Encephalomyelitis (Sleeping Sickness)

Three types of virus cause sleeping sickness in the horse: western equine encephalomyelitis (WEE), eastern equine encephalomyelitis (EEE), and Venezuelan equine encephalomyelitis (VEE). The virus is transmitted by mosquitoes and flies from wild birds and rodents to horses and people. In people, flulike symptoms that could cause death in the very young and old develop.

Symptoms of sleeping sickness usually start showing up in the late summer. Fever, inappetence (lack of desire or of craving), depression, elevated heart rate, diarrhea, paralysis, abnormalities in white blood cells, irritability, blindness, and death are commonly noted.

Control measures include controlling insect populations and vaccinating susceptible horses. Inoculation, which consists of two injections 7 to 10 days apart, is usually recommended before May. Immunity lasts about 6 months. There is no specific treatment for viral encephalitis. Supportive care includes administration of fluids, anti-inflammatory drugs, and anticonvulsants if necessary.

Equine Infectious Anemia (Swamp Fever)

Flies and mosquitoes carry swamp fever as well. The virus is related to the human immunodeficiency virus (HIV) but is not known to infect humans. Infection is by transfer of blood cells from an infected horse (contaminated needles). The virus persists in the white blood cells of all infected horses for life. Symptoms include intermittent fever, depression, progressive weakness, edema, weight loss, anemia, splenic lesions, labored breathing, pounding heartbeat, exhaustion, and death (see figures 11–57 and 11–58). Most affected horses die in a few days with the acute forms of the disease.

No specific treatment or approved vaccine is available in the United States. China has developed and uses a vaccine for equine infectious anemia (EIA). Control of horsefly, stable fly, and mosquito populations; improved sanitation; and isolation of infected animals help reduce the spread of the disease. The agar gel immunodiffusion (AGID) test, more commonly known as the Coggins test, has received wide acceptance as a method of determining carriers of the disease. A horse that has had swamp fever but has recovered becomes immune but serves to spread the problem because of intermediate carriers such as mosquitoes. The Coggins test involves drawing a blood sample and sending the serum portion (after clotting) to a federally approved diagnostic laboratory. Negative tests are required by many states for movement within the state or from other states. A negative Coggins test certificate is required at most of the large horse shows and race tracks as a part of the health papers.

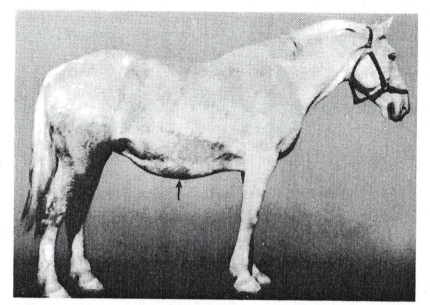

FIGURE 11–57 A horse affected with chronic infectious anemia (swamp fever). Courtesy of the U.S. Department of Agriculture.

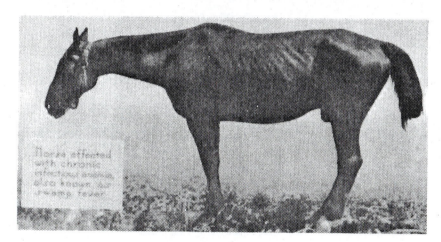

FIGURE 11–58 A case of the subacute form of infectious anemia showing extensive dropsical swellings on the abdomen. Courtesy of the U.S. Department of Agriculture.

Equine Protozoal Myeloencephalitis

Equine protozoal myeloencephalitis (EPM) is a neurological disease that was first recognized in horses in the early 1970s. It is caused by the protozoal parasite (single-celled organism) *Sarcocystis neurona*. It is found throughout all 48 continental states, southern Canada, and parts of South America. Horses ingest the organism, which is spread by opossums and scavenging birds. The horse is a dead-end host for the organism. (Carnivores eating an affected horse will not become infected.) The disease is not contagious between horses.

Clinical symptoms include asymmetrical and progressive incoordination, choppy or spastic gait, dragging of a hoof, knuckling over, stumbling, weakness while turning or backing, muscle atrophy, and paralysis. EPM can be diagnosed by testing cerebrospinal fluid (CSF) for evidence of antibodies to the parasite using the Western blot method. The primary treatment is with the antifolate drug pyrimethamine (Daraprim) in combination with sulfadiazine for at least 3 months. There are no vaccines currently available. Limiting wildlife exposure to horses' feed supply assists in the prevention of the disorder.

Founder (Laminitis, Fever in the Feet)

In founder, the horny laminae of the hoof become congested with blood, preventing a normal "cushioned gait." An obviously painful lameness ("walking on eggs") in the front feet, sometimes all four feet, suddenly appears followed by unusual growth of the hooves, which must be trimmed often. The foot is warm to touch. Swelling of the tissue between the foot bone and wall of the hoof may cause a rotation of the pedal bone in serious cases.

Founder (see figure 11–59) has been attributed to excessive consumption of grains, running on hard ground, certain drugs, infections, abortion, high fever, consumption of cold water after exercise, lack of exercise, metritis, and radical change in feed. Control measures include cooling out horses before giving them water, not overfeeding of grain or lush pasture, and improved management of the horses' excercise. Treatment may include standing in cold water to reduce swelling of the blood vessels, which will give some relief until the veterinarian arrives. A veterinarian should be consulted about corrective shoeing, diet modification, mineral oil, cold packs, hot packs (early), antihistamines, and the administration of digital nerve blocks to permit the horse to walk (increases blood flow to the area).

FIGURE 11–59
Founder (laminitis), a complex disorder affecting the feet, causes irreparable damage if the coffin bone rotates away from the hoof wall. Chronic laminitis or founder showing an unusual growth of hoof (b). Photos (a) and (c) show septic necrotizing laminitis in a more severe case. Courtesy of Dr. Jana Smith and Dr. L.S. Shapiro, L.A. Pierce College 240-acre teaching farm laboratory.

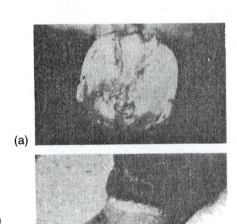

(a)

(b)

(c)

Heaves (Chronic Obstructive Pulmonary Disease)

Heaves is characterized by coughing and wheezing while at work. The horse goes into respiratory distress, has labored expiration, and experiences muscular hypertrophy. It is thought to be due to allergies to dusty or moldy feed, to air pollutants, or to hereditary weakness. Feeding clean forage with a little water sprinkled over it or all pelleted rations has shown good results in alleviating the condition. It is considered a serious fault in an otherwise healthy horse. Treatment consists of using a bronchodilator, mucolytic drugs, and corticosteroids. Affected horses must be kept in a dust-free environment.

Navel Ill

Navel ill, an infection of the navel cord of newborn foals, leads to an inflamed navel; hot, swollen joints; and lameness. It is caused by bacteria entering the bloodstream through the umbilicus. The navel cord should be dipped in a tincture of 10 percent iodine at birth as a simple preventive measure. Treatment by a veterinarian may include a blood transfusion and use of antibiotics and/or sulfa drugs.

Poll Evil (Fistula Withers)

Infection of a bursa in the poll or withers leads to ruptured, draining sores (see figure 11–60). It may be contagious, so infected animals should be isolated. The cause is generally bruises from poorly fitted saddles or headgear, possibly in combination with the same organisms that cause lumpy jaw and brucellosis in cattle. Treatment should always be under the direction of a veterinarian. Proper drainage and removal of dead tissue and/or use of caustic applications are common treatment methods.

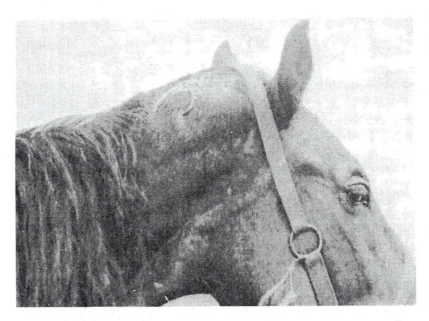

FIGURE 11–60 A horse affected with poll evil.
Courtesy of the U.S. Department of Agriculture.

Potomac Horse Fever (Equine Monocytic Ehrlichiosis)

Potomac horse fever (PHF) is an incurable disease caused by a rickettsiae organism, *Ehrlichia risticci*. It is spread by an arthropod vector (e.g., ticks, fleas, and lice). Its name comes from the first location where it was discovered in 1979, along the Potomac River. It is now found across the continental United States.

Symptoms include lethargy, anorexia, fever, colic, diarrhea, and laminitis. There is a 5 to 30 percent mortality rate with PHF. Treatment is usually not successful but includes intravenous fluid and electrolyte replacement. A vaccine that seems to control the incidence in problem areas has been developed.

Rhinopneumonitis (Equine Viral Rhinopneumonitis, Equine Abortion Virus)

Rhinopneumonitis is a viral disease similar to the flu. It is caused by one of two common equine viruses: equine herpesvirus 1 (EPV-1) and equine herpesvirus 4 (EHV-4). Transmission is via contact with nasal discharges, aborted fetuses, and placentas. Most of the outbreaks in young horses (weanlings) is caused by EHV-4. Symptoms for both viruses include pneumonia, death in newborn foals, abortion, fever, neutropenia, congestion, conjunctiva, malaise, cough, and inappetence.

Immunity results after natural infection. There is little cross immunity, however, between the two virus types. Modified Live Virus (MLV) vaccines are available in some countries but are controversial in nature. Immunity lasts only 2 to 4 months. Booster vaccinations must be given every 3 to 4 months. A killed vaccine for each type of virus is also available and is recommended for use in pregnant mares to prevent abortion. There is no specific treatment. Antipyretics and good animal nursing are all that are recommended. Antibiotics are sometimes given to fight off secondary respiratory infections.

Strangles (Equine Distemper)

Strangles, a contagious, bacterium-induced (*Streptococcus equi equi*) disease, almost always occurs when horses are between 6 months and 5 years of age. Symptoms include inflammation of the upper respiratory tract and swelling and abscessation of adjacent lymph nodes. High fever, inappetence, painful swallowing, cough, and discharge of pus from the nose are also noted. Morbidity levels reach nearly 100 percent; mortality levels are low, averaging only 2 percent.

Control methods include proper sanitation, quarantining new horses, and culturing their nasal discharges for *S. equi equi*. Treatment involves complete rest and extra care. Placing hot packs over abscesses and then incising and draining them are common treatment methods. Intravenous fluid therapy (see figure 11–61) and antimicrobials are also occasionally used.

Tetanus (Lockjaw)

Tetanus is caused by an organism that lives in the soil, manure, and dust—*Clostridium tetani*. The organism can only grow without air; thus puncture wounds are most likely to harbor it. The tetanus organism produces toxin (poi-

FIGURE 11–61
(STAT) large animal intravenous set with 10-foot polyurethane coil length allows simultaneous administration of drugs, plasma, blood, and electrolytes.
Courtesy International Win, LTD., and Mark Lorentz photographer.

son) 100 times as powerful as strychnine, causing muscle contraction. Horses are one of the most sensitive of all species.

Symptoms of tetanus include muscular spasms, respiratory failure, cardiac arrhythmias, rapid breathing, sensitivity to noise, stiffness and rigidity, folding of the inner eyelid over the eye, lockjaw, and death. Death may occur within 24 hours to 1 week. The mortality rate in horses is greater than 80 percent.

Treatments with muscle relaxants, tranquilizers, sedatives, tetanus antitoxins, and intravenous glucose injections are given but are seldom effective. Active immunization (tetanus toxoid) and good sanitation will prevent the disease. If no previous vaccination is given, use of a tetanus antitoxin will prevent tetanus when the animal is exposed (e.g., through castration, cuts, and nail punctures) to the organism.

Leg and Foot Problems

The variety of problems that affect the soundness of limbs in horses appears almost unlimited. A brief discussion of the most common conditions follows. See figure 11–62 for the location of specific unsoundness.

Bog Spavin. Bog spavin is a swelling of the joint capsule that occurs at the hock joint due to an accumulation of fluid (see figure 11–63). A large swelling is obvious at the anterior portion of the hock joint, although smaller protrusions may accompany it on either side. The condition does not usually result in permanent lameness. However, complete recovery is not to be expected in every case. Often the horse will appear lame until warmed up, especially during cold weather. The cause is usually faulty conformation—too straight in the hock joint (no cure), injury (for example, as a result of quick stops), or most

FIGURE 11–62
Anatomy and location of selected unsoundness in the horse. Reprinted with permission from J. Blakely and D.H. Bade, *The Science of Animal Husbandry,* 6th ed. (Englewood Cliffs, NJ: Prentice-Hall, 1994).

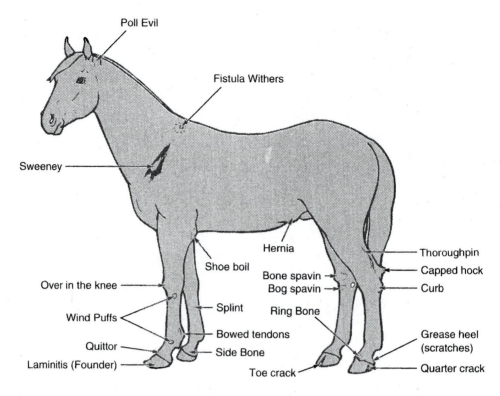

Poll Evil

Fistula Withers

Sweeney

Hernia

Thoroughpin

Shoe boil

Bone spavin

Capped hock

Bog spavin

Curb

Over in the knee

Ring Bone

Splint

Wind Puffs

Grease heel (scratches)

Bowed tendons

Quittor

Side Bone

Laminitis (Founder)

Toe crack

Quarter crack

FIGURE 11–63
Bog spavin is a bulge or swelling on the front inner side of the hock (leg on left). Although unsightly, it normally consists of an accumulation of fluid at the hock joint capsule, which usually does not cause lameness. It is thought to be inherited and is seen most often in straight-hocked horses.
Courtesy of Dr. Jana Smith and Dr. L. S. Shapiro, L. A. Pierce College 240-acre teaching farm laboratory.

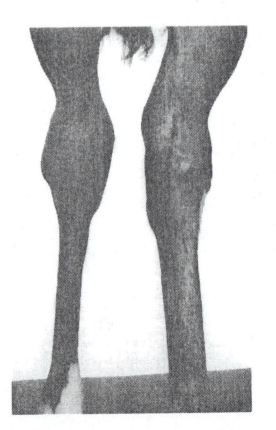

commonly *osteochondritis dissecans (OCD)*.[8] Horses with OCD can be diagnosed by X-ray examination. OCD is a disorder in young growing horses in which there is a progressive breakdown of the cartilage. Fragments of cartilage can be removed by arthroscopy. Deficiencies of vitamin A, vitamin D, calcium, or phosphorus alone or in combination can also produce bog spavin. In this case, a simple dietary supplementation seems to alleviate symptoms in 4 to 6 weeks. Horses 6 months to 2 years of age are most affected by nutritionally caused bog spavin.

Treatment of a trauma-induced condition varies. Two to three injections of a corticosteroid into the joint capsule given at weekly intervals is a common treatment by veterinarians. Horse owners have found rest and treatment of the horse with a stimulating liniment and daily massage for 2 to 3 weeks to be effective in some cases. Firing or blistering is not recommended.

Bone Spavin. To the layperson, bone spavin appears similar to bog spavin. However, this disorder is due to a bony enlargement and is located on the lower inside part of the hock (see figure 11–62). Bone spavin causes a reduction arc of the normal flight of the foot, reducing flexion to relieve pain and causing excessive wear of the toe. It is most common in horses used for quick stops, such as roping horses.

The cause is traced to hereditary weakness, bruises, sprains, or injuries. Rickets may predispose a horse to bone spavin as well. This serious condition affects the usefulness of thousands of horses. Complete recovery should not be expected, but a usable horse is possible with proper care and warm-up.

The most consistently effective treatment is surgery or firing by a veterinarian. Tissues around the bone are painful. Surgery to remove a portion of the cunean tendon and 60 days of rest usually relieve the pain and restore the majority of the soundness.

Firing (applying a hot iron or needle to a blemish), although of less value than tendon surgery, is sometimes prescribed. It is done to produce an acute reaction of a chronic inflammation in the hopes that it will undergo the natural process of healing. An effort is generally made to puncture the cunean tendon in firing to create the same effect as surgical removal.

Bowed Tendon (Tendinitis). Bowed tendons are generally brought about by severe strains, such as heavy training or racing. Poor conformation, poor conditioning and fatigue have also been associated with this disorder. The tendons behind the cannon bones (the flexor tendons) become enlarged in both the front and hind legs (see figure 11–64). Diagnosis is assisted with use of ultrasound. Treatment includes rest, application of cold packs and systemic anti-inflammatory agents. Some practitioners have used superior check ligament desmotomy with some degree of success to minimize the recurrence of tendinitis when the horse resumes training.

Capped Hock. Capped hock is caused by a trauma to the point of the hock (when a horse kicks a wall or trailer gate, for example). Symptoms include a thickening of the skin at the point of the hock. Treatment may include injection of corticosteroids.

---

[8]M.D. Markel, "The Musculoskeletal System and Various Disorders," in *UC Davis School of Veterinary Medicine Book of Horses,* ed. M. Siegal (New York, NY: HarperCollins, 1996).

FIGURE 11–64
Bowed tendon, also known as tendinitis, can cause lameness and can be a chronic condition.
Courtesy of Dr. Jana Smith and Dr. L.S. Shapiro, L.A. Pierce College 240-acre teaching farm laboratory.

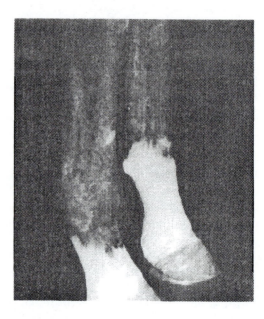

Contracted Heels.   Contracted heels can be an inherited disorder or may be acquired through improper shoeing. Contracted heels result when the frog or cushion of the foot is damaged and shrinks, allowing the heels to come together. The bottom surface of the foot becomes smaller in circumference than at the coronet band. Certain breeds of horses normally have a foot shaped in this fashion. Corrective shoeing can help in most cases. Proper trimming to allow frog pressure may promote foot expansion.[9]

Corns.   Corns are similar to the condition that occurs in humans (see figure 11–65). They are caused by bruising the sole of the feet, generally due to rough terrain or poor shoeing. Corns are difficult to heal; rest and qualified help may be necessary. Treatment usually consists of rest, special shoeing, poulticing, and strict sanitation.

Gravel.   Gravel, a lay term, supposedly is caused by the migration of a piece of gravel from the white line of the hoof to the heel area. Actually, a crack in the white line occurs first (see figure 11–62), allowing infection of the sensitive structures. Inflammation is the result, and since the gravel (or other material) blocks the logical route, drainage breaks out at the heel area.

Excessively dry feet, creating a crack in the white line or sole, is the most common cause, although founder and injury are also suspect. Lameness appears before drainage is visible, but it may go undetected until after drainage. Horses often modify their gait to obtain relief, and it may not be noticeable except to one familiar with the mount. If caught before drainage, recovery is almost certain. Recovery chances after drainage are good but less optimistic.

Treatment involves cleaning and proper draining of the foot. Soaking the foot in Epsom salts is recommended even if drainage has begun. Tincture of iodine and antiphlogistic pastes may be applied daily under a bandage for a week and once every 3 to 4 days afterward, provided that the foot stays dry. Treatment

---

[9]O. R. Adams, *Lameness in Horses*, 3rd ed. (Philadelphia, PA: Lea & Febiger, 1987).

FIGURE 11–65
Corns about the size of a quarter, on either side of the frog on the inner back side of the hoof, are the result of neglected feet, excessive trimming, improper shoeing, or other conditions that throw pressure on the feet. Corns are often discolored red or reddish yellow.
Reprinted with permission from J. Blakely and D.H. Bade, *The Science of Animal Husbandry,* 6th ed. (Englewood Cliffs, NJ: Prentice-Hall, 1994).

should continue until the foot is completely healed. Tetanus antitoxin should be administered.

Grease Heel (Scratches).   Pasterns and fetlocks become inflamed with a mangelike skin condition. It is caused by poor sanitation, although some horses are more susceptible than others. Treatment consists of cleaning the fetlocks with mild soap and water. The hair is clipped closely around affected areas, and astringent, antiseptic substances are then applied at regular intervals. Clean, sanitary quarters are a necessary part of the treatment routine.

Hoof Cracks.   Faulty conformation, brittle hooves, or foot injury can cause vertical cracks or splits from the coronet down (see figure 11–62). These cracks may be mild or very painful. Treatment usually involves trimming the hooves and special shoeing.

Navicular Disease.   Inflammation of the small navicular bone (see figure 11–66) in the front feet and the navicular bursa is one of the most important causes of lameness in horses. The exact cause is unknown, but hard work, small feet, and trimming heels too low may contribute to its development. Rest in the early stages produces improvement, but the condition reappears when training or work resumes. Both front feet may be affected, but usually only one foot is affected. Early signs are repeated lameness following a rest of a few days and pointing with the affected foot (shifting weight to the healthy foot and extending slightly the affected hoof). In riding, the toe of the foot strikes before the heel, shortening the stride and causing many owners to diagnose the injury as a shoulder problem.

The problem has been attributed to an inheritable condition resulting from upright conformation or a weak navicular bone. Although inheritance plays a part, it is not the actual cause and any horse may develop the problem. The chief aggravating factor is concussion of the front feet, especially if horses are worked on a hard surface or subjected to hard work such as racing, cutting, calf roping, and so on.

FIGURE 11–66
Radiograph of a horse's foot showing (from bottom toward top) the coffin bone, short pastern, long pastern, and small navicular bone between the coffin and short pastern.
Courtesy Dr. Craig Barnett, Bayer Agriculture Division.

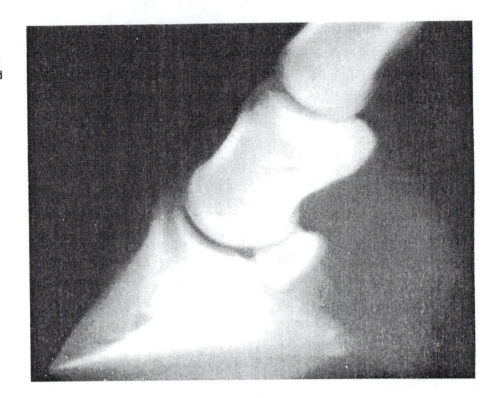

Treatment consists of special shoeing and/or injection of corticosteroids. Both are usually only temporary solutions. The only permanent solution that has been accepted is blockage of the nerves to desensitize that part of the foot. This procedure can add several years of useful service.

Quittor.   Quittor is a chronic inflammation of a collateral cartilage at the coronet (see figure 11–62). Swelling, heat, and pain over the coronary band are characteristic signs. It is manifested as a deep-seated running sore that is usually confined to the front feet but is occasionally seen in the hind feet. The infection may arise from a puncture wound or traumatic bruise that reduces circulation in the area.

Drainage and antiseptics have been used successfully, although surgery is usually recommended to remove necrotic tissue. Incisions are not sutured, but a bandage and a poultice such as Denver Mud (Demco Company, Denver, Colorado) are applied to speed healing.

Ringbone.   Ringbone is a new bone growth on the pastern bone of the front foot that is seen only occasionally on the hind feet. It causes lameness. The condition is rare in Thoroughbreds but quite common in other breeds. Two types afflict horses—high ringbone and low ringbone—both similar except for location of the new bone growth.

Trauma is the usual cause of ringbone. Pulling, bruising, or a direct blow to the phalanges leads to heat, swelling, and pain in addition to the new bone growth. Lameness is constant, and the horse will flinch when finger pressure is applied to the active area. X-rays are needed for positive diagnosis, however. Cold water bandages give some temporary relief. For more permanent relief, the nerve leading to the area is blistered, fired, or severed.

Sidebones.   Sidebones are quite common in horses, except Thoroughbreds. Lameness may or may not occur. Sidebones are more common in the forefeet and involve an ossification of the lateral cartilage just above and to the rear quarter of the coronet. The growth may occur on either or both sides of the foot.

The cause is probably concussion, traumatizing the cartilage. Poor shoeing or wire cuts have also been cited as causes. Unless pain and heat accompany sidebones, they do not often cause lameness, but mechanical interference with foot movement becomes a problem due to massive bone formation.

Treatment, if sidebones are definitely a cause of lameness, consists of cold water bandages for temporary relief. Corrective trimming and shoeing to roll the foot on the affected side are a more permanent solution. The hoof is also grooved to aid in expansion of the quarters of the foot, relieving pain.

Splints.   Splints (see figure 11–62) are new bony growths on the cannon bone, usually on the inside of the foreleg. When found on the rear legs, they are usually on the outside of the cannon. The condition is normally associated with hard training, malnutrition, or faulty conformation in young horses less than 2 years old. Horses 3 to 4 years old are occasionally affected. Trauma caused by slipping, running, falling, or jumping can induce the initial trauma that leads to ossification. In mild cases, the lameness is not evident in the walk. Other gaits exhibit the problem, but this condition does not always cause lameness.

Treatment of splints is more simple and effective than treatment of most leg disorders. Usually they heal of their own accord with 30 days of rest, but some owners have hastened healing with hot and cold applications to affected limbs, followed by application of an antiphlogistic pack to reduce swelling. Firing and blistering have been tried with some success but are generally thought to be unnecessary because time and rest will relieve symptoms eventually.

Stringhalt.   Stringhalt is an involuntary flexion of the hind legs, most evident when backing a horse. The foot is markedly jerked toward the abdomen and may actually hit the abdominal wall. The cause is unknown but it is speculated that nervous diseases, degeneration of nerves, and/or spinal cord disorders could be among the initial phases of the problems. Surgery is the only treatment and consists of removal of a portion of a tendon that crosses the lateral surface of the hock joint.

Thoroughpin.   Thoroughpin (see figure 11–62) is a puffy swelling that can be confused with bog spavin. It is found in the web of the hock and can be distinguished by its movement under finger pressure to the opposite side of the leg. The condition is not serious, and treatment with liniments and massage may reduce the swelling. However, drainage and an injection of corticosteroids is the most effective treatment.

Thrush.   Thrush is a degeneration of the frog of the foot caused by poor sanitation. It is most common in the hind feet. The first sign of trouble is a strong, offensive odor. Treatment consists of trimming away the infected part of the frog, sanitation, and the use of an antiseptic. Owners use many thrush remedy preparations including iodine, formalin, creolin, and carbolic acid.

Wind Puffs (Windgall).   A puff or windgall (see figure 11–62) is a harmless swelling of a bursa (fluid sac). The bursas just above the pastern joints in the fore and rear legs are affected. There is no heat or pain associated with them. The condition is attributed to working on a hard surface, racing, and so on. It is quite common in any hardworking horse, especially one that works on rough terrain. Treatment with cold packs and blistering agents is recommended. Liniments and massage may also be effective.

Parasites

Flies and Mosquitoes.   Flies and mosquitoes are not only irritating to animals but may also be vectors (carriers) of disease. Insecticide sprays are somewhat effective but are of short duration if the horse sweats and removes the protection. A clean stable and removal of manure heaps help prevent accumulation of flies. Spraying with an electric fogging machine or constructing a stable with good ventilation also helps.

Mites.   Microscopic creatures that burrow under the skin to lay eggs (causing mange) are called mites. If kept off the skin, they die within 1 hour to 2 weeks so control is easy with modern insecticides. Good grooming normally prevents them.

Stomach Bots.   The botfly (*Gasterophilus* spp.) (see figure 11–67) lays eggs on the hairs around a horse's fetlocks, knees, throat, jaws, and lips. The fly, some-

FIGURE 11–67
(a), The common botfly lays eggs in the spring on the leg hairs. (b), Eggs are hatched by the stimulation of the horse's lips, producing larvae that migrate to the stomach, where they attach and mature during the winter. Larvae are expelled with feces, pupate in the soil, and hatch out the following spring to produce another botfly in the continuing cycle.

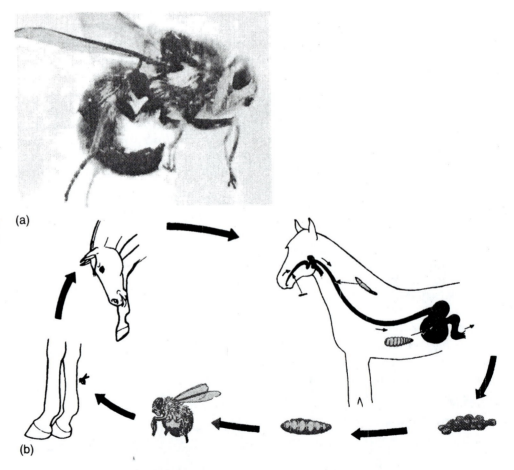

(a)

(b)

what smaller than and similar in appearance to the honeybee, does not sting the horse as popularly thought.

The eggs hatch in 7 to 10 days, and the young larvae enter the horse's mouth when the horse licks its forelimbs and shoulders where the fly deposited its eggs. The larvae molt in the horse's mouth and grow there for 2 to 4 weeks. Next the larvae migrate to the lining of the stomach and attach themselves, feeding on blood until maturity. They then release their hold, pass out with the droppings, enter the pupal or resting stage for 20 to 70 days, and eventually emerge as flies ready to lay eggs once again.

Signs of bot infestation are frequent digestive problems and even colic, emaciation, anemia, stomatitis, and lack of energy. The disease can be prevented by good grooming. Clipping of the attached eggs may be necessary in a control program. Treatment includes using an anthelmintic medication at least two but preferably four times per year. Ivermectin is effective against all stages of this parasite.

Strongyles (Bloodworms).   Strongyles (*Strongylus vulgaris*) are the most common and dangerous of parasites to the equine. They migrate along the arteries that feed the large intestine (see figure 11–68). Symptoms include diarrhea, emaciation, extreme weakness, anemia, and death. Treatment with anthelmintics (ivermectin, pyrantel, and benzimidazole) is effective (see figure 11–69). However, when choosing a dewormer and determining how often to administer it, you must first consider the horses' environment, stocking rate,

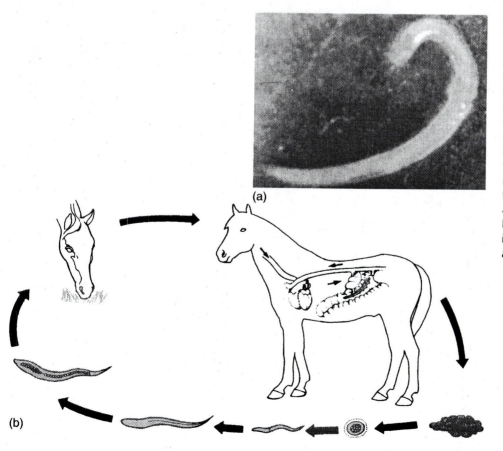

(a)

(b)

FIGURE 11–68
(a), Large strongyles are worms that inhabit the large intestine and lay eggs that mature to larvae in the excreted manure. (b), Larvae crawl up grass stems and are ingested by the horse to complete the cycle. Migration takes place through the stomach and circulatory system back to the large intestine, where eggs are laid.

FIGURE 11–69   A veterinarian administering an anthelmintic through a stomach tube.
Courtesy Dr. Craig Barnett, Bayer Agriculture Division.

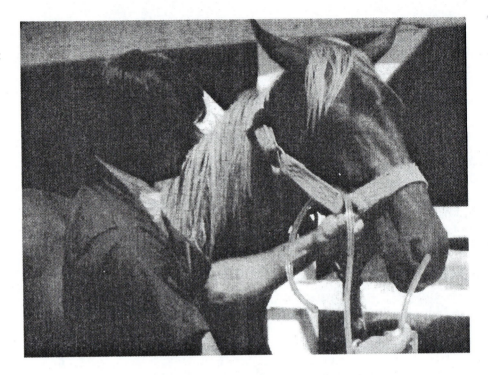

FIGURE 11–70
Ascarids (round-worms) in the intestines of this young foal caused a rupture, peritonitis, and death. The life cycle differs from strongyles in that damage is done in the liver and lungs. Larvae coughed up and swallowed mature in the stomach, produce eggs, and complete the cycle.

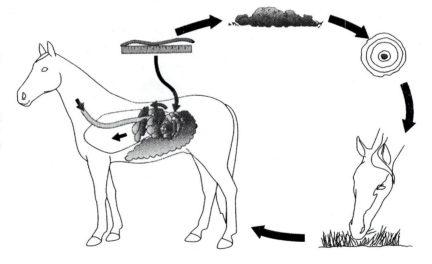

and location. Deworming programs should be individualized based on fecal counts performed at least once or twice a year.

Ascarids (large roundworms) (see figure 11–70) and stomach worms also infect horses. Signs include poor growth, anemia, dull coat, listlessness, and digestive problems including colic.

INJURIES OF HORSES

Teeth Wear

Any horse may wear its rear teeth or molars down, producing sharp points that make it painful to eat. Veterinarians **float** teeth (file off sharp points) to correct the condition.

Float to file off the sharp edges of a horse's teeth

Wounds

Loose nails, projecting boards, sharp objects, barbed wire, and the heels and teeth of other horses cause great damage every year to horses. Horses often panic in a critical moment, causing even further harm. Proper cleaning of wounds, suturing when needed, and tetanus shots should be considered.

Poisonous Plants

Poisonous plants could be considered a nutritional problem; a partial list of commonly found plants that are poisonous to horses follows:

1. Saint-John's-wort
2. Tansy ragwort
3. Buttercup
4. Fiddleneck
5. Foxglove
6. Jimsonweed
7. Milkweed
8. Nightshade
9. Water hemlock
10. Avocado
11. Azalea
12. Bracken fern

SUMMARY

The horse was one of the last farm animals to be domesticated. Horses were first used as food, then for war, sports, draft, and mining, and eventually domesticated for use as companion animals. It is estimated that the U.S. horse population ranges from 5 to 10 million head. There is no reliable count of the equine by the USDA. Most estimates are achieved by surveying equine veterinarians. In recent years, leading veterinary and other animal science institutions have spent more money to find improved methods of feeding, breeding, and managing U.S. equine.

EVALUATION QUESTIONS

1. _____ is a genetic disorder typically classified with the Arabian breed.
2. _____ are the most harmful internal parasites for horses.

3. The average gestation period of a mare is _____ .
4. The average length of the estrous cycle of a mare is _____ .
5. _____ is another name for swamp fever.
6. _____ is the number one killer (metabolic disease) in horses.
7. _____ is the most massive breed of draft horse.
8. _____ is a digestive disorder in the equine brought on by overeating, excessive drinking while hot, moldy feeds, and/or infestation of roundworms.
9. _____ is a respiratory defect in which the horse has difficulty completing the exhalation of inhaled air.
10. A female donkey is called a _____ .
11. Tests used to diagnose swamp fever in the equine are _____ or _____ .
12. Estrus in a mare lasts _____ .
13. Pointing with one foot in the horse is often a noticeable symptom of _____ .
14. The number of chromosome pairs for a horse is _____ , for a donkey is _____ , and for a mule is _____ .
15. There are _____ horses in the world, _____ horses in the United States, _____ horses in California, and _____ horses in Texas.
16. A young horse under the age of 1 year of either sex is called a _____ .
17. The _____ was once known as the horse of queens.
18. There are two types of quarter horses, the _____ and the _____ .
19. The Shetland pony was first used for _____ .
20. Draft horses exceed _____ hands and _____ inches.
21. The Thoroughbred originated in _____ .
22. Crossing a jack with a mare will produce a _____ .
23. A _____ – _____ cross will always produce the palomino color.
24. Iodine is used to treat the navel cord to prevent _____ .
25. To ensure the passage of meconium from the foal, a _____ can be administered.
26. _____ is the part of the horse's digestive system that allows for microbial breakdown of fiber.
27. _____ is the nutritional disorder in which the horse eats wood.
28. _____ is the most difficult muscle to breed on a horse.
29. _____ age when a horse has all of its permanent teeth.
30. The normal rectal temperature of a healthy horse is _____ .
31. Azoturia is brought on by irregular _____ without reducing _____ intake.
32. If colic is suspected a horse should be _____ to prevent twisted gut.
33. _____ is most commonly caused by osteochondritis dissecans.
34. Heaves can be an _____ reaction.
35. _____ is a running sore from the withers.
36. Degeneration of the _____ of the foot is characterized by an offensive odor.

37. The _____ fly lays eggs (on the horse's hair) that hatch into _____ that feed on _____ in the lining of the _____ .

38. To float a horse's teeth means to _____ them.

39. _____ is another name for bloodworms.

40. _____ is a bony enlargement that appears on the inside and front of the hind legs below the hock at the point where the base of the hock tapers into the cannon.

41. _____ is a male horse castrated before reaching maturity.

42. A hand is _____ inches of measurement.

43. _____ is a multicolored, spotted horse.

44. _____ refers to horse distemper.

45. _____ is the only draft breed of horse developed in the United States.

46. _____ is the breed of horse that became famous for pulling beer wagons.

47. During the nineteenth century U.S. Army troops killed large numbers of the _____ , a breed favored by the American Indian.

48. _____ is the largest of the pony breeds.

49. The age at which a horse reaches puberty is _____ .

50. Typically _____ occurs at the end of each teat prior to foaling.

51. _____ is also known as the first water bag.

52. _____ is also known as foal heat.

53. _____ and _____ are two common causes of abortion in mares not caused by microorganisms.

54. _____ (slinging the feet outward) and _____ (slinging the feet inward) are just two of several common defects seen in judging horses.

55. _____ groove appears on the surface of the upper corner incisor teeth of horses at _____ years of age.

56. The feeding of _____ oil meal is not recommended for horses due to its gossypol content.

57. The minimum forage requirement for an adult horse is _____ percent of its body weight per day.

58. With colder environmental temperatures horses should be fed increasing amounts of _____ (type of feed).

59. _____ is the preferred grain fed to horses.

60. _____ is the act of eating and sucking on fences due to boredom.

61. _____ causes the breakdown and release of myoglobin into the urine.

62. _____ are salvelike irritating substances placed on injured limbs to hasten the healing process.

63. _____ is also known as sleeping sickness.

64. _____ , a protozoal disorder, is spread primarily by the opossum.

65. _____ is a herpes viral disorder of horses that causes abortion and respiratory distress.

66. _____ is a leg disorder causing new growth of bone at the pastern joint.

DISCUSSION QUESTIONS

1. What are the three types of sleeping sickness?
2. Why is artificial insemination not widely practiced in the horse industry? What is the real reason?
3. Why is a horse born on December 31 a year old on January 1? How does this system affect the fertility of the mare? Is it the most efficient system?
4. Describe colic, navicular disease, and ringbone.
5. How might ivermectin affect the environment? What are the short-term and long-term effects?

RECOMMENDED READING

1. Siegal, M. (ed.). *UC Davis School of Veterinary Medicine Book of Horses: A Complete Medical Reference Guide for Horses and Foals.* (Harper Collins, 1996).
2. Hintz, H. F. *Horse Nutrition: A Practical Guide.* (NY: Arco, 1983).
3. Lewis, L. D. *Feeding and Care of the Horse,* 2nd ed. (Philadelphia, PA: Lea & Febiger, 1995).
4. Evans, J. W., A. Borton, H. Hintz, and L. D. Van Vleck. *The Horse.* 2nd ed. (Freeman, 1990).

C H A P T E R

12

The Rabbit Industry

OBJECTIVES

After completing the study of this chapter you should know:

- The history of rabbit domestication and current and future trends of the rabbit industry.
- Distinguishing characteristics of the most common breeds of commercially raised rabbits.
- How the Animal Welfare Act has changed the management of commercially raised rabbits.
- Specific management techniques for rabbits raised for meat rather than for pets.
- The use of rabbits for medical and product research, including the use of specific pathogen free and barrier-specific rabbits and their importance to the medical research industry.
- The benefits of coprophagy.
- Unique management concerns in feeding rabbits for each of the various rabbit subindustries (e.g., pets, meat, laboratory, and fur).
- Special reproductive management concerns and procedures of rabbits.
- The management of rabbits raised for their wool and grading, preparing, and cleaning of the wool.
- The most common diseases of rabbits and their diagnosis, symptoms (in brief), treatment, and prevention.

KEY TERMS

Cecotrophy	Hutches	Nosema
Coprophagy	Kindling	Oral papilloma virus
Eicosanoids	Mohair	Shope papilloma virus

INTRODUCTION

Although the last effort to have the rabbit accepted as a livestock animal by the U.S. Department of Agriculture (USDA) failed in 1972, rabbit production continues to develop into a significant agricultural enterprise in the United States.

In addition, rabbit production is a relatively important industry in several European nations (France, Spain, and Italy), where rabbit is regarded as a gourmet meat. Most rabbits are commercially raised for meat, fur or wool, and laboratory usage (medical research). Laboratories use nearly 600,000 rabbits a year for medical experiments and new product testing. Nearly 10,000 tons of Angora wool is consumed on the international market every year.[1]

A few rabbit breeds, however, were developed strictly for their aesthetic value and are kept as pets. These fancy rabbit breeds have very little meat or fur value. Various shows are held for fancy rabbit breeds, which are judged on aesthetic traits. Many of these breeds have traits such as variegated fur, extraordinarily long ears, and so on. These breeds are kept as pets in small numbers and are not covered in any great detail in this book. Although the general rabbit management principles and practices discussed in this chapter apply to such fancy breeds, our focus is on rabbitry for meat, fur, or wool.

Rabbits have many advantages as meat producers. They are excellent converters of feed to meat and naturally produce more meat per pound of female live weight than most other animals. For example, a sow that weans two litters of eight pigs each will produce 100 percent of its live weight in marketable pork. A 1,000-pound cow weaning a 500-pound calf per year produces 50 percent of its live weight as beef. A rabbit doe can wean 30 rabbit fryers weighing 4 pounds each per year, or about 1,000 percent of its live weight.

Today approximately 200,000 producers market 6 to 8 million rabbits annually in the United States.[2] U.S. citizens consume 8 to 10 million pounds of rabbit meat. Marketable meat from rabbits is produced in a short amount of time. Generally only 90 days are required from mating a rabbit doe (female) until fryer rabbits are ready for market.

Hutches rabbit pens or cages

In addition, rabbits can be kept easily, without large investments of money or equipment. Rabbitry ranges from a small backyard operation with three or four **hutches** (rabbit houses) to a large commercial operation of several hundred hutches. As a backyard project, rabbits are easily raised, do not make annoying noises, and occupy only a small area for production. Hence they are very popular as Future Farmers of America (FFA) and 4H beginning animal projects.

Total world rabbit meat production is estimated at 1.6 million tons of carcasses.[3] The largest producers of rabbit meat include Italy, France, Ukraine, China, Spain, and Russia (see table 12–1). The average world rabbit consumption is 0.66 pounds per person annually.

DOMESTICATION AND BREEDS OF RABBITS

The rabbit was initially domesticated in Africa, and it was first considered to be food in Asia about 3,000 years ago. In Europe, rabbits have been used for food for more than 1,000 years. Rabbits were brought to the United States from Europe in the early nineteenth century.

Like the other members of the class Mammalia previously discussed, rabbits bear their young live, have body hair, and produce milk to nurture their

---

[1]U.S. Department of Agriculture (USDA), 1999.

[2]*Small and Part-Time Farming Project,* Pennsylvania State University, January 11, 1999.

[3]M. Colin and F. Lebas, *Rabbit Meat Production in the World: A Proposal for Every Country, 8th World Rabbit Congress,* vol. 3 (Toulouse, France: World Rabbit Congress, 1996).

TABLE 12-1	Leading Nations in Rabbit Meat Production	
Country	Production (tons)	Consumption (kg/person)
China	120,000	0.069
France	150,000	2.756
Italy	300,000	5.587
Russia	100,000	0.673
Spain	120,000	3.152
Ukraine	150,000	2.886
United States	35,000	0.142
Total world	1,613,620	0.301

Source: M. Colin and F. Lebas, *Rabbit Meat Production in the World: A Proposal for Every Country, 8th World Rabbit Congress,* vol. 3 (Toulouse, France: World Rabbit Congress, 1996).

newborn. Rabbits were originally classified as rodents but are now placed in a separate order (Lagomorphs) because they have two more incisor teeth than rodents (six instead of four). The three rabbit genera all fall under the family Leporidae.

Rabbits have been around for more than 45 million years and appear to have originated in Asia. All breeds of domestic rabbits are descendants of the European wild rabbit. Contrary to the belief of many new rabbit producers, no domestic rabbit breeds in America developed from the native hare (jackrabbit, genus *Lepus*) or the North American cottontail (genus *Sylvilagus*). Each of these hares is of a different genus than the domestic rabbits (genus *Oryctolagus*), is quite different biologically, and has a different number of chromosomes and thus will not interbreed and produce young. Hence all breeds present in the United States today were either imported from other parts of the world or developed from imported breeds.

Rabbit breeds vary in size, use, color, and length of fur. Mature rabbit weights vary from 2 pounds to 20 pounds. Breeds with mature weights of 2 to 5 pounds are termed small breeds and are mostly fancy or pet rabbits. Breeds used for meat, fur or wool, or laboratory purposes are medium breeds (9- to 12-pound mature weight) and large breeds (over 13-pound mature weight). The color of fur varies from solids to spotted to variegated. The length of fur varies from shorthaired breeds to the Angora rabbit with hair 8 to 10 inches long.

Currently 45 breeds of rabbits are recognized in the United States. Each breed is recognized by the American Rabbit Breeders Association (ARBA) and is recorded in its *Standard of Perfection*. The *Standard of Perfection* is a book used by the rabbit judge as a guide for judging all the various breeds and by the breeder as a guide for learning what qualities are expected of top breeding animals.

Of the 45 recognized rabbit breeds, 18 are considered breeds of the commercial meat body type and size. Of these 18 breeds, 14 are actually suitable for the rabbit meat industry due to other factors (such as wool-bearing abilities and so forth). The 14 breeds are as follows:

Champagne d'Argent
Californian
Cinnamon
American Chinchilla

Crème d'Argent
French Lop
Hotot
New Zealand
Palomino
Rex
American Sable
Satin
Silver Fox
Silver Marten

The New Zealand White rabbit is the top breed for meat purposes. The Californian rabbit is also in high demand for this purpose.

American

The American rabbit originated in the city of Pasadena, California. It is of medium size and rarely seen outside of the United States. Coat colors include blue, silver, and white. Bucks weigh an average of 10 pounds , and mature does weigh an average of 11 pounds.

American Chinchilla

The American Chinchilla rabbit is of three varieties—the standard (mature weight of 6 to 7 pounds), the heavyweight (mature weight of 10 to 11 pounds), and the giant (mature weight of 13 to 15 pounds). All are used for both fur and meat production. It is a meaty rabbit of medium length and compares with other breeds in meat production, quality, and flavor. This breed was developed from the animals imported from France in 1919. Pelt (fur) color is an important selection trait for most breeders, with the ideal an overall grayish fur (chinchilla color) with a blue undertone.

Angora

The Angora rabbit comes from Ankara, Turkey. The Angora is a medium-sized, major wool (*mohair*) producer. Consequently, it requires a protein-rich diet for the wool to be strong. Angora rabbits are white, black, brown, or blue with red or blue eyes and are noted for long wool growth. Growth of the wool is usually about 1 inch per month (see section on rabbit wool) and is plucked every 3 months. Four types of commercial Angora rabbits are raised in the United States. The English Angora rabbits, which weigh 5 to 7 1/2 pounds, are the most popular Angora breed for show.[4] The French Angora weighs an average of 7 1/2 to 10 1/2 pounds. It includes both white and colored coats and has excellent wool for hand spinning. The Giant Angora weighs 8 1/2 to 10 pounds or more. Its wool is white only, with three hair types in the wool: underwool, awn fluff, and awn hair. The Satin Angora weighs an average of 6 1/2 to 9 1/2 pounds. Its coat consists of shiny wool fibers, both white and colored. This breed is the least popular of the Angora breeds.

---

[4]National Angora Rabbit Breeders Club, July 6, 1999.

Belgian Hare (Lapin-Lievre Belge)

The Belgian hare originated in Flanders, Belgium. It is called the racehorse of the rabbit industry. Its colors include a deep, rich red, tan, or chestnut shade. Mature bucks and does weigh 8 pounds.

Californian

The Californian rabbit was bred in Southgate, California, in 1923. It has broad shoulders, a meaty back, and a good dressing percentage. Californian rabbits have a white body with dark gray or black ears, nose, tail, and feet. The eyes are a bright pink. The ideal mature weight is 9 pounds for bucks and 9 1/2 pounds for does.

Champagne d'Argent

The Champagne d'Argent breed is also known as the French Silver breed. It is an old breed, grown in France for more than 100 years. Although also a good meat rabbit, its fur is one of the leaders for use in the garment industry. Champagne d'Argent rabbits are born black but take on a silver or skimmed milk color with a dark slate blue undercoat at 3 to 4 months of age. Mature weights are 10 pounds for bucks and 10 1/2 pounds for does.

Checkered Giant

The Checkered Giant rabbit is a very large breed that is preferred by some breeders as a meat producer but is used primarily in show and for fur. The breed is white with black markings down the back, the sides, and in circles around the eyes.

Dutch

The Dutch rabbit breed originated in Holland with English influence. Dutch rabbits are exhibition type; they are small but carry a lot of meat with little lost in dressing out. Their colors include black, blue, chocolate, tortoise, steel gray, and gray. Mature bucks and does weigh only 4 1/2 pounds.

Flemish Giant

Flemish Giant rabbits are the largest of the domestic rabbits, with mature weights in excess of 13 pounds. They are noted for size and quality of fur. One of their main uses is in crossbreeding programs with other breeds to increase meat production. Flemish Giants are found in a variety of colors ranging from steel gray or black to sandy, white, or blue. The popularity of the Flemish Giant for commercial meat production has diminished in recent years due to its semi-arched type, which results in an extremely long, unattractive, and bony carcass.

French Lop

The French Lop is one of four breeds of lop-eared rabbits: English, French, Holland, and Mini. They originated in France and are a cross of English Lop and Flemish Giant rabbits. French Lops have characteristic horseshoe-shaped ears, are 15 to 18 inches long, are heavily muscled, and have a good, thick pelt. However, they are noted for being bad tempered and untidy in their cages. French

Lops are primarily white with blue eyes, but all colors are found. Mature bucks weigh 11 pounds, and mature does weigh 12 pounds.

Havana

The Havana rabbit originated in Holland. It is a small, meaty-type rabbit with a lustrous and soft fur with lots of sheen. It is deep chocolate brown with dark brown eyes. Mature bucks and does weigh 5 1/4 to 5 1/2 pounds.

Himalayan

Many Himalayan rabbits are found in China and Russia along the Himalayan Mountains. They were further developed in England. Their long, slender body makes them an excellent exhibition rabbit. Colors include a white body and black muzzle, ears, front and hind limbs, and tail. The black coloring can be replaced by blue, brown, lilac, or chocolate. The eyes are red. The mature weight of both bucks and does is only 3 1/2 pounds.

Netherlands Dwarf

The Netherlands Dwarf originated in Holland. It is the smallest of all rabbits and has the greatest variety of colors and coat patterns. Mature bucks and does weigh only 2 pounds.

New Zealand

The New Zealand breed is the most popular breed in the United States today. This breed originated in California and Indiana. It is a very good meat and fur breed of medium size. New Zealand rabbits are either red or white. The red New Zealand rabbits have reddish bodies with cream bellies. Red New Zealand rabbits have brown eyes; white New Zealand rabbits have red eyes. Mature bucks weigh an average of 10 pounds, and does weigh 11 pounds.

Palomino

The Palomino originated in the state of Washington. It is described as a utility-type rabbit (meat-producing qualities with an attractive appearance) and is golden, bright orange with an undercoat. Mature buck weights average 9 pounds, and doe weights average 10 pounds.

Polish

The Polish rabbit comes from England. It is a small, compact rabbit with a short, fine, and dense fur. Its colors include black, blue, chocolate, white, and ruby-eyed white. Its eyes are red or blue. Mature buck and doe weights average 2 1/2 pounds.

Rex

The Rex rabbit's fur lacks the stiff guard hairs usually found on other rabbits. The fur has a plush-velvet, soft, silky feel and is desired by the garment industry. Colors include beaver, black, chinchilla, lilac, and opal with white or blue eyes. The Rex originated in Coulange, France. It is known as the pelt breed of rabbits but also has good meat-producing qualities. Bucks mature at 8 pounds, and does mature at 9 pounds.

Satin

The Satin rabbit is a good meat breed because of its type and size. Mature weights are 9 to 10 pounds. Nine colors are recognized; white is the most popular. Satin rabbits are noted for vivid colors and a sleek coat.

Tan

The Tan rabbit breed originated in England. Its two-colored body is solid and exhibits sheen and luster. The two colors found vary from black, blue, and chocolate to lilac with tan (reddish brown). Mature bucks weigh an average of 4 to 5 pounds, and does average 4 1/2 to 5 1/2 pounds.

SELECTION OF BREEDING RABBITS

Breeding stock are selected by rabbitry beginners and by producers who are selecting replacement does and bucks for the herd. A popular saying among rabbit breeders is that "anyone with enough money can buy a winning rabbit, but it takes skill and knowledge to breed and raise your own winners."[5] Beginners first select the breed or breed combination based on the intended use of the rabbit (meat, fur, or wool), the available markets for the end products, and personal preferences. Once the breed is determined, rabbits must be selected from the breed to serve as foundation stock. Foundation stock vary from 2 to 10 does and 1 buck to several hundred rabbits, depending on the size of the operation. This selection process is the same whether it is selection of replacement rabbits or selection of foundation stock of a future herd.

Breeding rabbits are selected based on their ability to produce plenty of offspring with compact, meaty bodies and thick hindquarters, loins, and shoulders. The body carcasses of the offspring should be meaty and solid, not flabby and loose.

As with other enterprises, production records are very valuable in the selection process. In selecting rabbits, individual doe, buck, and sibling records are used. The minimum records that should be considered are the average litter number and weaning weights. Some other selection criteria include the following:

1. *Good production:* Raises a uniform litter of more than eight fryers per litter with four litters per year
2. *Milking ability:* 10 or more teats; litter weights of over 8 pounds per litter at 3 weeks of age
3. *Mothering ability:* Easy breeding; high offspring survival rates
4. *Ability to grow:* Bunnies that are at least 1 1/4 pounds at 4 weeks of age and 4 pounds at 8 weeks of age

---

[5]P. Lamar, ARBA Commercial Department Committee, 1999.

5. *High-yielding carcass:* More than 50 percent dressing percentage
6. *Good fur quality*
7. *Good health and vigor*

As with other livestock selection, the more criteria used in the selection process, the more certain the success of the selected individual. In breeding, however, only one or two traits per breeding should be selected for at a time (review genetics selection in chapter 2). Selection should be from reliable breeders and based on more than just the eye appeal of a furry bunny.

For show rabbits the ARBA-sanctioned rabbit shows use the American system of judging. This system places only one first place animal in each class and awards only first through third place regardless of the size of the class (except at national competitions). The Danish system awards each exhibitor or rabbit a first, second or third place. When selecting rabbits, it is important to understand show place winning, the weight or meaning, and the system used.

PET RABBIT INDUSTRY

The Animal Welfare Act of 1966 (AWA) and its three amendments regulate professional pet rabbit suppliers.[6] Pet rabbit suppliers are defined as individuals who make more than $500 a year selling rabbits. The United States Department of Agriculture Branch of Animal & Plant Health Inspection Service (USDA/APHIS) then inspects the facilities of all licensed professional suppliers of rabbits, making sure the animals are raised in a humane and healthy manner. The main problem today is not the supplier or breeder of rabbits but the individual customer who buys a bunny and then does not know what to do with it after it ceases being "so cute." Many if not most of these rabbits die of neglect or are abandoned in the wild (a certain cruel death).

RABBIT MEAT INDUSTRY

Although rabbits are not considered livestock, they still must be slaughtered in federally or state-inspected (depending on the state) facilities if the meat is to be sold or even given away. The carcasses from backyard butchering of rabbits are not allowed to be sold or used in retail stores, restaurants, or supermarkets. Only USDA-inspected meats may be exported to other countries outside of the United States.

To be economically feasible, a commercial rabbit meat grower must maintain at least eight kits per litter. Crossbreeding is very popular in that it increases hybrid vigor in the fryers produced. As with the previously mentioned species of animals, hybrid vigor lasts only one generation. Thus, although disease resistance, growth rates, breeding consistency, litter sizes, and so forth improve with crossbreeding, this improvement is seen only in the first generation. Hence, only terminal crosses should include crossbreeding.

Rabbit meat is available as fryers, roasters, stewers, and capons. A fryer is a rabbit under 12 weeks of age and with an average carcass weight less than 3 1/2

---

[6]For a complete summary of the AWA, see L. S. Shapiro, *Applied Animal Ethics* (Albany, NY: Delmar, 2000).

pounds. Restaurants generally prefer slightly smaller fryers, whereas super-markets and butcher shops prefer larger carcasses.

Some processors demand younger fryers. At approximately 12 weeks of age, the rabbit reaches puberty and starts producing hormones that change the quality of the meat. The carcass becomes less tender, the testicles appear on the bucks, and the females begin to blossom. The highest prices therefore go for fryers.

Roasters are rabbits under 6 months of age that have not been sexually active. Stewer rabbits are those over 6 months old that have been sexually active and productive and have tough carcasses. Capons are castrated bucks. There is little use in commercial markets for capon, stewer, or even roaster rabbits. Some are processed into rabbit pepperoni sticks and the like.

The goal of commercial meat growers is to produce fryers that are 4 to 5 pounds by 8 weeks of age. Fryers should gain 1/4 pound per week and consume 4 pounds of pelleted feed for every pound of gain.

LABORATORY RABBITS

Approximately 600,000 rabbits are used in medical and product research and in educational training programs per year. The AWA ensures humane treatment, housing, and care of all rabbits used for laboratory research. It also protects the owners of pet rabbits from the theft of their animals by preventing their sale for or use in research.

Pressure from animal rights organizations has led to studies seeking alternatives to the use of animals in research. One example is the Draize eye test. Because of improved statistics and the development of partial-animal alternatives, the numbers of rabbits used in this and other research tests have been dramatically reduced in recent years.

Research facilities seek specific pathogen free (SPF) white New Zealand rabbits and barrier-specific rabbits. The SPF rabbits are guaranteed to be free of a specific pathogen. The more pathogens guaranteed, the more expensive the rabbit (and the costlier to produce). The barrier-specific rabbits are born by cesarean section and raised by hand to ensure no contamination between the doe and its kits. These rabbits, obviously, bring the highest prices.

REPTILE FEEDER MARKET

Snakes, crocodiles, alligators, and the like have to eat. Rabbits continue to be a choice of food for these animals in zoos and other operations. Chickens have proven to be too dangerous because of their talons and beaks. The most popular sizes for snake feeders are 3- to 5-pound rabbits. Endangered carnivores also have a fancy for rabbit; eagles, condors, and wolves being rehabilitated prefer rabbit meat.

RABBITRY MANAGEMENT

Success in rabbitry, like success in any other livestock enterprise, depends on the management of the rabbit herd. Management skills are always based on knowledge of the animal to be managed; rabbitry is no exception. The basic principles already learned in animal nutrition, reproduction, genetics, and health apply to rabbitry, along with those unique to the rabbit. In this chapter we deal with management of the rabbit herd, based on such knowledge.

Housing and Equipment for Rabbitry

Hutches.   Although the size of the rabbitry enterprise varies from two rabbits to thousands, all require the same basic housing and equipment for successful management. Rabbits are kept in cages called hutches for ease of handling, protection, and proper management. Selection of the type of hutch depends on the size of the operation, climate, investment potential of the owner, and desired ease of cleaning. Double-compartment hutches may be desirable for small backyard operations (see figure 12–1), whereas thousands of wire hutches kept in a building would meet the needs of larger operators. The climate dictates the type of hutch used. Adult rabbits can withstand very cold temperatures (below freezing) with a minimum of protection, but they cannot withstand very hot temperatures or prolonged direct exposure to sunlight. Rabbits need some indirect sunlight. A rabbit also cannot withstand prolonged exposure to drafts or dampness but needs some ventilation. Young rabbits are born without fur and are extremely susceptible to the environment. They need extra protection until they grow their fur. Hutches should provide indirect light and 8 to 10 changes of air per hour but protect the rabbit from exposure to strong drafts, rain, and prolonged sunlight. Hutches should be constructed to allow convenience and ease of rabbit handling for the owner, ease of cleaning and proper sanitation, and maximum comfort for the rabbit. Such hutches can be very simple in construction, keeping investments to a minimum.

Sanitation of hutches is a must for rabbit health. Ease of cleaning must be considered in design and placement of rabbit hutches. Hutches are constructed from wire, wood, or both. Wire is usually more expensive and does

FIGURE 12–1
Single-deck hutch.

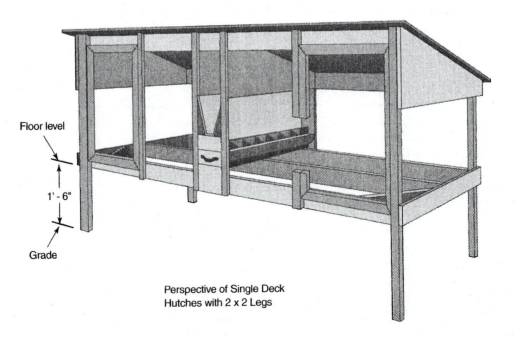

Floor level

1' - 6"

Grade

Perspective of Single Deck
Hutches with 2 x 2 Legs

not provide enough protection in adverse weather. Wood and wire hutches are easily constructed to meet the requirements based on the climate of the area.

Each compartment in a hutch should be large enough for one doe and its litter until weaning time or for 10 to 15 weaned bunnies. A rule of thumb is 1 square foot of floor space for each pound of rabbit. Usually 2 1/2 feet by 4 feet by 16 inches high is suitable for small breeds, and 2 1/2 feet by 6 feet by 20 inches high is suitable for large breeds. For ease of handling of the rabbit and cleaning, the door of the hutch should be large enough to move nest boxes, feeders, and so on in and out of the hutch easily (a minimum of 13 inches by 16 inches). Slotted wood (slots 1 to 1 1/2 inches wide with 5/8-inch spacing) or 5/8-inch wire mesh floors allow feces to fall through and aid greatly in cleaning the hutches. Only one doe or buck is housed per compartment. Does can be housed in adjacent hutches, but a solid wall should always separate a doe from a buck rabbit. One compartment can house 10 to 15 fryers (weaned bunnies) in cool weather or 10 to 12 fryers in hot weather. Two cages should be kept in a separate area for isolation of new or sick rabbits.

Nest Boxes.   A nest box provides shelter and protection of newly born bunnies and seclusion of the doe when it gives birth (kindles). It should be large enough to prevent crowding but small enough to maintain body heat. The size of the nest box is from 12 by 12 by 16 inches to 10 by 16 by 8 inches, with a 6-inch by 6-inch opening in the top or end (see figure 12–2). Ventilation holes should be placed in the sides of the nest box, and drainage holes should be placed in the floor. Nest boxes can be placed in the hutch or be part of the floor of the hutch. Boxes inside the hutch also provide the doe with a space to escape the pestering young bunnies. If the box top is level with the floor, an escape board (1- by 6- by 12-inch board placed 9 inches above the floor) should be provided for does with growing bunnies.

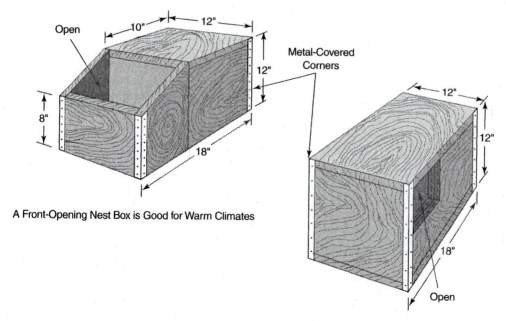

FIGURE 12–2 Nest boxes.

A Front-Opening Nest Box is Good for Warm Climates

A Covered Nest is Better in Cold Weather

Feeders and Waterers.   Feeders and waterers vary from various types of containers to self-feeder and automatic watering systems. Containers used for feed and/or water should be 3 to 4 inches deep, 6 to 8 inches in diameter, easy to clean, and heavy enough to prevent rabbits from tipping them over. Hay mangers should be about 6 inches off the floor and be filled from the outside of the hutch.

Other Equipment.   Other helpful equipment includes measuring cups to measure out feed in ounces, scales to aid in performance records, brushes and combs for grooming, and a clean carrying basket or cart for rabbit handling.

FEEDING THE RABBIT HERD

The rabbit's digestive system could be compared with that of the horse (chapter 4). It is a simple digestive system with an enlarged cecum and large intestine. This system allows for the intake and utilization of high-quality, leafy hays, grasses, and green vegetables. These items are broken down by bacteria in the lower digestive tract as in the horse. During the 1980s, researchers in France and elsewhere undertook considerable research aimed at improving rabbit nutrition in order to improve the rabbit meat industry.

■ **Coprophagy**
eating of feces

Unlike other animals, rabbits eat their feces (called **coprophagy**). This practice is quite normal for rabbits and is based on the construction of their digestive tract. Coprophagy is usually done late at night or early in the morning. Feces pellets that are light green and soft are eaten; those excreted in the day, which are brown and hard, are not eaten. The eating of the soft feces (cecotrophs) is actually called **cecotrophy**, a type of coprophagy. Cecotrophy

■ **Cecotrophy**
eating of the soft feces, or night feces, by rabbits

allows rabbits to make full use of the bacterial digestion in the lower tract—that is, conversion of forage protein to high-quality bacteria protein, synthesis of B vitamins, and breakdown of cellulose or fiber into energy. Cecotrophs contribute approximately 20 percent of the protein and 10 percent of the energy needs for maintenance as well as needed vitamins and minerals. Thus a rabbit should be allowed to practice coprophagy. Some older rabbits have difficulty reaching around to ingest their cecal pellets. Protein supplementation should therefore be considered when formulating their diets.

Like all animals, rabbits require the six basic nutrients—carbohydrates, fats, proteins, minerals, vitamins, and water. The amounts of each required depend on the age of the rabbit, desired production, and rate of growth. Clean water is a must. A lactating doe can drink 1 gallon of water per day. A salt spool is used to supply needed minerals. The protein content of rations is usually simplified by using two rations, one 16 percent and the other 14 percent. These rations are fed to rabbits with high or low requirements (see table 12–2). With rabbits raised

TABLE 12–2	Percent Composition of Rabbit Rations			
Rabbit Type	**Crude Protein**	**Fat**	**Fiber**	**Ash**
Sexually active buck, pregnant does, does with litters, growing bunnies under 3 months of age	14–18	3–6	15–20	12–14
Dry does, bucks, maturing bunnies	12–14	2–4	20–28	5–6

for wool production, dietary protein levels average 10 to 12 percent for short-haired adult rabbits and 13 to 15 percent for longhaired rabbits. Cecal pellets, which contain 25 to 28 percent protein, provide for at least 10 percent of the rabbit's amino acid needs.

Rabbits require 1 to 1 1/2 percent of their total caloric intake to be in fat. Two essential fatty acids, linoleic and linolenic acids, need to be balanced. **Eicosanoids** (prostaglandins and others) have the omega-3 and omega-6 fatty acids as their precursors and thus are necessary and must be supplied by the diet. Eicosanoids are essential for blood pressure control, blood clotting, muscle contractions, and memory.

A suggested ration is 40 percent rolled oats, 25 percent wheat bran, 15 percent rolled barley, 18 percent oil meal, 1 percent bonemeal, and 1 percent salt. This or a commercial ration should make up about 40 percent of the total feed; the rest should be hay or green feeds.

> ▤ **Eicosanoids** small lipids derived from arachidonic acid that are used to produce prostaglandins

Amount of Feed to Use

The amount of feed fed per day varies with the size of the rabbit and the production stage (see figure 12–3). Table 12–3 provides a guide for the amounts of feed to use. Green feeds or high-quality hays should be given along with the concentrate ration in table 12–3. Rabbits' rations should be changed gradually over a 7- to 10-day period. To change a ration gradually, a small amount of the new feed is mixed with the old, increasing the proportion of the new feed over the 7- to 10-day changeover period.

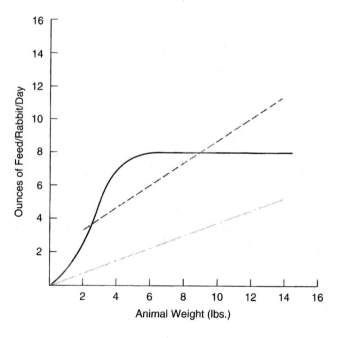

FIGURE 12–3
Feeding requirements for rabbits.
Reprinted with permission from J. Blakely and D. H. Bade, *The Science of Animal Husbandry,* 6th ed. (Englewood Cliffs, NJ: Prentice-Hall, 1994).

———————— Developing Young and Lactating Does (Full Feed)

– – – – – – Pregnant Does

——— · ——— Maintaining Bucks and Does

TABLE 12–3	Feeding Amounts for Various Rabbits
Rabbit Type	**Feeding Regime**
Young bunnies (birth to 3 weeks age)	Does' milk only
Bunnies over 3 weeks old	Whole grain (oats) free choice; some hay.
Weaned bunnies	Free-choice concentrate and hay ration.
Dry does, bucks	Full feed of hay plus 3 to 6 oz of concentrate per day if needed. Adjust to maintain proper weight. Do not let does become overfat.
Does at kindling	2 to 4 oz of concentrate first day. Gradually increase 1/2 to 1 oz/day to free choice in 5 to 7 days (4 to 7 oz/day). Feed 40% hay of the ration.
Lactating does	Free-choice concentrates, 40% hay.

Source: Reprinted with permission of J. Blakely and D. H. Bade, *The Science of Animal Husbandry*, 6th ed. (Englewood Cliffs, NJ): Prentice-Hall, 1994).

Feeding Schedules

The regularity of feeding is more important than the number of times per day the rabbits are fed. Nursing does should be fed twice daily with either grain and hay at both feedings or hay in the morning and grain and hay in the evening. Rabbits fed once daily (dry does, for example) should be fed in the late afternoon or evening because they eat mostly at night.

Types of Feeds

Concentrate or grain mixtures can be home mixed or commercial supplements. At least two different grains should be used in the mix to enhance intake. Grains, in order of preference by rabbits, are rolled oats, wheat, grain sorghum, and corn. Cottonseed meal should not be used as a protein supplement due to its gossypol content. Soybean meal, peanut meal, sesame seed meal, sunflower meal, and safflower meal are safe protein supplements for rabbits. Rapeseed meal should be avoided because it may cause problems in fertility and reproduction in does.

Hays that are used must be of high quality—high in nutrients, leafy, and clean. Legume hays, such as alfalfa, are preferred because they are high in protein, clean, leafy, and relished by rabbits. They can be used as the only feed for dry does, bucks, and older rabbits and up to 40 percent of the diet of lactating does. Grass hays are lower in quality and should not make up more than 10 percent of the diet.

Other green feeds, including garden greens, roots, and tubers, can be used if available. If used, they should be fed only to bunnies over 3 months of age. Greens include beets (roots and tops), cabbage, carrots (tops and roots), kale, potatoes, sweet potatoes (tubers and vines), turnips (tops and roots), and grass and legume leaves. Most commercial growers do not feed green feeds or root crops because of their lower nutrient density, high labor cost in feeding, and lowered efficiency in growth rate compared with more conventional feeds.

The fiber content of the rabbit's diet is extremely important. It is the best source of energy for cecal bacteria. Rabbits, like horses, convert this fiber into volatile fatty acids, providing energy for the animals. The fiber also protects rabbits from dietary-induced enterotoxemia, prevents blockages due to hair, and helps rabbits retain water in the digestive tract. Because of the aforementioned, fruits and vegetables should be fed at a minimal level due to their high water and low fiber content.

REPRODUCTIVE MANAGEMENT

Sexual Maturity and Mating

Although rabbit does reach puberty by 4 (in the small breeds) to 12 (in the large breeds) months of age, mating should be postponed for at least 1 to 1 1/2 months to allow increase in body size and maturity.

A doe's estrous cycle is different from that of other farm animals. Other farm animals have an estrous cycle in which the female has only a few days when it is in heat and will accept the male. The rabbit is almost the opposite. Does have a 15- to 16-day estrous cycle. Of these 16 days, they are in heat 12 days and will accept the buck when conception can occur. A doe is not in heat and will not accept the buck only the first and last 2 days of the cycle.

Many litters of bunnies can come from one doe in a year. Proper management usually allows for four to eight litters per year, depending on the production cycle. This system allows for proper weaning of a litter and proper conditioning of the doe prior to kindling (giving birth) of the next litter.

Young does are bred at 5 1/2 months of age to an experienced buck. If they do not mate, breeding is tried again every 10 days until the does are 6 1/2 months old; then young does are force mated. Force mating refers to manually restraining the doe for the buck to mate.

Does are rebred when the litter is 3 to 7 weeks old, depending on the number of litters desired per year. As discussed earlier, chances are good that they will accept the buck. If they do not, the producer should try to breed the does again in 3 to 4 days. Success usually follows. If the does still fight the buck, they may need to be force mated.

Does should be bred in the morning or afternoon before feeding, either by natural mating or by artificial insemination. Natural mating is accomplished by taking the doe to the hutch of the buck, never vice versa. Introduction of a buck into the doe's hutch will result in the doe fighting off the buck, not because the doe is not in heat but because it is possessive of its hutch. Mating usually occurs fairly rapidly. Because the doe will ovulate about 8 to 10 hours after the first service, some producers rebreed the doe at 8 to 10 hours after the first service. One buck is kept for 10 does, and a buck is not used more than once in 3 days. The reproductive life of a buck is 6 months to 2 years; the reproductive life of a doe is 2 to 3 years.

Approximately 10 to 18 days after breeding, the doe should be tested for pregnancy. Test mating and pregnancy diagnosis by palpation are the two primary means of verifying pregnancy. In test mating the doe is returned to the buck. If it refuses the buck, it is thought to be pregnant. Palpation of does is done 10 to 14 days after breeding. In palpation, the horns of the uterus are felt through the abdomen. Rabbits are easy to palpate, and with a small amount of practice palpation can be very accurate in diagnosing pregnancy. To palpate a doe, it is restrained by grasping a skin fold over its neck along with its ears with one

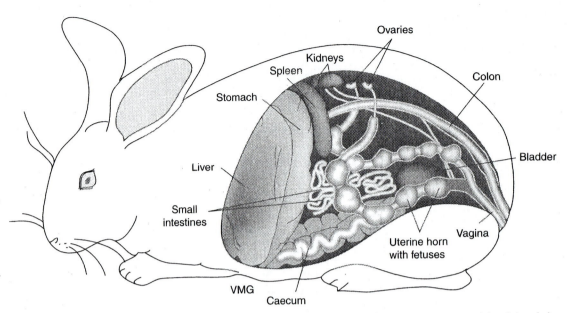

FIGURE 12–4 Pregnancy detection in the rabbit is relatively simple. With the thumb on one side of the abdomen and fingers on the other, a gentle sliding motion up to the rib cage can detect marble-sized fetuses.

hand. The other hand is then left free for palpation. The free hand, palm up, is passed under the abdomen to the hind legs. With the thumb on one side of the abdomen and fingers on the other, the finger and thumb are gently brought closer together and slid up to the rib cage. As this procedure is done, the horns of the uterus should pass through one's hand. The fetuses are the size of marbles (see figure 12–4). Any open does should be rebred.

Kindling

Kindling parturition in rabbits

Kindling (giving birth) normally occurs 31 to 32 days after breeding but can range from 28 to 35 days. Thus, does should be provided with a nest box no later than 27 days after being bred. Bunnies are born in litters of 2 to 15; most litters are 6 to 10. They are born in the nest box, usually during the night. They are born with their eyes shut and without fur. At 1 week, the eyes open, and the fur begins to appear. At about 3 weeks, the bunnies will come out of the nest box and begin eating with the doe. Bunnies are weaned at 7 to 8 weeks (depending on the breeding cycle of the producer). Management of the doe prior to, during, and after kindling involves the following:

1. Do not allow the doe to get too fat prior to kindling. Fat does have smaller litters and complicated kindlings.
2. At 21 to 28 days of gestation, disinfect the nest box, fill it two-thirds full with bedding, and place it in the doe's hutch. It will build a nest in it prior to kindling.
3. Prior to kindling, the doe will pull fur from its body and put it into the nest box. This process continues through kindling. Just prior to kindling, the doe loses its appetite (1 to 2 days prior).
4. A doe usually kindles during the late evening or early morning. Parturition usually is complete within half an hour. Take care to keep the doe calm during this time.

TABLE 12–4	Effect of Time of Rebreeding and Weaning on Number of Litters Produced per Year by Does		
Rebreeding[a] (days)	Weaning[a] (days)	Days between Litters	Litters/Doe/Year
14	28	45	8
21	35	52	7
28	42	59	6
42	56	73	5

Source: Reprinted with permission from J. Blakely and D. H. Bade, *The Science of Animal Husbandry,* 6th ed. (Englewood Cliffs, NJ: Prentice-Hall, 1994).
[a]Days after kindling.

5. Examine the nest box after kindling to record births and to remove any dead bunnies.
6. If more than eight bunnies are born in a litter, foster extra bunnies to other litters by rubbing the bunny with the foster mother's fur.
7. Remove the nest box when all the young have left it, about 3 weeks after kindling.
8. Cull does that lose or destroy their litters twice or lack natural mothering ability.

Management of Weaned Bunnies

Bunnies can be weaned from doe milk as early as 4 weeks after birth by placing them in another pen to be fed until slaughter. Usually, such early weaning results in smaller and less meaty bunnies than those weaned at 7 to 8 weeks of age. However, earlier weaning allows for more litters per year per doe (see table 12–4). Regardless of the weaning strategy, bunnies are usually slaughtered for fryers at 8 weeks of age (56 days).

RABBIT FUR OR WOOL INDUSTRY

The Rex rabbit is raised for its pelt because it produces a soft, luxurious fur. Top-quality tanned and primed Rex pelts sold for $12 to $14 per pelt in 1999. Pelts are commonly used for trimmings on doll clothes and on craft items. The quality pelts (prime) are made into garments and accessories. The four Angora rabbit breeds are raised for their wool (mohair), which is made into quality luxury garments.

Mohair may be sheared, plucked, combed, or clipped from the rabbit depending on individual market preferences and usage. Raw wool is generally sold to spinners, and the spun wool can be sold at a higher price to yarn shops and to knitters. Prices vary according to the quality, length, texture, and manner of gathering.

The amount of wool produced per rabbit per year can average anywhere from 8 ounces to 3 pounds or more. Growers wanting higher yields of wool should raise the Giant Angora or the German Angora rabbits, which produce considerably more wool per animal. The German Angora is used much more in Europe than in the United States. In 1999, Angora wool was selling for $4.00

per ounce for clipped wool (cut from the rabbit with scissors) and $4.50 to $8.00 per ounce for plucked wool.[7]

Angora rabbits require constant grooming and preventive maintenance for wool block. Wool block is basically the same thing as fur block in other rabbit breeds or as hairballs in cats. It is much more prevalent, however, in Angora rabbits due to the length of wool (see later section "Wool Block").

RABBIT HEALTH MATTERS

There are dozens of common health disorders in rabbits. We touch on the most common or serious ones in this section.

Broken Back

Rabbits should never be picked up by their ears. Trying to escape a handler not holding them properly is a common cause of broken back with inexperienced rabbit growers. In addition, soft radio music should be played at all times to reduce fright from sudden noise in the rabbitry. Rabbits can break their backs on their own in a cage simply trying to escape sudden noises. For humane reasons paralysis lasting 1 to 2 weeks warrants euthanasia.

Coccidiosis

Two common types of coccidiosis affect rabbits, intestinal and liver. The best preventive measure is using a wire brush on the cage floor daily to remove droppings stuck to the wire. The coccidia require a time outside of the rabbit before they can become infective. Treatment of the liver form with sulfaquinoxolone or amprolium has proven effective.

Ear Mite

Two common mites infect rabbits. *Psoroptes cunniculi* are common ear mites; they cause a shaking of the head, flapping of the ears, and scratching of the ears. Treatment includes placing mineral oil in infected ears for 3 days, repeating 10 days later.

Mange mites (*Sarcoptes scabiei*), also known as fur mites, can be treated with malathion or rotenone. These destructive external parasites cause a tremendous loss of fur to the rabbit.

Encephalitazoonosis

Encephalitazoonosis, also known as *nosema*, is a protozoan parasite (*Encephalitozoon cuniculi*) that causes brain and kidney infections. It is spread through the urine. Prevention is through improved sanitation. There is no known treatment.

---

[7]C. Haenszel, Fuzy Farm, Indiana, 1999.

Enteritis Complex

Enteritis complex is the most common cause of death in rabbits. It is associated with one or a combination of the following enteric disorders: mucoid enteritis, enterotoxemia, Tyzzer's disease, and coccidiosis. Tyzzer's disease is caused by *Clostridium piliforme,* a gram-negative, spore-forming bacillus. Symptoms include rapid weight loss, profuse watery diarrhea, fecal staining of the hindquarters, rough appearance, sunken eyes, and death within 12 to 48 hours. The disorder is associated with poor sanitation and stress (overcrowding).

Infectious Myxomatosis

In 1859, a single pair of rabbits was introduced and released in Australia by an English colonist. By 1890, more than 20,000,000 rabbits inhabited the colony. Rabbits became a serious economic problem. After trying to beat tens of thousands of rabbits to death the Australians tried a more scientific approach—biological warfare, so to speak. In 1950, myxoma virus (a poxvirus) was introduced into Australia to kill off large numbers of rabbits. Symptoms of this fatal disorder include a fever of 108°F, listlessness, purulent conjunctivitis, anorexia, labored breathing, and death. Today, myxomatosis is found in rabbits in California and Oregon and is spread primarily by mosquitoes and fleas.

Mastitis (Blue Breasts)

The chief causative organism of rabbit mastitis is *Staphylococcus* spp. Poor sanitation increases the spread in the rabbitry. The mammary gland becomes hot and swollen and if untreated becomes cyanotic (blue bag). Treatment consists of intramuscular administration of 200,000 units of penicillin G per 10 pounds per day for 3 days. Extra care in monitoring treated does is required because penicillin can cause diarrhea in rabbits.

Pasteurella multocida (Snuffles)

Although SPF rabbits can be purchased for snuffles, large numbers of rabbits infected with snuffles are still found, especially in the pet population. *Pasteurella multocida* is the primary organism that causes snuffles. *Bordetella bronchiseptica* and staphylococcal infections have been associated with snuffles as well. Symptoms include pneumonia, abscesses, weepy eyes, pyometra, orchitis, wry neck, sneezing, and nasal discharge. It is extremely contagious. Experimental vaccines have shown some success. Treatment with antibiotics placed in the drinking water reduces the spread through the rabbitry.

Papillomatosis

Papillomatosis is a condition of small warts or tumors forming in or on the rabbit. The two main types of warts are the *oral papilloma virus,* found under the tongue and on the floor of the mouth, and the *Shope papilloma virus,* found in the horny warts on the neck, shoulders, ears, and abdomen of the rabbit. Isolation and culling are the two main control methods of papillomatosis.

Pinworms

Pinworms (*Passalurus ambiguus*) are small, white worms found in rabbit droppings. Fecal analysis should be used to positively identify the parasite. Treatment consists of monthly treatment with piperazine or phenothiazine in the feed.

Ringworm

Ringworm is much more common in Europe than in the United States. Hair loss can be seen on the face and on the feet of rabbits infected with this fungal disorder. Treatment with griseofulvin mixed with strawberry jam and fed at a rate of 25 milligrams per kilogram for 15 to 25 days has proven effective.

Sore Hocks (Ulcerative Pododermatitis)

Sore hocks is also known as pododermatitis in domestic rabbits. It is caused by excess or continued pressure on the skin from bearing weight on the wire-floored cages or by scraping the skin with an item infected with *Staphylococcus aureus*, which is usually present in the hutch. The disease does not involve the hock but the plantar surface of the metatarsals (hind foot) and sometimes the volar surface of the metacarpal (forefoot) region. The best treatment is to cull and breed for rabbits with big, well-furred feet. Rubber door mats seem to help prevent this disorder.

Tapeworm

Rabbits are an intermediate host for dog and cat tapeworms. There are no visible signs of infection in rabbits.

Viral Hemorrhagic Disease

Viral hemorrhagic disease (VHD) is a viral disorder with an incubation period of 3 days and a mortality rate approaching 100 percent. In 1989, Mexico had a severe outbreak of VHD that killed 50,000 rabbits outright and caused another 100,000 exposed rabbits to be destroyed. The virus is found throughout Europe but has avoided American rabbitries so far. The virus causes hemorrhage in the lungs and intestines, with acute bloody nose and bloody stools. No vaccine has been approved for use in the United States, although England has developed one for its rabbitries. To prevent the spread of this disorder, all purchased rabbits should be quarantined. Certificates of negative tests on all imported rabbits should be required.

Vent Disease (Rabbit Syphilis, Venereal Spirochetosis)

Vent disease is caused by *Treponema cuniculi*. Clinical signs include blisters, ulcers, and scabs in the genital areas, as well as the lips and eyelids. Treatment with penicillin has eliminated the disease over a 3-week period.

TABLE 12–5	Comparison of Nutritional Content of Meat			
Edible Portion Uncooked	Protein (%)	Fat (%)	Calories per 100 grams	Cholesterol (milligrams per 100 grams)
Beef	13.7–16.3	28–39	317–406	95–125
Chicken	20	4.8–11	126–179	60–90
Lamb	13–15.7	27.7–39.8	312–410	—
Veal	18.5–19.1	12–16	184–218	100–140
Turkey	20.1–23.2	4.6–24	139–177	—
Salmon, Pacific	17.4	16.5	218	60
Tuna	24.75	3–5.2	126–146	149–172
Rabbit	20.8	10.2	175	50

Source: Modified from U.S. Department of Agriculture (USDA) Circular No. 549, "Proximate Composition of American Food Materials," 1952 and J. S. McLester and W. J. Darby, *Nutrition and Diet in Health and Disease,* 6th ed. (W. B. Saunders, Philadelphia, PA).

Wool Block

Long, luxurious wool on a rabbit can prove to be a serious health hazard. The longer the wool is left on the rabbit, the more likely some of it will loosen and be ingested as the rabbit grooms itself. The ingested wool forms a wool ball in the digestive system and can kill the rabbit in a matter of days. Fresh pineapple juice (10 milliliters), administered daily for several days, will partially digest hairballs and has been successful in treating mild cases.

On a final note about raising rabbits, some research shows healthful benefits of consuming rabbit meat compared with other animal products. Rabbit meat is relatively high in protein and low in fat and cholesterol compared with traditional red meats (see table 12–5).

SUMMARY

Although most American consumers consider rabbits mostly as pets and secondarily as research animals, there has been considerable development of rabbit production as an agricultural enterprise. Rabbits raised for meat, rabbits raised for wool (Angora wool), rabbits raised for their pelts, and rabbits raised for medical research have created this small but energetic industry.

New rabbit breeds continue to be developed to meet this changing need. Profit margins per animal are small, and thus most rabbit producers are either hobbyists or use this venture to supplement their income. Large commercial-type operations use the same or similar management techniques adopted by other livestock enterprises.

EVALUATION QUESTIONS

1. _____ is the soft feces eaten by the rabbit.
2. _____ is the practice of consuming feces.
3. Parturition in a rabbit is called _____ .

4. A young rabbit is called a _____ .
5. _____ is a rabbit breed from Turkey.
6. _____ is the racehorse of the rabbit industry.
7. The smallest of all rabbits is the _____ .
8. _____ is the major meat-producing rabbit breed.
9. _____ is the French rabbit breed valued for its pelt.
10. The gestation period of a doe is _____ .
11. _____ is the most common cause of mastitis.
12. Rabbit syphilis is referred to as _____ .
13. *Pasteurella multocida* causes _____ in rabbits.
14. _____ is the most common cause of death in rabbits.
15. _____ is a condition of small warts.
16. _____ is a fatal viral disorder purposely given to rabbits in Australia to thin out their population.
17. Lapine is to rabbit as _____ is to sheep.
18. Rabbits are of the _____ order.
19. The dressing percentage of the rabbit is _____ .
20. A rabbit kept for meat consumption before the age of 12 weeks is called a _____ .
21. Pregnancy testing of rabbit does by palpation can be done as early as _____ after breeding.
22. Doe rabbits are usually rebred when the litter is _____ .
23. A young rabbit doe should first be bred at about _____ months of age.
24. A 4-pound rabbit doe should produce _____ pounds of marketable meat per year.
25. From mating to weaning of rabbit fryers requires about _____ (days/months).
26. _____ is a highly contagious disorder caused by *Pasteurella multocida* in conjunction with *Bordetella bronchisepticum*.
27. _____ is the largest rabbit breed.
28. The rabbit's digestive system is much like that of a _____ .
29. Rabbit mohair is produced by the _____ breed.
30. Bunnies can be weaned from doe milk as early as _____ of age.
31. _____ (number) rabbits are slaughtered annually in the United States for meat.
32. The nation _____ is the number one producer of rabbit meat.
33. _____ are rabbits over 6 months old that have been sexually active and productive and have tough carcasses.
34. A _____ is a castrated buck.
35. _____ rabbits are guaranteed to be free of a specific pathogen.
36. _____ rabbits are born by cesarean section and raised by hand to ensure no contamination between the doe and its kits.
37. Rabbits are kept in cages called _____ for ease of handling, protection, and proper management.
38. _____ are also called night or soft feces.
39. _____ hormones have the omega-3 and omega-6 fatty acids as their precursors.
40. Cecal pellets contain _____ percent protein.
41. The amount of wool produced per rabbit per year averages anywhere from _____ to _____ .

42. The _____ breed of rabbit is most commonly afflicted with wool block.
43. _____ and _____ are the two types of coccidiosis that affect rabbits.
44. _____ is a viral disease that killed more than 50,000 rabbits in Mexico and forced the slaughter of another 100,000 rabbits in 1989.

DISCUSSION QUESTIONS

1. List the climate conditions a rabbit cannot withstand and must be protected from for good production.
2. Describe how to palpate a doe.
3. What are the advantages of rabbits for meat production? For meat consumption?

13

The Camelid (Lamoid) Industry

OBJECTIVES

After completing the study of this chapter you should know:

- The historical, current, and future trends of the camelid industry in the United States.
- Any unique anatomical features and management in feeding llamas.
- The importance of colostrum in disease prevention of crias.
- Reproductive management of the llama and any special concerns compared with other livestock species, including male berserk syndrome.
- The most common diseases of llamas and their symptoms (in brief), diagnosis, treatment, and prevention.
- The prevalence of plants poisonous to llamas in and around their normal surroundings.
- How llamas have been used successfully as guardian animals.

INTRODUCTION

Llamas continue to increase in popularity in the United States. The International Lama Registry lists 20,248 members of its organization, with most (19,052) from the United States. Oregon, California, Washington, Texas, and Ohio lead the nation in Lama farms and ranches (see table 13–1). The average herd size for the United States is 7.09 llamas per owner; in Canada the average herd size is only 3.9 llamas per farm. Total llama numbers have increased from an estimated 60,000 in 1993[1] to more than 126,000 in 1999.[2] In addition, there are more than 12,000 alpacas and 150 guanacos in the United States. In 1994, there were more than 8 million llamas and alpacas in the world (mostly in South America). More than 3 million llamas roam the Andes today (2 million in Bolivia and 1 million

---

[1]H. W. Leipold, T. Hiraga, and L. W. Johnson, "Congenital Defects in the Llama," *Veterinary Clinics of North America* 10(2):401–420, 1994.

[2]International Lama Registry, July 1, 1999.

TABLE 13–1	Leading States in Llama Numbers	
State	**Llamas**	**Llama Owners**
California	11,489	1,879
Colorado	6,488	951
Ohio	5,806	1,062
Oregon	13,977	1,693
Texas	7,634	1,163
Washington	10,684	1,576
Total U.S. population	126,750	19,052

Source: International Lama Registry, July 1, 1999.

in Peru). Although Australia has 15,000 alpacas, 99 percent of the world's alpacas are found in Peru, Bolivia, and Chile.

Llamas (*Lama glama*) are members of the family Camelid, order Artiodactyla, and suborder Tylopoda; the family also includes camels, vicunas (*Lama vicugna*), guanacos (*Lama guanacoe*), and alpacas (*Lama pacos*). The genus *Lama* (one *l*) includes the llama, alpaca, and guanaco. The vicuna, because of certain structural differences, is sometimes placed in its own genus, *Vicugna vicugna*. Guanacos (gwuh-NAH-koze) and vicunas (vye-KOON-yuhs) are wild lamoids that live in small bands of females, usually led by one male. Vicunas are an endangered species.

It is generally believed that the domestic camel, both the one-humped Dromedary or Arabian camel (*C. dromedarius;* native to India, North Africa, and the Near East) and the two-humped Bactrian camel (*C. bactrianus;* native to Central Asia), originated on the central plains of North America. Some 3 million years ago camels began to migrate to Asia and Africa, while llamalike animals headed south throughout South America. Camels completely disappeared from North America at the end of the last ice age, 10,000 to 12,000 years ago. Llamas and alpacas were domesticated before the Inca Indian civilization some 5,000 to 6,000 years ago in Peru. Today there are no llamas or alpacas in the wild state.

In the late 1800s both llamas and camels were reintroduced to North America. The camels died out in the deserts of the Southwest. The llama has gradually increased in numbers and is showing promise as an additional North American livestock species.

The South American Indians of Argentina, Bolivia, Chile, Ecuador, and Peru maintain most of the world's llama herds. Llamas have been used as beasts of burden and for their meat since the time of the Incas. Today the South American llama is used as a pack animal, as a producer of fiber for rugs and ropes, as a meat animal, and as a producer of dung for fuel. In the United States, llamas are used for light draft, pack animals, fiber production, show, and guard animals and are kept as companion animals (see figure 13–1). The smaller alpaca (averaging 35 inches at the withers compared with 47 inches for the llama) is used for its fine fiber production and for show.

Mature llamas weigh anywhere from 250 to 500 pounds. They mature by 4 years of age and usually live 15 to 20 years. A mature alpaca weighs 135 to 180 pounds. All of the South American camelids have 74 chromosomes and can thus interbreed and produce fertile offspring. The average gestation of a llama is 350 days. Llamas may be solid, spotted, or marked in a variety of patterns. Their wool colors range from white to black and many shades of gray, red, roan, and beige.

FIGURE 13–1
Guardian llama and its emu friend. Both are used to protect goats from predatory coyotes and dogs that frequent this college farm laboratory. Photo by author.

LLAMA NUTRITION

Llamas and other camelids are sometimes referred to as pseudoruminants because they have three and not four stomachs. The omasum is absent in all camelids. Digestion of fibrous feeds, fermentation, and regurgitation of ingesta are very similar, however, to those of other ruminants.[3] The dental formula of a llama is 1-1-2-3/3-1-1-3, or 30 teeth in all. Unlike other ruminants, llamas have canine teeth.

Llamas prefer more coarse, dry vegetation than alpacas, which prefer moist and lush forage. Llamas eat grass, forbs, browse, shrubs, and trees. They are much more flexible in their choice of forages and are generally considered more efficient than sheep or cattle at converting plant material into usable protein and energy. They are both grazers (eat grasses and forbs) and browsers (eat shrubs and trees).

Llamas eat 1.8 to 2.0 percent of their body weight per day. For a mature 300-pound llama this total is equivalent to 3 to 6 pounds of roughage a day for maintenance alone. Unless the animals are growing, in late gestation, in early lactation, extremely athletic, or exposed to prolonged cold temperatures, they should not be fed free-choice forages or be supplemented with concentrates.

Water requirements average 1 to 3 1/2 gallons a day per adult llama (or 5 to 8 percent of body weight per day). Water should be offered free choice and not be put near feeders or dung piles. It is important to keep the water trough clean to prevent several common microbial disorders in llamas. Llamas are capable of doing well on the trail on one good drink a day. Decreased water intake usually decreases feed intake, which negatively influences animal performance.

Loose iodized salts should be offered free choice in a sheltered container. Some llama breeders have found that combining the salt with minerals encourages their consumption and yet minimizes overconsumption. Calcium,

[3]P. J. Van Soest, *Nutritional Ecology of the Ruminant* (Ithaca, NY: Cornell University Press, 1994).

phosphorus, and selenium are the minerals of major concern in llama diets. As with other animals mentioned previously, an imbalance in calcium and phosphorus levels causes poor bone growth, decreases milk production, and contributes to crooked legs, osteoporosis, rickets, and urinary stones. The recommended calcium–phosphorus ratio is 1.2–2.0 to 1, which is commonly found in many grass hays and pastures. Too high a level of selenium can bring about the nutritional disorder known as blind staggers or alkali disease, whereas too little can cause white muscle disease or interfere with normal growth, reproduction, and lactation.

With normally cured hay, llamas receive all of their required vitamins A, D, and K and produce adequate amounts of their B complex requirements. If stored forage loses its green color, vitamin A should be added to the diet. Vitamin E is usually destroyed in cured forages, and some evidence shows benefit to its supplementation.

Llamas are usually segregated according to their dietary requirements. Llamas can be grouped into four categories: age, stage of pregnancy, mother with cria, and obese llamas:

Age

- Weanlings up to 1 1/2 years old
- Males 1 year or older
- Older females still breeding

Stage of pregnancy

- Pregnant females (1 to 9 months) with no baby at side
- Pregnant females within 2 months of parturition

Mothers with babies at their sides
Obese females

When feeding obese llamas, only 1 percent of their body weight per day should be fed. No grain and only grass hay is usually fed. The offspring of obese mothers should be weaned later, and no free-choice feeding (including grazing) should be allowed.

Colostrum feeding is extremely important to the overall health of the new cria. It is imperative that only the first-milking colostrum be given to a cria during its first meal. Because colostral antibodies are not species specific, initial protection can come from other milk sources. Beginning the second day, the cria should receive at least 10 percent of its body weight per day in goat milk if llama milk is unavailable. Second choice is lamb milk replacer (nonmedicated and diluted one part milk replacer to six parts water).

Colostrum is rich in immunoglobulins, specifically immunoglobulin G (IgG). This protein protects the cria against viral and other infectious agents that invade the blood. In humans, IgG passes readily through the placenta into the fetus, providing the infant with protection against diseases to which the mother was previously exposed. In horses, cattle, sheep, and camelids, IgG does not pass through the placenta and thus the newborn is without protection until it consumes the antibody-rich colostrum. Passive transfer of immunoglobulins occurs over the first 24 hours of life. Within 6 hours of birth, transfer is reduced by one-third, and by 24 hours only 11 percent of the immunoglobulins can be absorbed.

When a cria does not ingest enough colostrum or when the IgG level is below normal, failure of passive transfer (FPT) occurs. In llamas, approximately

10 percent of newborns are afflicted with FPT, leading to weak or dead newborns.[4] Several tests have been developed to measure IgG levels in colostrum. Because the cria cannot manufacture adequate amounts of IgG for about the first 6 weeks of life, it is critical to provide this passive transfer to the newborn. At 24 hours after birth, cria average 1,657 milligrams per deciliter of IgG (numbers range from 200 to 4,500). The amount of IgG needed to prevent disease depends on the amount of infectious agents present in the birthing area and the amount of colostrum ingested and absorbed. It is estimated that a healthy cria needs a minimum of 800 milligrams per deciliter to be considered capable of providing a strong immune response. Lowered IgG levels have been associated with increased susceptibility to parasites such as giardia and Eperythrozoondsis (EPE) and to chronic respiratory problems, lethargy, and diarrhea.

If after testing the cria, FPT is identified, a veterinarian should be consulted about providing IgG either intravenously or intraperitoneally. Donor llamas have blood drawn into a sterile receptacle. It is then centrifuged to separate the plasma from the red blood cells. The plasma can be used immediately or frozen for up to 1 year. If administered intraperitoneally, crias weighing 18 to 22 pounds should receive 300 to 500 milliliters of plasma; those weighing 23 to 30 pounds should receive 500 to 1,000 milliliters. The IgG titer of the donor should be high to minimize the quantity of transfer plasma required.

Kent Laboratories (23404 NE Eighth Street, Redmond, WA 98053) has developed a llama-specific IgG test (radial immunodiffusion [RID] test) currently used by several U.S. veterinary teaching hospitals and large llama ranches. Other, less specific colostrum quality tests are also on the market.

LLAMA REPRODUCTION

Male llamas reach sexual maturity at 2 1/2 to 3 years of age. Female llamas are sexually mature at 1 to 3 years of age and can conceive at an average of 18 months. It is best to wait to breed females until after they reach 24 months of age. Females should be at least 60 to 70 percent of their mature adult weight prior to breeding to reduce the incidence of birthing complications and ensure a more fully developed cria at birth.

Llamas do not have a heat cycle; they are induced ovulators. They will ovulate 24 to 36 hours after breeding and can breed at any time of the year. Breeding lasts from 5 to 45 minutes; the average is about 20 minutes long. The male breeds the female while she is sitting down (in the kush position).

If the female becomes pregnant it will not be receptive to the male 21 days later. Instead it will spit at the male, signifying that it is no longer interested. This is not a sure pregnancy test method but is used early on. Blood progesterone levels can be monitored day 21 after breeding. Anything over 2 nanograms per milliliter may indicate pregnancy. Ultrasound methods after day 20 and repeated after day 40 (confirmation that the cria is still alive) are becoming common practice.

Most pregnancies last 340 to 345 days, but the normal range is from 330 to 375 days. Signs of approaching birth include enlargement of the mammary

---

[4]D. Jorgensen, Triple J Farms, Redmond, Wash., May 1994.

glands and the softening and release of the cervical plug. Prelabor getting up and down and trying to nest or find a comfortable position takes approximately 6 hours. This period is followed by breakage of the water bag and the delivery of the cria. Like most other mammals discussed, the cria comes out in a diving position: the front two legs first with the head in between them. Any other position is considered abnormal or dystocia. The actual labor generally takes less than 2 hours.

Once the cria is breathing normally, it should be helped by wiping the mucus away from its nose and mouth and dipping the umbilicus with a tamed iodine solution. The cria should be up and standing and encouraged to nurse within the first hour to hour and a half. The placenta should be expelled within 4 to 6 hours. If it is not completely expelled within 12 hours, a veterinarian should be called.

Most crias are born from a standing mother and normally do not require assistance. Twinning is rare in llamas. Normal birth weights range between 20 and 35 pounds. Crias are usually weaned at 4 to 6 months of age.

MALE BERSERK SYNDROME

Baby llamas, or crias, are very cute. Consequently, humans want to cuddle and pet and provide extra tender loving care for them. Bottle-feeding baby male llamas and providing them with this extra attention results in abnormal behavior known as male berserk syndrome. When males reach puberty they do not see the difference between their human caretakers and other llamas. They become very territorial and very aggressive. This syndrome is permanent, and affected animals must usually be destroyed (because they attack humans). To prevent this problem, owners are strongly recommended to allow the bonding to occur between mother and cria and not cria and human. They should interface with the cria as little as possible. Castrating the llama early (between 2 and 6 months of age) helps but does not guarantee against this disorder.

HERD HEALTH

A veterinarian who specializes in or has experience with camelids should be consulted in setting up a herd health program. Each area of the country has its own particular needs and problems. Each farm dictates a different management and herd health program to successfully raise its animals. The veterinarian is the person most capable of making herd health decisions.

Individual records of each animal will help the veterinarian manage the animals. A sample record for llamas can be found in figure 13–2. This record is currently used on a college farm laboratory. A more detailed record may be needed or desired for a commercial farm.

Most crias are vaccinated before 4 weeks of age, when maternal antibodies may still interfere with an immune response. Booster vaccinations are thus required much sooner, as they are with any young mammal. With adult llamas, booster vaccinations are required annually. As previously discussed, the veterinarian should be consulted as to schedule, vaccines, and methods used to prevent, diagnose, and treat disorders. Most llamas, however, are vaccinated for tetanus and *Clostridium perfringens*. Depending on the area in

L.A. Pierce College Farm Laboratory
Individual Llama Record

_____ name _____ registration number _____ sex

_____ DOB _____ farm of origin

_____ sire _____ sire's registration number _____ sire's farm

_____ dam _____ dam's registration number _____ dam's farm

_____ birth weight _____ weaning weight _____ yearling weight

_____ mature weight _____ date

Fiber (quality) _____ (length) _____ (date) _____

_____   _____   _____

_____   _____   _____

Vaccinations
Tetanus   ____ ____ ____ ____ ____ ____ ____

Clostridium perfringens
   Type C and D ____ ____ ____ ____ ____ ____ ____

Clostridium septicum ____ ____ ____ ____ ____ ____ ____

Clostridium chauvoei ____ ____ ____ ____ ____ ____ ____

Rabies   ____ ____ ____ ____ ____ ____ ____

Leptospirosis   ____ ____ ____ ____ ____ ____ ____

_____   ____ ____ ____ ____ ____ ____ ____

_____   ____ ____ ____ ____ ____ ____ ____

Parasite Control

Fecal Analysis   Eggs Identified _____ Date _____

                 Eggs Identified _____ Date _____

                 Eggs Identified _____ Date _____

                 Eggs Identified _____ Date _____

Worming Medication Used _____ Date _____

                        _____ Date _____

                        _____ Date _____

                        _____ Date _____

External Parasites Identified _____ Date _____ Treatment _____

                             _____ Date _____ Treatment _____

Miscellaneous Health Record Information

_____

_____

Conformation evaluation: Performed by _____ (date) _____

_____

_____

FIGURE 13–2   Sample individual llama record.

which the llamas reside, additional vaccines for leptospirosis, anthrax, and rabies may be warranted.

The veterinarian may ask for some vital signs prior to his or her arrival. Temperature, pulse, and respiration rates can give the veterinarian an idea of the urgency of a more complete diagnosis and treatment. The average rectal temperature of an adult llama is 99 to 101.8°F (37.2 to 38.7°C); normal cria temperature is 100 to 102.2°F (37.8 to 39°C). Normal resting heart rate is 60 to 90 beats per minute; normal resting respiration rate is 10 to 30 breaths per minute. The respiration rate should be checked from a distance first so as not to excite the animal. Then the heart rate and rectal temperature are checked.

Injections should only be performed on restrained animals. For intramuscular injections, use of a 16- to 22-gauge needle of 1 1/2 inches (depending on the viscosity of the product injected) should be used. Muscles in the neck region should not be used. The muscles in the hind leg are usually preferred because they have less fiber. However, avoidance of the sciatic nerve is important. If frequent injections are needed, the site should be rotated to prevent muscle soreness. For subcutaneous injections, a 16- to 22-gauge needle of 1 1/2 inches is also used. The preferred site is the skin in the region of the withers, where there is also less wool.

Collecting Blood Samples

The veterinarian may need to take a blood sample for a more complete diagnosis in an ill llama. With practice, llama breeders can easily learn how to take samples from the tail vein or from the ear vein (e.g., small samples needed for pregnancy testing). For larger samples, the jugular vein (in the neck) is used. Because many vital structures are located in this region, the veterinarian or his or her trainee should perform this task. If the veterinarian is using an automated cell counter, adjustments need to be made to obtain an accurate reading. Llama red blood cells are smaller and shaped like those found in reptiles (elliptical).

Worming

Many llama breeders simply worm their animals on a regular basis (twice per year) and practice a rotation of worming medication. However, it is very easy and better medicine to find out what worms are present and to what degree before simply trying to treat blindly. In addition, a periodic fecal examination for parasites allows caretakers to see if their current treatment plan is actually working. Some of the most common internal parasites of llamas include stomach worms (*Haemonchus, Ostertagia,* and *Trichostrongulus* spp.), lungworms (*Dictyocaulus* spp.), thread-necked strongyles (*Nematodirus* spp.), nodular worms (*Oesophagostomum* spp.), whipworms (*Trichuris* spp.), liver flukes (*Fasciola* spp.), and meningeal worms (*Paralaphostrongylus* spp.). Coccidiosis (*Eimeria* spp.) may also be a significant problem in crias, contributing to diarrhea and dehydration.

The most common external parasite affecting llamas is lice (*Mallophaga* spp. [chewing lice] and *Anaplura* spp. [sucking lice]). Unthriftiness, wool loss, and scratching are usually signs of lice infestation. Topical treatment with carbamate powder is very effective in most cases. Ticks and mites occasionally infest llamas but are not as common as lice.

Poisonous Plants

Following is only a partial list of poisonous plants that commonly occur in areas inhabited by llamas in the United States. Prevention includes scouting out pastures prior to letting the animals out. In addition, most llamas will not voluntarily consume these plants unless they are hungry; by not overgrazing pastures, preventing roadside eating, and preventing ingestion of unknown plants, most poisonings can be prevented.

1. Black locust, false acacia
2. Japanese or English yew
3. Kentucky coffee tree
4. Ornamental rhododendron and azalea
5. Box (ornamental hedge, *Buxus sempervirens*)
6. Ohio buckeye
7. Trumpet vine
8. Lantana
9. Oleander
10. Tobacco
11. Tree tobacco
12. Water hemlock
13. Black nightshade, bull nettle, Carolina nettle
14. Milkweeds
15. Jimson weed, Jamestown weed
16. Death camus, sandcorn
17. Pokeweed, poke
18. Mountain laurel, ivy bush, bank laurel
19. Sheep laurel, wicky, sheepkill
20. Evergreen rhododendron, great laurel
21. Western azalea
22. California rosebay
23. Japanese pieres, mountain fetterbush, staggerbush
24. Labrador tead
25. Black laurel, Mt. laurel
26. Dog laurel, fetterbush

Nail Trimming

The simple act of maintaining proper nail length can prevent a more serious problem that may lead to lameness. Most overgrown llamas' nails will grow to the side and interfere with their normal gait. Llamas used as pack animals walk long distances and will not need their nails trimmed as often (or not at all). Sheep- or goat-cutting nippers used to cut away excess growth will maintain a healthy hoof for the llama.

USES OF LLAMAS

Fiber

Llama fiber is used for knitwear, woven fabrics, rugs, and rope (see figure 13–3). The fiber is classified by fiber length as short (classic), medium, or heavy wool fiber. Weavers and spinners also classify the wool as fiber type,

FIGURE 13–3 This show llama has excellent fiber quality sought after by many weavers and spinners. Photo by author.

hair type, or mixed fiber and hair and by relative quantity of thicker guard hair (lots, some, or little).

Llamas are normally sheared every 2 years, each yielding about 7 1/2 pounds of fleece. In 1999, llama ranchers were marketing clean llama fleeces at $2.25 per ounce. Alpaca fleeces were selling for $2 to $5 an ounce to local artisans. A well-kept alpaca can yield as much as 5 to 8 pounds of fleece per year.[5]

Guardians

Llamas are territorial by nature and are instinctively suspicious of dogs and coyotes. Most guard llamas will stand guard at the birth of lambs and protect sheep from dangerous canines by herding. The best sheep guards are gelded llamas over the age of 18 months; they respect humans and can be easily haltered and led. Guard llamas need to be alone (not doubled with another llama) for bonding with the sheep to occur. The primary means of defending the sheep is by herding them together and by charging, striking, or stomping at the canines. Although proven very effective against one or two dogs or coyotes, llamas are no match for a mountain lion or for large packs of dogs or coyotes.

Other Uses of Llamas

In the United States commercially raised llamas are primarily used for packing. Most llamas can carry 45 to 60 kilograms (99 to 132 pounds) 35 kilometers a day for 20 days. Hunters, equestrians, and hikers regularly use llamas to carry their

---

[5]Alpaca Owners and Breeders Association (AOBA), July 1999.

gear up in the western mountains of the United States. Llamas are also used for cart pulling; in animal-facilitated therapy; as exhibition animals in shows, fairs, and parades; and as companion animals.

SUMMARY

Llamas continue to increase in popularity. It is estimated that there are more than 138,000 llamas and alpacas in the United States—a 100 percent increase in just one decade. The llama is used as a packing animal, guardian animal, and companion for the family. Because of the increase in popularity, it is important for the veterinary student, in particular, to learn the physiology, medicine, and management of this beautiful animal.

EVALUATION QUESTIONS

1. _____ is the normal rectal temperature of a llama.
2. _____ is the average gestation period of a llama.
3. Llamas are native to the high mountain region of _____.
4. The llama was domesticated by the _____.
5. There are _____ llamas and _____ alpacas in the United States.
6. The average respiration rate is _____ and the average pulse is _____ in llamas.
7. The chromosome number of a llama is _____.
8. _____, _____, and _____ are the states with the three largest populations of llamas in the United States.
9. _____ is a wild member of the South American camelid family that is an endangered species.
10. The average weight of a mature llama is _____.
11. The average weight of a mature alpaca is _____.
12. _____ are animals that eat grasses and forbs, whereas _____ are animals that eat shrubs and trees.
13. A mature adult llama will eat _____ percent of its body weight per day in dry matter.
14. When feeding obese llamas, only _____ percent of their body weight per day should be fed.
15. By the second day of life, the cria is consuming _____ percent of its body weight per day in colostrum.
16. The immunoglobulin _____ is measured to determine the quality of colostrum.
17. _____ is abbreviated FPT.
18. Female llamas should not be bred until they have reached at least _____ percent of their expected mature body weight.
19. Llamas are bred in the _____ position.
20. An average cria weighs _____ pounds at birth.
21. Crias are weaned at age _____.
22. _____ is the most common external parasite affecting llamas.
23. Llamas can easily carry _____ pounds on long packing trips through the mountains.

DISCUSSION QUESTIONS

1. List the common diseases of llamas and their prevention and treatment.
2. How are reproduction, breeding, and infertility different in llamas than in other domestic animals?
3. List common external and internal llama parasites and their prevention and treatment.
4. How is feeding llamas different than feeding sheep, cattle, or horses?
5. Is a llama a true ruminant? Explain.
6. What are some common poisonous plants that llamas might come across in the western United States? How might one prevent such a poisoning?
7. What is male berserk syndrome? What are its causes, prevention, and treatment?
8. Describe an ideal guardian llama. What is it used for?

RECOMMENDED READING

1. Fowler, M. E. *Medicine and Surgery of South American Camelids.* (Ames, Iowa: Iowa State University Press, 1989).
2. Hoffman, C., and I. Asmus. *Caring for Llamas. A Health and Management Guide.* 2nd Edition. (Grand Junction, Colorado: Rocky Mountain Lama and Alpaca Association, 1996).
3. Hume, C. "On Feeding Llamas." *Llama World* 1:6–9, 1983.
4. Johnson, L. "Llama Medicine." *Veterinary Clinics of North America, Food Animal Section* (Philadelphia, PA, 1989).

14

Ostriches

OBJECTIVES

After completing the study of this chapter you should know:

- The historical, current, and future trends of the U.S. ostrich industry.
- Unique reproductive characteristics and management concerns of ostriches.
- The marketing of meat, hide, and feathers in the ostrich industry.
- The special considerations used in feeding commercially raised ostriches.
- Diseases of ostriches and their symptoms (in brief), diagnosis, treatment, and prevention.

INTRODUCTION

Ostriches *(Struthio camelus)* belong to the ratite (raft without keel) family, a family of birds that cannot fly and do not have a keel. Other members of this family include rheas, kiwis, emus (see figure 13–1), and cassowaries. Ostrich skeletons and fossils that date back over 120 million years—to the time of the dinosaurs—have been found.

The ostrich industry started more than 100 years ago in South Africa. In the late 1980s, ostriches gained popularity in the United States. Today there are more than 2,000,000 ostriches being raised in over 50 countries throughout the world. In the United States there are more than 2,000 members of the American Ostrich Association, and there are more than 500,000 ostriches on their farms and ranches. Over half are raised in Texas, California, and Arizona. The birds are raised for their hide, which can produce as much as 14 square feet of leather; their meat (the carcass yields about 70 pounds of meat); and their feathers (about 3 to 4 pounds per bird). As previously mentioned, ostriches are part of the ratite family, a group of flightless birds with a flat breastbone but without the keel-like prominence characteristic of most flying birds. Ostriches are the largest bird in the world, standing 7 to 9 feet tall and weighing 250 to 350 pounds (see figure 14–1). They can live 70 to 80 years and reproduce for more than 40 years.

FIGURE 14–1
Ostrich teenagers
located on a large
ostrich ranch in
Southern California.
Photo by author.

Ostriches are known to run at speeds exceeding 50 miles per hour and with strides as long as 20 feet or more. They can maintain a speed of 40 miles per hour for at least 30 minutes. They cannot fly, however. Ostriches are the only bird with only two toes. When defending themselves, ostriches flail their legs and kick. Their sharp toenails can severely injure or even kill a person or predator such as a dog, a coyote, or even a small lion. Ostriches do not bury their heads in the sand as depicted on so many cartoons and comics.

REPRODUCTION

In the wild ostriches reach sexual maturity at between 4 and 5 years of age. Wild ostriches stay in small flocks, and the males scratch out the nests for the hens. The less colorful hens sit on the eggs during the day, and the males take over during the night. Both parents brood the young chicks. However, in captivity, hens begin to lay at 2 to 3 years of age and lay 30 to 100 eggs per year. Males usually reach maturity at around 2 1/2 years of age.

During the mating season, the males develop a red coloring on their beaks, around the eyes, and on the skin around the leg bones. Ostriches are polygamous. Domestically raised birds are mated one male to every two to several females. A breeding pair in the 1980s sold for as much as $50,000. In 1999 a breeding pair could be purchased for $2,000 to $5,000. In an average year, a hen lays 40 eggs, of which 95 percent are fertile; 75 percent of these fertile eggs will hatch. Hatching to full-grown slaughter weight of 250 pounds takes 9 to 12 months. The reproductive anatomy of the ostrich is very similar to that of the chicken previously discussed. The hen has a single left ovary and oviduct consisting of the same five parts: infundibulum, magnum, isthmus, uterus, and vagina. The male has two intraabdominal testicles, near the kidneys.

MEAT, HIDE, AND FEATHERS

A 250-pound bird yields a 100-pound carcass. Meat cuts are classified into prime cuts (steak and fillet) and trimmings or offcuts. In 1997, more than 100,000 U.S. birds were processed. At that time nearly 20 pounds of the best quality meat in a bird sold for $10 a pound, another 20 pounds for $4 to $7 a pound, and another 60 pounds (made primarily into jerky and ostrich sticks) for much lower prices. A live 250-pound ostrich selling for slaughter purposes usually brings in $400 to $500. Ostrich meat is much more like beef than like poultry. It is a lean red meat, high in protein but low in fat (see table 14–1). In 1992, only two fine restaurants offered ostrich meat on the menu in the United States. In 1999, there were several hundred restaurants found nationwide.[1] Some restaurants now sell ostrich burgers for as little as $3.50 a burger compared with the $20 to $30 per entrée of the early 1990s.

One ostrich egg equals the content of 24 chicken eggs. Even though the egg is large in comparison to the chicken egg, the ostrich egg is the smallest proportionally to the size of the adult bird of any egg laid. Usually, the cost of the ostrich egg in the United States is prohibitive for egg production. However, scientists believe that the ostrich could be developed to lay an egg a day, making the eggs more economical. Ostriches are photoperiod dependent. During the typical breeding season of March through September, an average ostrich hen will lay an egg every other day if the eggs are removed each day for artificial incubation. If the eggs are not removed the hen usually stops laying after a clutch of 12 to 22 or more eggs are laid. Ostrich eggs incubate for 42 days. The newly hatched chick stands 10 inches tall and weighs about 2 pounds. Ostrich chicks grow more than 10 to 12 inches a month for the first 6 months and reach slaughter size and weight at around 12 months of age.

The hide from a mature ostrich can be made into three full boots. One company in Dallas, Texas, the John Mahler Company, has a $10 million import-export hide business involving the purchase of ostrich skins from South Africa. According to Dr. Marjorie Mahler, president of the firm, the ostrich hide is one of the toughest yet most pliable skins available. South Africa still dominates the world's ostrich leather supply market.

TABLE 14–1	Nutrient Comparison of Ostrich and Other Commonly Consumed Meats				
Meat	**Protein (%)**	**Fat (g)**	**Calories (kcal)**	**Iron (mg)**	**Cholesterol (mg)**
Ostrich	26.9	2.8	140	3.2	83
Chicken	28.9	7.4	190	1.2	89
Turkey	29.3	5.0	170	1.8	76
Beef	29.9	9.3	211	3.0	86
Pork	29.3	9.7	212	1.1	86
Veal	31.9	6.6	196	1.2	118
Deer	30.2	3.2	158	4.5	112

Source: Texas Agricultural Extension Service, Texas A & M University, 1999.

---

[1] American Ostrich Association (AOA), July 1999.

The full quill area (where feathers are stuck into the skin) brings in about $400 to $450 tanned and $200 green. The plumage of the male is quite attractive and is used for a variety of purposes in home decorating and sometimes in clothing. The feathers are used as household feather dusters, shoes, clothes, plumed headpieces, and more recently computer dusters. The feathers leave no dust and eliminate static electricity.

DANGEROUS BIRDS

Ostriches must be confined behind a very high (at least 6-foot) fence. Ostriches can be very dangerous. They defend themselves by flailing their legs and kicking. Their toenails are extremely sharp and can severely injure or kill a person if they attack. They have been known to knock down lions in their native Africa. Caretakers should walk behind them rather than in front. Ostriches kick forward and down. They have been seen disemboweling full-grown bulls.

NUTRITION

Ostriches are grazers (hindgut fermentors); they consume grass, leaves, and other forbs in addition to insects and arthropods. It is important to maintain a relatively high fiber content in their diet. Young ostriches need from 6 to 9 percent fiber in their diet, and adults require a minimum of 15 to 18 percent. Young ostriches tend to become obese on reduced-fiber rations. Obesity is one of the major causes of decreased production in birds. Three sample rations developed by Oklahoma State University are shown in tables 14–2, 14–3, and 14–4.

Ostriches need to have free-choice water to prevent impaction. Ostriches also need grit. In the wild, ostriches pick up pebbles and other hard objects,

TABLE 14–2	A Typical Ostrich Starter Ration
Ingredient	**Amount (lb.)**
Soybean meal	26.50
Corn grain, yellow	25.20
Alfalfa (dehydrated)	18.60
Peanut hulls	9.53
Sorghum	8.50
Corn screening	5.50
Calcium carbonate	1.84
Dicalcium phosphate	1.84
Nutrablend vitamins	1.25
Fat (animal)	0.80
Salt	0.60
DL methionine	0.07
Total weight	100.00

Source: J. Berry, Oklahoma State University Extension Facts No. 3988, Extension Poultry Specialist, Oklahoma State University, 1999.

TABLE 14-3	A Typical Ostrich Grower Ration
Ingredient	**Amount (lb.)**
Soybean meal	26.00
Corn grain, yellow	22.90
Alfalfa (dehydrated)	15.00
Sorghum	15.00
Wheat midds	6.50
Fish meal (menhaden)	5.00
Peanut hulls	4.80
Calcium carbonate	1.47
Nutrablend vitamins	1.25
Dicalcium phosphate	1.23
Fat (animal)	0.50
Salt	0.30
DL methionine	0.04
Total weight	100.00

Source: J. Berry, Oklahoma State University Extension Facts No. 3988, Extension Poultry Specialist, Oklahoma State University, 1999.

TABLE 14-4	A Typical Ostrich Breeder Ration
Ingredient	**Amount (lb.)**
Soybean meal	27.40
Corn grain, yellow	21.00
Corn screenings	10.20
Sorghum	10.00
Alfalfa (dehydrated)	9.80
Meat and bonemeal	8.18
Calcium carbonate	4.90
Peanut hulls	4.80
Nutrablend vitamins	1.25
Dicalcium phosphate	1.20
Fat (animal)	0.60
Salt	0.60
DL methionine	0.06
Total weight	100.00

Source: J. Berry, Oklahoma State University Extension Facts No. 3988, Extension Poultry Specialist, Oklahoma State University, 1999.

which they retain in their gizzard to help grind the food they consume. Oyster shells (3 percent by weight) are commonly fed for this purpose. A protein level of 15 to 24 percent, a calcium–phosphorus ratio of no greater than 2.25, and an average energy value of 2,100 to 2,300 kilocalories of metabolizable energy per kilogram of diet should be fed.

DISEASES AND OTHER HEALTH-RELATED DISORDERS

Poor sanitation in incubators, hatching rooms, chick rooms, and brooders is the leading cause of health-related problems in ostriches. Overcrowding is also a major problem. Ostriches need at least 250 to 300 square feet per bird up until 6 to 8 months of age, when the growth rate starts to slow down. Many ostrich breeders vaccinate for salmonella, *Escherichia coli,* and the various clostridial diseases as an additional preventive measure.

Diarrhea

Diarrhea is the most common clinical sign in young ostrich chicks. Chicks can develop diarrhea from a sudden change in diet, a new diet, or pathogenic reasons. Treatment for nutritionally caused diarrhea includes bismuth subsalicylate or simply allowing the chicks to adjust to new or changed diets gradually. Ostrich chicks are coprophagic, and thus transmission of many bacterial infections is quite rapid. The most common bacterial causes include *E. coli, Salmonella* spp., *Pseudomonas* spp., *Campylobacter jejuni, Klebsiella* spp., *Clostridium perfringens, C. colinum, Mycobacterium* spp., *Streptococcus* spp., and *Staphylococcus* spp. Culture and antibiotic sensitivity tests should determine treatment of bacterial diarrhea.

Egg Binding

Ostrich hens can become egg bound. There may be a genetic predisposition to egg binding, and it may be complicated by poor nutrition, obesity, metritis, or environmental factors. Ultrasound and radiology are generally used to make a positive diagnosis of this disorder. Treatment includes the injection of vitamins A, D, and E, along with calcium and oxytocin. Increasing the bird's ambient temperature seems to help as well. Breaking the egg in the uterus should be avoided, because there is a high risk of lacerations from fragments of the shell.

Oviduct Infections

Oviduct infections are quite common in ostriches. Metritis can cause abnormal shell development to no egg production at all. Salpingitis is also a common bacterial infection in the hen.

Common isolates on culture that cause pneumonia, air sacculitis, and/or peritonitis in the ostrich include *E. coli, Pseudomonas* spp., and *Acinitobacter* spp. Papilloma viruses have also been isolated in the ostrich. Hens with this virus show malformed eggs, sudden stop in egg production, and/or odoriferous eggs. Hens may also have a discharge below the cloaca with a peculiar odor. The white blood cell count ranges from 20,000 to 100,000.

Both ultrasound and X-rays have been helpful in assessing the amount and consistency of exudate in the oviduct of the hen. Treatment must be based on isolate culture and sensitivity tests.

Poxvirus Infection

Poxvirus infection is a viral disorder spread by insects. It is a self-limiting, low-mortality disease that usually affects young chicks. It produces the typical pox lesions on the face, ears, and neck.

Yolk Sacculitis

The incidence of yolk sacculitis increases markedly when ostrich owners try to assist the chick at hatching. This practice often results in the yolk sac becoming contaminated with gram-negative bacteria.

Parasites

External parasites of the ostrich include lice, ticks, and quill mites. Biting lice cause skin and feather damage. Feather mites live in the vein on the underside of the feather and feed on the blood. Treatment for ticks and mites is primarily with ivermectin at 0.2 milligrams per kilogram at 30-day intervals. Lice are controlled by dusting at 14-day intervals with carbaryl dust.

Internal parasites of the ostrich include wireworms (nematodes), tapeworms (cestodes), ascarids, and strongyloides and various protozoa (*Hexamita, Giardia, Trichomonas, Cryptosporidium,* and *Toxoplasma* spp.). Ivermectin at 0.2 milligrams per kilogram is effective at treating nematode infestations. Metronidazole at 10 milligrams per kilogram can be used to treat most ostrich protozoa. Tapeworms are best treated with fenbendazole. A veterinarian should be consulted on the culture, diagnosis, and treatment of all internal parasites.

For further information, readers are encouraged to contact

American Ostrich Association
3950 Fossil Creek Blvd., Ste. 200
Fort Worth, TX 76137
aoa@flash.net

For additional information on ostrich diseases, readers may consult

D. M. Allwright, "Ostrich Diseases," in *Practical Guide for Ostrich Management and Ostrich Products,* ed. W. A. Smith (Matieland, South Africa: Alltech, 1995).

SUMMARY

Ostriches belong to the ratite family. They are birds but cannot fly. The ostrich industry began in South Africa more than 100 years ago. In the United States, increasing numbers of farms have experimented with and some have profited by this large meat- and hide-producing bird. More than 100,000 ostriches are processed for meat every year in the United States alone. Hundreds of restaurants are now offering ostrich meat as an alternative to beef hamburger and steak.

EVALUATION QUESTIONS

1. _____ is a group of flightless birds that have a flat breastbone without the keel-like prominence characteristic of most flying birds.
2. The number of ostriches currently thought to exist in the United States is _____.
3. _____ is an example of another member of the ostrich family.
4. Ostriches are native to _____ (country).
5. At slaughter, an ostrich will weigh _____ pounds.
6. It takes _____ (time) for an ostrich to reach slaughter weight.
7. A domestic ostrich hen will start laying eggs at age_____.
8. _____ is where feathers are stuck into the skin.
9. In a good year, the average ostrich hen will lay _____ eggs.
10. An ostrich egg equals the content of _____ chicken eggs.
11. _____, _____, and _____ are the three leading ostrich-producing states in the United States.
12. The ostrich can run as fast as_____ miles per hour for a duration of _____ minutes or more.
13. The typical breeding season of an ostrich is during the months of _____ and _____.
14. _____ is the most common nutritional cause of decreased production in ostriches.
15. In the wild, ostriches consume _____ needed to help grind up food in their gizzards.
16. _____ is the most common clinical sign in young ostrich chicks.

DISCUSSION QUESTIONS

1. Why has the general American public not accepted the consumption of ostrich products as readily as chicken, turkey, and beef?
2. What do you perceive the future of ostrich farming to be in the U.S.?

Glossary

Abomasum Fourth or true stomach of a ruminant.

Abortion Premature expulsion of the fetus before it has matured enough to survive.

Absorption Process by which digested food particles pass through the digestive tract into the circulatory system.

Acid detergent fiber (ADF) Measure of cellulose and lignin in forages and feeds analyzed by boiling the sample in an acid detergent solution; it consists primarily of cellulose, lignin, and silica, which are the least digestible parts of the plant; as the plant matures ADF increases and digestibility decreases.

Acidosis Disorder resulting from lower than normal pH in the body.

Aerobic Requiring the presence of oxygen to live and reproduce.

Aflatoxin Carcinogenic fungal toxin produced by improperly stored or damaged feeds.

Agriculture Utilization of biological processes, on farms, to produce food and fiber useful to humanity.

Ahimsa Hindu belief in noninjury.

Albumin White of the egg.

Allele Genes that are located at the same point (locus) on each of a pair of chromosomes but that affect the same trait in a different or alternative manner.

Alveoli Individual grapelike structures of the mammary gland that secrete milk.

Amino acids Building blocks of proteins; they contain oxygen, carbon, hydrogen, and nitrogen.

Anaerobic Without oxygen.

Anemia Deficiency in the quantity or quality of blood.

Anestrus Phase or period of sexual quiescence.

Animal rights Philosophy that animals have the same basic rights as human beings; against speciesism.

Animal unit Generalized unit for describing the stocking density or carrying capacity of feed on a given area of land; one animal unit is the amount of feed a 1,000-pound cow with calf will consume.

Animal welfare   Philosophy that humankind has dominion over animals and, as such, has responsibility for their well-being.

Anion   Negatively charged ion. In dairy nutrition, important anions in the dry cow rations are chloride and sulfate.

Anorexia   Without appetite.

Anthelmintic   Drug used for treating internal worms in animals.

Antibiotics   Class of drugs produced by living microorganisms that inhibit the growth of other microorganisms.

Antibody   Immunoglobulin molecule produced by lymphoid tissue following exposure to an antigen. Antibodies help the body by neutralizing the specific antigens that induced their formation.

Antigen   Protein that stimulates production of a specific antibody.

Antiphlogistic   Paste or poultice that counteracts inflammation and fever.

Antitoxin   Antibody that specifically neutralizes the effects of a toxin.

Ascarids   Roundworms.

As-fed   Refers to feed as consumed by the animal.

Ash   Mineral matter present in feed.

Autosomes   Chromosomes excluding the X and Y chromosomes.

Babcock test   Test for butterfat quantity in milk and milk products.

Backgrounding   Practice of raising and preparing a beef animal from weaning until it is ready for placement in the feedlot.

Bacterin   Vaccine made with killed or attenuated bacteria.

Balling gun   Long, metal or plastic (for young calves) instrument with a cup-like depression at one end used for placing a bolus or magnet in the back of the mouth of cattle so that it can be swallowed without being chewed.

Band   Operational unit of sheep on the range, numbering from 1,000 to as many as 3,500 head.

Bang's disease   Same as brucellosis or undulant fever. A highly contagious abortion disease.

Barred   Term used to describe striped markings on poultry.

Barrow   Castrated male swine.

Beta-carotene   Precursor of vitamin A.

Bile   Fluid produced by the liver that emulsifies fats in the small intestine.

Binomial nomenclature   System developed using two names—a genus and a species—in classifying all life forms.

Biosecurity   Management procedures that minimize the risk of disease transmission.

Blistering   Controversial treatment in which salvelike irritating substances are placed on the leg of a horse to bring more blood to the part, hastening nature's reparative process.

Bloat   Metabolic disorder in ruminants in which gas is produced in excess compared to what can be removed; also known as ruminal tympany.

Blood horse   Pedigreed horse; term is synonymous with the Thoroughbred breed.

Bloodworms   Strongyles.

Boar   Noncastrated male swine.

Body condition score   System that reflects the amount of subcutaneous fat stored in the body; helpful in determining energy levels needed when balancing rations.

Bolus   Large pill for dosing animals; sometimes used interchangeably with *cud.*

Bovine   Referring to cattle.

Bovine viral diarrhea   Viral disease of cattle characterized by diarrhea, abortion, and immunosuppression.

Bracket pedigree   Three- to five-generation diagrammatic record of an individual's ancestors.

Break joint   Cartilage at the lower end of the shank bone, that can be snapped when butchering lambs and that is used to determine the age of the lamb.

Breed   Animals of common origin and having characteristics that distinguish them from other groups within the same species.

Broiler   Young chicken under 13 weeks of age (of either sex); it provides tender meat and has a soft, pliable, smooth-textured skin and flexible breastbone cartilage; also known as a fryer.

Brood   Group of baby chickens.

Brooder   Heat source for a recently hatched chick.

Broodiness (Broody)   Behavior of a hen when it stops laying eggs and wants to sit on a nest to hatch them.

Browse   Leaves, shoots, and twigs of brush plants, trees, or shrubs fit for grazing.

Buck   Male goat, deer, or rabbit.

Bull   Noncastrated male bovine.

Bummer   Orphan lamb.

By-product   Secondary product produced in addition to the primary product (e.g., corn oil versus corn grain).

Cabrito   Meat from young goats.

Calf   Young bovine of either sex.

Calorie   Unit of heat energy defined as the amount of heat needed to raise the temperature of 1 gram of water from 14.5°C to 15.5°C.

Candling   Process of examining an intact egg by holding the egg in front of a strong beam of light while checking for shell soundness, interior quality, and stage of embryonic development.

Capon   Castrated male chicken less than 8 months of age.

Caprine   Pertaining to the goat.

Carbohydrate   Nutrient consisting of carbon, hydrogen, and oxygen with the general chemical formula $(CH_2O)_n$; includes sugars, starches, cellulose, and gums.

Carding   Process in which loose, clean, scoured wool is converted into untwisted strands ready for spinning.

Carnivore   Animals that eat primarily flesh (meat).

Carrier   Animal that serves as a source of infectious organisms to other animals.

Carrying capacity   Potential stocking rate that will achieve a target level of animal performance on a particular grazing unit under a specified grazing plan.

Cation   Positively charged ion. In dairy nutrition, used in the calculation of dietary cation–anion difference (DCAD). A negative DCAD during the later dry period helps reduce the risk of milk fever.

Cecotrophy   Eating of the soft feces, or night feces, by rabbits.

Cecum   Blind pouch located at the junction of the small and large intestines. It is a necessary and functional part of the digestive tract in horses and rabbits.

Cellulose   Carbohydrate that forms the skeleton of most plants.

Chaff Glumes, husks, or other seed covering together with other plant parts that are separated from the seed in threshing or processing.

Challenge feeding Gradual increase in the level of concentrates fed to challenge cows to reach their maximum genetic potential for milk yield and to prevent ketosis; also known as lead feeding.

Chevon Meat from goats slaughtered at weaning or older.

Chick Young poultry.

Chimeras Animal produced by the mixing of two or more cell populations or by the grafting of an embryonic part of one animal onto the embryo of another.

Chromosome Rod-shaped bodies that occur in pairs in the nuclei of cells and that carry the genetic makeup of the animal in the form of genes arranged along the chromosome.

Classification System of placing life forms in order based on similar physical characteristics.

Clinical Later stages of a disease in which an abnormality in body function is readily observed.

Cloaca Common junction for the outlets of the digestive, urinary, and genital systems through the vent.

Clone Genetically identical offspring produced by natural or artificial asexual reproduction of a single organism or cell.

Clutch Eggs laid by a hen on consecutive days.

Cockerel Young male chicken.

Coggins test Test used to diagnose equine infectious anemia; also known as the agar gel immunodiffusion (AGID) test.

Cold blooded Horse of draft horse breeding with no Eastern or Oriental blood. These animals are generally easygoing, quiet, and not easily excitable.

Colostrum By legal definition, the lacteal secretion 15 days before and 5 days after parturition; it is rich in the antibodies needed to provide passive immunity to the newborn.

Colt Young male horse (usually under 3 years of age).

Comb Fleshy crest on the top of the head of poultry.

Complete feed Sole ration fed to an animal (without any additional substances, other than water, consumed) and meeting the animal's nutrient needs.

Concentrate Feed that is relatively high in energy and low in fiber (<18 percent crude fiber).

Contagious Capable of being transmitted from one animal to another.

Coprophagy Eating of feces.

Corpus luteum Also known as yellow body; temporary structure formed on the ovary after ovulation that is needed to maintain pregnancy in most species and that secretes progesterone.

Cotyledon Fetal membrane that joins with the maternal caruncles of the uterus, forming placentomes (connection of the placenta).

Cow Female bovine that has had one or more calves.

Cow hocked Condition in which the hocks are close together and the fetlocks are wide apart; term is used to describe both cattle and horses.

Cowper's gland Also known as bulbourethral gland; an accessory sex gland in the male that produces the seminal fluid needed to clean the urethra and neutralize its acidic environment.

Creep feeding Feeding young animals extra feed in an area separated from their dams; the opening to the area is too narrow or too low for the dams to enter, but the young animals can enter freely.

Cribbing   An undesirable behavior in which a horse places its upper teeth on the edge of a fence post, feeder or stall door, arches its neck, inhales, and grunts. It is seen more in stabled or confined horses and believed to be due to boredom.

Crimp   Natural waviness of wool fibers.

Crimped   Mechanical method of processing grain products with rollers that have a corrugated surface.

Crop   Area of the digestive tract in poultry that is located between the esophagus and proventriculus. Its purpose is to store ingested feed and moisten and soften it prior to its entering the proventriculus.

Crossbreeding   Mating of animals of different breeds.

Crude protein (CP)   Estimated chemically derived protein content of a feed based on its percentage of nitrogen.

Crutching   Practice of clipping the wool from the rear quarters of a ewe before it lambs.

Cryptorchidism   Hidden testicle; a condition in which one or both testicles are retained in the abdominal cavity; if both are retained the animal is sterile.

Cud   Bolus of regurgitated feed that a ruminant animal chews during rumination.

Cull   To remove from the flock or herd an animal that is genetically or phenotypically inferior.

Cutability   Yield of trimmed boneless meat.

Cuticle   Thin coat of albumen-like material deposited over the shell of the egg while in the vagina.

Dairy Herd Improvement Association (DHIA)   Dairy record-keeping association under the direction of the National Cooperative Dairy Herd Improvement Programs and the USDA designed to measure milk and component production as well as genetic, financial, and management of the individual animal and herd.

Dam   Female parent.

Debeaking   Trimming of the lower one-third of the beak in growing and mature birds to prevent cannibalism.

Dehorning   Removal of horns or their developing cells from cattle, sheep, and goats.

Dewlap   Pendulous skin fold hanging from the throat of cattle.

Digestible energy (DE)   Energy that was absorbed by the animal after accounting for the energy lost in the feces.

Digestible protein (DP)   Protein actually absorbed (determined by subtracting the protein content of the feces from the crude protein of the feed).

Digestion   Process of breaking down feed into particles small enough to be absorbed.

Diploid   Having two sets of homologous chromosomes; somatic cells are diploid.

Disaccharide   Union of two molecules of simple sugars.

Dock   That part of a sheep at which the tail was cut. To cut the tail from a sheep or cow.

Doe   Female adult goat, deer, or rabbit.

Dominant   Gene that hides or masks the effect of another gene in the same allelic series.

Double muscling   Abnormal condition in which muscle cells multiply or increase out of proportion to normal cells thus giving the animal excessive muscle development.

Draft animal   Animal used for pulling loads.

Drake   Male duck.

Drench   Administration of liquid medicine to an animal via the mouth.

Dressing percentage   Percentage of live weight of an animal that is carcass at slaughter.

Dry cows   Cows not producing milk; they are usually made to go dry 60 days prior to the next calving and lactation.

Duodenum   First segment of the small intestine, following the stomach.

Dyspnea   Difficult or labored breathing.

Dystocia   Difficult or abnormal parturition.

Eastern   Refers to horses of Arab, Barb, or similar blood.

Eicosanoids   Small lipids derived from arachidonic acid that are used to produce prostaglandins.

Elastrator   Bloodless means of docking tails or castrating using rubber bands to cut off the circulation, causing that part of the body to wither and then slough off.

Electrolyte   Electrically charged dissolved substance.

Embryo transfer   Transfer of an embryo from one animal to another.

Endemic   Disease (usually with low mortality rate) that is present in a region for a long time.

Endocrinology   Study of the endocrine system and its hormones.

Enteritis   Inflammation of the intestines.

Entropion   Genetic disorder in which the eyelids are turned inward.

Environmental mastitis   Inflammation of the mammary gland caused by organisms that survive in the environment (outside the animal), such as *Escherichia coli*.

Enzyme   Organic catalyst that speeds up the digestion of feed without being used up in the process.

Epididymis   Tube that carries sperm from the seminiferous tubules to the vas deferens; final place for maturation of sperm in the male.

Epinephrine   Hormone produced by the adrenal medulla that causes vasoconstriction. It is also known as the flight-or-fight hormone and counteracts the effects of oxytocin. It is released when the animal is afraid or under stress.

Epithelium   Tissue lining the internal cavities and covering the external surfaces of the body.

Equine   Pertaining to a horse.

Equus   Latin word for horse.

Eructation   Act of belching.

Esophagus   Muscular tube from the mouth to the stomach.

Essential amino acids   Amino acids that cannot be synthesized in the body at a rate needed for normal growth.

Estrogen   Major female sex hormone produced by the ovaries; responsible for the outward signs of heat or estrus.

Estrous   Entire cycle of reproductive changes in the nongravid female; period from one estrus to another.

Estrus   Also called heat; period during which the female is receptive to the male for purposes of breeding.

Ewe   Female sheep.

Expected progeny difference (EPD)   Estimate of the genetic worth of an animal in passing on its traits to its offspring.

Facing   Practice of clipping the wool above and below the eyes of a sheep to prevent wool blindness.

Fallopian tube   Also called oviduct; site of fertilization in most species.

Farrier   Person who shoes, trims, and cares for horses' feet.

Farrow   To give birth for a swine.

Fat-corrected milk (FCM)   Adjustment of a quantity of milk with different fat percentages to equivalent amounts based on an energy basis.

Fecundity   Ability to produce offspring.

Feedstuff   Any material used as food by an animal.

Fell   Thin membrane that covers the carcass.

Feral   Wild or nondomestic.

Fertilization   Union of a sperm and an ovum to form a zygote.

Filly   Female horse under 3 years of age.

Fimbria   Hairlike projections on the infundibulum that help guide the ovulated ovum into the fallopian tube.

Finish   Degree of fatness or fattening of an animal.

Firing   Applying a hot iron or needle to a blemish.

Fisking   Method of forcing the hands between the pelt and fell to remove the pelt from the carcass.

Fistula   Artificial window, surgically placed into the rumen.

Fleece   Wool from one sheep.

Float   To file off the sharp edges of a horse's teeth.

Flock   Small group of sheep.

Flushing   Nutrition procedure used to increase the ovulation rate by increasing the weight gain just prior to breeding. Involves increasing the amount of grain or quality of forage 2 weeks prior to and 1 week following breeding.

Foal   Young horse of either sex, usually less than 1 year of age.

Foal heat   First heat occurring after foaling in a mare, usually on the 9th day; also known as 9-day heat.

Fodder   Entire aboveground portion of a plant in the fresh or cured form.

Follicle   Ovarian structure that houses the ovum.

Fomite   Something that physically transports an infectious agent from one place to another.

Forage   Any feed material greater than 18 percent crude fiber.

Forbs   Nongrass broadleaf plants, including sage, shinoak, and saltbrush, commonly found on range and eaten by livestock; many livestock producers call forbs weeds.

Freemartin   Infertile heifer born cotwin to a bull.

Freshen   To give birth and begin lactating.

Frog   Wedge-shaped structure at the bottom of a horse's foot that helps pump blood through the foot by expanding as the foot strikes the ground.

Furstenberg's rosette   Sphincter muscle that controls the release of milk through the streak canal of each teat.

Gait   Manner or style of locomotion; a distinctive rhythmic motion of the animal's feet and legs.

Galvayne's groove   Longitudinal depression on the surface of the upper corner incisor teeth of horses 10 to 29 years of age; used to assist in age determination.

Gamete   Mature reproductive cell; sperm or ovum.

Gander   Mature male goose.

Gelding   Castrated male horse.

Gene   Hereditary units located on a chromosome.

Genotype   Genetic makeup of an animal.

Genus   Taxonomy category composed of a group of similar species.

Gestation   Period of pregnancy.

Gilt   Young female swine that have not farrowed.

Gizzard   Also known as the ventriculus; this is the muscular stomach of the bird.

Glycogen   Animal starch.

Goose   Mature female goose.

Gosling   Young goose or geese of either sex.

Gossypol   Compound found in cottonseed feeds that is toxic to simple-stomached animals.

Graafian follicle   Mature follicle, ready to ovulate.

Grade   Animal that possesses the distinct characteristics of a particular breed but at least one of its parents cannot be traced to the registry (herd book).

Grass tetany   Metabolic disorder caused by deficiency of magnesium in the diet; also known as grass staggers and hypomagnesemic tetany.

Gravid   Pregnant.

Grease wool   Wool as it comes from the sheep and prior to cleaning; it contains the natural oils and impurities present in the fleece.

Green chop   Forages freshly cut and chopped in the field and hauled to livestock for consumption without further processing or drying.

Grit   Small particles of granite added to poultry rations to help in their grinding of feed in the gizzard.

Groats   Grain from which the hulls have been removed.

Gross energy (GE)   Total heat of combustion of a feed burned in a bomb calorimeter.

Gubernaculum   Ligament that extends from the inguinal region to the tail of the epididymis.

Gullet   Another word for esophagus.

Hand   Measurement used to describe the height of a horse at its withers; 1 hand is equal to 4 inches.

Hank   Measurement of the fineness of wool; a hank is 560 yards of yarn. More hanks of yarn are produced from fine wools than from coarse wools.

Haploid   Number of chromosomes found in the gametes, which is half the normal number of chromosomes found in the somatic cells (diploid condition).

Hardware disease   Caused by the ingestion of small pieces of baling wire, nails, screws, and so forth and their puncturing and/or scratching of the walls of the digestive system in ruminants; also known as traumatic gastritis.

Hay   Aerial portion of forage stored in the dry form for feeding to animals.

Haylage   Low-moisture silage; usually with 45 to 55 percent moisture.

Heart girth   Measurement taken around the body just behind the shoulders; used to estimate body weight in cattle.

Heat increment (HI)   Heat produced by an animal associated with nutrient digestion and utilization.

Heifer   Female bovine that has not had a calf.

Hen   Adult female poultry.

Herbivore   Animals that subsist on feed of plant origin.

Heterosis   Improvement in production, disease resistance, stamina, and so forth in the offspring usually seen when crossbreeding or crossing of strains in both animals and plants; also called hybrid vigor.

Heterozygous   Condition in which the two alleles at a given locus are not the same.

Hinny   Cross between a stallion and a jennet.

Holandric inheritance   Genes carried on the portion of the Y chromosome that has no homologous portion on the X chromosome; traits represented by these genes are thus passed only from father to son.

Homogenization   Process in which milk fat is broken up into small particles so that they will not rise, allowing them to stay evenly spread throughout the milk.

Homozygous   Pertaining to the condition in which both genes at a particular location, on homologous chromosomes, are the same allele or are identical (either dominant or recessive).

Hot blooded   Of Eastern or Oriental blood. These horses tend to have higher temperaments, are more sensitive, and are more active than cold-blooded horses.

Hothouse lamb   Lambs born in the fall or early winter and finished for market between 6 and 12 weeks of age and weighing from 30 to 60 pounds.

Hulls   Outer covering of grain or other seed.

Hutch   Rabbit pen or cage.

Hydrogenation   Chemical process that saturates fats with hydrogen ions and increases the trans-fatty acids in a feed in an attempt to prevent the oxidation of unsaturated fatty acids.

Hypocalcemia   Low blood calcium level.

Ileum   Third portion of the small intestine.

Immunoglobulin (Ig)   Protein that contains a portion of the serum antibodies. Immunoglobulins are synthesized by B lymphocytes and plasma cells.

Inbreeding   Breeding of animals more closely related than the average relationship in the population.

Incomplete or partial dominance   Case in which one gene does not completely mask the effect of its allele.

Incubation   Process of holding eggs under controlled heat and moisture conditions and thus permitting fertile eggs to hatch.

Infundibulum   Funnel-shaped opening of the oviduct close to the ovary.

Insulin   Hormone secreted by the pancreas that is involved with the regulation and utilization of blood glucose.

Inter-   Prefix meaning between.

Interstitial cells   Also known as Leydig cells; located in the spaces between the seminiferous tubules in the testes, they produce the male hormone testosterone.

Intra-   Prefix meaning within.

In vitro   Within an artificial environment or outside the living body.

In vivo   Within the living body.

Ionophores   Class of antibiotics that are used extensively as feed additives for cattle.

Isthmus   Part of the oviduct where the shell membrane is added to the egg in poultry.

Jack   Male donkey.

Jejunum   Middle portion of the small intestine.

Jennet   Female donkey.

Jowl   Meat from the cheeks of a hog.

Judging   Attempt to rank or place animals in the order of their excellence in body type.

Ked   Wingless fly that lives off the blood (from the skin) of sheep.

Keel   Breastbone of a bird.

Kemp   Large, chalky white hairs found in the fleece of some breeds of goats (Angora).

Ketosis   Metabolic disorder caused by sudden need for energy and the excessive formation of ketone bodies in the blood; also known as acetonemia, pregnancy disease, and twin-lamb disease.

Kid   Goat under 1 year of age.

Kilocalorie (Kcal)   1,000 calories.

Kindling   Parturition in rabbits.

Kosher slaughter   Slaughter and handling of meat in accordance with the laws of the Torah (first five books of the Bible) according to Jewish law. The person who performs the slaughter is called a *Shochet*.

Labia   Inner (labia minor) and outer (labia major) folds of the vulva.

Lactose   Milk sugar.

Lamb   Young sheep (under 1 year of age). Act of giving birth (for a ewe).

Lameness   Deviation from the normal gait.

Laminitis   Inflammation of the laminae between the hoof wall and coffin bones leading to lameness.

Lanolin   Refined wool grease.

Lard   Rendered fat from pork carcasses.

Legumes   Plants that have nodules formed by bacteria on their roots. The bacteria are capable of fixating nitrogen from the air and passing it on into the plant, increasing its protein content. Examples are members of the pea family, peanuts, alfalfa, clover, soybeans, and so forth.

Leydig cells   Also called interstitial cells; cells that produce testosterone between the seminiferous tubules in the testes.

Libido   Sex drive.

Lice   Small, flat, wingless parasitic insects.

Lignin   Indigestible compound that forms a portion of the cell walls of mature plants and trees.

Linebreeding   Breeding an animal to a common ancestor; not as closely related as inbreeding.

Locus   Location on a chromosome where a particular gene is found.

Loin   Portion of the animal's topline in the region of the lumbar vertebrae (between the back and rump regions).

Maggot   Larvae of a fly.

Magnum   Part of the oviduct in poultry that secretes the albumen, or white of the egg.

Mane   Long hair on the top of the neck of a horse.

Mange   Reportable contagious disease caused by mites; also called scabies.

Marbling   Intermingled fat in the lean muscle that contributes to its tenderness and palatability.

Mare   Female horse over 3 to 4 years of age or one that has produced a live foal.

Mastitis   Inflammation of the mammary gland.

Mature equivalent (ME)   Standardization of milk records to the level that would have been produced had that animal been a mature cow on the day of test.

Meconium   First stools passed by a newborn.

Megacalorie (Mcal)   Equivalent to 1,000 kilocalories or 1 million calories.

Meiosis   Cell division that produces gametes, which contain the haploid number of chromosomes.

Metabolizable energy (ME)   Energy absorbed and utilized by the animal after subtracting for energy lost in the feces, urine, and gases.

Metritis   Inflammation of the uterus.

Milk fever   Potentially fatal metabolic disorder caused by a sudden need of calcium following birth and initiation of lactation; also known as parturient paresis.

Milk letdown   Expulsion of milk from the alveoli and small ducts into the gland cistern of the udder.

Mitosis   Cell division of somatic cells that produces identical diploid daughter cells.

Mohair   Fleece of the Angora goat.

Molting   Process of losing feathers from the wings and body.

Monogastric   Simple-stomached or nonruminant animal.

Monotocous   Normally giving birth to one young each gestation period.

Morbidity   Rate of being sick after being exposed to a pathogen.

Mortality   Death rate.

Moufflon   Wild-type short-tailed sheep.

Mule   Cross between a male donkey (jack) and a female horse (mare).

Multiparous   Females having had two or more pregnancies that resulted in live offspring.

Mustang   Wild or native horse of the western plains, descended from Spanish horses.

Mutation   Chemical change in a gene such that it functions differently than before the change; is passed on to each succeeding generation.

Mutton   Meat from a mature sheep.

Necropsy   Postmortem examination of an animal's internal organs.

Net energy (NE)   Most commonly used energy term for determining lactating dairy cows' energy needs; the energy available after removing energy lost in feces, urine, gas, and heat increment (heat produced during digestion and metabolism).

Neutral detergent fiber (NDF)   Measures the cell wall of a plant (cellulose, hemicellulose, and lignin) and has an inverse relationship with dry matter intake; a minimum level of ADF is required, however, to prevent ruminal disorders such as acidosis, rumen parakeratosis, and low milk fat percentage.

Nitrogen free extract (NFE)   Readily available carbohydrates of a feed (primarily the sugars and starches).

Nulliparous   Having never given birth.

Nutrient   Chemical substances in the feed that are used to support life, growth, reproduction, and lactation.

Offal   Organs, trimmings, and tissues that are removed at slaughter.

Omasum   Third compartment of the ruminant stomach; also known as the manyplies or bible stomach.

Omnivore   Animals that eat both plants and meat.

Open   Not pregnant.

Outbreeding   System of mating animals less closely related than the average for the breed.

Ovine   Refers to sheep.

Ovulation   Release of the ovum from the ovary.

Ovum   Female reproductive cell or gamete; commonly called an egg.

Oxytocin   Hypothalamic hormone responsible for milk letdown and contractions of the uterus both at birth (labor) and during mating.

Paleolithic   Old Stone Age.

Parasite   Organism that lives on or in an animal and at the expense of the living animal.

Parenchyma   Functional layer.

Parturition   Act of giving birth.

Passive immunity   Transfer of antibodies to prevent disease.

Pastern   Joint between the first and second phalanx of each foot. Also refers to the long pastern bone (first phalanx) and the short pastern bone (second phalanx).

Pasteurization   Process of heating milk to a temperature and holding time sufficient to kill pathogenic (disease-causing) bacteria.

Pedigree   Diagram of an animal's ancestry.

Pelt   Skin of a sheep, rabbit, or goat as removed at slaughter; contains the hide with its hair, wool, or fur.

Peristalsis   Rhythmic contractions of the digestive tract that propel its contents from the esophagus to the anus.

Phenotype   Physical characteristics and performance of an animal.

Photoperiod   Amount of light or day length.

Photosynthesis   Formation of carbohydrates and oxygen by the action of sunlight on the chlorophyll (green pigment) in a plant.

Piebald   Horse with a black coat with white spots.

Piglet   Young pig.

Placentome   Combined cotyledon and caruncle "button" attaching the membrane that encloses the fetus.

Pleiotropic effect   Results when one gene controls or contributes to the phenotype of more than one trait.

Polled   Naturally hornless animal.

Polyestrus   Having many estrous cycles throughout the year.

Polytocous   Litter bearing.

Porcine   Pertaining to swine.

Postpartum   After birth.

Poultice   Moist, mealy mass that is applied hot to a sore or inflamed body part for healing purposes.

Poults   Young turkeys of either sex.

Predicted transmitting ability (PTA)   Estimate of a sire's or dam's potential to transmit yield productivity to its offspring.

Preen   Combing the feathers with the beak.

Primiparous   Having given birth once or in the process of first birth.

Probiotics   Microbes used as feed additives.

Progeny   Animal's offspring.

Protozoa   One-celled animals, some of which are pathogenic to domestic animals.

Proximate analysis Chemical evaluation of a feedstuff.

Puberty Period when the reproductive system becomes mature in form and function.

Puffs Windgalls, bog spavins, or thoroughpins.

Pullet Young female chicken.

Punnett square Diagrammatic representation of gene segregation and their recombination at fertilization.

Purebred Animal with two purebred parents that is hence eligible for inclusion in the breed registry.

Qualitative traits Traits controlled by one or a few pairs of genes and whose phenotypes (e.g., coat color, and size) are clearly seen.

Quantitative traits Traits usually controlled by many pairs of genes and whose phenotypic expressions (e.g., milk production and fertility) are more varied and influenced more profoundly by environmental and other factors than are qualitative traits.

Quarantine Segregation of susceptible animals to prevent the spread of infectious diseases.

Rabbitry Place where rabbits are raised.

Ram Sexually intact male sheep.

Raw wool Wool shorn from live sheep and unprocessed; same as grease wool.

Recessive Gene whose expression is hidden by a dominant gene and only expresses itself when in the homozygous state. Recessive alleles are represented by lowercase letters.

Reclaimed wool Wool that is salvaged by reworking new or old fabrics.

Registered Animal whose name is recorded in the breed registry.

Reticulum Second compartment of the ruminant stomach; also known as the honeycomb or hardware stomach.

Roaster A young chicken (usually 6 to 8 lbs.) that is tender-meated, with soft, pliable and smooth textured skin. The breastbone cartilage is less flexible than that of a broiler or fryer and the roaster is of heavier weight.

Rolled Mechanical method of processing grain products by compressing them between smooth rollers resulting in flake form.

Rooster Mature male chicken; also known as a cock.

Roughage Feedstuff of greater than 20 percent crude fiber and less than 60 percent in total digestible nutrients.

Rumen First and largest compartments of the ruminant stomach; also known as the paunch.

Rumination Cud chewing.

Saturated fat Completely hydrogenated fat.

Scouring Treating wool with a soap and soda solution removes the soluble impurities, which include the yolk (yellow coloring), suint (salts caused from sweating that produce the characteristic odor), and gland secretions that serve to protect the fibers while they are on the sheep.

Scours Diarrhea.

Scrotum Skinlike sac that houses and protects the testicles.

Scurs Small rounded growths of horn.

Seminiferous tubules Convoluted tubules located within the testes that are responsible for the production of spermatozoa.

Septicemia Disease caused by the presence of bacteria and/or their toxins in the blood; also known as blood poisoning.

Sex limited Phenotypic expression of a gene limited to one sex.

Sex linked Genes that are carried on the nonhomologous portion of the X chromosome.

Shearing Removal of the fleece, either by hand or by power clippers.

Sickle hocked Hind legs set too far forward, giving the impression of a sickle shape when viewing the animal from the side.

Sigmoid flexure S curve in the penis of a bull, ram, or boar that allows for elongation or straightening of the penis during copulation.

Silage Feed (usually the entire aerial portion of the plant) cut, chopped, and stored under anaerobic conditions and allowed to ferment before being fed. Normal moisture content is 65 to 70 percent.

Sire Male parent.

Soilage Same as green chop.

Sow Mature female swine or female swine that have farrowed.

Species Taxonomy category that is subordinate to a genus.

Speciesism Denotes having a preference in the treatment of one species over another.

Spermatozoa Male gamete that fertilizes the ovum.

Spinning count Measurement of fineness of a fleece based on the length of yarn (expressed in hanks) that can be spun from 1 pound of scoured wool.

Splay footed Toes turned out.

Springer Heifer within 2 to 3 months of calving.

Stag Male chicken under 10 months of age that shows developing sex characteristics intermediate between those of a roaster and those of a rooster; term is also used to describe livestock that have been castrated after reaching sexual maturity.

Staple Length of wool fibers.

Steer Castrated male bovine.

Stocker Beef animal in the phase between being weaned and ready for finishing in the feedlot.

Stocking rate Ratio of males to females; also refers to the number of animals per acre in a pasture or range setting.

Strain Families or breeding populations within a variety that possess common traits.

Subclinical Early stages of a disease in which clinical symptoms are not readily apparent.

Suint Salts of perspiration in the raw wool fleece usually removed when it is scoured.

Switch Brush of hair at the end of a bovine's tail.

Syndactylism Genetic disorder in Holstein cows in which one or more of the hooves are solid in structure rather than cloven; also known as mule foot.

Tagging Shearing the dung tags or locks of wool and dirt from the dock and around the udder.

Tankage Animal residue left after rendering fat in a slaughterhouse and used for feed or fertilizer.

Taxonomy Science of classifying life forms into different categories.

Testosterone Male sex hormone.

Tom Male turkey.

Total digestible nutrients (TDN) Method of estimating amount of energy in a feed by adding up the digestible crude protein, crude fiber, nitrogen free extract, and ether extract $\times$ 2.25 (fats).

Toxoid   Detoxified toxin (type of vaccine) used to stimulate formation of antitoxin in an animal's body.

Traumatic gastritis   Also known as hardware disease.

Tripe   Walls of the reticulum and sometimes the rumen (from cattle) used as human food.

Trocar   Sharp-pointed instrument equipped with a cannula that is used to puncture the body wall and rumen to relieve the gaseous pressure in bloat.

Twitch   Instrument of restraint that pinches the upper lip of the horse so that it can be held still while undergoing treatment.

Type   Physical conformation of an animal.

Unsaturated fat   Fat containing double bonds and thus not completely hydrogenated.

Urea   Synthetically made nonprotein nitrogen used to cheapen the feed protein in ruminants. Also the primary end product of protein metabolism in mammals.

Uropygial gland   Found in many birds and aids in the waterproofing of their feathers; also known as preen gland.

Vagina   Birth canal and area of semen deposit during natural breeding in most species.

Variety   Subdivision of a breed used to describe poultry.

Vas deferens   Tube that transports mature semen from the tail of the epididymis past the accessory sex glands to the urethra.

Veal   Calves fed for early slaughter (usually under 4 months of age).

Vent   Common area for evacuation of feces and eggs in poultry.

Vestibule   Area common to both the urinary and reproductive systems in mammals.

V muscle   Muscles of the chest of a horse that give the appearance of an inverted V.

Volatile fatty acids (VFAs)   Referring to acetic, propionic and butyric acids produced in the rumens of ruminants and the ceca of horses and rabbits; provide extensive energy to the animal.

Wether   Castrated male sheep or goat.

Wintered   How an animal is fed through the winter months.

Withers   Area on the top of the shoulders.

Wool blind   Growth of wool on the face and around the eyes in a manner that obstructs the vision of the sheep.

Worsted   Manufacturing process of wool using the longer wool fibers (2 inches or more in length) and in which the wool is carded, combed, and drawn into a thin stand of parallel fibers spun together to form a strong but thin yarn.

Xenotransplant   Transplant of organs between species (e.g., between animals and humans).

Yolk   Natural grease of wool that keeps the wool in good condition; lanolin is a common product used in cosmetics that is refined from this grease.

Zoonotic   Capable of being transmitted between animals and humans.

Zygote   Cell formed by the union of a sperm and an ovum.

Index

Abdomen, palpation of, 50
Abdominal pain, 489–490
Abnormalities, in sheep, 290–291. *See also* Health management
Abomasums, 73, 124–125
Abortion
  bovine neospora, 162
  chlamydial, 185
  enzootic, 185
  enzootic abortion of ewes (EAE), 293
  in horses, 465
  vibriosis, 298
  See also Enzootic abortion
AB 611 (California statute), 234, 236
Absorption, 70
Accessory organs, 78
Accessory sex glands, 34
Accuracy, 242
Acetic acids, 74
Acetonemia, 125–126, 295
Acid detergent fiber (ADF), 83
Actinobacillosis, 155
Actinomycosis, 155
Additives, 216–217
Adjusted 205-day weight, 241
Adjusted 365-day weight, 241
β-adrenergic agents, 221
African geese, 423
Ages, recording, 246
Aging, 209
Ahimsa, 209–210
Airflow meter, 141
Airsacculitis, 401
Albumen, 46
Allantois, 45
Alleles, 20
Alpacas, 533. *See also* Camelid industry
Alveoli, 137
American albino, 448
American buckskin, 448
American Buff, 419
American Chinchilla, 512
American Cormo, 258

American Cream draft horse, 443
American LaMancha, 174
American Landrace, 316
American Mammoth Bronze, 417
American Poultry Association, Inc., 381
American quarter horse, 434–435
American rabbit, 512
American Rabbit Breeders Association (ARBA), 511
American saddlebred, 435
American Standard of Perfection, The (American Poultry Association, Inc.), 381, 383
Amerifax, 194
Amino acids, 79, 81
Amnion, 45
Amylase, 78
Anaerobic, 462
Anaplasmosis, 147–148
Anatomy
  of chickens, 388–391
  female, 37–42
  male, 31–36
Anemia, 361
Angora goats, 176–177
Angora rabbits, 512, 525–526
Angus, 195
Animal & Plant Health Inspection Service (APHIS), 516
Animal Damage Control (ADC) program, 300
Animal Improvement Programs Laboratory (AIPL), 136
Animal Medicinal Drug Usage Clarification Act (AMDICA), 234, 236
Animal rights, 209–210
Animals
  in North America, 3
  in United States, 3–7
Animal science, 1, 15
  future of, 12–15
  global politics, numbers, and economics, 8–11

origins of, 2–3
taxonomy, 7–8
Animal unit, 218
Animal welfare, 210–211, 373–374
Animal Welfare Act (AWA), 516
Ankina, 195
Ankole-Watusi, 195–196
Annular ring, 137
Anthrax, 149–150
Antibiotics, 394
Antioxidants, 394
Anus, 70
Apoplexy, 292–293
Appaloosa, 448–449
Appointment births, 50
Arabian camels, 533
Arabian horses, 435–436
Arachidonic fatty acids, 84
Arginine, 81
Arsenicals, 394
Artificial insemination, 51–60
  dairy cattle, 133
  horses, 466
  sheep, 286–287
  swine, 347–348
Ascaridia galli, 408
Ascarids, 369, 504, 550
Aspergillosis, 401–402
Atrophic rhinitis, 361–362
Aujeszky's disease, 362–363
Autosomes, 23
Avian digestive systems, 66, 76–78, 390
Avian influenza, 402
Avian leukosis, 405
Avian pneumoencephalitis, 406–407
Aylesbury, 419
Ayrshires, 105
Azoturia, 488–489

Babcock, S. M., 99
Babcock test, 99
Baby pig disease, 367
Baby pig scours, 365
Bacillary hemoglobinuria, 157–158

Back, broken, 526
Backfat, 326
Bactrian camels, 533
Bakewell, Robert, 19, 267
Bang's disease, 151
Barbados Blackbelly, 274
Barn itch, 164, 370
Barnes dehorner, 231
Barrows, 331
Barzona, 196
Beef breeds, 194–205
Beef carcass, 205–217
Beefmaster, 197
Beef production, 189–190, 249
  breeds, 194–205
  carcass, 205–217
  fattening cattle, 218–221
  husbandry, facilities, and equipment,
    222–236
  selection, 236–249
  stocker systems, 217–218
  U.S. regions of, 190–194
Behavioral problems, in poultry,
    408–409
Belgian draft horse, 443–444
Belgian hare, 513
Belted Galloway, 197
Berkshire, 316
Big liver disease, 405
Bile, 70
Billy chevon, 181
Binomial nomenclature, 8
Birth. See Parturition
Birth weight, 241
Birth weight EPD, 242
Black East Indie, 419
Blackleg, 150, 291
Black Welsh Mountain, 276
Blistering, 489
Bloat, 74, 123, 291
Blonde d'Aquitaine, 197
Blood, 27, 48
Blood samples, from llamas, 539
Blood spot, 392
Bloodworms, 503–504
Bloody scours, 365
Blowflies, 299
Bluebag, 291
Blue breasts, 527
Blueface Leicester, 276
Bluetongue, 184, 292
Boars, 331
Body, maintenance of, 87
Body condition scoring, 116
Boer goats, 177
Bog spavin, 495, 497
Bolus, 74
Bomb calorimeter, 89
Bone spavin, 497
Borden, Gail, 99
Border Leicester, 267
Bordetella bronchiseptica, 527
Bouncing, 48
Bovine genital campylobacteriosis, 160
Bovine herpesvirus I (BHV-1), 161–162
Bovine leukocyte adhesion deficiency
    (BLAD), 136
Bovine neospora abortion, 162

Bovine respiratory syncytial virus, 162
Bovine somatotropin (BST), 146–147
Bovine spongiform encephalopathy,
    155–156
Bovine viral diarrhea, 162
Bowed tendon, 489, 497
Bracket pedigree, 25
Braford, 197–198
Brahman, 198–199
Brahmanstein, 108
Brahmousin, 199
Branding, 231–232, 245
Brangus, 199
Breeder cattle production, 193
Breeding and genetics, 18–19, 27
  cells, 19–20
  dairy cattle, 112–113, 133–137
  hereditary defects, 24–25
  horses, 459, 460
  incomplete, partial, or codominance,
    22–23
  mating systems, 25–27
  Mendel's experiments, 20–22
  sex determination, 23
  sex-limited genes, 23
  sex-linked traits, 23–24
  and sheep nutrition, 279
  swine, 359
  See also Reproduction; Selection
Breeding rabbits, 515–516
Breeding sheep, 303–304
Breeding swine, 342
Breeds
  beef cattle, 194–205
  dairy cattle, 104–108
  defined, 383
  goats, 172–178
  horses, 433–457
  poultry, 382–388
  rabbits, 510–515
  sheep, 258–278
  swine, 316–322
Broad Breasted Bronze, 417
Broad Breasted White, 417
Broad tail pelts, 277
Broiler industry, 410–416. *See also*
    Poultry industry
Broken back, 526
Brooding, 422–423
Brown stomach worm, 165
Brown Swiss, 105–106
Browse, 180
Brucellosis, 151, 184, 298, 363
Buff geese, 423
Buff Orpington, 419
Building ventilation, 374
Bulbourethral glands, 34
Bulk tank, 143
Bull-testing stations, 243
Bummers, 279, 289
Burdizzo pincer, 230
Bureau of Land Management (BLM),
    217
Butyric acids, 74
By-products
  beef, 216–217
  pork, 335
  sheep, 309

Cabrito, 181
Calf
  bouncing, 48
  dairy, 109–112
Calf diphtheria, 151
Calf-rearing systems, dairy, 112
Calf scours, 124
Californian rabbit, 513
California Reds, 274
California Variegated Mutant, 276
Calorie system, 89
Calving ease, 242
Camelid industry, 71, 532–533, 542
  herd health, 537–540
  llama nutrition, 534–536
  llama reproduction, 536–537
  llama uses, 540–542
  male berserk syndrome, 537
Camels, 533. *See also* Camelid
    industry
Canadian geese, 423
Candling, 416
Cannibalism, 408–409
Cannula, 74
Capon, 412
Capped hock, 497
Caprine arthritis encephalitis, 184
Caprine pleuropneumonia, 184
Caracul pelts, 277
Carbohydrates, 66, 81–84
Carbonizing, 304
Carcass
  beef, 205–217
  sheep, 308
  swine, 330–341
  See also Slaughtering process
Carcass merit, 339
Carcass yield, 332–334
Carding, 304
Cardiorespiratory systems, of chickens,
    391
Carnivores, 68
Caseous lymphadenitis, 292
Cashmere goats, 178
Casting, 228
Castration, of sheep, 290
Castration equipment, 228–230
Cattle, in United States, 3–4. *See also*
    Beef production; Dairy industry
Cattle grubs, 163–164
Cayuga, 419
Ceca, 77
Cecal worms, 408
Cecotrophy, 520
Cells, 19–20
Cellulose, 82
Certified litter (CL), 347
Certified mating (CM), 347
Certified meat hogs, 329
Certified meat sire (CMS), 347
Cervical plug, 40
Cervix, 39, 40
Cestodes, 550
Cestodiasis, 165
Challenge feeding, 119
Champagne d'Argent, 513
Charbray, 199
Charolais, 200

Checkered Giant, 513
Chemical castration, 230
Chemical dehorning, 231
Chemical digestion, 69
Chemotherapeutic agents, 221
Chester Whites, 317
Cheviot, 261
Chevon, 181
Chewing lice, 539
Chianina, 200
Chickens, anatomy and physiology of, 388–391. *See also* Poultry industry
Chimeras, 57–58
Chinese geese, 423
Chlamydial abortion, 185
Chlorophyll, 66
Cholera, 363–365
Chorioallantoic membrane, 47
Chorioallantois, 45
Chorion, 45
Chromosomes, 19
Chronic obstructive pulmonary disease, 493
Chronic respiratory disease, 405–406
Chronic specific enteritis, 153–154
Chyme, 70
Circling disease, 292
Class, 8, 383
Classification, 7
Claw, 142
Clay, Henry, 458
Clinical mastitis, 144
Clitoris, 40–41
Cloaca, 34
Cloning, 59, 60, 101
Clun Forest, 276
Clydesdales, 444–445
Coccidiosis, 165, 402–403, 526, 539
Coccidiostats, 394
Cock, 412
Cockle, 298
Codominance, 22–23
Colibacillosis *Escherichia coli,* 365
Colic, 489–490
Color registries, 448–451
Color sexing, 392
Colostrums, 109
Columbia, 271–272
Columbus, Christopher, 4, 6, 205
Comanches, 449
Commercial beef cattle, 193–194
Compudose, 221
Concentrate, 78
Condensed milk, 99
Confinement finishing, 346
Conformation, 19
Connemara, 451–452
Consumer grades, for eggs, 416
Contagious ecthyma, 184–185, 296–297
Contagious pustular dermatitis, 296–297
Contracted heels, 498
Cooling, 208
Coop, Ian, 268
Coopworth, 267–268
Coprophagy, 520
Corneal nerve block, 230

Cornish game hens, 387, 412
Corns, 498
Coronado, Francisco Vásquez de, 6
Corpus luteum (CL), 38
Corrals, 223–227
Corriedale, 272
Cortés, Hernando, 4, 6, 449
Cotswold, 268
Cotyledon, 40, 47
Cotyledonary placental attachment, 40
Cowper's glands, 34
Cows. *See* Beef production; Dairy industry
Creep feeding, 194
Crested duck, 419
Crias, 537
Cribbing, 486
Crop, 76
Crossbreeding, 26, 344
Cross contamination, 382
Crude protein, 80
Crutching, 288
Cryptorchidism, 33, 290
Cryptosporidiosis, 162–163
Cud, 74
Curled-toe paralysis, 403
Cuticle, 42
Cutout data, 244
Cuts
  beef, 215–217
  pork, 334, 335
Cystine, 81
Cytoplasm, 19

Dairy breeds, 104–108
Dairy calves, 109–112
Dairy concentrates, 116
Dairy Cow Unified Score Card, 131–132
Dairy farms, income and expenses, 103
Dairy genetics, 134–136
Dairy goat management, 178–179
Dairy Herd Improvement Association (DHIA), 100
Dairy industry, 97–98, 166–167
  breeding, selecting, and judging, 128–137
  current state of, 101–102
  dairy breeds, 104–108
  farm income and expenses, 103
  feeding and general management, 109–123
  herd health disorders, 147–165
  history of, 98–101
  housing, 166
  metabolic disorders, 123–127
  milk secretion and milking machines, 137–147
Dairy rations, 121–122
Debouillet, 259
Degraded intake protein (DIP), 122
Dehorning equipment, 230–231
Dehorning saw, 231
Delaine-Merino, 259
DeLaval, Gustav, 99
Demodectic mange, 186
Deoxyribonucleic acid (DNA), 19
Dermatophytosis, 164–165

De Soto, Hernando, 5, 6
Devon, 200
Dexter, 200
Diarrhea, 549
Diestrus, 43
Diet. *See* Nutrition/feeding
Differentiation, 46
Digestibility, feed, 88
Digestible energy (DE), 89
Digestion, 69
Digestive systems, 68–69
  avian, 76–78
  chickens, 390
  functional cecum, 70–71
  monogastric, 69–70
  ruminant, 71–76
Diploid, 20
Dipping, 233–234
Disaccharide, 81–82
Disbudded, 183
Disease, defined, 401. *See also* Health management
Diseases of Dairy Cattle (Rebhun), 147
Displaced abomasums, 124–125
DNA. *See* Deoxyribonucleic acid
Dolly (cloned sheep), 60
Domestication, of rabbits, 510–515
Dominant genes, 20
Donkeys, 457
Dorper, 274
Dorset, 262–263
Double-coated breeds (sheep), 274–275
Down, 45
Draft horses, 433–434, 443–447
Drenching, 232
Dressing, 331
Dressing percentage, 208–209, 332–334
Dried poultry litter (DPL), 123
Dromedary camels, 533
Drugs, new regulations for, 234, 236
Dry cows, 113
Dual-purpose breeds (sheep), 271–274, 278
Duck plague, 403
Ducks, 397, 417–419. *See also* Poultry industry
Duck virus enteritis, 403
Duodenum, 70
Duroc Jersey, 318
Dutch Belted, 108, 201
Dutch rabbit, 513
Dwarfism, 290–291
Dysentery, 365

Ear mites, 526
East Friesian, 272
Easter lambs, 289
Economics, global, 8–11
Egg binding, 549
Egg Grading Manual, Agricultural Handbook No. 75, 416
Eggs, 392–393
  hatcheries, 415
  incubation of, 414
  production expectations, 415–416
Egyptian geese, 423
Eicosanoids, 521

Eisenhower, Dwight D., 382
Elastrator, 230
Electric dehorner, 231
Electrolytes, 124, 394
Embryonic vesicle, 47
Embryo splitting, 56–57
Embryo transfer, 54–56
Emden, 423
Emulsifier, 70
Encephalitazoonosis, 526
Encephalitis, 292
Endocrine, 70
Endocrinology, 31
Endometrial cups, 40
Energy, feed, 89
Enteritis complex, 527
Enterotoxemia, 185, 292–293
Entropion, 291
Environmental Protection Agency
      (EPA), 300
Enzootic abortion, 185
Enzootic abortion of ewes (EAE), 293
Enzootic pneumonia, 368
Enzymatic digestion, 69
Epididymis, 31, 34
Equine abortion virus, 494
Equine digestive system, 66, 70–71
Equine distemper, 494
Equine encephalomyelitis, 490
Equine industry, 427–432, 505
   color registries, 448–451
   diseases, 487–504
   donkeys, 457
   draft horse breeds, 433–434, 443–447
   injuries, 505
   light horse breeds, 433, 434–442
   mules, 457–458
   nutrition, 474–487
   pony breeds, 434, 451–455
   reproduction and management,
         458–466
   selection, 467–472
   sport horse breeds, 455–457
   taxonomy, 432
   teething, 472–474
   in United States, 6–7
Equine infectious anemia, 490
Equine monocytic ehrlichiosis, 494
Equine protozoal myeloencephalitis,
      491–492
Equine viral rhinopneumonitis, 494
Equipment
   beef cattle, 222–236
   rabbit, 520
Erysipelas, 366
Escherichia coli, 145, 217
Escherichia coli complex, 293
Esophageal groove, 75–76
Esophagus, 76
Essential amino acids, 81
Essential nutrients, 78–87, 394
Estrogen, 38
Estrous cycle, 42–43
Estrus, 39, 42, 459, 465
Estrus synchronization, 286–287
Ether extract, 84
Ewe breeds, 278
Exocrine, 70

Expected progeny difference (EPD),
      242
Expenses, dairy farm, 103
Exudative epidermitis, 367
Eye appraisal, of swine, 323–326

Facilities, beef cattle, 222–236. *See
      also* Shelter/housing
Facing, 288, 291
Fallopian tube, 38–39
False rabies, 362–363
Family, 8
Farrowing, 342, 344–345
Farrow-to-finish operations, 346
Fats, 84–85
Fat-soluble vitamins, 84, 87
Fattening, 218–221
Fatty acids, 84
Feathers
   chicken, 391
   ostrich, 546–547
Feather sexing, 392
Fecal contamination, 382
Federal Meat Grading Service, 211
Feed additives, 216–217
Feed digestibility, 88
Feed energy, 89
Feeder cattle grades, 212–213
Feeder cattle production, 193
Feeder pigs, 341–342
Feeders, rabbit, 520
Feeding. *See* Nutrition/feeding
Feed nutrients, 87–91
Fell, 308
Female anatomy, 37–42
Fenbendazole, 550
Fences, for pastures and corrals,
      223–225
Fertilization, 39, 44–46
Fetal membrane slip, 47
Fever in the feet, 492
Fiber, 540–541
Fibropapillomas, 160
Fiddleneck poisoning, 127
Fifth quarter, 216
Fimbria, 39
Fine-wool breeds (sheep), 258–259
Finished market hogs, 331
Finishing, swine, 345–347, 358
Finnish Landrace, 272–273
Finnsheep, 272–273
First generation, 21
First limiting amino acid, 349
Fisking, 308
Fistulated, 75
Fistula withers, 493
Fjord horse, 436
Flagellum, 44
Fleece, 305
Flemish Giant, 513
Flies, 502
Float, 505
Fly strike, 369
Foal heat, 464–465
Foaling, 461–464
Foot-and-mouth disease, 152, 367
Foote, W. C., 275
Foot problems, in horses, 495–502

Foot rot, 152–153, 185, 293
Forage
   dairy cattle, 116, 119
   goats, 180
Forbs, 180
Founder, 492
4H, 510
Fowl. *See* Poultry industry
Fowl pox, 403
Fowl typhoid, 403–404
Fox, Michael, 60
Frame size, 212–213
Franklin, Benjamin, 420
French Alpine goats, 173
French Lop, 513–514
Fructose, 81
Fryers, 410, 413. *See also* Broiler
      industry; Poultry industry
Functional cecum digestive system,
      70–71
Funnel, 41
Fur industry, rabbits, 525–526
Furstenberg's rosette, 140
Future Farmers of America (FFA), 510

Galactose, 81
Gallbladder, 70
Galloways, 201
Galvayne's groove, 474
Gander, 420
Gas gangrene, 156
Geese, 397, 419–423. *See also* Poultry
      industry
Gelbray, 201
Gelbviehs, 201
Genes, 19
   defined, 20
   dominant, 20
   recessive, 20
   sex-limited, 23
Genetic abnormalities/disorders
   dairy cattle, 136–137
   sheep, 290–291
   See also Health management
Genetic engineering, 100
Genetics. *See* Breeding and genetics
Genotype, 19
Genus, 8
Germinal disc, 45
Gestation, 46–50
   dairy cattle, 112–113
   horses, 460
   sheep, 279, 288–289
   swine, 358–359
Gilson, Warren, 133
Gilts, 331, 358
Gizzard, 77
Gland cistern, 137
Glandular stomach, 73, 77
Glickman, Dan, 14
Glucose, 81
Glycogen, 78
Goat industry, 171–172, 187
   breeds, 172–178
   dairy goat management, 178–179
   feeding, 180–181
   herd health, 184–186
   kid management, 183–184

lactation, 181
milk characterisitics, 179–180
parturition, 182–183
reproduction, 181–182
Goat milk, 179–180
Goat pox, 185
Goiter, 125
Gonadotropin-releasing hormone (GnRH), 38
Goose, 420
Goslings, 422–423
Graaff-Reinet disease, 294–295
Graafian follicle, 38
Grade animal, 27
Grades
beef carcass, 211–214
eggs, 416
pork carcass, 334–335
poultry, 412–413
Grading, 208
beef carcass, 211–214
pork carcass, 334–335
Grandin, Temple, 210
Grass staggers, 125, 293–294
Grass tetany, 125, 293–294
Gravel, 498–499
Gravid, 47–48
Grease heel, 499
Grease wool, 304
Greasy pig syndrome, 367
Grit, 77, 394
Grooming equipment, 236
Gross energy (GE), 89
Groundsel poisoning, 127
Growth-promoting substances, 220–221
Guanacos, 533. See also Camelid industry
Guardians, llamas as, 541
Gubernaculums, 31
Guernseys, 106
Gulf Coast Native, 276
Gullet, 76

Hair breeds (sheep), 274–275
Hairy foot warts, 160
Halving, 207
Ham-loin index, 334
Hampshire pigs, 319
Hampshire sheep, 263
Hand mating, 347–348
Hand milking, 140
Hanoverians, 455
Haploid, 20
Hardware disease, 73, 127
Hardware stomach, 73
Hatch Act, 99
Hatchery, 415
Havana rabbit, 514
Health management
camelids, 537–540
dairy cattle, 147–165
goats, 184–186
horses, 487–504
ostriches, 548–550
poultry, 400–409
rabbits, 526–529
sheep, 290–300
swine, 360–373, 374

Heape, Walter, 54
Heat detection, 349
Heaves, 493
Heifers, springing, 113. See also Beef production; Dairy industry
Hemorrhagic septicemia, 158–159, 296
Hens, 412. See also Poultry industry
Herbivores, 69
Herd health. See Health management
Hereditary defects, 24–25
Hereford beef cattle, 201
Hereford swine, 319
Hermaphroditism, 182
Heterozygous, 20
Hexoses, 81
Hides, ostrich, 546–547
Himalayan rabbits, 514
Hindu religion, 209–210
Histidine, 81
Hogs. See Swine industry
Holandric inheritance, 24
Holsteiner, 456
Holsteins, 107–108, 136–137
Homozygous, 20
Honeycomb, 73
Hoof cracks, 499
Horizontal transmission, 401
Hormonal assays, 48
Hormones, 221
Horses, in United States, 6–7. See also Equine industry
Hothouse lambs, 289
Housing. See Shelter/housing
Hulet, Clarence, 273
Husbandry, beef cattle, 222–236
Hutches, 510, 518–519
Hybrids, turkey, 417
Hydrochloric acid, 70
Hydrogenation, 85
Hydrolysis, 79
Hyperkeratosis, 127
Hyperperistalsis, 70
Hypocalcemia, 126
Hypodermosis, 163–164
Hypoglycemia, 125–126, 367
Hypomagnesemic tetany, 125
Hypoperistalsis, 70
Hysteria, 408–409

Icelandic sheep, 276
Identification, of beef cattle, 231–232
Ileum, 70
Immunologic diagnosis, 49–50
Inbreeding, 26
Income, dairy farm, 103
Incomplete dominance, 22–23
Incubation, of eggs, 414
Independent culling level, 248–249, 330
Indian Runners, 419
Infectious bovine keratoconjunctivitis, 156–157
Infectious bovine rhinotracheitis, 161–162
Infectious bursal disease (IBD), 404
Infectious conjunctivitis, 156–157
Infectious coryza, 404
Infectious keratitis, 156–157

Infectious myxomatosis, 527
Infertility, in sheep, 298
Infundibulum, 38, 41
Inguinal canal, 31
Inherited abnormalities, in sheep, 290–291. See also Health management
Injuries, of horses, 505
Insulin, 78
Intense work, 475
Interdigital phlegmon, 152–153
Interstitial cells, 33
Intradermal test, 154
Intramammary injections, 145–147
Involution, 40
Ionophores, 220
Isoleucine, 81
Isthmus, 41
Ivanoff, E. I., 51
Ivermectin, 550

Jacks, 457
Jacob, 276
Jennets, 457
Jerseys, 106–107, 137
Jewish law. See Kosher slaughter
Johne's disease, 153–154, 185, 294
John Mahler Company, 546
Judging
dairy cattle, 128–133
sheep, 301–305
steers, 238–240
swine, 323–326

Karakul, 276–277
Katahdin, 275
Ked, 298
Kemp, 176
Ketosis, 119, 125–126, 185, 295
Keystone, 231
Khaki Campbell, 418
Kidding, 182–183
Kidney stones, 297
Kids, management of, 183–184
Kindling, 524–525
Kingdoms, 8
Kosher slaughter, 210–211

Labia majora, 41
Labia minora, 41
Laboratory rabbits, 517
La Bouhite, 294–295
Lactation
goats, 181
sheep, 279
swine, 359
Lactose, 81, 82
Lamb dysentery, 294
Lambing jug, 288
Lamb joint, 307
Lambs
orphan, 279, 289
wether, 302
See also Sheep production
Laminitis, 492
Land Grant Act, 99
Lanolin, 304
Lapin-Lievre Belge, 513

Laptop measurements
  beef cattle, 242–243
  sheep, 306
  swine, 327
Large intestine, 70, 78
Larkspur poisoning, 126–127
Lasalocid, 220
Layer industry, 410–416. *See also*
      Poultry industry
Layers, 413, 416
Least-cost ration, 91
Leghorns, 384
Leg problems, in horses, 495–502
Leptospirosis, 154–155, 298, 367–368
Leucine, 81
Libido, 289
Lice
  dairy cattle, 164
  goats, 186
  llamas, 539
  poultry, 404
  sheep, 298
  swine, 369
Light horse breeds, 433, 434–442
Light-induced estrus, 465
Lighting, for egg growth, 415
Light work, 475
Limber legs, 137
Limosins, 201
Lincoln, 268
Linear score, 144
Linebreeding, 26
Liners, 143
Linoleic fatty acids, 84
Linolenic fatty acids, 84
Lipase, 78
Lipids, 84
Listerellosis, 292
Listeriosis, 185
Liver, 70
Liverflukes, 539
Llamas, 533, 534
  nutrition, 534–536
  reproduction, 536–537
  uses of, 540–542
  See also Camelid industry
Lobule, 137
Lockjaw, 297, 494–495
Locus, 19
Loin-eye area, 333–334
Loin muscle meat quality, 339
Longhorn, 205
Long-wool breeds (sheep), 266–271
Lot mating, 347–348
Lumen, 137
Lumpy jaw, 155
Lungworms, 369, 539
Lymphoid leukosis, 405
Lysine, 81
Lysocellin, 220

Machine milking, 140–143
Machine sexing, 392
Mad cow disease, 155–156
Mad itch, 362–363
Maedi, 294–295
Maggots, 299
Magnum, 41

Mahler, Marjorie, 546
Maine-Anjou, 202
Male anatomy, 31–36
Male berserk syndrome, 537
Malignant edema, 156
Malignant hyperthermia, 370
Mallards, 418
Malpositions, 46
Malpresentation, 46
Malta fever, 184
Maltose, 81–82
Mammary gland, in dairy cattle, 137
Management
  dairy cattle, 109–123
  goats, 183–184
  horses, 458–466
  rabbits, 517–520
  swine, 341–347
Mange, 164, 186, 299, 369
Mange mites, 526
Manyplies, 73
Marbling, 85, 211
Marchigiana, 202
Marek's disease, 405
Market hogs, 331
Marketing
  pork, 331
  sheep, 301–307
Marsh's progressive pneumonia,
      294–295
Masticates, 73
Mastication, 70
Mastitis, 143–147, 186, 527
Maternal breeding value (MBV) ratio,
      242
Maternal Sow Line National Genetic
      Evaluation Program, 338
Mating systems, 25–27. *See also*
      Breeding and genetics;
      Reproduction
Maturity, 211
Measurements. *See* Performance
Meat and Poultry Acts, 382
Mechanical dehorner, 231
Mechanical digestion, 69
Mechanical refrigeration, 99
Meconium, 464
Medium-wool breeds (sheep), 259,
      261–266
Medium work, 475
Meiosis, 20
Melengestrol acetate (MGA), 221
Mendel's experiments, 20–22
Meningeal worms, 539
Merck Veterinary Manual, The, 147
Metabolic disorders
  dairy cattle, 123–127
  sheep, 291–298
  See also Health management
Metabolizable energy (ME), 89
Metastrongylus elongates, 369
Metastrongylus pudendotectus, 369
Metastrongylus salmi, 369
Metestrus, 42
Metestrus bleeding, 42
Methane inhibiters, 220
Methionine, 81
Metronidazole, 550

Microbial disorders, in sheep, 291–298.
      See also Health management
Miduterine artery, 48
Milk, 48
  condensed, 99
  goat, 179–180
  residual, 139
Milk cup, 143
Milk EPD, 242
Milk fever, 126, 186
Milk hose, 143
Milking machines, 140–143
Milking phase, 140
Milking process, 140–147
Milking shorthorns, 108
Milk letdown, 139
Milk pump, 143
Milk replacer system, 112
Milk secretion, physiology of, 138–139
Milk tank, 143
Milk urea nitrogen (MUN), 123
Minerals, 85–86, 486
Miniature horses, 455
Minor breeds (sheep), 275–278
Missouri Fox Trotter, 437
Mites
  goats, 186
  horses, 502
  poultry, 405
  rabbits, 526
  sheep, 299
  swine, 369
Mitosis, 19–20
Mohair, 172, 512, 525
Molting, 391
Monday morning disease, 488–489
Monenisn, 220
Monogastric digestive system, 66,
      69–70
Monosaccharides, 81
Monotocous, 38
Montadale, 263
Morab, 437
Morgan horse, 437
Morgan, Justin, 437
Mosquitoes, 502
Mouth, 76
Mule foot, 136
Mules, 457–458
Multiple farrowing, 345
Murray Grey, 202
Muscles, of chickens, 389–390
Muscle thickness, 213
Muscovy, 418
Mutation, 24
Mutton-type sheep, 301–304. *See also*
      Sheep production
Mycoplasma mastitis, 145
Mycoplasma pneumonia, 368
Mycoplasmosis, 405–406
Myiasis, 369

Nail trimming, 540
Narasin, 220
National Association of Animal
      Breeders (NAAB), 132
National Association of Swine
      Records, 329

National Pork Producers Council (NPPC), 338, 373
National Poultry Improvement Plan (NPIP), 383
Natural breeding/mating
  dairy cattle, 133
  sheep, 287
  See also Breeding and genetics; Reproduction
Navajo-churro, 277
Navel III, 493
Navicular disease, 499–500
Necropsy, 401
Necrotic laryngitis, 151
Negative energy balance, 122–123
Nematodes, 165, 550
Nest boxes, 519
Net energy (NE), 89
Netherlands Dwarf, 514
Neutral detergent fiber (NDF), 83
New Hampshire chickens, 386
New Zealand rabbits, 514
Newcastle disease, 382, 406–407
Nez Perce Indians, 448–449
Nidation, 46
9-day heat, 464
Nitrogen free extract (NFE), 84
Nodular worms, 539
Nonruminant herbivore digestive system. *See* Functional cecum digestive system
North America, animals in, 3
North County Cheviot, 263
Nosema, 526
Nubians, 172
Nuclear transfer, 59, 60
Nucleus, 19
Nurse cow system, 112
Nutrients
  essential, 78–87, 394
  feed, 87–91
Nutrition/feeding, 65–68, 92
  beef cattle, 218–221
  dairy cattle, 109–123, 143
  digestive systems, 68–78
  essential nutrients, 78–87
  feed nutrients, 87–91
  goats, 180–181
  horses, 474–487
  llamas, 534–536
  ostriches, 547–548
  poultry, 393–400
  rabbits, 520–523
  sheep, 278–285
  swine, 347–360, 374
Nutritional muscular dystrophy, 297
Nymphs, 164

Oberhasli goats, 175
Occupational Safety and Health Administration (OSHA), 407–408
Offal, 308
Ohio Improved Chester, 319
Oil, 84
Older goat, 181
Omasum, 73
Omnivores, 69

Oral papilloma virus, 527
Order, 8
Orf, 184, 296
Organic, 13–14
Orphan lambs, 279, 289
Osteochondritis dissecans (OCD), 497
Ostriches, 544–545, 550
  dangers from, 547
  health management, 548–550
  meat, hide, and feathers, 546–547
  nutrition, 547–548
  reproduction, 545
Outbreeding, 26
Ovaries, 38
Overeating disease, 292–293
Overo horses, 450
Oviduct infections, 549
Ovine progressive pneumonia, 294–295
Ovulation, 38
Ovum, 38
Oxford, 263

Packing, 541–542
Paint horse, 449–450
Palomino horses, 450–451
Palomino rabbits, 514
Palpation, of abdomen, 50
Panama sheep, 273
Pancreas, 70
Papillae, 72
Papillomatosis, 527
Papillomatous digital dermatitis, 160
Parakeratosis, 350
Paralytic myoglobinuria, 488–489
Parasites
  goats, 186
  horses, 502–504
  ostriches, 550
  poultry, 400–409
  sheep, 298–299
  swine, 368–370
  See also Health management
Paratuberculosis, 153–154
Parental generation, 21
Parrot mouth, 291
Partial dominance, 22–23
Parturient paresis, 126
Parturition, 40, 50
  goats, 182–183
  horses, 461–464
  sheep, 288–289
  See also Breeding and genetics; Reproduction
Pasteurella multocida, 527
Pasteurellosis, 296
Pasture finishing, 346
Pastures
  farrowing sows on, 344–345
  fences for, 223–225
Paunch, 72
Pediculosis, 186
Pedigree
  beef cattle, 241
  sheep, 305–306
  swine, 326–327
Pegging down, 325
Pekin ducks, 419

Pellet binders, 394
Pencil-and-pad measurements
  beef cattle, 242–243
  sheep, 306
  swine, 327
Penis, 31, 34
Pens, 225–227
Pentoses, 81
Pepsin, 70
Peptide bonds, 80
Percheron, 445–446
Peren, Geoffrey, 269
Perendale, 269
Performance
  beef cattle, 241–244
  sheep, 306
  swine, 327–329, 338–341
Performance testing, 244
Permanent markings, 245. *See also* Branding
Perosis, 407
Persian lamb pelts, 277
Peruvian Paso, 439
Pet rabbit industry, 516
Phenotype, 19
Phenylalanine, 81
Photoperiodism, 286
Photosynthesis, 66
Phylum, 8
Physiology, of chickens, 388–391
Phytoestrogen-induced infertility, 279
Piedmontese, 202
Pigs. *See* Swine industry
Pilgrim, 423
Pinkeye, 156–157, 295
Pintos, 449–450
Pinworms, 528
Pinzgauer, 202
Pizzle rot, 298
Placentation, 47
Placentome, 40
Plants, poisonous, 126–127, 505, 540
Pleiotropic effect, 182
Plymouth Rocks, 384
Pneumonia, 157, 296
Poisonous plants, 126–127, 505, 540
Poland China, 320
Polish rabbit, 514
Politics, global, 8–11
Poll evil, 493
Poloxalene, 74
Polydome, 166
Polyestrus, 286
Polypay, 273
Polysaccharides, 82
Polytocous, 38, 347
Ponies, 434, 451–455
Pony of the Americas (POA), 452
Porcine reproductive and respiratory syndrome (PRRS), 370
Porcine stress syndrome (PSS), 370–371
Pork, soft, 85
Pork carcass, 330–341
Pork Industry Handbook, 339
Postdip, 143
Postfoaling care, 464
Posthitis, 298

Potomac horse fever, 494
Poult enteritis mortality syndrome, 407
Poultry industry, 379–382, 423
    broiler and layer industry, 410–416
    chicken anatomy and physiology, 388–391
    diseases and parasites, 400–409
    ducks, 417–419
    female reproductive anatomy, 41–42
    fertilization and development, 45–46
    geese, 419–423
    male reproductive anatomy, 34–35
    nutrition, 393–400
    reproductive system and eggs, 392–393
    taxonomy and breeds, 382–388
    turkeys, 417
Poultry Products Inspection Act (PPIA), 382
Poults, 417
Poxvirus infection, 549
Predators, of sheep, 300
Predicted transmitting ability (PTA), 135
Predip, 143
Preening, 391
Pregnancy diagnosis, 47–50
Pregnancy disease, 125–126, 185, 295
Pregnancy examination, for horses, 465–466
Pregnancy toxemia, 185, 295
Pregnant mare serum gonadotropin (PMSG), 40
Prepuce, 34
Prespermal fraction, 348
Probability, 20
Probiotics, 220
Procurement grades, for eggs, 416
Production expectations
    eggs, 415–416
    poultry, 412
Production registry (PR), 329, 346
Production testing
    beef cattle, 244–248
    sheep, 306–307
    swine, 329
Proestrus, 42
Profit maximization, 91
Progeny testing, 244
Progesterone, 38
Prolactin, 38
Propionic acids, 74
Prostaglandin, 38
Protein, 79–81
Protein quality, 349
Protozoa, 550
Proventriculus, 77
Proximate analysis, 82
Pseudorabies, 362–363
Psoroptic mange, 164, 186
Psoroptic scab, 299
Puberty, 36
    dairy cattle, 112
    goats, 183–184
    poultry, 392
    sheep, 286
Pullorum disease, 407–408

Pulpy kidney disease, 292–293
Pulsator, 142
Punnett square, 21
Purebreds, 27
Purebred swine operations, 346–347
Pygmy goats, 177

Qualitative traits, in poultry, 393
Quality grades, for beef carcass, 211–212
Quantitative traits, in poultry, 393
Quittor, 500

Rabbit industry, 509–510, 529
    domestication and breeds, 510–515
    feeding, 520–523
    fur or wool industry, 525–526
    health management, 526–529
    laboratory rabbits, 517
    management, 517–520
    meat industry, 516–517
    pet industry, 516
    reproduction, 523–525
    reptile feeder market, 517
    selection, 515–516
Rabbit syphilis, 528
Radiography, 48
Ralgro, 221
Rambouillet, 259
Ram breeds, 278
Rancidity, 85
Range paralysis, 405
Ration balancing, 90–91
Rations
    dairy cattle, 121–122
    swine, 359–360
Rebhun, W. C., 147
Rebreeding, 464–465
Receiving vessel, 143
Recessive genes, 20
Recombinant DNA, 60
Records, production testing, 246
Rectal palpation, 47
Rectal prolapse, 291
Rectovaginal constriction, 137
Rectum, 70
Red Angus, 202–203
Red Brangus, 203
Red nose, 161
Redpoll, 108
Reduction division, 20
Red water disease, 157–158
Refrigeration, mechanical, 99
Registered animals, 27
Registered program, 193
Releaser, 143
Replacement gilts, 358
Reproduction, 30–31, 60–61
    artificial insemination, 51–60
    estrous cycle, 42–43
    female anatomy, 37–42
    fertilization, 44–46
    gestation, 46–50
    goats, 181–182
    horses, 458–466
    llamas, 536–537
    male anatomy, 31–36
    ostriches, 545

    poultry, 392–393
    rabbits, 523–525
    sheep, 286–289
    swine, 347–349
    See also Breeding and genetics
Reptile feeder market, 517
Residual milk, 139
Resting phase, 140
Restraining equipment, 228
Retail cuts
    beef, 215–217
    pork, 335
Reticular groove, 75–76
Reticulum, 73
Rex rabbit, 514–515
Rhinopneumonitis, 494
Rhode Island Reds, 384, 386
Rickets, 127
Ringbone, 500
Ringworm, 164–165, 186, 370, 528
Roasters, 412
Rock Cornish game hen, 412
Roman geese, 423
Romanov, 275
Romney, 270–271
Roosters, 412
Ross, 387–388
Roughage, 78
Roundworms, 165, 369, 408, 504
Rumen, 72–73
Rumensin, 220
Ruminal tympany, 74, 123
Ruminant digestive system, 66, 71–76

Saanens, 173–174
Sable goats, 175
St. Croix sheep, 275
Salers, 203
Salinomycin, 220
Saliva, 48
Salmonella pullorum, 407–408
Salmonella, 298
San Miguel sea lion virus disease, 372–373
Santa Gertrudis, 203
Sarcoptic mange, 164, 299
Satin rabbit, 515
Saturated fatty acids, 84–85
Scabies, 164, 299
Scales
    sheep, 306
    swine, 327
Scoring, body condition, 116
Scottish Blackface, 277
Scouring, 304
Scours, 70, 124, 294
Scrapie, 295–296
Scratches, 499
Scrotal circumference, 242
Scrotum, 31, 33
Seasonal breeders, 36
Sebastopol, 423
Secondary ducts, 137
Secondary sex glands, 34
Second generation, 21
Segregated early weaning, 371
Selection, 515–516
    beef cattle, 236–249

horses, 467–472
sheep, 301–307
swine, 323–330
See also Breeding and genetics
Selection index, 249, 330
Semen, swine, 348–349
Seminiferous tubules, 33
Septicemia, 401
Serological test, 154
Sex control, 58–59
Sex determination, 23
Sex-limited genes, 23
Sex-linked traits, 23–24
Sex preselection, 100
Sexual maturity, in rabbits, 523–524
Sheath, 34
Sheath rot, 298
Sheep nose botfly, 299
Sheep production, 253–255, 309–310
    diseases, 290–300
    dual-purpose breeds, 271–274
    fine-wool breeds, 258–259
    hair and double-coated breeds, 274–275
    long-wool breeds, 266–271
    medium-wool or meat breeds, 259, 261–266
    minor breeds, 275–278
    nutrition, 278–285
    predators, 300
    reproduction, 286–289
    selection, wool care, and marketing, 301–307
    slaughtering process, 307–309
    tail docking and castration, 290
    in United States, 4–5
Sheep scab, 299
Shehitah laws, 210
Shell, 143
Shelter/housing
    beef cattle, 222
    dairy cattle, 166
    eggs, 413–414
    poultry, 411–412
    rabbits, 518–520
    swine, 374
Shetland ponies, 452
Shetland sheep, 277
Shipping fever, 158–159, 296
Shires, 446
Shope papilloma virus, 527
Shorthorns, 203
Show animals, swine, 326
ShowRing, 326–327
Shrinkage, 304
Shropshire, 264
Sidebones, 501
Sigmoid flexure, 34
Silver, L. B., 319
Simmental, 203
Simple-stomached digestive systems. *See* Monogastric digestive system
Skeleton, of chickens, 388–389
Skin, of chickens, 391
Skinning, 207
Slaughter cattle production, 193
Slaughtering process

beef cattle, 205–211
sheep, 307–309
swine, 331
See also Carcass
Slaughter sheep, 302–303
Sleeping sickness, 490
Slipped tendon, 407
Small intestine, 77
Smith, Jim, 133
Snuffles, 527
Soest, Van, 82–83
Soft pork, 85, 371
Somatic cells, 19–20
Sonoray, 243–244
Sore hocks, 528
Sore mouth, 184, 296–297
Southdown, 265–266
Sows, 331, 338, 339. *See also* Swine industry
Spallanzani, L., 51
Spanish goats, 178
Species, 8
Speciesism, 210
Specific pathogen free pigs, 371
Sperm-poor fraction, 348
Sperm-rich fraction, 348
Sphincter muscle, 137
Spider syndrome, 291
Spinose ear tick, 299
Splints, 501
Spool joint, 308
Sport horses, 455–457
Spotted swine, 321
Spraying, 233–234
Springing heifers, 113
Stags, 331, 412
Standardbred, 440–441
Standard of Perfection (American Rabbit Breeders Association), 511
Staphylococcus aureus, 145
Starch, 82
Steers, judging, 238–240. *See also* Beef industry
Stewing chicken, 412
Sticking, 207
Stiff lamb disease, 297
Stocker systems, 217–218
Stocking rate, 35
Stomach, ruminant, 74–75
Stomach bots, 502–503
Stomach worms, 504, 539
Strains, 383
Strangles, 494
Streak canal, 137
Streptococcus agalactiae, 145
Streptomyces fungi, 220
Stringhalt, 501
Strip cup, 143
Strongyles, 503–504
Strongyloides, 550
Stunning, 205
Subclinical mastitis, 144
Suburethral diverticulum, 41
Sucking lice, 539
Sucrose, 81
Suffolk draft horse, 446–447
Suffolk sheep, 266

Suint, 304
Superior, 231
Surfactant, 74, 123
Swamp fever, 490
Swine classes, feeding, 357
Swine dermatosis, 350
Swine dysentery, 365
Swine fever, 363–365
Swine industry, 314–315, 374
    animal welfare, 373–374
    breeds, 316–322
    health management, 360–373
    history, 315–316
    management, 341–347
    nutrition, 347–360
    pork carcass, 330–341
    reproduction, 347–349
    selection, 323–330
    taxonomy, 316
    in United States, 5–6
Swine-testing stations, 328–329
Syndactylism, 136
Synovex H, 221
Synovex S, 221

Tail docking, 290
Tall fescue toxicosis, 126
Tamworth, 321
Tandem selection, 248, 330
Tan rabbit, 515
Tapeworms, 408, 528, 550
Targhee, 273–274
Taurine, 81
Taxonomy, 7–8
    horses, 432
    poultry, 382–388
    swine, 316
Teasing, 459
Teat cistern, 137
Teat cup assembly, 142
Teething, 472–474
Teeth wear, 505
Temperature, and sheep reproduction, 286
Tendonitis, 489, 497
Tennessee walking horse, 441
Termination cross, 193
Tertiary ducts, 137
Testes, 31, 33
Testosterone, 33
Tetanus, 186, 297, 494–495
Texas Longhorn, 205
Texel, 266
Texture, 211
Thoroughpin, 501
Thread-necked strongyles, 539
Threonine, 81
Thriftiness, 213
Thrush, 501
Tobiano horses, 450
Toggenburgs, 173
Total digestible nutrient (TDN), 89
Toulouse, 423
Toxoplasmosis, 297
Traits
    poultry, 393
    sex-linked, 23–24
Trakehner, 456

Transmissible gastroenteritis, 372
Transport myopathy, 370
Traumatic gastritis, 73, 127
Trematodes, 165
Trichinella spiralis, 369
Trichinosis, 369
Trichomoniasis, 159, 408
Trocar, 74
Trypsin, 78
Tryptophan, 81
Tube dehorner, 231
Tuberculin testing, 99–100
Tuberculosis, 159
Tunis, 266
Turkeys, 394, 396–397, 417. *See also* Poultry industry
Twin-lamb disease, 295
Tying-up syndrome, 488–489
Type, 19
Tyrosine, 81
Tyzzer's disease, 527

Ulcerative pododermatitis, 528
Ultrasonic device, 243–244
Ultrasound, 48
Undegraded intake protein (UIP), 122–123
Unified Meat Type Hog Certification Program, 346
United Kingdom Farm Animal Welfare Council, 374
United States
animals in, 3–7
beef production in, 190–194
U.S. Department of Agriculture (USDA)
Animal & Plant Health Inspection Service, 516
Animal Damage Control program, 300
Animal Improvement Programs Laboratory, 136
beef carcass grades, 211, 214
Dairy Herd Improvement Association, 100
egg grades, 416
and National Poultry Improvement Plan, 383
and organic foods, 13–14
poultry grades, 412–413
and Poultry Products Inspection Act, 382
U.S. Food and Drug Administration (FDA), 234
U.S. Forest Service, 217, 421
U.S. procurement grades, for eggs, 416
Unsaturated fatty acids, 84
Uremic poisoning, 297

Uric acid, 78
Urinary calculi, 297
Urine, 48
Urine scald, 298
Urolithiasis, 297
Uropygial gland, 391
Uterine milk, 40
Uterus, 39–40, 41

Vaccinations
beef cattle, 232–233
llamas, 537, 539
sheep, 300
See also Health management
Vacuum controller, 141
Vacuum level, 141
Vacuum pump, 141
Vacuum reserve, 142
Vacuum tank, 141
Vagina, 40, 42
Vaginal biopsy, 48
Valine, 81
Variety, 383
Vasa deferentia, 34
Vas deferens, 34
Vegetarianism, 210
Venereal spirochetosis, 528
Vent, 42
Vent disease, 528
Ventilation, 374
Ventriculus, 77
Vertical transmission, 401
Vesicular exanthema, 372
Vesicular stomatitis (VSV), 159–160
Vestibule, 41
Vibriosis, 160, 298
Vicunas, 533. *See also* Camelid industry
Vietnamese Potbellied pigs, 321
Villi, 70
Viral hemorrhagic disease, 528
Virgin Island White, 275
Virus pneumonia, 368
Visceral lymphomatosis, 405
Vital amine, 87
Vital signs, of chickens, 391
Vitamins, 84, 87, 486
Volatile fatty acids (VFAs), 71
Vulva, 40–41

Warbles, 163–164
Warts, 160
Washington, George, 458
Water, 78–79
Water belly, 297
Waterers, rabbit, 520
Water-soluble vitamins, 87

Wattles, 173
Waxing, 461
Weaning
dairy calves, 110–111, 112
rabbits, 525
Weaning EPD, 242
Weende system of proximate analysis, 88
Welfare. *See* Animal welfare
Welsh Harlequin duck, 419
Welsh ponies, 452, 454
Wether lambs, 302
Whipworms, 539
White Beltsville, 417
White muscle disease, 127, 297
White Park, 205
White scours, 124
Whole milk system, 112
Wholesale cuts
beef, 215–217
pork, 335
Wholesale grades, for eggs, 416
Wiltshire Horn, 277–278
Windgall, 501–502
Wind puffs, 501–502
Wintered, 194
Wireworms, 550
Wooden tongue, 155
Wool blind, 288
Wool blindness, 291
Wool block, 529
Wool care, for sheep, 301–307
Woolen, 305
Wool industry, rabbits, 525–526
Wool maggots, 299
Wool-type sheep, 304–305. *See also* Sheep production
Worms, 408, 539. *See also specific types*
Worsted, 305
Wounds, of horses, 505

Xanthophylls, 394
X disease, 127

Yearling weight EPD, 242
Yield, pork carcass, 332–334
Yield grades, beef carcass, 211, 214
Yolk, 304
Yolk sacculitis, 549
Yorkshire, 322

Zero tolerance, 382
Zoonotic disorder, 99–100
Zwoegersiekte, 294–295
Zygote, 20